The Origin of the Domestic Animals of Africa
Volume I

THE ORIGIN OF THE
DOMESTIC ANIMALS
OF AFRICA

H. EPSTEIN
The Hebrew University of Jerusalem

Revised in collaboration with
I. L. MASON
Institute of Animal Genetics, Edinburgh

Volume I

AFRICANA PUBLISHING CORPORATION
New York · London · Munich

Published in the United States of America 1971
by Africana Publishing Corporation
101 Fifth Avenue, New York, N.Y. 10003

Great Britain
Meier and Holmes Ltd.
18–22 Inverness Street, London, N.W. 1
Meier and Holmes SBN 900253 00 2

Germany
Internationaler Universitäts-Buchhandel
Landwehrstr. 31, 8 Munich 15

Copyright © 1971 by H. Epstein
All rights reserved.

No part of this publication may be reproduced, stored in a retrieval system, or transmitted, in any form or by any means, electronic, mechanical, photocopying, recording or otherwise, without the prior permission of Africana Publishing Corporation.

Library of Congress Catalog Card № 76-136491

ISBN 0-8419-0066-3

Printed in Germany (East) by Edition Leipzig

Preface

This work is the outcome of over 40 years of continuous study and research, beginning in 1927 when the author took up stock farming in the Western Transvaal and became interested in the history of the Africander breed of cattle. At that time very little was known on the subject, and it was with difficulty that the few scattered facts could be gathered. It soon became apparent that to trace the path of the ancestral stock of the Africander cattle and of the Hottentots, their former breeders, back to their earlier home in East Africa, it was necessary also to pay attention to the other domestic animals of the Hottentots. And since the origin of the animals of the Hottentots could not be fully understood without a comparison of their racial characteristics with those of the domestic breeds of the Bantu, the theme gradually expanded until it finally embraced the various types and breeds of domestic animals throughout the whole of Africa.

During 1933, the year before his departure from South Africa, the author worked in co-operation with Professors H. H. Curson and J. H. R. Bisschop, whose publications constitute outstanding contributions in this field of research and to whom the writer is indebted for the encouragement and sympathy generously accorded to his work. In 1934 a grant awarded by the Research Grant Board of the Union of South Africa from funds provided by the Carnegie Corporation of New York made it possible for the author to travel in East Africa and the countries of the Middle East and study the local breeds of livestock. Thereafter his work was continued in Palestine, save for a period in 1936 spent at the British Museum (Natural History). During the years 1942–1948 the writer served as Technical Adviser (Meat Division) to the Government of Palestine, and in the course of this work he handled large numbers of cattle, buffaloes, sheep, goats, pigs and camels. The animals were purchased in Cyrenaica, Egypt, the Sudan, Arabia, Turkey, Iraq, Syria, the Lebanon, Transjordan and Palestine. This opportunity added considerably to his practical knowledge of numerous types and breeds of unimproved stock and furnished valuable material for this book. From 1948 to 1960 the author continued his work as Technical Adviser (Meat Division) with the Government of Israel, and in 1950 he was appointed to the chair of animal breeding at the Hebrew University of Jerusalem.

In 1958 he visited the Kladrub stud and several experiment stations in Czechoslovakia and also sheep farms and breeding stations in the German Federal Republic and in Scotland. In 1963 he went to Iran, Burma and Thailand and spent four months in the Chinese People's Republic, eager to discover in the livestock of the eastern part of the Old World any parallel traits with those of the domestic fauna of Africa. The outcome of this study was his monograph on the 'Domestic Animals of China'. In the same year he also visited the Mongolian People's Republic and the Union of Soviet Socialist Republics, again with a view to becoming acquainted with the livestock of these regions. In 1968 he went twice to the U.S.S.R. and India and spent two months in Nepal, always for the purpose of studying the local domestic fauna and obtaining a wider background for his principal interest and work on 'The Origin of the Domestic Animals of Africa'.

By 1949 the manuscript had been completed. However, the author's attempts to find a publisher met with difficulties. In view of the large number of illustrations and the high cost of production a subsidy was found necessary. For his purpose the author approached the South African Council for Scientific and Industrial Research, and in 1950 he was informed that a subsidy of £ 1000 would be granted on condition that the Council "be allowed to have the manuscript edited by competent persons of its own choice to ascertain whether the volume can be reduced without in any way interfering with the subject matter". In due course the Council succeeded in obtaining the consent of Mr. I. L. Mason of the Institute of Animal Genetics, Edinburgh, to undertake this task.

Mr. Mason read the manuscript in the 1950's and the revised manuscript in 1963. Along with the author, he also read the proofs. In 1958, 1963 and 1968 the author went to Edinburgh, each time for one month, to discuss with him questions which had not already been dealt with by letter. On each occasion, Mr. J. P. Maule, Director of the Commonwealth Bureau of Animal Breeding and Genetics, placed the Commonwealth Bureau's large library and other working facilities at the disposal of the author; in addition, he entrusted the Bureau's collection of photographs of African livestock to him for publication in this work. The author's sincere gratitude is due to Mr. Maule for his generous assistance.

The author wishes to express his gratitude to the South African Council for Scientific and Industrial Research for their choice of Mr. Mason as editor. No serious writer interested in the accuracy of his work and the soundness of his arguments could have had a better critic—severe but open-minded and just. With his thoroughness and keen intelligence he guided the author away from the pitfalls of some long-held but mistaken pet ideas. The author wishes to express his deep gratitude to Mr. Mason for all he has done in these many years of scientific collaboration.

The author is also indebted to Mr. M. L. Ryder, Animal Breeding Research Organisation, Edinburgh, who read and commented upon the chapter on sheep.

In 1955 the manuscript, which with the continuous accumulation of new material was gradually expanding into two volumes, was again offered to publishers, but owing to the large investment involved and the limited demand likely for such a specialised work these were reluctant to undertake publication. In 1963 the author met Professor Lothar Hussel of Karl-Marx-University, Leipzig, whom he told of his problem. Professor Hussel promised to find a publisher, and after a short time the author was informed by VEB Edition Leipzig, Verlag für Kunst und Wissenschaft, that they were prepared to publish his work in collaboration with the Africana Publishing Corporation of New York.

During all these years the South African Council for Scientific and Industrial Research was kept informed by Mr. Mason and the author of their various endeavours to find a publisher and of the need for a larger subsidy. In response, the South African Council for Scientific and Industrial Research raised the amount to 5770 Rand, almost three times the original sum. To this the South African Friends of the Hebrew University of Jerusalem, on the initiative of Judge Felix Landau of Jerusalem, added another 1500 Rand. The author wishes to record his grateful acknowledgement to the South African Council for Scientific and Industrial Research and the South African Friends of the Hebrew University of Jerusalem for their generous support. It is his hope that the two volumes as they now appear in print will not disappoint their sponsors.

H. Epstein

Political designations of countries are given in their official names. Wherever other designations have been used, these refer to geographical descriptions. As far as possible, geographical names have been spelled according to J. Bartholomew (ed.), The Times Atlas of the World, Comprehensive Edition, 1967.

Source of Illustrations

Volume I

Figures	by courtesy of
514, 515	Adamantides
99	Ashmolean Museum
223, 596	W. G. Beaton
337, 354, 614, 662, 663	Berlin Museum
392–395, 401	P. Bonelli
538, 541, 544, 549, 586	J. C. Bonsma, Pretoria
2	G. Budich, Berlin
155, 156, 172, 182, 190, 192, 195, 196, 198, 199, 206, 207, 247, 248, 256, 290, 293, 369, 498, 601, 654	British Museum (Fig. 290 Photograph by K. Liebscher)
89, 90, 245, 298	Cairo Museum
194	Cambridge
171	Candia Museum
525	J. B. Condy
21, 23, 48, 71, 98, 158, 163, 164, 189	C. M. Cooke, Ealing
410, 420	A. W. Chalmers
422	Credit P.R.O., Khartoum
556	H. H. Curson
481, 482, 598	G. Doutressoulle
114	Dummerstorf Collection
246	Editions des Musées Nationaux, Paris
376, 584	The Farmer's Weekly
208, 225, 300, 301, 327, 387, 416, 430, 431, 435–437, 439, 445, 446, 460–462, 464, 473, 497, 558, 565, 595, 597	D. E. Faulkner
157	Fogg Art Museum, Harvard University, Cambridge, Mass.
563	J. Ford
291, 292	The Director of Geological Survey of India
389, 390, 396, 403, 404	Y. S. Goor
476	G. M. Gates

415, 417, 418, 421	H.T.B. Hall
10	Heimpel, Raabs – Austria
523	J.H.N. Hobday
653	Indian Museum, Calcutta
64, 647	Institut für Kulturmorphologie, Frankfurt a.M.
424, 425	Institut National pour l'Etude Agronomique du Congo
280 (left), 282, 356	Iraq Museum
443, 444	Kenya Information Service
205	Kühn Museum, Halle
56	LIFE photograph by N.R. Farbman Copyright Time Inc. 1947
150, 151, 178, 246, 280 (right), 283	Louvre Museum
504	M. Maricz
441, 442, 484, 573	I.L. Mason
334, 426, 428, 469, 505, 511	J.P. Maule
419, 500	E.A. McLaughlin
15	E. Mendelssohn
92, 106, 281	Metropolitan Museum
134	Moscow Zoological Park
161, 646	National Geographic Magazine
472	T.C. Okoro
318, 455, 467, 560, 568, 644	Onderstepoort Collection
216, 477, 478	J. Pagot
302	D.S. Rabagliati
405, 406	J.L. Read
554, 557, 561, 562	Rhodesian Indigenous Cattle Breeders Society
588, 589	Rijksmuseum Amsterdam, Foto-Commissie
557	D.A. Robinson, Causeway, SR
305, 310, 313, 316, 345, 346, 579, 626–628	J.E. Rouse, Saratoga
397–399, 402, 432, 491, 492	R.H.D. Sandford, FAO
466	Service Général de l'Information de Madagascar
465, 559	J.M. da Silva
14	G. Strauß
592	E. von Uechtritz
176	Vatican
296, 297	The Director of Veterinary Services, Cairo
542	The Director of Veterinary Services, Onderstepoort
354	Vorderasiatisches Museum, Berlin
569, 570	J. Walton
531, 533, 534, 536	F.H.B. Watermeyer
325	G. Williamson
42	A. de Witt, Potchefstroom
566, 567	E. Wyatt-Sampson

Contents

Preface	v
Source of Illustrations	viii
Chapter I Dog	**1**
I. A Survey of Wild Canidae	3
II. Studer's Classification of Prehistoric Dogs	18
III. Distribution and Characteristics of the Dogs of Africa	28
i. The Pariah Dogs of Africa	28
ii. The Greyhounds of Africa	58
iii. The Mastiffs of Ancient Egypt	71
iv. The Hounds of Africa	73
IV. Origin and Descent of the Dogs of Africa	83
i. The Influence of Domestication on the Canine Cranium	83
ii. The Origin of the Pariah Dogs of Africa	110
iii. The Origin of the Greyhounds of Africa	147
iv. The Origin of the Mastiffs of Ancient Egypt	171
v. The Origin of the Hounds of Ancient Egypt	181
Chapter II Cattle	**185**
I. On the Classification of Cattle	187
II. The Humpless Longhorn Cattle of Africa	201
i. Recent Humpless Longhorn Cattle in Africa	201
ii. Ancient African Longhorn Cattle	213
iii. Origin and Descent of the African Longhorn Cattle	226
III. The Humpless Shorthorn Cattle of Africa	259
i. Distribution and Characteristics of the Shorthorn Cattle of Africa	259
ii. The Geographical Origin of the African Shorthorn Cattle	288
iii. The Phylogeny of the Shorthorn Cattle	307

IV. The Humped Cattle of Africa 327
    i. On the Classification of the Humped Cattle of Africa 327
    ii. The Zebu Cattle of Africa 340
    iii. The Sanga-Fulani Group 409
    iv. Origin and Descent of the Humped Cattle of Africa 505
    v. Summary—Classification, Distribution and Origin of African Cattle . 555

Chapter III   Buffalo   557

I. The Classification of Wild Buffaloes 559
II. Distribution and Characteristics of the Domestic Buffalo of Egypt 564
III. The Origin of the Domestic Buffalo of Egypt 567

Volume II contains the following chapters:
        Chapter IV   Sheep
        Chapter V   Goat
        Chapter VI   Pig
        Chapter VII   Ass
        Chapter VIII   Horse
        Chapter IX   Camel
        Bibliography
        Indexes

IV. The Horned Cattle of Africa

i. [Introduction]
ii.
iii. The Sanga Cattle Group
iv.
v.

Chapter III. Buffalo

I. The Classification of Old Buffaloes
II. Distribution and Characters of the Domestic Buffalo of Egypt
III. The Origin of the Domestic Buffalo of Egypt

Chapter I

DOG

I. A Survey of Wild Canidae

Among the numerous members of the family Canidae, only three species, i.e. the coyote and its relatives (Lyciscus), the true jackals (Thos oken = Canis aureus), and the wolves (Canis lupus), show in the most important cranial and dental features a close resemblance to the domestic dog. The quest for the latter's ancestors may therefore be limited to these three species. The coyotes, jackals and wolves are characterised by the marked development of the frontal sinuses and the powerful dentition, more especially the broad strong canines and strong flesh-teeth. The pupils of the eyes are round, and the tail is of moderate length. The cranial superstructure is generally well developed in relation to the size of the skull, while the supraorbital processes are always convex and laterally bent down. In the larger representatives the frontal ridges or crests are joined into a single sagittal (parietal) crest which frequently reaches a considerable height, whereas in the smaller forms the fronto-parietal ridges remain apart.

The Coyote of North America

The coyote or prairie wolf, Canis (Lyciscus) latrans Say., of North America, comprising a large number of geographical races, ranges from Salvador and western Costa Rica including the peninsula of Nicoya to the lower parts of Hudson's Bay (Nelson, 1932). The coyote is considerably smaller than the common wolf from which it is also distinguished by a more bushy tail and thicker and longer fur. Owing to the considerable length of the splanchnocranium, the skull has a greyhound-like appearance. The coyote is excluded from the ancestry of the domestic dogs of the Old World for geographical reasons.

The True Jackals

The jackals proper, i.e. excluding the side-striped and the black-backed jackals of Africa, are divided into the golden jackals of Asia and the grey jackals of Africa. Both groups display considerable racial variability, although the typical subspecies (Canis aureus typicus L.) is exceptional in its comparatively uniform cranial conformation.

1. Grey wolf (left) and coyote (right) (from "The National Geographic Magazine")

2. Asiatic jackal (Canis aureus) (photograph: G. Budich)

The bushy tail is relatively shorter than in the wolves. As the number of caudal vertebrae in jackals and wolves is equal, the shortness of the jackal tail is due to the short caudal vertebrae (Klatt and Vorszeher, 1923).

The Asiatic Jackal

Canis aureus Linnaeus, 1758, occurs in nearly the whole of southern Asia, and extends north to the Caucasus and the steppes around the Caspian Sea, and west into the southern parts of the Balkan peninsula. The range includes Romania, Greece; western and southern Turkmenia, Tajikistan, the whole course of the Amu-darya; Persia,

Iraq, Asia Minor, Afghanistan, Syria, Palestine, Arabia; Baluchistan and Sind, south through peninsular India to Ceylon, eastwards to Nepal, Assam, Burma and Thailand (Ellerman and Morrison-Scott, 1951).

The Asiatic jackal reaches a body length of 65–80 cm, tail length of 22–30 cm, and a shoulder height of 45–50 cm. The eyes are light brown in colour and have round pupils (Brehm, 1922). The ears are small and triangular in shape, their relative size being comparable to that in the Pomeranian. The general colour of the coat varies from a pale isabelline to a pale rufous. In the male the basic colour is golden-yellow, with a more reddish brown hue between the ears, on the back of the nose, and on the hindlegs above the hocks. The colour of the female is slightly duller, the reddish brown regions on the head and hindlegs in particular showing a lack of lustre. A line of black hair extends from below the eye to the ear; but the outside of the ears and the nape of the neck are devoid of black. In general, the coat of the female shows less black than that of the male. The hair at the throat, near the tip of the nose and the lips is of a dull greyish white colour; but nowhere is this clearly marked off from the golden-yellow of the main areas. The lower part of the neck is whitish, broken by a broad black stripe extending from shoulder to shoulder, with a parallel duller line in front. The belly displays the basic golden colour of the coat, but in a somewhat lighter hue. The inside of the legs is of the same greyish white colour as the throat, while the feet are light yellow. On the back the coat shows a great deal of black which begins behind the nape of the neck and reaches its greatest development behind the middle of the back. The sides are also characterised by a fair amount of black hair which forms about five or six indistinct stripes lower down. A conspicuous dark line runs from the root of the tail across the upper thigh to the knee. The tail is golden-yellow broken by black;

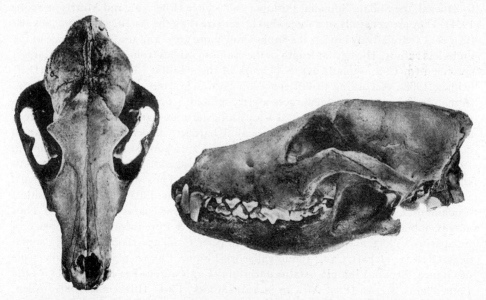

3. Skull of female Canis aureus typicus, frontal and lateral views (after Hilzheimer)

but there is no distinct basal spot. Generally, the hair has a white base followed by a black and a yellow zone, and ending in a black point. The underfur has a grey base changing into a golden colour towards the tips. The coat is thick and coarse, forming a kind of mane above the shoulders (Hilzheimer, 1908).

The most characteristic cranial feature of Canis aureus is the profile. Owing to the great height of the aboral part of the nasals, this ascends with hardly any stop in an almost uninterrupted line from the tips of the nasals to the highest point of the neurocranium. Viewed from above, the splanchnocranium appears short, broad and plump. The neurocranium is strongly developed and distinguished by vertical walls and the slight vaulting of the roof. The narrowest part between the temples is well marked. From thence the distance to the supraorbital processes is short. Neither the sagittal crest nor the supraorbital ridges are well developed. The frontal region, bordered by the squamosal ridges, the postorbital processes, the maxillae, and the nasals, is very broad, while the postorbital processes are long, strong and considerably bent down. The stop in the region of the nasal roots, which are situated far behind the end of the maxillary processes, is hardly perceptible. The mandible is strong, with the posterior edge of the ascending ramus straight, and the horizontal ramus gradually tapering towards the incisive part (Hilzheimer, 1908).

Canis aureus has repeatedly been crossed with the domestic dog. Hilzheimer obtained a triple hybrid, wolf—jackal—domestic dog (Lotsy, 1922).

The North African Jackal

The North African or grey jackals range over the whole of North Africa south to about 5° N, including Egypt, Libya, westwards to Morocco, Rio de Oro, thence southwards to Senegal, the Sudan, Somalia, Ethiopia and Kenya (Ellerman and Morrison-Scott, 1951). They are generally of a somewhat larger size than the Asiatic or golden jackals; their ears are relatively longer, the flanks show more grey, and the outer surface of the limbs less rufous. The basilar length of the cranium ranges from 187 mm in the largest specimen (of C. a. soudanicus) to 117 mm in the smallest (of C. a. riparius) (Hilzheimer, 1908). According to differences in body size, coat colour and cranial conformation several geographical races are recognised:—Canis aureus lupaster Hemprich & Ehrenberg, 1833 (in place of Canis lupaster and Canis sacer); Canis aureus algirensis Wagner, 1841 (in place of Sacalius barbarus H. Smith, C. a. tripolitanus, C. lupaster grayi Hilzh. and C. studeri Hilzh.); Canis aureus maroccanus Cabrera, 1921 (in place of C. lupaster maroccanus Cabr.), Mogador, Morocco; Canis aureus anthus Fréd. Cuvier, 1820 (in place of C. anthus and Thos senegalensis) of Senegal and the western Sudan; Canis aureus soudanicus Thomas, 1903 (in place of C. anthus soudanicus, C. thooides, C. doederleini Hilzh. and C. variegatus Cretzschm. = Thos aureus nubianus); Canis aureus riparius Hemprich & Ehrenberg, 1832 (in place of C. riparius, C. hagenbecki, C. mengesi, C. lamperti, C. mengesi lamperti and C. somalicus) (Red Sea jackal); Canis aureus gallaensis Lorenz, 1906 (in place of C. gallaensis) of Guinea, Arussi, Ethiopia—status uncertain; Canis aureus bea Heller, 1914, of the Loita Plains, Kenya (East African jackal) (Studer, 1901; Hilzheimer, 1908; Allen, 1939; Ellerman and Morrison-Scott, 1951).

4. Canis aureus lupaster (after Brehm)

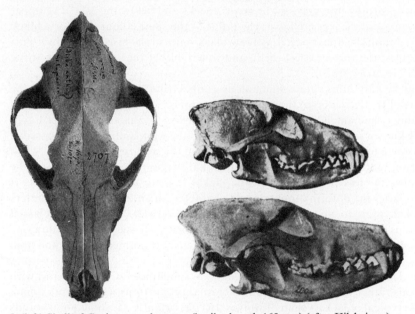

5. (left) Skull of Canis aureus lupaster (basilar length 165 mm) (after Hilzheimer)

6. (right) Crania of Canis aureus L. from Asia Minor (above) and Canis aureus lupaster H. & E. from Egypt (below) in equal reduction (after Antonius)

The Egyptian jackal, Canis aureus lupaster, of Egypt, Libya and southern Palestine, is a slender leggy animal with a shoulder height of about 50 cm (Brehm, 1922). It has a relatively large head, long pointed canines, and large, broad and pointed ears. The coat is characterised by a dirty sand colour, light yellow legs, and general lack of grey. The dorsal crest is rufous, each hair having a bright white ring. The skull differs from that of the Asiatic Canis aureus by greater size, stronger teeth with longer and more pointed canines, a concave splanchnocranial profile, and stronger development of all superstructures. The neurocranium is moderately large, with a prominent sagittal crest and long occipital protuberance.

Canis aureus algirensis, which ranges from Tripolitania through Tunisia to Morocco, is distinguished by a long slender skull, devoid of sharp edges, with a flat frontal region and a long shallow stop. The soft coat is of a very light sandy grey colour; the chest is yellow, and the belly sand-coloured. The dorsal line is black mixed with white; a more or less distinct lateral line runs along each flank, curving down to the side of the thigh (Anderson and De Winton, 1902). The area between the dorsal and lateral stripes is lighter in shade, although exhibiting a fair amount of black hair.

The type specimen of Canis aureus anthus from Senegal is a small slender jackal with a dorsal height of 41 cm. The back and flanks are dark grey with a few yellowish spots, the hair being composed of black and white zones in addition to a few fawn ones. The grey colour is not even, but is lighter or darker in different areas. The neck is greyish fawn, turning into grey on the head, cheeks and below the ears. The upper part of the snout and the legs are fawn; the tail is of the same colour save for a black longitudinal spot on the upper part and a few black hairs in the brush. The lower part of the mandibles, the throat, chest, belly and inner side of the legs are dirty white (St. Hilaire and Cuvier, 1820).

The Nubian jackal, Canis aureus soudanicus, ranges from Upper Egypt through the northern part of the Republic of the Sudan to Kordofan. Hilzheimer (1908) wrote of this jackal under the name Canis doederleini as follows:—"This is the largest Egyptian wild dog known so far. In size it is not inferior to the wolf, and it is not impossible that it was known to the earlier authors, who believed the wolf to occur also in North Africa." Brehm (1863) described the Nubian jackal in the following words:—"The wolf dog shows far less resemblance to the jackal than to our European wolf. It resembles the latter in conformation, size and character; in all these respects it differs from the jackal, even to an untrained eye." Antonius (1922), who likewise classed C. a. soudanicus (C. doederleini) with the wolf, described a live specimen as follows:— "It was a small, tawny yellowish grey dog, with rather short pointed canines and high legs."

Hilzheimer (1908) based his classification of C. a. soudanicus (C. doederleini) with the grey jackals rather than the wolves on its cranial proportions which approximate more to those typical of jackals than of wolves. In coat colour C. a. soudanicus stands also nearer to the jackals than to the wolves in which the coat shows either more white, more black or more fulvous, especially on the extremities.

It appears that Brehm and Antonius classed C. a. soudanicus with the wolves chiefly because of its large size. But this is an unsafe criterion; for there is one particularly striking feature about C. a. soudanicus: its weak dentition. As the extreme races of a species—large and small—are frequently characterised by a disproportionately weak or strong dentition, we may assume that C. a. soudanicus represents but a larger evolutionary form of the grey jackals of North Africa.

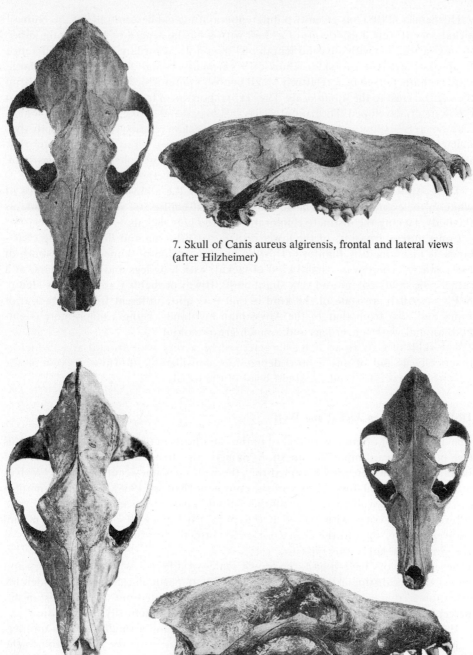

7. Skull of Canis aureus algirensis, frontal and lateral views (after Hilzheimer)

8. Crania of Canis aureus soudanicus, basilar length 187 mm, frontal (left) and lateral (right bottom) views; basilar length 138 mm, frontal view (right top)

Hilzheimer (1908) has given two different craniological descriptions of the Nubian jackal: one (Canis doederleini) of a skull with a basilar length of 187 mm, the other (Canis variegatus) with a basilar length of 138 mm. In accordance with the difference in absolute size, the two skulls show different neurocranial development. The large skull is characterised by a relatively small neurocranium which slopes gradually from the sagittal crest to the squamosals. The cranial portion in front of the temporal region is powerfully developed. The frontal area is very broad and strong, while the splanchnocranium appears low and weak. The dentition is weak in relation to the large body size, much weaker than the dentition, in relation to body size, of Canis aureus lupaster. The smaller skull is distinguished by a relatively larger and more spherical neurocranium, shorter and less prominent sagittal crest, and weaker occipital protuberance.

The coat of C. a. soudanicus is distinguished by the almost complete absence of white in the even grey-brown basic colour. The extremities are of the same shade as the body, lacking the distinctive coloration of the grey jackals.

Canis aureus riparius, of the lower districts of Ethiopia and the adjoining coastlands, is mentioned by Blanford (1870) in his description of Annesley Bay, south of Massawa:—"There was a small kind of jackal, with long legs and longish ears, of a rather pale sandy colour and very slight build. It was probably Canis riparius Hemp. & Eh. ... All I can state of the kind is that it is quite different from the jackal of India and also from that of the Abyssinian highlands, being a much more slight-built animal, with longer legs and a much greyer colour."

The skull of C. a. riparius is characterised by a very wide frontal area, which is practically devoid of any central depression, considerable downward slope of the postorbital processes, and the slight bend of the nasals.

Distinction between Jackal and Wolf

The difference of opinion with regard to the classification of Canis aureus soudanicus (see p. 8) raises an important question, namely, the distinction between wolves and jackals. It is now generally accepted that the wolf does not occur in Africa. Wolves are absent from the fossil record of the continent (Romer, 1938). In Werth's map of the range of the wolf (p. 12) the southern limit skirts Africa in the north and east. Flower (1932) writes with regard to Egypt:—"No real wolf occurs in Egypt; all 'wolves' sent as such to the Giza Zoological Garden and Museum were specimens of the Greater Jackal, Canis lupaster."

Several authors have tried to lay down scales of differences by which wolves and jackals may be distinguished. Studer (1901; 1903), in particular, has paid considerable attention to this subject. But Hilzheimer (1908) has pointed out that there is an imperceptible transition from the jackals to the wolves in practically every feature.

In size there is no difference between a large jackal and a small wolf. An uninterrupted line of skulls can be arranged, according to size, from the largest wolf to the smallest jackal. Nor is there any fundamental difference in the ratio of neurocranial to splanchnocranial length, as Studer (1901) believed. For although in jackal skulls the neurocranium is never shorter than the splanchnocranium (owing, probably, to the relatively small body size of jackals and the correspondingly large brain case), many

wolves are also characterised by this peculiarity. As Klatt (1913) plainly remarked:—
"Since the jackal is but a miniature edition of the wolf, it is evident that it must possess a relatively larger and more forward extending brain; this is a physiological necessity."

The condition of the bullae is also significant. In jackal skulls the largest bullae exceed those of wolves; but small jackal bullae and large wolf bullae overlap in size. In the shape of the bullae there seems to be a slight difference between jackals and wolves. In the latter these are generally flatter and broader, the roof gradually passing into the bony part of the external auditory canal. In the former the bullae are very high and narrow, while the bony part of the auditory passage is strongly marked off from the roof of the bullae.

Studer (1903) mentioned the larger and peculiarly shaped orbits in jackal skulls as distinguishing features. Yet the larger orbits are due to the relatively larger eyes which distinguish the smaller from the larger canidae.

While the molars are generally larger in jackals than in wolves, the difference is so slight, and there are so many exceptions, that it cannot be considered as of any value for specific distinction. The same applies to differences in the absolute size of the canines and premolars.

While Hilzheimer (1908) concluded his comparison between wolves and jackals with the remark that "these examples may suffice to prove that there is no sharp line of demarcation between wolves and jackals", it would appear that this inference is not acceptable in its generalised form. For jackal skulls are distinguishable from those of wolves by the smaller size of the carnassials as compared with the molars behind them. Jackals differ from wolves also in the form of their first upper molar tooth. In the jackals this tooth has a platform, the cingulum, running around the labial side of the crown; in the wolves the platform is reduced or wanting in the middle of its length. This condition of reduction of the cingulum must be regarded as a specialisation in the wolves, since a complete cingulum is present in Alopex, Vulpes and other Cynoids (Miller, 1920; Jones, 1921).

The Wolf

With the exception of a very few breeds of domestic dog, the wolf is the largest living member of the family. True wolves are restricted to the northern hemisphere where they form a compact circumpolar group. Their range is very extensive, embracing Europe, the greater part of Asia as far east as Japan, and nearly the whole of North America. In Western Europe they are now extinct, except in Portugal, Spain, Italy, Sicily, Sweden and (occasionally) Norway. They are widely distributed in the U.S.S.R. The western limit, though fluctuating considerably in north-west Russia, may be taken as a line running from Sweden to Finland, and then along the eastern borders of the Baltic Soviet Republics, Poland and Czechoslovakia; thence through Romania to Yugoslavia and Bulgaria, with occasional extensions into northern Greece and Turkey. The Asiatic range includes Soviet Asia, except various northern islands; Sinkiang, Mongolia, Manchuria; Kansu in northern China, eastwards to the Gulf of Po Hai (Chihli), and adjacent parts of central China; Korea, Japan, Tibet; in Pakistan and India, from Baluchistan and Kashmir south at least to Dharwar, and

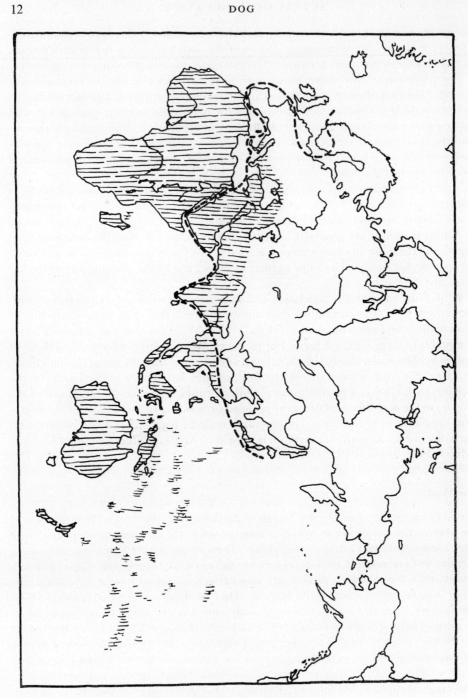

9. – – – Southern Range Limit of Old World Wolves
  ⌊⌊⌊ Distribution of Pariah Dogs (after Werth)

eastwards to Bengal; in south-west Asia, from Iran to Iraq, Asia Minor, Jordan and Arabia. Wolves are widely distributed in North America (Ellerman and Morrison-Scott, 1951).

In accordance with their extensive range, wolves display great variability in size and coloration. In certain localities distinct geographical races have evolved, which are distinguished by the shape of the skull and the colour of the coat; in other areas no particular geographical races have appeared (Antonius, 1922). In the Scandinavian countries small and large wolves, slender and coarse animals, some long-faced and others short-faced, are found side by side (Brinkmann, 1921). Studer (1901) thought that the skull of the wolf is the most variable among the skulls of all wild mammals. But

10. Running wolf, Poland (after Heimpel)

this is probably an exaggeration, considering that, in addition to the wolf, numerous wild species, far from being genetically uniform, contain a vast reservoir of overt or concealed variability, including recessive mutants, isoalleles and the huge array of morphs involved in balanced genetic polymorphism, both phenotypically conspicuous and cryptic (Mayr, 1963). Iljin (1941) has suggested that the causes of variation of several cranial characters of the wolf are of a genotypical nature.

Adult wolves range from 40–50 kg in weight, with a recorded maximum of 78 kg. The animal reaches a shoulder height of 85 cm and a length of body from the tip of the nose to the base of the tail of 125 cm, with a tail length of approximately 50 cm. The female is slightly smaller than the male. The basilar length of the skull varies between 164 and 236 mm (Hilzheimer, 1926). Nehring has described a huge wolf skull with a basilar length of 272 mm (Studer, 1901). The largest wolves are found in the northern parts of the range, although not in the extreme north; the smallest races occur in the south.

The wolf is remarkable for its sharp-cut hindquarters, its long flexible spine, firm set of the shoulders and keel-shaped chest with lack of depth (Hilzheimer, 1932). In the front view the chest is almost invisible behind the leg joints, the "elbows" being

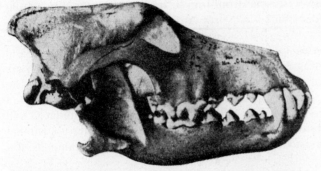

11. Male wolf skull from Galicia (after Antonius)

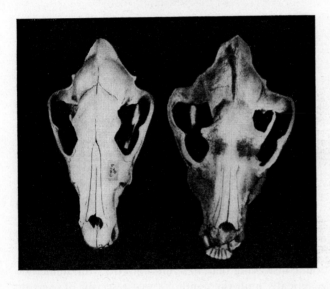

12. Wolf skulls, left with narrow, right with broad cheek-bones (after Iljin)

turned inwards and pressed to the chest. The legs are straight, with the pads slightly turned outwards. The straight log-shaped tail is carried low (Iljin, 1941).

The prevailing colour of the fur is zonargrey (i.e. black, brown or blue with one or several yellow or white bands or zones in the upper part of the hair), rufous or yellowish grey, more or less mingled with black, on the upper part of the body, with whitish underparts. The woolly underfur is slate-brown. The tail may be tipped with black. From this ordinary type of coloration there may be variations owing to the development of a more or less marked grey or red tinge. In some cases the fur is much paler than usual, and even pure white, as in Canis lupus albus Kerr of the tundra and forest-tundra of U.S.S.R., and in others nearly or quite black. The so-called "blue" wolves occasionally met with in the northern region of Siberia are zonarblack with a dilute black band, giving the impression of an ashen blue shade. In Europe the light-coloured varieties are characteristic of northern, and the dark of southern regions, black wolves being not uncommon in Spain. A black race occurs also in Tibet; in 1963 the author saw a family of three local black wolves at Changchun, Kirin province, north-east China. Sporadic cases of black wolves have been frequently reported in zoological literature; a juvenile black wolf, shot near Chkalov, is exhibited in the Moscow Zoological Museum. Iljin (1941) considers the occasional appearance of such black wolves as the result of matings of two wolves heterozygous for black (to be more precise, in non-zonarity), zonar grey (wild grey) in the wolf being dominant to black.

The Wolves of America

Great individual variation in size, cranial conformation and coat colour is found also in the wolves of North America where at one time twenty-three races of them lived, ranging from Newfoundland in the east, to Vancouver Island, British Columbia, in the west, and from the plateau of middle Mexico to approximately 440 miles from the North Pole (Young, 1944). As in Europe and Asia, the wolves in America are smaller, slenderer, leggier and darker in colour in the south than in the north (Lydekker, 1893–96). Wallace (1889) has stressed the great cranial variability in North American wolves inhabiting the same region.

Small Southern Races

Everywhere the small southern races, such as Canis lupus hodophylax Temm. of the southern islands of Japan, C. l. pallipes Syk. of India, C. l. arabs Pocock of southern Arabia, and C. l. deitanus of southern Spain, pass almost imperceptibly into the northern representatives of the species, from which they differ little except in size and occasionally in colour. Individual variability within these races is very high.

The Indian Wolf

The Indian wolf, Canis lupus pallipes Syk., is regarded by several authors as distinct from the common species (Noack, 1915; Lydekker, 1916; Brinkmann, 1921). It occurs

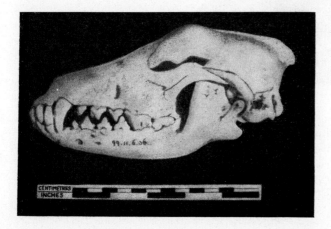

13. Cranium of Canis lupus arabs (after Clutton-Brocks)

14. (left) Head of female Canis lupus pallipes (photograph: G. Strauss)
15. (right) Female Canis lupus pallipes

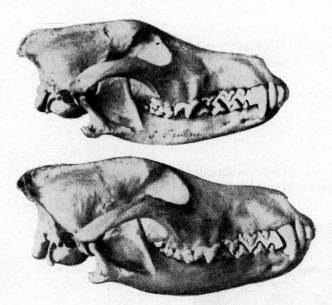

16. Crania of Canis lupus pallipe from West Pakistan (above) and of a female Persian wolf (below) (after Antonius)

on the right bank of the Indus, whence its range extends west into Baluchistan and thence to Iraq, the Jordan river and northern Arabia. On the left bank of the Indus it is replaced by the ordinary wolf which, although widely distributed over Asia, is not found east of the Bay of Bengal.

Like most southern wolves, Canis lupus pallipes is smaller in size and slighter in build than the northern races. According to Huxley (1880), it "more nearly approaches the jackals than any other Old World Wolf...". The fur is greyish fulvous in colour, usually with a brownish tinge, sometimes with more or less black, and occasionally with a marked rufous tint. The skull is distinguished by a relatively broad vaulted neurocranium, low position of the splanchnocranium, slender nasal part, well developed bullae, and weak dentition; but individual variability in these features is considerable. Considering that the average basilar length of the pallipes skull, in conformity with the smaller body size, is below that of the typical Canis lupus, 192 as against 217 mm (Brinkmann, 1921), most of the distinctive cranial features of the Indian wolf should be attributed to the relatively larger size of the brain, which is a consequence of the small body size. Therefore, the genetic differences between Canis lupus pallipes and the northern wolves may be reducible to different ranges of variation in body size, which renders it doubtful if the Indian wolf is entitled to rank higher than a geographical race.

II. Studer's Classification of Prehistoric Dogs

Studer's systematics of prehistoric dogs in accordance with cranial conformation, and his classification and genealogy of recent dogs on the basis of their cranial resemblance to prehistoric prototypes have become obsolete with our growing understanding of the biological causes of the basic cranial differences among ancient and recent dogs. Yet the legitimate criticism directed against his taxonomy, the establishment of more scientific systems of canine classification, a new nomenclature, and several recent finds of prehistoric dog skulls that have shattered Studer's theory of the phylogeny of his prehistoric prototypes, have not diminished the importance of his work. It is still practically impossible to discuss the origin of domestic dogs without reference to Studer's views, while the terms he employed for different cranial types have retained their general usefulness to this day. For these reasons our description of African dogs and the discussion of their descent and origin are prefaced by a summary of Studer's classification.

Studer (1901) divided the prehistoric dogs and their derivatives into two main groups: palaearctic and southern dogs. The link connecting the palaearctic and southern dogs is represented by the prehistoric Canis familiaris poutiatini, described by Studer in a later work (1906).

Canis Familiaris Poutiatini

The type specimen of C. f. poutiatini, consisting of a complete skeleton, was found on the southern shore of Lake Vysokoye, near Moscow, in strata believed to belong to the Russian Mesolithic; but Gandert (1930) considers a neolithic age as likely. In life, C. f. poutiatini was about the size of a Collie. Antonius (1922) suggested that but for its slightly longer legs it resembled the Australian dingo. The skull has a basilar length of 169 mm. The neurocranium is long, vaulted in the parietal, and markedly constricted in the temporal, region. The forehead is high and broad, slightly dished, with sloping supraorbital processes. The temporal ridges fuse below the coronal suture into a high sagittal crest which extends aborally into a prominent occipital protuberance. The splanchnocranium is narrow from the Foramina infraorbitalia orally, the lateral walls descending steeply to the alveolar portion. The stop is shallow, the zygo-

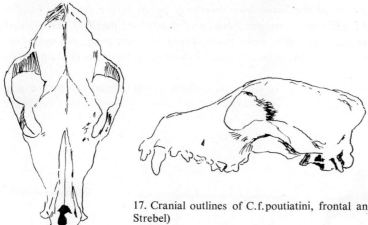

17. Cranial outlines of C.f.poutiatini, frontal and lateral views (after Strebel)

matic arches are short, and the orbits wide. In profile the height of the forehead, from which the neurocranium and splanchnocranium slope in opposite directions, is prominent, the aboral descent of the neurocranium being considered as a major domestication feature in the canine skull.

Palaearctic Dogs

Studer's group of palaearctic dogs comprises five different types:—1) Canis familiaris palustris Rütimeyer, 2) Canis familiaris inostranzevi Anuchin, 3) Canis familiaris leineri Studer, 4) Canis familiaris intermedius Woldřich, 5) Canis familiaris matris optimae Jeitteles.

Canis Familiaris Palustris

The palustris or turbary type in domestic dogs was first described by Rütimeyer (1861) from neolithic lake dwellings in Switzerland, about 3500–2000 B.C. Dogs of the same general type, although occasionally of a somewhat larger size, have been found also in neolithic sites in southern Britain (Windmill Hill, Whitehawk Hill, Trundle). The Swiss lake dwellings, however, do not harbour the oldest palustris type in Europe, as still earlier and more primitive representatives of this group occur in mesolithic sites in northern Europe. The typical C. f. palustris has a small cranium with a basilar length of 130–145 mm, a rounded capacious neurocranium with large orbits and only slightly developed superstructure; the parietal ridges are joined into a moderate sagittal crest which extends aborally into a short occipital protuberance. The slightly dished forehead and the zygomatic arches are of moderate width, the splanchnocranium is pointed and relatively short, the neurocranium considerably exceeding the splanchnocranium in length, and the stop is well defined.

The Ladoga form of C. f. palustris, i.e. C. f. palustris ladogensis Anuchin, is distinguished from specimens of the Swiss lake dwellings by its larger size (the basilar length measures 145–155 mm), stronger bone and dentition, markedly developed sagittal crest, greater parietal width, more pointed splanchnocranium, and shallower stop; the neurocranium is relatively less developed than the splanchnocranium (Gandert, 1930).

The dwarf form of C. f. palustris, i.e. C. f. spaletti Strobel, from the bronze age terramare of Upper Italy, the neolithic turbary station of Bodman, Lake Überlinger

18. Skull of C. f. palustris from the lake dwelling of Font (after Antonius)

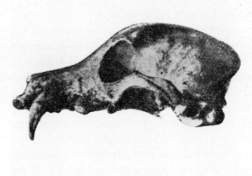

19. Skull of a recent Scotish Terrier of palustris type, frontal and lateral views (after Antonius)

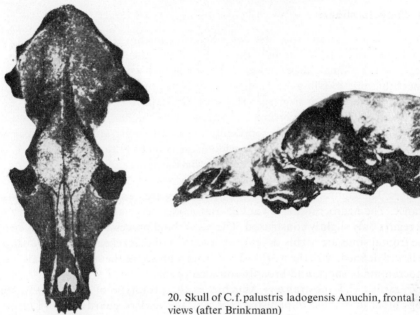

20. Skull of C.f. palustris ladogensis Anuchin, frontal and lateral views (after Brinkmann)

(L. of Constance), the lake dwellings of Ripač in Bosnia, and several sites in Germany (Sipplingen, Baden), Switzerland (Egolzwil) and Austria, has a basilar skull length of only 125–130 mm; the neurocranium is spherical and devoid of a sagittal crest, and the mandible is short and markedly curved.

Studer regarded the Pomeranian, primitive terrier (prior to the cranial changes brought about in the modern breeds), the Chow, the Battak dog of the highlands of Sumatra, etc. as recent representatives of the typical palustris form, and some of the arctic dogs as descendants of the larger Ladoga type.

21. Nenets (Samoyed) (C.f. palustris ladogensis type)

Canis Familiaris Inostranzevi

Canis familiaris inostranzevi was first described by Anuchin (1882) from crania recovered by Inostranzev from late neolithic strata of the Lake Ladoga moors. A cranium of this type, although larger in size and resembling a wolf's skull, was found in the late neolithic pile dwellings of Font at Lake Neuenburger; another one, dating probably from the bronze or iron age (Hallstatt), in Lake Bieler; Wettstein has described a similar skull from the bronze age settlement of the Zurich Alpenquai, where C. f. palustris is absent. The earliest representatives of the inostranzevi type occur in mesolithic sites in Sweden and Denmark, which date from 6500 to nearly 8000 B.C.

The skull of C. f. inostranzevi Anuchin has a basilar length of 177 mm. It is characterised by a long powerfully developed sagittal crest, and correspondingly short parietal ridges. The sagittal crest markedly slopes backwards, ending in a long occipital protuberance. The neurocranium is vaulted and moderately narrows orally, with the temporal region only slightly constricted. The postorbital processes are markedly bent down, the frontal sinus are highly developed, and the broad forehead is deeply dished. The stop is well defined, with the forehead towering high above the straight nasals. The splanchnocranium is short, and broadly rounded orally.

Studer regarded C. f. inostranzevi Anuchin as the prototype of the mastiffs, the arctic dogs other than those of palustris type, and the herders' guard dogs of Europe. The term "herders' guard dog" or, simplified, "guard dog", also called "shepherd-dog" (Hilzheimer, 1932; Epstein, 1969) — in German cynological literature "Hirtenhund" (Hilzheimer, 1926) — is employed with reference to a peculiar cranial type within the range of variation of C. f. inostranzevi, and should not be confused with sheepdog. In cynology the term mastiff is commoly applied to, and includes, breeds

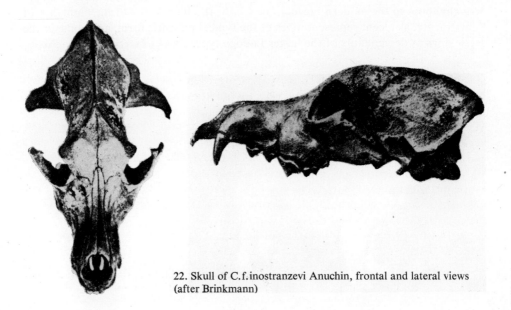

22. Skull of C.f.inostranzevi Anuchin, frontal and lateral views (after Brinkmann)

23. Norwegian Elkhound
(C. f. inostranzevi type)

of mastiff as well as of herders' guard dog type. Indeed, the border lind between these two inostranzevi types is arbitrary, for the guard dogs merely represent a more primitive or generalised, and the mastiffs a more specialised stage of the same general type. According to Brinkmann (1925), the original C. f. inostranzevi form still survives in the Elkhound of Norway (Fig. 23). The tendency to the mastiff type is indicated by a mandible of C. f. inostranzevi from Kassemose, Denmark; this is so strongly curved that the dog to which it belonged must have had the facial profile of a mastiff.

Canis familiaris leineri

The type specimen of Canis familiaris leineri Studer is represented by a skull recovered from the late neolithic turbary station of Bodman on Lake Überlinger. Studer has pointed out that in its slender form this skull resembles a greyhound's, and in its straight facial profile that of C. f. matris optimae. The basilar length is 200 mm, and the total skull length 236 mm. Both the neurocranium and splanchnocranium are long, the sagittal crest is straight and moderately developed, and the occipital protuberance very prominent. The lateral walls of the parietal region are vaulted. In the temporal region the skull is markedly constricted, so that the pear-shaped parietal and occipital parts are clearly marked off from the broad forehead. The latter is flat and high, and only slightly dished. The occipital triangle is high, the base of the skull broad, and the tympanic bullae are very large. The neurocranium passes without a stop into the long splanchnocranium which gradually narrows orally. The constriction of the splanchnocranium at the anterior end of the zygomatic arches, which appear considerably extended above the temporal fossae, gives the region at the posterior end of the nasals and the portion of the frontal processes of the maxillae flanking the nasals the appearance of being distended upwards, and causes their sharp angulation with the steeply descending walls of the maxillae. The dentition is strong.

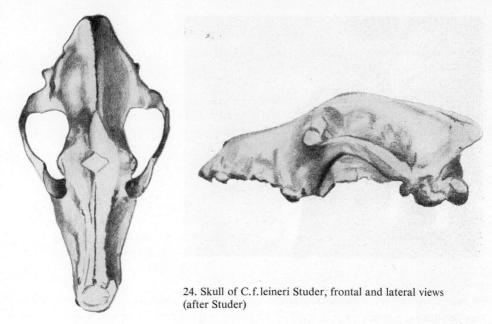

24. Skull of C.f.leineri Studer, frontal and lateral views (after Studer)

Skulls of the same general type, but larger and more powerful, with a slightly blunter splanchnocranium, stronger zygomatic arches and highly developed sagittal crest, have been found in Ireland. Studer regarded C. f. leineri as the ancestral form of the Irish Wolfhound, Scottish Deerhound, and the larger greyhounds of Europe. But Hauck (personal communication, quoted by Menzel and Menzel, 1960) has pointed out that the type specimen of C. f. leineri in conformation and dimensions is not a greyhound skull; therefore, it does not represent the ancestral form of the northern greyhounds, such as the Irish Wolfhound and Scottish Deerhound (Figs. 24, 123 and 124), whose rough wiry coat is attributed to an admixture of rough-coated guard dogs to the greyhound type.

Canis Familiaris Intermedius

The cranial type specimen of the hounds of Europe, recovered from a bronze age tomb in Lower Austria, has been described by Woldřich (1882a) with the name Canis familiaris intermedius. In Switzerland the earliest remains of C. f. intermedius Woldřich date from the most recent neolithic period. These are more primitive in conformation than the bronze age type specimen, linking the intermedius type with C. f. palustris. C. f. intermedius has since been found also in a neolithic context at Most in C.S.S.R. (Zeuner, 1963a). In North Germany it is already represented in mesolithic sites. In Europe it survived unchanged until the La Tène period. The original cranial type occurs among the primitive hounds of Europe, such as the Norwegian Harehound, to this day.

The cranial type specimen has a basilar length of 164 mm; it is characterised by a large, high and vaulted neurocranium, only slightly dished frontal area, strongly developed frontal sinuses, markedly constricted temporal region, short sagittal crest, and long, highly arched temporal ridges. The splanchnocranium is relatively short and broad, and blunt orally; the stop is well defined. C. f. intermedius differs from C. f. poutiatini principally by a larger, wider and higher neurocranium which is less constricted in the temporal region, a higher forehead, shorter, broader and less pointed splanchnocranium, and more clearly defined stop due to the height of the forehead.

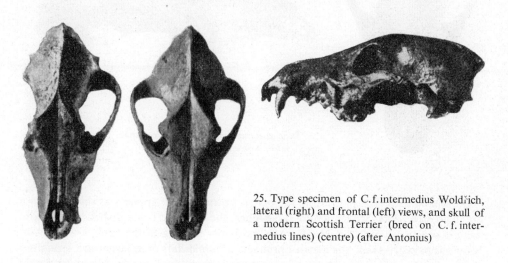

25. Type specimen of C.f.intermedius Woldřich, lateral (right) and frontal (left) views, and skull of a modern Scottish Terrier (bred on C. f. intermedius lines) (centre) (after Antonius)

Hilzheimer (1926), following Antonius (1922), regarded C. f. intermedius as a specialised derivative of C. f. poutiatini, the relation of these two types being paralleled by that of C. f. palustris of the Swiss lake dwellings and its earlier Ladoga form (Ebert, 1924–32). From C. f. palustris, C. f. intermedius is distinguished by the greater length of the cranium, more especially of the nasal part, greater width of the maxillae, lesser cranial height, and stronger development of the superstructure; its forehead is wider and flatter, the oral part more broadly rounded, and the stop less pronounced. In the shape of the neurocranium C. f. intermedius resembles C. f. palustris (Studer, 1901).

Canis Familiaris Matris Optimae

The type specimens of the primitive sheepdogs have been described by Jeitteles (1877) with the name Canis familiaris matris optimae. As it was formerly believed that the matris optimae type made its first appearance during the bronze age, it has occasionally been called "bronze dog". In reality it occurs already in mesolithic strata in North Germany. The basilar length of C. f. matris optimae skulls measures 170–190 mm. Typically the neurocranium exceeds the splanchnocranium in length; but in recent sheepdogs, grouped with C. f. matris optimae, the neurocranium and splanchnocranium are occasionally of equal length (Hauck, 1950).

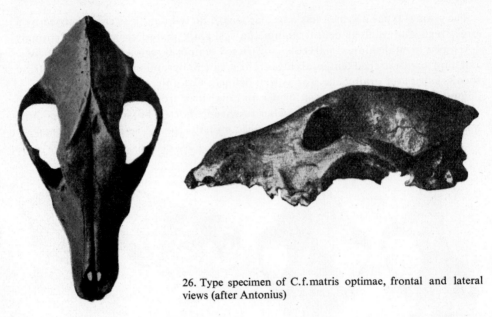

26. Type specimen of C.f.matris optimae, frontal and lateral views (after Antonius)

Compared with C. f. poutiatini, the matris optimae skull appears as though drawn out in all directions (Studer, 1906). It is higher; the forehead is not dished but flat, and there is practically no stop, the frontals gradually sloping to the roof of the nasals. From the palustris skull, the matris optimae cranium differs in its larger size, straighter profile, longer and more pointed splanchnocranium, and regular development of the sagittal crest. From the intermedius skull it is distinguished by the relatively smaller maxillary width, longer nasals and lesser neurocranial height. C. f. matris optimae is most closely related with C. f. leineri; indeed, many recent crania of the general matris optimae type are more greyhound-like than the skulls of greyhounds proper or the cranium of C. f. leineri. But the type specimens of C. f. matris optimae are smaller and more compact than the leineri skull, the occipital protuberance is shorter, the occipital triangle lower, the forehead is relatively longer and narrower, the splanchnocranium shorter, wider aborally and more pointed orally, the nasals are shorter, the tympanic bullae smaller, and the parieto-frontal crests unite into the sagittal crest more aborally.

Southern Dogs

The southern dogs, represented by the dingo-pariah group, have been classed by Studer (1901; 1906) into three sections in accordance with his assumption of three distinct phases in their state of domestication:—1) The dingo of Australia and the Tengger dog of Java, regarded as the wild ancestral stock of the southern group of dogs; 2) the pariah dogs of Africa, southern Asia and the south-eastern Balkans, still in a semi-wild state; 3) the southern greyhounds and the Tibetan Mastiff, fully domesticated.

Studer considered the dingo, Tengger and pariah dogs as cranially akin to C. f. poutiatini, the Tibetan Mastiff as of a similar cranial type, though much enlarged through crossing with the wolf, and the southern greyhounds as derivatives of the pariah type, with the splanchnocranium drawn out in length.

Studer's theories on the wild (as distinct from feral) state of the dingo and Tengger dog, the semi-wild (as distinct from semi-feral) state of the pariah dogs, and the origin of the Tibetan Mastiff from an outcross of pariah dogs with the wolf are erroneous, as will be shown in our subsequent discussion (for the status of the dingo and Tengger dog see pp. 114–120; for the origin of the Tibetan and other mastiffs pp. 172–181.

III. Distribution and Characteristics of the Dogs of Africa

i. The Pariah Dogs of Africa

1. Distribution and Characteristics of the Pariah Dogs of Africa

Name and Distribution

The term pariah is derived from the Tamil "paraiyan"; literally meaning "drummer", it is applied to members of a low caste in southern India. As applied to domestic dogs, it denotes a group whose individual members are generally distinguished from the majority of other domestic dogs in that they are not attached to human masters, frequently not even to certain households; they are not bred, reared or protected by man, but eke out a miserable existence scavenging on whatever they can pick up in the streets and outskirts of towns and villages.

The range of the pariah dogs extends from the eastern shores of the Atlantic through Africa, the southern parts of the Balkan peninsula, the Caucasus and southern Asia to China and the islands of Japan. Pariah dogs occur in numerous islands of the Pacific, also in Australia and New Zealand (see Map. p. 12). Again, dogs of pariah type and habits are found in many parts of South and Central America. Until comparatively recently they were also present in southern Spain (Vesey–FitzGerald, 1957).

Werth (1944) employs the term Shenzi (Swahili for wild or uncultivated for the whole pariah group. The term pariah, he reasons, refers to an outcast, whereas these dogs were originally not outcasts, but served their owners as food. Like the pig, goat and hen, used for the same purpose, they were not fed, but had to find their own sustenance.

In the present description the term pariah has been retained to denote a group of African dogs of a primitive, generalised racial type, more especially in cranial conformation, irrespective of whether the individual representatives conform to the general habits of pariah dogs—as the great majority actually do—or whether some breeds[1] or individuals have risen to a more companionable state in their association with human society, — i.e. have become domestic dogs proper.

[1] Here and in the following the term "breed" is employed either with reference to groups of domestic animals kept in geographical, at any rate genetic isolation, or to such groups which are not genetically isolated and in which individual animals are not easily distinguishable from those of similar neighbouring groups, but which represent populations that would statistically differ from those of adjacent regions in certain taxonomic characters, somewhat analogously to wild geographically continuous subspecies with free gene flow in their contact areas.

Studer's Description of the Pariah Skull

According to Studer (1901), the pariah has a narrow skull, with a long neurocranium, vaulted in the parietal region, a narrow, low and dished forehead with sloping supraorbital processes, and very narrow between the orbits. The sagittal crest is well developed, and the occipital protuberance markedly drawn out aborally. The Bullae osseae are fairly large, bluntly keeled, and the occipital triangle is low. The low position of the forehead, due to the weak development of the brain cavity, is responsible for the obliquity of the orbital plane. At the lower edge of the temporal ridges, behind the orbital processes, the frontals are markedly distended. The splanchnocranium is of limited width, gradually narrowing orally, the stop is little pronounced, the profile at the posterior end of the nasals is concave, the back of the nasals gradually sloping orally. The maxillary portion at the attachment of the jugals is narrow, resulting in only a slight constriction in front of the infraorbital foramen. From the narrow back of the nasals the lateral walls of the maxillae slope steeply, in the anterior part nearly vertically. The zygomatic arches are strong but not very wide; commonly the neurocranial portion is as long as, or longer than, the splanchnocranial.

The Pariah Dogs of Egypt

Dogs can be seen in every part of Egypt and Sinai wherever there are human inhabitants. They are commonly of the pariah type—according to Lydekker (1893–96), of a single race, but according to Flower (1932), of several different breeds. Their size is about that of a sheepdog or collie, but they are of slightly stouter build, with a broader head and long bushy tail. The shoulder height averages 55 cm (Lydekker, 1893–96). The ears are short, erect and pointed. In young pariahs the ear is pendulous; with the strengthening of the muscles between the third and sixth month it changes into the common prick ear (Siber, 1899). The coat is usually lemon-and-white or pure white in colour, sometimes all lemon or all brown. In some individuals the brown tends to grey, and in others to yellow. The legs and underline are often white. Occasionally black or tawny animals may be observed, with white feet and tip of the tail (Jarvis, 1936). Flower (1932) says that black dogs are common.

27. Egyptian Pariah Dog (after a drawing by R. Hartmann)

According to Gaillard and Daressy (1905), the Egyptian pariah is more slender in conformation than are the pariah dogs of Istanbul. It has a long heavy head with straight rather short prick ears. The thick tail is carried low, reaching down to the heels. The forefeet are furnished with five toes, the hindfeet with four, lacking in a dewclaw. The coarse rough coat is commonly dark red in colour, occasionally a light yellow; black-spotted specimens are rare.

Murray (1935) says that "there are almost as many strains in the dog tribes of Egypt as among the Bedouins, their masters". The pariahs act as watchdogs, preferring to stay about the tents on guard rather than follow their masters about the country. They make quite good sheepdogs, if brought up with the flock when young. The Hanadi, a western tribe settled in Sharqiya, when hunting jackals, always take one or two pariahs with them, as the saluki will not close with a jackal whereas a good pariah will when the greyhounds have brought it to bay.

The watchdogs of the Awlad 'Ali, a tribe of the north coast of Egypt from Salum to Alexandria, are big curly-tailed dogs, often with a great deal of white on them. They strongly resemble the dogs of Tripolitania, possibly because the Awlad 'Ali, until about 150 years ago, lived for a long time in the "Green Mountain" of Cyrenaica. In the west the Awlad 'Ali have sheepdogs of pariah size and the chestnut colour so common among pariahs (Murray, 1935).

Cranial Conformation in Egyptian Pariah Dogs

Cranial variability among the pariah dogs of Egypt is considerable, and some pariah skulls markedly deviate from Studer's description of the type specimen (see p. 29). Noack (1907) found considerable conformational variation, indicating a complex mixture of types, in four pariah skulls collected in the vicinity of Cairo and Saqqara. In the narrow splanchnocranium and the width of the zygomatic arches these skulls show a marked resemblance to C. f. palustris; but in size and profile three of the skulls approach C. f. intermedius, while the fourth differs by its smaller size, convex parietal region, greater neurocranial width, less acute angulation below the orbits, and shorter splanchnocranium. The well marked parieto-frontal crests are joined into a moderately high sagittal crest, remaining separate only in the smallest of the four skulls. The occipital crest is well developed, the zygomatic arches are wide, and the nasals slope steeply laterally.

'African' Hairless Dogs

The occurrence of a hairless variation in African pariah dogs, characterised also by the lack of the upper and lower premolars, has been reported from Egypt as well as from West Africa (Reade, 1863), Central Africa (Pechuel-Loesche, 1893) and South Africa (De Bylandt, 1897). Darwin (1868) was informed by Dr. Bowerbank of a hairless Berber bitch and dog of black skin colour. A hairless dog, reputed to have come from Egypt, was kept in the Zoological Collection in London in 1832. However, this specimen may have been wrongly assigned to Egypt; for Sonnini (1801) stressed that he had not encountered any hairless dogs in Egypt or Turkey.

The idea that hairless dogs came from Egypt is traceable to Linné's "Systema Naturae", Editio X (1758), where this dog is listed as "Canis aegyptius, nudus absque pilis" (Plate, 1929–30). This error has been repeated by numerous authors for 150 years. Strebel (1905) wrote about hairless dogs:—"They were formerly called Africans, in reality they originate from Central and South America." Hairless dogs have been observed in China, southern India, Ceylon, the West Indies and South America; the Mexican variety is recognised as a toy dog breed by the American Kennel Club (Hubbard, 1946).

The Pariah Dogs of North Africa

Small, primitive, terrier-like pariah dogs of a mean weight of approximately 5 kg occur in large numbers in Libya and the Atlas countries. In Libya they are particularly numerous in the coast region, the Berber towns of the Jebel and the town of the Jewish cave dwellers of Tigrinna; they are also common in the oases of the Sahara. In general, they are characterised by prick ears, occasionally with drooping tips, a short smooth coat with somewhat longer hair on the underside of the tail, either black-and-white or brown-and-white, more rarely self-coloured. Peters (1940) recorded the following measurements in two Libyan pariah dogs weighing 4.75 and 5.5 kg respectively (cm):

Measurements	I	II
Height at shoulder	30.2	25.9
Length of body	32.8	37.5
Height of chest	14.7	14.9
Heart girth	38.0	38.0
Length of tail	16.2	22.5
Length of head	13.1	15.4
Width of head	8.1	8.6
Interorbital distance	3.7	3.5
Width at mandibular angles	5.6	4.3

Berber Dogs

The Berber dogs of Libya, Algeria and Morocco show an improvement on the common pariah, but are much rarer. They are fairly large, strong watchdogs, with a bushy tail and dirty white coat, occasionally black-and-white or with a few red spots on the body, head or tail, more rarely brown or black. Generally the Berber dogs of the Atlas countries differ from the common pariah in the prevalence of sheepdog features. According to Fitzinger (1876) they are nearly related to the Pyrenean Mountain dog, but are smaller, with a flatter forehead, shorter neck, more compact body, and longer hair on the neck, chest, posterior part of fore- and hind-legs, and lower part of the tail. The latter is sickle-shaped, but carried low and nearly straight when the animal is running. Two types of coat are distinguished: one moderately long and the other short. Peters (1940) writes that the Berber dogs belong to the same type as the pariah

28. North African pariah dog (after Peters)

29. Tripolitanian dog (after Hilzheimer)

dogs which formerly occurred in Istanbul, but they also resemble Eskimo dogs. Frank (1965), who separates the Berber or Kabyle dog racially from the African pariah and greyhound groups, describes it as of sheepdog size, prick eared, with a thick white or yellow-and-white coat and bushy tail. "It resembles the Chow, also Arctic and sheep-dogs." The Chow, as Epstein (1969) has pointed out, is a derivative of South China pariah stock; prima facie the derivation of the Berber dog from North African pariah stock appears therefore to be most likely.

30. Algerian dog (after a drawing by Pierre Mégnin)

31. Head of a Berber dog with cropped ear tips (after Peters)

Cranial Conformation in Berber Dogs

In six crania of pariah dogs collected in the vicinity of Mogador, Morocco, Noack (1907) has recorded a combination of the C. f. palustris and C. f. intermedius types, with the frontal conformation approximating to C. f. matris optimae. Several features are suggestive of the Egyptian pariah, and some of the Moroccan jackal (sic!). One of the six skulls, considerably larger than the rest, has a narrow wolf-like neurocranium, wide zygomatic arches, and a stop nearly as shallow as in the greyhound. In the other five skulls, which are of a fairly uniform type, the neurocrania are egg-shaped, at the posterior end broader than in the centre; the flat fronto-parietal ridges extend considerably backwards, forming low sagittal crests in three of the skulls, but

32. North African pariah skull (basilar length—173 mm), front and lateral views (after Hilzheimer)

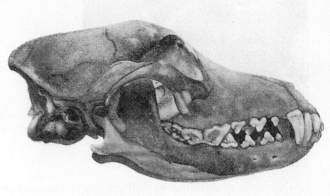

33. North African pariah skull (basilar length—151 mm), front and lateral views (after Hilzheimer)

remaining separate throughout in the others. The forehead is moderately rounded, slightly dished in the centre, and only little elevated above the frontal sinuses. The nasals slope obliquely sideways. The posterior edge of the Foramen infraorbitale is markedly distended by the anterior root of the upper carnassial, resulting in a clear demarcation between the moderately long nasal region and the posterior part of the maxilla. The interorbital width varies in accordance with the width of the nasals. In profile the crania show a marked resemblance to the skull of C. f. intermedius, though the stop is less pronounced than in the typical intermedius skull. The dentition, again, approaches that of the Egyptian pariah. Noack concluded that the Berber dogs represented a complex mixture of various racial types; nonetheless, he considered their descent from Egyptian pariah dogs as possible, although there were certain conformational differences. Antonius (1922) also regarded them as a pure pariah breed improved by selection.

The Armenti

A similar dog, named after the village of Armant, on the left bank of the Nile, occurs in Egypt where the breed is valued for its strength, courage and intelligence. Siber (1899) has described it as larger in size than the common pariah, with a broad head, pointed muzzle, erect ears of medium size, strong well-placed legs, a bushy tail carried low but with the tip slightly curled up, and a long rough coat usually greyish yellow in colour with a black muzzle, occasionally black throughout. However, this description and Strebel's drawing of an Armenti (Fig. 35) are not quite correct; the Armenti is actually lower in the legs, and has a wider and blunter muzzle, and short drooping ears which are partly covered by the long hair of the head. The height at the back is approximately 55 cm, and the weight about 24 kg. Hubbard (1946) describes the colour of the Armenti as black, black-and-white, tan-and-white or grizzle-and-white, with the topknot, muzzle and brisket white.

It has been suggested that the Armenti is not an indigenous breed, but owes its origin to European dogs left in the vicinity of Armant by Napoleon's soldiers during

34. (left) Armenti, front view (after Hubbard)
35. (right) Strebel's drawing of an Armenti

the Egyptian campaign. Hartmann (1864) believed the Armenti to go back to a guard dog which a Russian traveller left at Armant (Siber, 1899). Considering however that dogs of a similar type occur throughout Morocco, Algeria, Tunisia and Tripolitania, and that no canine breed evolved in such an haphazard manner and subjected to a complete lack of attention and selection could have retained its identity for so long a period, these theories of the Armenti's descent appear to be quite unfounded. As Murray (1935) says:—"The fierce Armenti watchdogs of Upper Egypt so closely resemble the Pyrenean sheep-dog, that they are nowadays said to be descended from the dogs brought over by Napoleon's army. This is unlikely, since Wilkinson, writing

36. Pariah skull of C.f.matris optimae type from the Upper Nile (after Antonius)

only thirty years after the French invasion, calls them Hawara dogs, after the well-known Berber tribe of Upper Egypt." A similar view has been expressed by Peters (1940) who regards them as a long-haired mutation of the Berber dog.

Menzel and Menzel (1960) refer the Armenti to the C. f. matris optimae cranial type. The occurrence of this type among the dogs of Upper Egypt is indicated also by a pariah skull from the Upper Nile (Fig. 36), in which the resemblance to the cranium of the early prototype of the Collie (C. f. matris optimae) is striking (Antonius, 1922).

Hilzheimer (1908) has given a short description (without photograph) of a powerful male canine cranium which Schweinfurth had collected in Egypt. It is distinguished by strongly developed superstructure and very large frontal sinuses in relation to which the brain case appears small. The parietal region is relatively narrow, the constriction of the skull in the temporal region, half-way between the postorbital processes and the weak occipital protuberance, only slight, and the frontal region very broad and vaulted. The stop is well defined, and the lateral walls of the broad splanchnocranium are parallel. Hilzheimer concluded that the skull obviously belonged to an animal of the mastiff group "descended from wolves", adding, however, that it might be the skull of an Armenti. This would indicate that the Armenti, and possibly also the nearly allied Berber dogs, occasionally display a tendency to the mastiff type.

East African Pariah Dogs

Dogs of a similar type to the Armenti are kept in the highlands of Ethiopia to guard the herds and flocks (Siber, 1899). In order to make them fierce, they were locked up in dark pits for months (Hildebrandt, 1874). The present author has met similar dogs occasionally in Syria and Palestine, besides the common pariah; they are usually held in a superior state as guardians of tents, buildings and flocks. They are common in Anatolia. Jarvis (1936) also encountered them in the Libyan, Egyptian and Sinai deserts. In his discussion of the pariah he says:—"There is a variety that appears to be a distinct breed and this is a leggy savage beast all black in colour, and these are valued by their owners as being particularly useful for guarding gardens or flocks."

South of Egypt proper the pariah dogs are slightly lighter in build and finer in bone than those of the Lower Nile valley. They stand about 40–50 cm at the shoulder and are usually white, yellow, red or sand-coloured, sometimes fawn or greyish white. The coat is short, the tail long and straight, rarely curled, and the ears are fairly long and erect (Siber, 1899).

In Eritrea and Ethiopia the pariah is practically the only type of dog to be found, save for a few salukis bred in the vicinity of the Red Sea coast (Marchi, 1929). Hilzheimer (1908) has recorded the measurements of two crania of small domestic dogs from Ethiopia, which are similar to the skull measurements of Pomeranians. In type the two skulls belong to the palustris group, although the relatively large dentition shows them to be rather more primitive than are the European representatives of that cranial group. The primitive character of the Ethiopian pariah dogs is expressed also in their low brain weight, which ranges from 80–86 g, with an average of 83.5 g,

37. White Nile pariah (after C. Keller)

whereas in European dogs of a similar body size the brain weighs about 10–15 per cent more (Klatt, 1921).

In some parts of Africa, as among the Wakindiga hunters and gatherers and the Wapare in the Bantu sphere of East Africa, the dog is absent, and no evidence is available that it had formerly been present there (Hahn, 1896; Kroll, 1929). The recent absence of the animal among the Wachagga and several other Bantu tribes is probably due to reasons of taboo.

The original type of pariah dog of the Bantu of East Africa is about the size of a Pomeranian, of slender build, with a pointed head, triangular prick ears carried horizontally, and a slightly twisted tail; the coat is yellow or yellowish brown in colour,

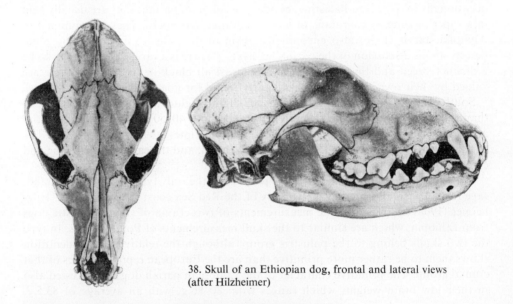

38. Skull of an Ethiopian dog, frontal and lateral views (after Hilzheimer)

39. Pariah dog of the Tindiga, East Africa (after Werth)

40. Baganda (Mutesa's) dog (after a sketch by Speke)

41. Baganda (Kamaraviona's) dog (after a sketch by Speke)

with white spots on the legs (Werth, 1915). Stuhlmann's (1909) description differs in some respects from Werth's; he recorded that the pure East African pariah was a rather small dog of compact build, with a pointed muzzle, large, triangular and erect, rarely drooping, ears, a coarse-haired sickle or ring tail with the tip turned to the right, and a yellow or brownish coat, sometimes black broken by tan above the eyes, on the legs and below the tail.

Around the partly excavated ruins of the ancient Arab town of Gedi, south of Malindi on the Kenya coast, Mason (1965—personal communication) observed the Giriama dog, which is very common on the coast but rarely encountered in other parts of Kenya. It resembles the Basenji (see below), but it is usually black with a tan face, a little shorter in the leg and with a less curly tail. All native dogs near Gedi are of this type, in striking contrast with the miscellaneous packs seen elsewhere in Kenya.

Among the Baganda and Bahima, lacustrian peoples of the Eastern Bantu group, dogs were formerly held in high esteem. One of the Baganda clans has adopted the dog as its totem animal, another clan uses the dog bell as a totem, for the Baganda, in common with many other Bantu tribes, provide their dogs with wooden bells when taking them on the hunt (Kroll, 1929). At the court of the Baganda king Mutesa, Speke, on his expedition to the Victoria Nyanza in 1859–63, found a considerable variety of dogs, some carrying a strain of greyhound blood, similar to the dog of Rumanica, ruler of the Karagwe (Fig. 80), and others of pure or mixed pariah stock. Speke drew several sketches of these dogs belonging to Mutesa and his adviser, Kamaraviona. They were red or white, very rarely of two colours; "the pet dog of Mutesa", Johnston (1902) recorded, "appears to have been white". Smaller in size than the common pariah, they were thin-boned, over-fat, with long thin muzzles and drooping or semi-prick ears, betraying saluki influence or some other mixture of stocks. They were seldom heard to give tongue. Wilson and Felkin (1882) wrote that the Uganda dogs were generally of a tan colour, and resembled smooth English

42. Elgumi dog (after Johnston)

43. (left) Pariah dog from the Lake District of East Africa (after a drawing by Stanley)
44. (right) Tail of Maviti (Mahenge) dog (after Siber)

terriers more than any other dogs. Under the influence of Mohammedan (Arab) traders from Zanzibar and the Sudan, to whom Speke's expedition opened the way to Uganda, dogs seem to have lost their former popularity among the lacustrian tribes. For only two decades later Stanley and the missionaries Wilson and Felkin observed that the esteem in which dogs were still held among the Baganda in Speke's time was rapidly decreasing (Siber, 1899).

The Elgumi people (sometimes called Wamia) between Mount Elgon and the Nzoia valley, north-east of Victoria Nyanza, have terrier-like dogs (Fig. 42) (Johnston, 1902). Stanley (1880) found typical pariah dogs all over Unyamvezi, Ujiji, Urua and Manyuema, in the vicinity of Lake Tanganyika, small, rough-coated, with long pointed noses and thin curved tails. The pariah dogs of the Tindiga of East Africa (Fig. 39) are of a fawn colour (Werth, 1944).

The Maviti of the Mahenge plateau, to the east of Lake Nyasa, a branch of the Angoni who fled north after revolting from the great Zulu chief, Chaka, bred pariah dogs for the sake of their meat, while the skin of the tail severed from the live dog served to attach the blade to the shaft of the spear. Siber (1899) has described the Maviti dogs as strong muscular animals, compact and short-coupled, with a broad rump, straight well-placed legs, a wolf-like head, strong dentition, long prick ears, and a bushy tail curved at the tip in a manner similar to the tail of the Bagirmi dog (Fig. 44).

Congo Dogs

The pariah dogs of the Congo belt are generally fawn, yellow, red or white, more rarely variegated. The tail is commonly carried close to the ground; occasionally it is curled up. These dogs usually howl, but do not bark. In villages in which they form a source of meat their condition approaches frequently that of a well fed pig; thus, the Ekoi, Yaunde, Banyangi, Babunda and Achewa of equatorial Africa castrate the males to speed up the fattening process (Kroll, 1929). But among Bantu and other peoples not keen on their flesh, they are usually starved and neglected. Exceptional in this respect are the dogs of the Batwa of the Congo. The Batwa are hunters and gatherers devoid of domestic animals save dogs and occasionally a few hens. The dogs are used in packs of three or four in the chase; they are treated better and show considerably higher intelligence than the common pariah dogs of the cultivators (Von Wissmann, 1883; 1890).

The dogs of the Mangbattu and Azande (Nyam-Nyam), true African peoples of the Ubangi-Uele basin, in the Eastern Province of the Congo, are distinguished by a small stature, straight legs, a curled tail, and short smooth coat which is red or yellow in colour with the neck commonly white (Schweinfurth, 1918). They have long prick ears, a pointed nose, and well defined stop. Cranially the Azande and Mangbattu dogs have been classed with C. f. palustris (Antonius, 1922).

The dogs of the Bongo, closely related to those of the Nyam-Nyam, are slightly larger in size, possibly owing to the influence of pariahs from the Upper Nile. They have fawn coats and long bushy tails carried close to the ground. Schweinfurth (1918) noted the peculiar erectability of the dorsal hair in the Bongo dogs, which is exhibited at the slightest emotion.

45. Pariah dog from the Congo (after Brehm)

46. (left) Pariah dog of the Babwende, Congo (after Stanley)

47. Nyam-Nyam dog (after Schweinfurth)

While the dog of Congoland is nearly always of the fawn pariah type, in Cameroun and in Lunda and Kioko, Angola, large black dogs occur; "their origin is undoubtedly Portuguese" (Johnston, 1907). The larger types of northern Congo, on the other hand, may be derived from the Sudan. "Nowhere in Negro Africa or in Upper Egypt is there a trace of the handsome Eskimo-like, Chow-like dog which is so characteristic a feature in the life of the Berbers and Tuareg of Northern Africa and the Sahara Desert, nor of the primitive greyhound type (slugi), also found in that region." The Ituri pygmies have no domestic animals except—and this not everywhere—prick eared fox-yellow dogs (Johnston, 1902); but they do not use them in the hunt

48. Basenji dog
49. (left) Babangi dogs, western equatorial Congo (after Johnston)

(Werth, 1944). Stuhlmann (1909) noted the very small size of the dogs of the Aka (Akka) and Ituri pygmies.

The Basenji, a Western Bantu people, and kindred tribes of the lower Congo, have dog sof typical pariah conformation and the size of Fox Terriers (Tudor–Williams, 1946). The height at the back averages 40 cm, and the weight ranges from 8 to 12 kg. These dogs are used by native hunters for beating game. The head is of moderate length, broad between the ears, flat across the top of the clean-cut skull, slightly dished between the orbits, with the stop little defined. Between the eyes and upon the forehead the skin is wrinkled. The eyes are rather small, fairly deep-set, and coloured hazel or brown; the ears are pointed, placed far apart, and carried stiffly erect, of medium size and open to the front; the muzzle is fairly sharp but not snipy, with level teeth and a black nose. The neck is rather long and well muscled; the body is compact and flexible, with a deep chest, fairly short and level back, and a moderately lifted loin with strong couplings. The legs are fairly long and springy, straight and of good bone, with small narrow feet having well arched toes. The tail is characteristically curled very tightly over the set-on, the tip generally resting to one side. The coat is short and smooth on a remarkably pliant skin; its colour is a rich chestnut-red, yellowish brown or black with white legs, underline and collar, the white extending from the neck to the nasal part of the head and thence in a thin median line into the forehead. This colour pattern is frequent in pariah dogs, not only in Africa but also in Asia. Occasionally Basenji dogs are cream-coloured or black-and-tan.

Normally these dogs do not bark, but they are capable of very nearly all the sounds common to a dog, with the exception of the yap of a terrier and the bay of a hound (Hubbard, 1946; 1948). Basenjis have an annual seasonal mating cycle in contrast with the semi-annual non-seasonal cycles of other breeds (Fuller and DuBuis, 1962).

The Congo basin is the central area of esteem of the dog. In Africa the eating of dog meat had originally a ceremonial character. The custom of sacrificing and eating dogs is mainly found in the area of West African hoe-cultivators, i.e. in the western and central Sudan, Upper Guinea and the Congo basin. Here the dog is regarded as the mythical conveyer of culture, in particular of the fire. In the area of the Zambesi, Angola, east and north-east Africa this custom has survived only sporadically, having been suppressed by more recent waves of culture, more especially of cattle-breeding peoples (Frank, 1965).

West African Pariah Dogs

The dogs of West Africa are of the common pariah type found throughout the Continent, cross-breeding with animals brought from the coast having hardly affected their conformation except occasionally in the carriage of the ears and tail and the length of the head. The West African pariah is distinguished by a conically shaped head, about 20 cm long and 13 cm wide, with a flat forehead displaying hardly any superstructure, a weakly pronounced stop, slightly overshot upper jaw, approximately 11 cm long prick ears, and a wrinkled skin over the forehead; occasionally, under the influence of European dogs introduced from the coast, the ears are semi-pendulous. The slender body is about 67 cm long; the length of the forelegs is 29 cm, of the hindlegs 40 cm, and of the tail 33 cm. The coat is smooth, yellow or fawn in colour with white patches on various parts of the body; occasionally the main colour is white (Pierre, 1906).

The Bagirmi and other true African tribes near Lake Chad have pariah dogs of a similar type to those found in Ethiopia,—short-coated animals with erect ears and a thin curled tail (Fig. 51).

The Malinke, Wassulonke and the majority of Bambara, occupying most of the region between the Atlantic and the upper Niger as far south as about latitude 9° N., in addition to a few Dahomey tribes, such as the Nagot, Baribas and Pila-Pila, breed dogs for their meat. The Dan of Liberia use their dogs in the hunt and also train them to guard their infants as well as the huts used by toddlers who are still too small to go into the bush. In this way the dogs become closely attached to their little providers and remain their faithful guardians (Himmelheber, 1958).

In southern Cameroun dog flesh is the food of the warriors (Werth, 1944). The Bakosi of Cameroun breed two distinct pariah types: a slender dog for the hunt and a coarser one for meat. The former stands about 35–40 cm at the shoulder. The head is broad and of moderate length, the spoon-shaped ears are carried laterally, and the tail, approximately 35 cm long, is either curled or straight. The hair is short and smooth; yellow, brown, black or dark grey in colour, nearly always variegated with white. The coarser type, bred for meat, is in the majority, and the males are commonly

DISTRIBUTION AND CHARACTERISTICS

50. (left) Pariah from (Mali) (after Pierre)
51. (right) Bagirmi dog (after Siber)

castrated to facilitate fattening. Owing to the slenderness of conformation and the great length of leg of early castrated dogs, these may be mistaken for greyhounds. In addition to these two types, a few dogs of small size are encountered among the Bakosi; these are distinguished by a long coat, bushy tail and pendulous ears; their colour is usually grey or greyish brown (Staffe, 1938; 1941).

South African Pariah Dogs

In the savanna belt of southern Africa many Bantu peoples train their dogs for hunting purposes; they have developed a light, slender and leggy type, approaching the

52. Tswana dog from the Kalahari (after Schultze)

greyhound in conformation. Schultze (1907) observed such dogs among the Tswana of the Kalahari (Fig. 52); and Holub found similar animals among the Bashukolumbwe, a Barotse subtribe in the river district of the Luanga, north of the Zambesi line. Njambo, chief of the Bashukolumbwe, is said to have possessed 7000 head of cattle and 70 dogs. The original hunting dogs of the Ama-Xosa, in the eastern part of Cape Province, were of a similar type: strong, well proportioned, of active appearance, nearly as slender as greyhounds, with erect, sharply pointed ears. These dogs were highly valued by their owners and kept in a sleek and well fed condition (Soga, 1931). But those Bantu tribes that did not keep their dogs either for hunting or for the sake of their meat had usually small scraggy beasts in a continuous state of semi-starvation. In the Chobe district, the pariahs, "little prick-eared, jackal-looking dogs", received no other food but fish (Selous, 1881). Bryant (1949), who arrived in Zululand in 1883, wrote that most of the Bantu dogs, including those of the Zulu, were of jackal-pariah type. But besides these, the Zulu possessed another entirely different breed which was smallish in size but stout in build, with a shorter and squarer muzzle, short light-brown hair, fierce in temperament, and carrying a slight mane along the neck and spine. Though occasionally met with in the Zulu kraals towards the end of the last century, it has since disappeared.

Cranial Types in South African Pariah Dogs

Petters (1934) has given a description of 22 crania of South African native dogs collected at Kailan, a Bantu reserve in the district of Kovimwaba, eastern South Africa. The basilar length of the skulls ranges from 193–121 mm; 193–136 mm in animals over 5 years of age, and 165–121 mm in those under 5 years. In one old dog the basilar length is less than that of a one-year-old animal, indicating that the differences in basilar length are not only due to age. The skulls display a considerable variety of racial characteristics. Petters distinguishes between three different basic types: one characterised by a slender conformation, the second by plumpness, and the third by small cranial size, in addition to several skulls of intergrading types.

The first group, regarded as the original type of Bantu dog, is characterised by a slender skull (Fig. 53), in which the supraorbital processes steeply slope from the narrow dished forehead. A short distance before the coronal suture, the temporal ridges fuse into a moderately developed parietal crest which extends backwards to the pronounced occipital protuberance. The frontals are markedly distended behind the temporal ridges, and constricted in the temporal region. The occipital triangle is low. The bullae appear considerably inflated. The palate is long and very narrow so that the premolars stand widely apart. In front of the Foramina infraorbitalia the long narrow splanchnocranium is but little constricted. The zygomatic arches are not prominent. The profile evenly slopes from the forehead backwards and forwards without a conspicuous stop. The neurocranium is only slightly longer than the splanchnocranium. The skull is distinguished by several primitive characteristics, notable especially in the frontal, parietal and occipital regions, and in the length of the splanchnocranium, while the narrow palate and the slight superstructural development are characteristic of a higher specialisation. Although this skull displays several grey-

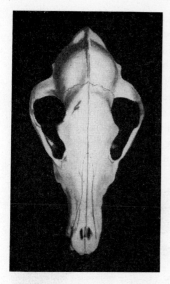
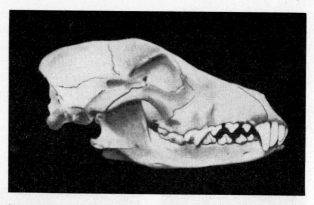

53. Cranium of a South African Bantu dog with greyhound features, frontal and lateral views (after Petters)

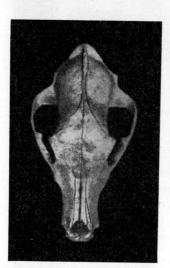
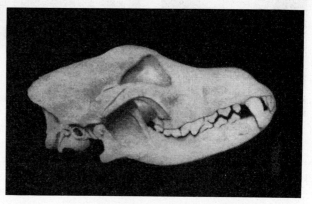

54. Cranium of a South African Bantu dog of inostranzevi type, frontal and lateral views (after Petters)

hound features, Petters (1934) classed the group of which it is typical with the pariah dogs, since it differs from the greyhounds in the considerably greater width of the frontal region between the inner orbital margins, and generally complies with Studer's definition of the pariah (see p. 29).

In the typical skull of the second group (Fig. 54) the parietal region, sloping roof-like orally and aborally, is much narrower at the parietal suture than in the auricular region. The supraorbital processes are bent steeply downwards, and the temporal ridges unite immediately before the coronal suture into a low parietal crest which extends backwards to the well developed occipital protuberance. The occipital triangle is low. The lateral walls of the neurocranium are nearly parallel from the temporal region to a short distance in front of the supraorbital processes, so that this portion has a pipe-like appearance. The forehead is very broad, dished, generally high, distended behind the temporal ridges, and only moderately constricted at the height of the temporal fossae. The zygomatic arches are little extended. The splanchnocranium is very wide at the anterior end of the zygomatic arches, narrowing orally so that P_3 stands oblique, and is rounded off in front. The profile shows a moderately defined stop at the posterior end of the nasals. The bullae are of medium size. The ratio of

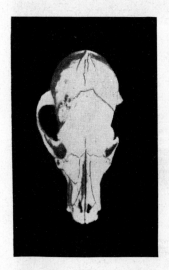
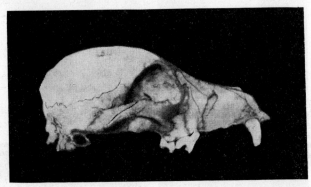

55. Cranium of a South African Bantu dog of palustris type, frontal and lateral views (after Petters)

neurocranial to splanchnocranial length is 10:9. The cranium of this dog holds an intermediate position between the inostranzevi and typical pariah types, the wide frontal region and the conformation of the splanchnocranium resembling those in C. f. inostranzevi, while the moderate concavity of the facial profile, the conformation of the neurocranium, the shape of the bullae, and the dentition are more typical of the common pariah.

The third group (Fig. 55) is the most distinct. The neurocranium is wider in the parietal than in the auricular region. The well developed temporal ridges include a lyre-shaped area, and fuse into a low crest only at the interparietal. The neurocranium is strongly vaulted, and the occipital protuberance is low and little developed. Above the Foramen occipitale, which is dorsally extended and pear-shaped, the high occipital triangle shows an oval elevation, typical of C. f. palustris. The bullae are moderately developed. The forehead is only slightly dished, and moderately constricted in the temporal region. The splanchnocranium is short, relatively wide, pointed orally, and slightly constricted in front of P_3. The skull reaches its greatest height at the intersection of the coronal suture and the median line, whence the profile descends steeply backwards. The stop is inconspicuous, and the zygomatic arches are little extended. In conformation and measurements the skull shows considerable similarity to typical C. f. palustris skulls, although the dentition and the poorly defined stop are pariah-like.

Petters (1934) considered a spontaneous evolution of the pariah dogs of the South African Bantu in the direction of the palustris and inostranzevi types as improbable, although he believed such evolution to have produced the palustris type in the Battak dog of Sumatra, the Tengger dog of eastern Java, and in the dogs of the Mangbattu and Azande of north-eastern Congo. Against the evolution of the palustris type from the primitive pariah he argued that a people showing little interest in their dogs can hardly be credited with the production of distinct racial types, and against the development of the inostranzevi form from pariah dogs that this is dependent on the introduction of wolf blood. Petters therefore believed that the palustris and inostranzevi strains in the dogs of the South African Bantu are due to the introduction of European breeds.

While the assumption that several racial peculiarities in the dogs of the Southern Bantu are ascribable to the recent introduction of European dogs cannot be disproved, it should be borne in mind that a high degree of variability characterises the majority of African pariah dogs. This applies also to regions where an influence of modern European breeds cannot possibly be suggested. Thus the poutiatini and palustris types are widespread; in several African pariah populations a tendency towards the slender greyhound type is observed. Among North African pariahs many resemble primitive sheepdogs; and in Egypt Schweinfurth collected a skull of mastiff type. Again, the theory that the evolution of the inostranzevi type presupposes the introduction of a wolf strain, although postulated by Studer (1901) and repeated by several later authors, is quite unproved.

Hottentot Dogs

The original dogs of the Hottentots of South West Africa seem to have been of pariah stock, possibly with a strain of greyhound blood carried south from East Africa, the Hottentots' original home, or bred on slender lines similar to the dogs of the Kalahari Tswana (Fig. 52). Kolb (1719) wrote that they had a small head with a very sharp muzzle and erect pointed ears. They were seldom higher than one ell (45 cm) and barely one-third longer. The coat was ash-grey. They were very faithful when their masters were in danger, and for this reason much valued by the European settlers as well as by the Hottentots. A Hottentot dog is represented in a painting by Daniell (1804–05) (Fig. 590). Andersson (1856) and Fritsch (1868) described the Hottentot dogs as half-starved mongrels with the ribs sticking out of the skin. Meyners d'Estrey (1891), while praising its courage and devotion as a watchdog, called the Hottentot dog one of the ugliest members of the canine family, with rough bristling hair, dirty-grey in colour, a pointed muzzle, prick ears and broad feet. Schinz (1891) described the dogs of the Hottentots as of medium size, with short hair, long snouts and drooping ears, von François (1896) as coarse and ugly, and varying in size and colour, while others have referred to them as lean hungry-looking mongrels, half-starved and savage-tempered. Schlettwein (1914) regarded the Hottentot dog as a particularly useful breed, hunting not by scent but by sight:—"One can see at a glance that these dogs are descended from greyhounds." According to Theal (1910), it was an ugly creature, its body being shaped like that of a jackal and the hair on the spine turned forward; but a faithful and serviceable animal of its kind. Some Hottentots still had dogs with ridges on their backs until fairly recent times. Jim Anderson, who lived in South Africa at the beginning of this century, wrote to George Gilky that there were about a dozen ridged dogs in the possession of Hottentots at Naauwpoort (merely a village then). An officer of the Scots Guards, who was in command at Naauwpoort in 1901, had two of these ridged dogs in his hunting pack; one of them was sandy-red with a black muzzle, and the other a dirty cream with a dark stripe down the back (Hawley, 1957).

Bushman Dogs

The Bushmen of South West Africa did not originally have any domestic animals. Many groups of present-day Bushmen have no dogs at all, though a few of the half-starved wandering tribes that eke out a miserable existence in the interior of the Kala-

56. A Bushman with his dogs

57. Masked hunters with dog. Bushman painting from South Africa (after Orpen)

hari desert do possess dogs. They first acquired these from the Hottentots who entered their hunting grounds from the north. But the Naron tribe procured its dogs from Tswana, Bergdama or Europeans (Schapera, 1930).

Burchell (1811) wrote that the dogs most common among the Bushmen were a small species entirely white, with erect pointed ears. Green (1952), travelling in the Kalahari in 1936, found that the finest type of Bushman hunting dog, a light brown ridgeback mongrel with a dark brown stripe and a trace of the greyhound in its appearance, was on the verge of extinction. One of the most typical of the dogs which he saw stood about 35 cm at the shoulder, with a body length seemingly out of proportion to that height. It had a broad forehead, sharp muzzle, upright ears and long drooping tail. Hawley (1957) reports that Mr. J.T. Robinson, palaeontologist of the Transvaal Museum, Pretoria, saw ridged dogs with Bushmen in south-east Angola in 1954, and that Mr. B.J.G. de la Bat, game warden in the Etosha Game Reserve, South West Africa, in 1955 destroyed some Bushman dogs, two of which had prominent ridges.

Bushman dogs are quite silent on the spoor and never betray the presence of the hunters by barking, though they will snarl and sometimes bark if one approaches an encampment. They display a considerable variability in colour and conformation, some having prick or semi-prick ears, while in others the ears are completely drooping. The majority show a strong tendency to the greyhound type, so that it is difficult to decide whether the Bushman dogs should be classed with the latter or with the pariahs with which they are here grouped in view of their original descent from the pariah dogs of the Hottentots.

2. Ancient Egyptian Pariah Dogs

Early Evidence of the Dog in Africa

The claim for the earliest presence in north-west Africa of a "probable Canis familiaris" is found in a report from a mid-palaeolithic site near Algiers, which contains Mouste-

58. Jackal with lions and prey (rock engraving at Kef el Msauora in the Sahara)

rian implements (Wulsin, 1941). Romer (1928; 1938) referred this canid to the jackals, for he did not find any unquestioned remains of dogs in the same area until well into the neolithic stage; in the reported instances the remains proved to be jackals rather than dogs.

There has been no claim for the presence of the dog in the neolithic Fayum or in Tasian or Badarian sites of Egypt (for chronology see pp. 213–214) (Reed, 1960). But three dogs are stated to have been found at Merimda beni Salama (Junker, 1929), invalidating Moustafa's assertion that no definite skeletal evidence of the dog exists in Egypt until the late Gerzean period. Several bowls with painted dogs are known from Amratian times (Massoulard, 1949; Kantor, 1953). The skeletal remains of dogs from the Gerzean of Maadi and Heliopolis have a wolf-like appearance (Moustafa, 1952; 1955).

Rock Drawings of Dogs in the Western and Eastern Deserts of Egypt

In the North African deserts records of domestic dogs appear first in rock drawings. In the very heart of the Sahara, where today not a beast nor a tree is to be seen, are paintings of wild bulls, oryx and Barbary sheep as well as human figures and dogs (Childe, 1934). In the Western desert the most ancient rock drawings are quite distinct from those of a later date. They are the works of hunters (Earliest Hunters[1]) whose minds were occupied with animals, footprints of game, traps and geometrical designs. They hunted with the bow and arrow; and as suggested by the crocodile in their drawings, they lived in close contact with the Nile. They had no domestic animals except the dog which occurs in some of their pictures. "The predynastic dog looks like that depicted at Alpera," Spain (Childe, 1934). It seems to have been a very primitive type, similar, possibly, to the present Berber dog; in the drawings it has a long body, long legs, prick ears, and the tail carried high over the back (Fig. 59). The elevation of the tail appears to be one of the first effects of domestication; already in wolves and jackals born wild but reared in captivity the raising of the tail is observed as an expres-

---

[1] The terms "Earliest Hunters", "Autochthonous Hamitic Mountain Dwellers", "Eastern Invaders" and "Early Nile Valley Dwellers" are employed by Winkler (1938–39) with reference to different styles in the ancient rock drawings of southern Upper Egypt.

59. (left) Domestic dog and various wild animals. Rock drawing by Earliest Hunters, Western Desert (after Winkler)

60. (right) Hunter with dog and various wild animals. Rock drawing by Hamitic Mountain Dwellers, Eastern Desert (after Winkler)

61. Hunters with dog, wild asses and boat. Rock drawing by Eastern Invaders, Eastern Desert (after Winkler)

sion of excitement and expectation, while in the wolf a slight lateral twist of the tip of the tail is also noticeable.

We do not know when the hunters of the Western desert received the dog. Apparently it was imparted to them by early immigrants, possibly Natufians or an intermediary hunting community, in a similar manner in which several thousand years later the Bushman hunters of South West Africa obtained their first dogs from Hottentots.

The rock drawings of the early hunters are followed by those of Hamitic pastoralists

(Autochthonous Hamitic Mountain Dwellers), spread all over the Western and Eastern deserts of Egypt. These possessed two different canine breeds: a rather large dog of pariah type, probably similar to the dog of the early hunters, and the prick-eared greyhound.

At an unknown date the country between the Nile and the Red Sea was invaded by a foreign sea-faring people (Eastern Invaders) who seem to have been in communication with Mesopotamia at one time. Their rock drawings show the elephant, hippopotamus, giraffe, ibex, antelope, Barbary sheep, wild cattle, wild ass, stag (?), lion (?), ostrich, lizard and dog; in one drawing a dog is seen on a lead. It differs from the dogs of pariah type drawn by the early hunters and Hamitic pastoralists, being smaller and stockier in build, with short legs, a short neck, long prick ears, and a short tail carried

62. (left) Man and dog hunting ostrich. Rock drawing by Eastern Invaders, Eastern Desert (after Winkler)

63. (right) Rock drawing of pariah-like dog and antelopes from southern Upper Egypt (after Winkler)

64. Rock Drawing of guard dogs from Zalad el Hamad, Libya

erect. In addition to this breed, which may have resembled a Pomeranian, greyhounds occasionally occur in the rock drawings of these people.

On the western frontier of the country between the Nile and the Red Sea rock drawings in Gerzean (Early Nile Valley Dwellers) sites show also two kinds of dog: the prick-eared greyhound and an ordinary pariah (Winkler, 1938–39).

In rock drawings of dogs at Aswan, dated to the period of the Old Kingdom, Schweinfurth (1912) identified three different breeds: the Tesem (greyhound), a Pomeranian-like dog, and the common pariah dog of Egypt (Fig. 27).

Skeletal Remains of Ancient Egyptian Pariah Dogs

In Pharaonic times pariah dogs were common in Egypt. The skeleton of a mummified dog of pariah type in the Cairo Museum is characterised by slender bones. The length of the spinal column from the first dorsal vertebra to the posterior point of the ischium measures 45 cm, and the height at the dorsal spinous processes 43 cm. The femur is longer than the tibia, a phenomenon characteristic also of the wolf, whereas in the majority of domestic dogs the tibia exceeds the femur in length. The number of sacral vertebrae varies in different specimens between three and four. The cranium has a long moderately vaulted neurocranium and relatively short splanchnocranium. The forehead is high, broad and slightly dished, the zygomatic arches are fairly short and high, and the stop is very shallow.

The skull of this dog, which is representative of many crania of ancient Egyptian pariah dogs, conforms to Studer's description of the type specimen of C. f. poutiatini (see pp. 18–19). Of the predynastic dogs depicted on rock surfaces in the deserts of

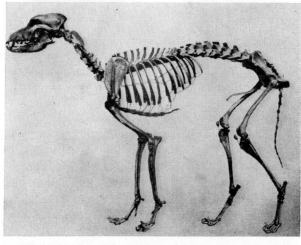

65. Skeleton of (another) Ancient Egyptian pariah dog (after Gaillard and Daressy)

66. Skull of Ancient Egyptian pariah dog (after Gaillard and Daressy)

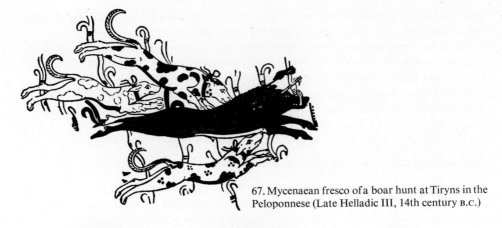

67. Mycenaean fresco of a boar hunt at Tiryns in the Peloponnese (Late Helladic III, 14th century B.C.)

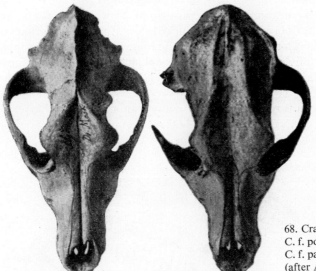

68. Crania of Ancient Egyptian dogs of
C. f. poutiatini (left) and
C. f. palustris (right) types
(after Antonius)

Egypt many probably belonged to this canine type. Hilzheimer (1932) has pointed out that dogs of apparent poutiatini type are frequently represented in ancient vase and mural paintings, as in the Mycenaean fresco of a boar hunt at Tiryns (Fig. 67), and they may at one time have ranged over the whole eastern Mediterranean region. Hutchinson (1962) refers the dogs depicted at Tiryns to the saluki type; but this is doubtful in view of their strong massive bodies which contrast with the greyhound's slenderness.

In addition to the poutiatini type, the palustris type is frequent in skulls of mummified dogs from ancient Egypt (Fig. 68). The small Pomeranian-like dogs represented in ancient rock drawings of hunting scenes in the Eastern desert of Egypt (Figs. 61 and 62) may be of this type.

But poutiatini and palustris are not the only cranial types found in the pariah dogs of ancient Egypt. In a collection of skulls of mummified dogs from Abydos, Thebes and Asyut, Hauck (1941) established the following types:—A large pariah, with an occasional tendency to the matris optimae type, and a small pariah; C. f. palustris as well as medium, small, and dwarfed Pomeranian-like skulls; crania of greyhound type of ordinary and of small size; and C. f. inostranzevi. Hilzheimer (1908) found among

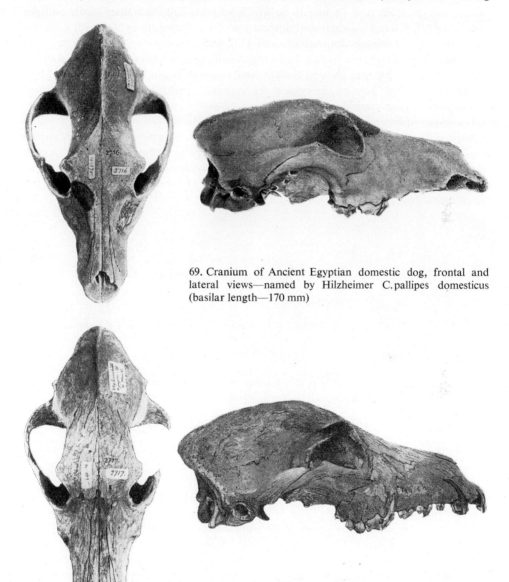

69. Cranium of Ancient Egyptian domestic dog, frontal and lateral views—named by Hilzheimer C. pallipes domesticus (basilar length—170 mm)

70. Cranium of Ancient Egyptian domestic dog, frontal and lateral views—named by Hilzheimer C. hadramauticus? (sacer?) domesticus (basilar length—164 mm)

several skulls of ancient Egyptian dogs only two of pariah type; these resembled the crania of mummified dogs described by Gaillard and Daressy (1905) as "chiens errants". The majority belonged to different types, more highly specialised than the pariah, such as the greyhound (which Gaillard and Daressy also found mummified), the hound, and a sheepdog-like breed. Other individual skulls were regarded by Hilzheimer as type specimens of breeds on which he bestowed the names Canis pallipes domesticus (Fig. 69), C. doederleini domesticus, C. hadramauticus? (sacer?) domesticus (Fig. 70), C. lupaster domesticus, and C. studeri domesticus, believing these to have been derived from the Indian Canis lupus pallipes as well as from various African grey jackals to which they showed certain similarities in cranial conformation. "Most of the canine breeds kept by the ancient Egyptians," Hilzheimer (1908) wrote, "seem still to survive in present-day Africa."

The large number of different types thus encountered among a relatively small number of skulls illustrates that the cranial variability characteristic of recent pariah populations already existed in ancient times. Moreover, similar to recent pariah populations, the different types of ancient Egyptian pariah dogs, established on the basis of a few individual skulls, are merely phases of an intergrading assemblage that forms a real continuum (see also pp. 120–122).

ii. The Greyhounds of Africa

1. Distribution and Characteristics of the Greyhounds of Africa

The greyhounds are generally characterised by a long narrow muzzle, nearly straight facial profile, slender body, long neck and limbs, and the habit of hunting by sight, and not by scent.

The Saluki

Greyhounds occur throughout the desert regions of North Africa, enjoying esteem, consideration and affection where the pariah is regarded as unclean and treated with contempt. The dry hot steppes on the fringe of the Sahara and the oases of the desert, where antelopes and desert hares abound, as in the Eastern Territories of Algeria, are the real home of the saluki, as the North African greyhound is called. The finest specimens are bred by the bedouin of the steppe and desert regions, such as the Hamyans, Aulad-Sidi-Cheikh, Harrar, Arbaa and Aulad Nail.

Salukis are commonly of medium size, the shoulder height of males varying between 58 and 70 cm and the weight between 23 and 25 kg, while females are often considerably smaller, which is a characteristic of this breed. The head is long and narrow, the skull moderately wide between the ears and not domed, the muzzle long and thin; the teeth are strong, and the stop is not pronounced. The ears are of moderate length, mobile, hanging close to the skull, and covered with long silky hair. The eyes are large and oval in shape, but not prominent. The neck and shoulders are slender but well muscled, the chest is very deep, the abdomen thin, the belly drawn up, the loin slightly arched, the hip bones are set well apart, and the tail is long, set on low, and carried in a curve, either rat-like or with silky hair hanging from the underside. The legs are long and slender, particularly in the upper part, with clean sinews, hard feet, and long arched toes. Murray (1935) writes:—"There is nothing of the cat's paw about the saluki's, but a good splay foot, to meet the sandy ground that he has to run over. The Arabs keep

71. Saluki

his foot in condition with henna, which tans the pad, and also reduces inflammation caused by cuts and bruises." According to the standard of points adopted in Europe and America, however, the toes should not be splayed out, though the saluki must not be cat-footed.

There is a smooth variety as well as a longer-haired type, the coat of which is soft and of silky texture. The longer-haired variety has a slight feather on the forelegs, the back of the thighs, and sometimes a slight woolly feather on thighs and shoulders. The smooth saluki has no feathering. The colours are white, cream, fawn, golden, red-grizzle and tan, occasionally white-black-and-tan or black-and-tan. But the most common coloration is that of the sand of the desert—a light fawn, with the head and muzzle of a darker, and the belly and inner sides of the legs a lighter hue.

The Saluki Skull

Studer (1901) observed that large greyhounds employed for hunting, such as the saluki of northern Africa and the Borzoi of Russia, have the skull very much drawn out,

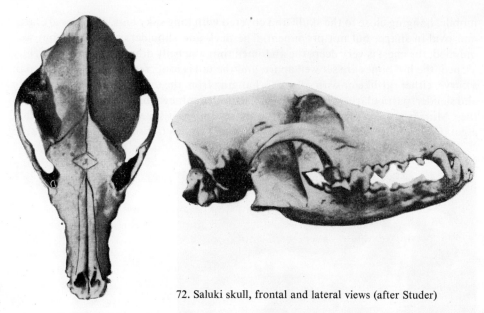

72. Saluki skull, frontal and lateral views (after Studer)

the cranial superstructure, more especially the sagittal crest, considerably heightened, the forehead low, and the stop very slightly or not at all pronounced. The portion in front of the first superior premolar is particularly long, high and narrow, and the outer wall of the maxilla forms a right angle with the narrow nasals. Owing to the prolongation of the maxilla, the premolars are placed far apart; in some instances there is an accessory premolar.

Two saluki skulls from Shtida and Boulawan, Morocco, described by Noack (1907), are characterised by the backward slope of the occiput, weak development of the occipital crest, shallow stop, and the absence of a sagittal crest, the parieto-frontal ridges remaining separate throughout. In comparison with the cranium of the Borzoi, the saluki skulls are shorter, wider between the orbits, and more robust, the olfactory tubes are better developed, and the nasals vaulted in consequence. The superior development of the olfactory tubes is correlated with the fact that many salukis are capable of hunting by scent.

Greyhounds in North Africa

On his wanderings in North Africa, Hamilton (1856) observed a pure saluki breed in Cyrenaica, which he described as follows:—"There is a race of greyhounds peculiar to this country generally of pale fawn colour, with very short hair and limbs almost as fine as those of the Italian pet greyhound." To this Murray (1935) has added that the Hanadi, a western tribe of Arabs settled in Sharqiya, but originally living in Beheira Province, possess a similar kind of greyhound, with no feathering and looking very much like a diminutive European hound.

In the coastal regions of Tunisia, Algeria and Morocco the saluki was usually not as well bred as among the desert tribes. Variability in size was considerable, the larger hounds, with a shoulder height exceeding 70 or 75 cm, being employed in hunting the wild boar, jackal and antelope, and the smaller, standing 60–65 cm at the shoulder, the desert hare. The coat colours commonly encountered were grey streaked with yellow, black, fawn, and variegated (Siber, 1899).

Greyhounds are bred also in Kordofan and by the Northern Hamitic Tuareg who inhabit the desert from Tuat to Tombouctou and from Fezzan to Zinder. Among the Tibu (believed to be the direct descendants of the ancient Garamantes), who occupy the Tibesti massif and who inhabited, until they were ousted by the Senussi, also the oasis of Kufra, a very poor type of saluki is found, starved and much neglected; for the Tibu have no passion for the hunt, although they occasionally catch antelopes in pitfalls and snares.

South of the Sahara desert greyhounds are only occasionally met with. In the Northern Provinces of Nigeria, between the Niger and Benue, Rohlfs found a breed of small greyhounds, apparently of degenerated saluki stock (Siber, 1899).

Greyhounds in East Africa

In East Africa greyhounds occur over a wide area. Particularly well bred specimens are found among the Middle Nilotic Shilluk. The Shilluk greyhound is of a fairly robust type, slightly below the size of an average pointer, and commonly fox-red in colour with a black muzzle, occasionally grey spotted with black. The coat is short and smooth, the tail long and thin, and the ears droop at the tips. In common with other North African greyhounds and the majority of African pariahs, the Shilluk greyhound has no dewclaws on the hind-feet. The speed and agility of these greyhounds is tremendous; they hunt an antelope down without difficulty (Schweinfurth, 1918).

The Low Nilotic Dinka, neighbours of the Shilluk, keep a coarser type of greyhound, considerably influenced by pariah blood and practically always of a fawn colour. The Dinka frequently castrate the males, believing this to enhance their speed and hardiness. Millais (1924) wrote that the Dinka hounds were of the greyhound-like pariah type, and usually in a state of semi-starvation. The Nuer, belonging with the Dinka to the Low Nilotic group, and roaming the country between the Bahr el Jebel and the Sobat, south of the Shilluk, keep a similar greyhound, short-coated and fawn or red in colour. The Shukurieh, occupying the region between the lower Atbara and the Nile from Berber to Khartoum, and the Kababish and Hassanieh, inhabiting the Bayuda steppe, on the left bank of the Nile, have beautiful greyhounds said to be descended from those of the Shilluk. Greyhounds of a similar type are bred also by the Berun of Sennar, between the Blue and the White Nile, and their eastern neighbours at Dinder and Gedaref. In fact, all Eastern Hamites, comprising a group of nomad pastoral people known as Beja, who inhabit the Eastern desert of Egypt, the Red Sea province of the Sudan, and extend through Eritrea to reach Ethiopia, are in possession of greyhounds used in hunting antelopes and smaller game. The Beni Amer, one of the Beja peoples, used to employ their greyhounds even against the rhinoceros (Siber, 1899).

73. Saluki bitch from Tripolitania (after a drawing by Pierre Mégnin)

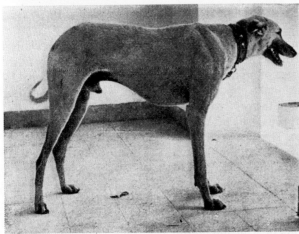

74. Libyan Saluki dog (after Peters)

75. Tunisian Saluki (after Siber)

76. (left) Shilluk Greyhound (after C. Keller)
77. (right) Dinka Greyhound (after a sketch by Millais)

78. Greyhounds from the Northern Sudan (Nubia) (after a drawing by Jean Bungartz)

Bisharin Greyhounds

Murray (1935) writes that the Bisharin, another of the Beja peoples, have a peculiar greyhound with prick-ears and curly tail, which he considers a descendant of the Tesem of the ancient Egyptians, "itself with a skull very like that of Canis lupaster, the largest of the three varieties of the Egyptian jackal". In the Red Sea hills of Etbai, the Bisharin feed their hounds on milk with very little solid food for most of the year so that few of them are equal to the running down of a fast hare alone, though they may do this sometimes when the hare is caught napping in a bush and gets a very bad start, or if more than one hound is in the hunt. Siber (1899) mentioned an allied breed of grey-

79. Greyhound from the Lake District of East Africa (after Leutemann)

hound with prick ears and a short coat, isabelline to reddish brown in colour, in the region between the Bahr el Azraq and the Sobat.

It is doubtful if the Bisharin greyhound may be regarded as a pure descendant of the Tesem. Its prick ears and curly tail are probably secondary acquisitions derived from a pariah crossing of the saluki. A similar cross between the saluki and pariah is found among the Sinai Arabs who call it Dirra.

Greyhounds of the Eastern Bantu

Among the half-Hamitic pastoralists of Kenya and Uganda dogs are widely used for hunting (Murdock, 1959). In the territories of the pastoral Bahima of Buganda and the Watusi of Rwanda-Burundi, greyhounds, closely resembling those of the Dinka and Shilluk, are found in addition to the much smaller pariah. They are characterised by a long pointed head, erect ears with folded tips, a long slender body, deep chest, slender legs, a nearly straight and only slightly hairy tail, and short, yellowish brown, white, variegated or, rarely, dark coat (Stuhlmann, 1909). Greyhounds considerably influenced by pariah blood are bred by some of the Eastern Bantu tribes, extending from Uganda in the north through Kenya, Tanzania, Zambia, Malawi and Mozambique north of the Zambesi river. The coastal Swahili and the Ubena, north of Lake Nyasa, hunt with dogs which show the influence of saluki blood in their pendant ears and rat-like tails,—in marked contrast with the prick-eared pariahs with their commonly curled tails.

In the kennels of Rumanika, ruler of the Karagwe, one of the Lacustrian Eastern Bantu tribes on the western shore of Lake Victoria, Speke, the discoverer of Victoria Nyanza and the source of the Nile, found, in 1862, a dog with a somewhat coarse body, but the long and slender head of a greyhound. Speke's drawing of this animal (Fig. 80) resembles the sketch of a greyhound after a wall painting from the tomb of Roti, one of the rulers of the XIIth Dynasty of Egypt (Fig. 81). It is notable that both Rumanika's and Roti's greyhounds have the ears cropped. Cropping of the ears, especially of pariah dogs, is still widely practised in East Africa and south-west Asia where it is regarded as a prophylactic against rabies. Kroll (1929), referring to this practice among the Low Nilotic Dinka, Eastern Bantu Akamba and Western Bantu Bakongo and Bashilange, erroneously thought that it was carried out for beauty's sake.

Dogs of Greyhound Type in the Congo

Among the dogs of the Congo region, commonly of the pariah type, some approximate to the greyhound in general conformation and the character of the skull. Frobenius (1907) has recorded that the dogs of the Baluba, a Western Bantu people in the southern part of Lulua, are more leggy than the common Congo pariah, with the head approaching the greyhound type (Fig. 82).

Greyhounds in South Africa

The occurrence of greyhounds among several tribes of the Southern and Western Bantu groups has repeatedly been reported. Holub (1879) encountered a breed of strong

80. (left) Greyhound of the Karagwe, Victoria Nyanza (after a drawing by J. H. Speke)
81. (right) Ancient Egyptian Greyhound (after a wall painting from the tomb of Roti, XIIth Dynasty)

82. Head of a Baluba bitch (after Frobenius)

83. Prick-eared Greyhound from Zanzibar (after a drawing by Pierre Mégnin)

well-shaped greyhounds among the Bashukolumbwe, a Western Bantu tribe of Zambia, occupying the country between the Nueko and Kafue rivers, north of the Zambesi. It seems, however, doubtful that the Bashukolumbwe were in possession of greyhounds proper, brought along on their southward migration or subsequently obtained from an Eastern Bantu people; rather their dogs appear to have been light pariahs bred and trained for hunting purposes, similar to the dogs of the Kalahari Tswana (Fig. 52). This is confirmed by Petters' (1934) description of South African pariah skulls with greyhound features (see p. 46).

A Greyhound from Zanzibar

A prick-eared greyhound from Zanzibar, where the small number of greyhounds encountered are commonly characterised by ears with folded tips, was described by Mégnin (1897). This greyhound, which had been taken to France, stood 68 cm at the shoulder and was distinguished by a rough red-and-white coat, strong body and the absence of dewclaws; it was intelligent and keen-scented. Referring to the robust conformation, rough coat, prick-ears, high intelligence and fully developed scent of this dog, Siber (1899) remarked that it was typical of African crossbred dogs derived from saluki and pariah stocks.

2. The Greyhounds of Ancient Egypt and North Africa

The Tesem

The earliest record of greyhounds in Egypt is represented by a painting of a man with four hounds led on leashes on a pottery bowl (the famous "Golenischeff bowl") from the Amratian period (4th millennium B.C.). From their build Hilzheimer (1932) inferred that the animals could only be greyhounds, of Whippet size, with the ring tail characteristic of the ancient Egyptian Tesem and teeth and jaws exaggerated. Greyhounds of a similar type are depicted also in the rock drawings of the early pastoral peoples in the Eastern and Western deserts (Winkler, 1938–39). In several scenes the greyhounds are shown hunting in couples—male and female, a practice still common in Afghanistan.

Records of greyhounds from the period of the dynastic conquest and protodynastic times show an animal of medium size, with a long slender head ending in a thin pointed muzzle, and a moderate stop. The prick ears are long and narrow, the neck and body long and thin, the chest is flat, the thighs are well muscled, and the legs are strong with

84. Hunter with four greyhounds from the Amratian period of Egypt

broad feet. The coat is short and reddish-yellow in colour, with a white underline. The tail is tightly curled like that of a pug, only occasionally twisted. "According to Egyptian ideas," Hilzheimer (1932) says, "this curly tail must have formed a sign of good breeding, for it is always carefully represented in all works of art, often to an exaggerated degree. Obviously the Egyptians were especially proud of this characteristic quite unnatural in a greyhound, who needs a long tail to guide himself; it must therefore have been extremely difficult to obtain by breeding."

Skeleton and Skull of the Tesem

The dog buried about 3000 B.C. at Saqqara in the tomb of Queen Her-neit was apparently of this type (Hutchinson, 1962). Gaillard and Daressy (1905) have described the skeleton of an ancient Egyptian mummified greyhound of the tesem type from Asyut,

85. Gerzean rock drawing of greyhounds at Hôsh (after Schweinfurth)

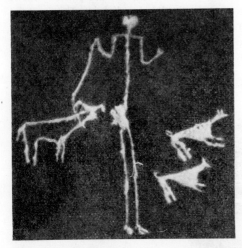 

86. (left) Gerzean rock drawing of hunter and greyhounds, southern Upper Egypt (after Winkler)

87. (right) Greyhound and game on gaming disc of black steatite inlaid with alabaster, from Saqqara (Ist Dynasty)

88. Tamed African hunting dogs (Lycaon pictus) (above) and Greyhounds (below), from the tomb of Ptahhetep, Saqqara (about 2500 B.C.)

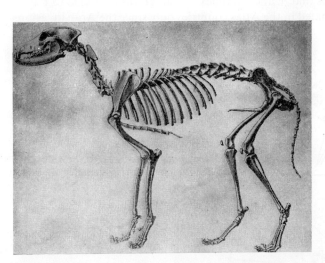

89. (left) Skeleton of an Ancient Egyptian greyhound from Asyut
90. (right) Skull of an Ancient Egyptian greyhound from Asyut

Upper Egypt (Figs. 89 and 90). It is distinguished by a long, slightly arched spinal column, narrow ribs, long extremities, and the absence of dewclaws on the hindlegs. The length of the spine from the first dorsal vertebra to the posterior point of the ischium measures 57 cm, and the height at the dorsal spinous processes 56 cm. The cranium is long and slender, with a well developed sagittal crest, pronounced occipital protuberance, narrow intraorbital region, dished forehead, and moderate stop.

Late Dynastic Greyhounds

In wall paintings from the later dynasties a change in the conformation of the greyhound is apparent. First the ring tail, strangely anomalous in a greyhound, gives way to the rat tail, as in the red greyhound from the hunting kennels of Roti, represented in a fresco at Beni Hasan together with a hound of light brown colour with dark patches (Fig. 91). Still later the long prick ears are lost, and the greyhound with small ears drooping at the tips appears nearly in the form in which it occurs in North Africa today (Fig. 92).

The Tesem and Saluki in North Africa

The range of the tesem did not remain restricted to the Nile valley and the Eastern and Western deserts. At an early period prick-eared greyhounds also reached the Atlas countries. In a rock drawing from Ghat, Fezzan, two men are seen hunting a Barbary ram, with five greyhounds showing the beautiful slender forms, prick ears and ring tail characteristic of the Tesem (Fig. 93).

Przezdziecki (1954) believes that in ancient times the saluki was restricted to Egypt, and that it has been disseminated in the Atlas countries only since the Arab invasions

91. Hound and prick-eared greyhound with a rat-like tail (wall painting from Beni Hasan)

92. Lion hunt—black and red ink on limestone, late XVIIIth Dynasty (c. 1350 B.C.)

93. Hunters with greyhounds. Rock drawing from Ghat, Fezzan (after Frobenius)

of the 11th century A.D. While the arabization of North Africa undoubtedly contributed to the saluki's diffusion, it is most unlikely that prior to this event only the Tesem should have spread into the Atlas countries, but not the more highly specialised saluki, which is so well adapted to the hunting of the gazelle and Barbary sheep in the desert.

iii. The Mastiffs of Ancient Egypt

Mastiffs are powerfully built, muscular animals with large massive frames. They are generally characterised by a large, elongated, flat or slightly convex neurocranium; the superstructure is strongly developed, and the occipital protuberance is prominent. The muzzle is short, broad and deep, usually blunt and cut off square, but in a few modern breeds undershot. The stop is pronounced, the chops or flews are well developed, and the small ears are set wide apart high on the sides of the head.

Ancient Egyptian Records of Mastiffs

Mastiffs may have reached Egypt in small numbers already during the protodynastic period. At Hierakonpolis, Quibell (1900) found a serpentine vase with handles in the shape of animal heads, the mountain sign below each. Quibell regarded them as bulls' heads, whereas Hilzheimer (1926) described them as those of mastiffs. From the same period (Hierakonpolis I) dates a mace head of grey steatite with alternating lions and large mastiffs, each of which attacks the lion before him with teeth and claws.

It cannot be ruled out that these records may represent mastiffs not actually bred in Egypt at that time, but carved from models introduced from Asia. For the type of design, a circular interlocking of the individual figures, is characteristic of Mesopotamia and occurs on numerous cylinder seals, on a silver vase of Entemena, and on the mace head of Mesilim of Kish (Frankfort, 1951). However, Hilzheimer (1932) was convinced that Egypt was "the most southerly point reached by this mastiff-type in prehistoric times. We can trace them in the period of the First Dynasty, about 3000 B.C. They probably accompanied nomad foreign tribes thither, and soon became extinct for lack of further stock. They must have resembled the huge Babyloni-

94. Procession of mastiffs and lions, protodynastic mace head from Hierakonpolis I (After Quibell)

an dog, from which they were most probably descended, and were not much smaller than the lions often depicted with them."

A pictorial record of the mastiff in Egypt is known also from a later period. "There is a coloured painting from the tomb of Redmera at Thebes, which depicts the tribute from various parts of Asia. Eight dogs form a part of the tribute, among them a greyhound, a mastiff, and a large spotted type not unlike the Dalmatian hound" (Gwatkin, 1933–34).

Into the Phoenician colony of Carthage mastiffs were imported from Assyria in the 7th century B.C. (Tschudi, 1926).

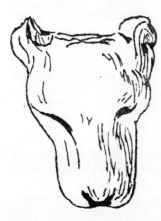

95. Mastiff (bull?) head, protodynastic vase handle from Hierakonpolis I (after Quibell)

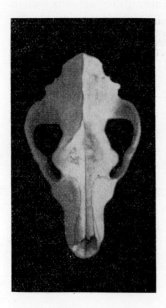
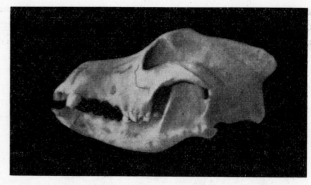

96. Skull of a mummified Egyptian dog of C.f.inostranzevi type, frontal and lateral views (after Hauck)

Cranium of a Mummified Dog of C. F. Inostranzevi Type

Among 38 skulls of mummified dogs from Abydos, Thebes and Asyut, in the Vienna Museum of Natural History, one cranium, of C. f. inostranzevi type, resembles a mastiff skull. It measures 194 mm in basilar length and 221 mm in profile length, and is characterised by a wide, deeply dished forehead which steeply descends orally, convex frontals with the clearly marked frontal ridges joining in front of the intersection of the sagittal and coronal sutures, and moderately long, aborally narrow nasals. The sagittal crest is moderately high, sharp-edged, and slightly wavy laterally. The orbits are fairly large, the preorbital depressions deep, and the postorbital processes strong and blunt. The Foramen magnum is oval horizontally, the basi-occipital broad and short, and the tympanic bullae are flat, furrowed and oblique; the preoccipital processes are short and broad, the postglenoid processes wide and curved. The mandible is strongly bent, with the heavy horizontal branch moderately high, and the symphyseal part narrow and slender. The teeth are small (Hauck, 1941).

From the small number of records it would appear that mastiffs were only occasionally imported into Egypt, possibly because they did not thrive in the Nile valley.

iv. The Hounds of Africa

The hounds are generally characterised by a large skull, straight and deep muzzle, full and well defined stop, large, low-set, drooping ears, fairly heavy flews, and a slightly curved stern set on rather high. All hounds hunt by scent.

1. The Rhodesian Ridgeback

Characteristics of the Rhodesian Ridgeback

In Rhodesia a breed of dog has been evolved which is characterised by a peculiar lay of the hair along the spine. The Rhodesian Ridgeback or Lion Dog, as it is called, is a strong, powerful hound with a large broad skull flat across the top and a well defined stop; the muzzle is rather long, deep and exceedingly powerful; the ears are set rather high, of medium size, tapering from wide bases to rounded tips, and carried folded close to the sides of the skull; the eyes are of moderate size, round, and generally dark in colour. The neck is fairly long, strong and entirely free from throatiness, leading to a body which is muscular, fairly broad, firm and symmetrical. The chest is deep and capacious, though not barrelled; the back is straight and of moderate length, but with a slight arch over the croup; the loins are fairly lifted and the couplings powerful. The legs are straight and well muscled and boned, with fairly large, round, compact and hard-padded feet. The tail is long, set rather high, thick at the root but tapering to a point carried low to about the hocks. The coat is dense, short, harsh in texture and waterproof; it is distinguished by the "cowlick" forming either a fiddle-shaped ridge or a razor back. The ridge of hair, growing in the opposite direction from the rest of the coat, runs from the set-on to the shoulders in a gradually widening furrow until at its end it flourishes into a decided burr or knot. In some of the older types the various cowlicks on the back, resulting from irregular placement of a number of crowns, produced untidy ridges; fiddle ridges were then also permitted by the breed standard. But now clear ridges with two crowns symmetrically placed within the third of the ridge nearest to the withers are demanded. A few Ridgeback dogs are furnished with manes. The coat may be fawn, wheaten, dark tan, brindle, tawny or of other main colours, interspersed with a little white; the desired colour is now light to red wheaten. The male reaches a height of 65 cm at the shoulder and a weight of 35 kg; the female is a little smaller than the male (Hubbard, 1946, 1948; Hawley, 1957).

Origin of the Rhodesian Ridgeback

The breed owes its existence to a cross of the original hunting and watch dogs of the Cape Boers, introduced into Rhodesia by mining prospectors, transport riders and hunters, with mastiffs and hounds, possibly retrievers, brought over from Europe (Lloyd, 1937). Some authorities (Young, 1944; Vesey–FitzGerald, 1957) believe it to have been developed by a series of crossings between native dogs and the modern Bloodhound and Great Dane.

Origin of the Ridgeback Pattern

The ridgeback feature, which, until the beginning of this century, seems to have been widely, though sparsely distributed throughout South and South West Africa, was probably imparted to the Rhodesian Ridgeback by Boerhounds, many of which were distinguished by it. The Farmer's Weekly, Bloemfontein, of 2.11.1938, mentions a letter addressed by Mr. F. E. B. Bedford, of Crawford, to the "Cape Times", discussing

97. Rhodesian Ridgebacks (after Lloyd)

the origin of the Ridgeback. Mr. Bedford wrote that he saw the first dogs with ridges on their backs on the family's farm "Beestkraal" at Hanover about 1887–88 when he was a boy. His father, like nearly every farmer, kept steekbaard (stickbeard, i.e. rough coated) or vuilbaard (dirtybeard)—honde, about the size of greyhounds, dirty-white in colour, very vicious, good runners and fighters. But these dogs had no ridge on their backs. The first ridge dogs were a litter of pups born on Mr. Bedford's farm out of a pure-bred English Bulldog bitch by a steekbaard sire. The steekbaard dog that had fathered

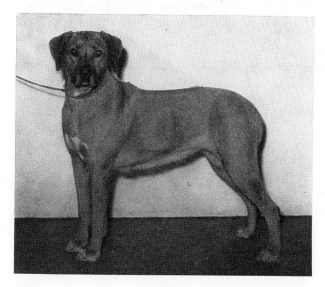

98. Rhodesian Ridgeback bitch

them was a descendant of steekbaard dogs brought from Swellendam by Mr. Bedford's great-grandfather, H. van Zyl, when the trek to the Colesberg district took place.

Gwatkin (1933–34) recorded a number of letters giving information on the origin of the Ridgeback. A resident in the Northern Transvaal wrote:—"In the northern parts of the Bushveld, along the Palala and Pongola rivers, about 15 years ago, quite a number of the original settlers had these dogs. The Burkes, van Vuurens, van Ecks, and Viviers were some of the names mentioned to me." Another correspondent stated that about the time of the Boer War he first saw Ridgebacks in the possession of Jacob Coetzee, a postcart contractor between Dordrecht and Jamestown, C.P. Mr. A. Giese, of Wankie, pegger of the Wankie Colliery and one of the settlers of Rhodesia, gave information that he had met these dogs along the Crocodile river in 1889, but he had not seen one in Rhodesia until 1910. He also stated that the hunter, van Rooyen, always had some of these dogs with him. The most conclusive proof that the Ridgeback originally came from the Cape was furnished in a letter from Mrs. Lovemore of Rhodesia:—"In 1875 my parents (Rev. Charles and Mrs. Helm) trekked by wagon up from Swellendam to Bulawayo. On their way up someone gave them a pair of dogs with a ridge down their backs. I remember, as a child, having these dogs, and later on, when

Mr. van Rooyen went on his hunting trips, he used to borrow some from us and he found them very brave when hunting lions and leopards. About 20 years ago Mr. Selmes, Mining Commissioner, imported a dog with a ridge and a short tail from Ceres district."

Hawley (1957) records that, about 1906, children coming to the farm school at Braamspruit, a few miles out of Aliwal North, had a ridged dog. He also learnt from a friend of his, J. P. Louw, in Bushmanland, that in the early thirties, i.e. before Rhodesian Ridgebacks were introduced into those parts, he saw six ridged dogs in the region of Brandvlei. Another ridged dog belonged to a station master near Port Nolloth just prior to 1920. "Brandvlei and Port Nolloth are right on the old beaten track of the Hottentots."

The Ridgebacks brought from the Cape to Rhodesia must not be thought to have been identical in type with the modern Ridgeback. The original Ridgeback has been crossed, standardised and improved out of all recognition. But the reversed trend of the spinal hair has remained the standard feature of the breed.

This peculiar factor carried by the steekbaardhonde of the Cape Boers seems to go back to a breed of dog present in South Africa at the time of the Dutch East India Company's first settlements at the Cape. Mr. F. E. B. Bedford in his above-quoted letter to the "Cape Times" stated that his father had been told by his great-grandfather that the steekbaard was a cross between a Russian hound and a greyhound. But this view is quite unfounded. There is a significant passage in Theal's (1910) description of the domestic dog of the Hottentots:—"He was an ugly creature, his body being shaped like that of a jackal, and the hair on his spine being turned forward..."; and ridgeback dogs have survived among Hottentots and Bushmen also into recent times (see pp. 50–51).

Descent from Hottentot Dogs

In the early days of the European settlement at the Cape of Good Hope, the Boers were very keen on acquiring cattle and sheep from the Hottentot tribes, organising numerous expeditions into the interior for this purpose. There are no records that they showed any particular liking for the Hottentots' dogs as well. But since the Boers got their cattle and sheep from the Hottentots, and as both their steekbaard dogs and the dogs of the Hottentots shaved the rather rare ridgeback character, we may assume that not only cattle and sheep but also dogs occasionally passed into the hands of European farmers, or that some of their bitches were served by Hottentot dogs.

The Hottentot dog, through lack of interest or perhaps open hostility on the part of the early settlers, is now extinct, and no skeletal material has been preserved in any museum. Whereas with the likewise extinct Hottentot cattle at least their descendants, the Africander, remain for study and comparison, the Rhodesian Ridgeback, although partly descended from the Hottentot dog and preserving its most characteristic feature, cannot be taken to fill this place; for, as Gwatkin (1933–34) has pointed out, "the Rhodesian Ridgeback has undoubtedly been crossed with other breeds introduced by Europeans and certainly in a most haphazard manner." Indeed, the ridgeback dogs that were brought to the Bulawayo Kennel Club Show in 1922, where the establishment of the Rhodesian Ridgeback breed was initiated, were of all types and sizes, from an undersized Great Dane to a small Bull Terrier; all colours were represented, but reds and brindles predominated (Hawley, 1957).

Origin of the Hottentot Dogs

The means of arriving at a sufficiently well founded theory on the origin of the Hottentot dogs are therefore limited to the conclusions that may be drawn from the existing pictures of these animals, such as the painting by Samuel Daniell (Fig. 590), and from the descriptions furnished by Theal and others (see pp. 49–50). Since the Hottentots themselves appear to have originated somewhere in the lake district of central East Africa, where they acquired their cattle and sheep, it is reasonable to assume that they came into possession of their dogs in the same region. These dogs belonged to the pariah type ubiquitous in Africa and southern Asia, although Schinz's reference to their long snouts and drooping ears suggests pariah dogs influenced by greyhound (saluki) blood.

The ridgeback feature has given rise to a strange theory on the origin of the Hottentot dog. Gwatkin (1933–34) discovered that the latter was not the only breed with the hair on the spine bristling forward, but that it shared this peculiarity with a dog from the small island of Phu-Quoc, off the coast of Cambodia, in the Gulf of Siam (Thailand). This dog is described as follows:—"A long head with powerful jaws, erect ears, reddish eyes, with a savage expression, somewhat coarse body, neck very long and flexible, shoulders sloping, belly drawn up, loins broad and strong. Straight and lean legs, stifles rather straight with muscular thighs, longish feet with hard pads. Coat, on the whole body and legs very short and dense, on the back the hair is growing the wrong way, towards the head, and is much longer and harder. Colour, reddish fawn with black muzzle, the hair on the back being darker. Height about 21 inches, weight about 40 lbs."

In the absence of a craniological analysis of the Phu-Quoc dog, the description points to a typical southern Asiatic pariah, but distinct in the forward growing hair along the spine. From the similarity of this feature in the Hottentot and Phu-Quoc dogs, Gwatkin (1933–34) argued that "the Hottentot dog, if not actually the same as the Phu-Quoc, is definitely derived from the same Asiatic stock that originated the Phu-Quoc."

While the African and southern Asiatic pariah dogs are probably descended from the same wild stock, Gwatkin's argument does not prove it. For had the Hottentot or the Phu-Quoc dog normally directed hair on the back, this would not invalidate the theory of their monophyletic origin. The range of major inheritable variations arising under domestication is limited; even entirely different species and genera occasionally show peculiar parallel features. Such parallel variations may be induced or brought to light by similar or different circumstances. In the case of the ridgeback feature common to the Hottentot and Phu-Quoc dogs, inbreeding may be a contributory factor, since both breeds are insular forms, as it were; for the Hottentot dog was long removed from the influence of other breeds, as the Hottentots for a considerable time roamed country that was devoid of people in possession of domesticated animals (Epstein, 1954a).

The ridge character, is of course, inherited. Hare (1932) has described a related congenital abnormality of the skin of Rhodesian Ridgeback dogs, which consists of a narrow and deep invagination of the skin in the mid-dorsal line on the neck, withers or croup. It seems also to be inherited, for an affected dog mated to his normal sister sired four normal and two affected pups.

The ridge factor occasionally occurs in dogs belonging to breeds in which it has not become a standard feature. Thus, a Labrador bitch bred in England from non-ridged stock had a small ridge of about 7–8 cm on the neck; one of her litter sisters was similarly marked. Many dogs, especially Bulldogs and Boxers, have the hair on the sides of the shoulders fanning out from a crown, cow-lick fashion and finishing up in the pattern of a comma. Two German Weimeraners had ridges about 20 cm long on their necks, extending from behind the ears to nearly at the shoulders. Such mutations can easily be fixed by inbreeding (Hawley, 1957).

The parallelism in the character of the coat in the Hottentot and Phu-Quoc dogs proves therefore nothing with regard to their genealogy; and it is fallacious to believe that this feature could not have arisen in Africa spontaneously, but that the Ridgeback, "favourite of the navigating Easterners", must have been introduced by sea from the original home of the Phu-Quoc. Gwatkin went so far as to state that as the Chinese and other eastern peoples had been navigating the Indian Ocean for a considerable period before the 10th century A.D., it was from this source doubtless that the Hottentots received their dogs. In support of this theory he even referred to Mongolian characteristics in the Hottentots themselves. Hubbard (1948), who shared Gwatkin's view on the origin of the Rhodesian Ridgeback from the Phu-Quoc dog, asserted that the latter had been introduced into Africa by Phoenician traders.

2. Ancient Egyptian Hounds

Early Records of Hounds in Egypt

Hounds made their appearance in Egypt at the end of the predynastic period. They are represented on a slate palette from Hierakonpolis, and by four ivory figurines from the tomb of Narmer-Mena, first ruler of the Ist Dynasty. The artists of ancient Egypt possessed a very intimate knowledge of the character and peculiarities of the hounds. The barking bitch behind the broken down ostrich (Fig. 102) is remarkably true to life and equally well observed is the attitude of the hound giving tongue on the scent (Fig. 103).

Different Types of Hound

The ancient records display a considerable variety of types in the hounds employed by the Egyptians. There are antelope hounds, harriers and trail hounds adapted to different forms of the hunt, and there are hounds hunting in packs. On their first appearance on the monuments of ancient Egypt, the hounds exhibit already so high a degree of specialisation that many centuries of selection and human endeavour must have preceded their introduction into the Nile valley.

DISTRIBUTION AND CHARACTERISTICS

99. (left) Hounds and game on a predynastic slate palette from Hierakonpolis

100. (right) Outline of an ivory statuette of a hound from the tomb of Narmer-Mena

101. (above) Hound hunting Dorcas antelope
102. (centre) Bitch hunting ostrich and hare
103. (below) Hound howling on the scent

104. (left) Sitting hound in hollow stone relief from Ancient Egypt (c. 2000 B.C.)
105. (right) Head of an Ancient Egyptian trail hound

106. A mechanical toy in the form of a hound, late XVIIIth Dynasty, Egypt

107. Pack of Ethiopian hounds sent to Thotmes III (c. 1500 B.C.)

On one of the monuments a hunter is seen leading two hounds on the leash, with a Beisa antelope slung over his shoulders; the hounds somewhat resemble heavy pointers, particularly in the carriage of their sterns. On a hollow relief in stone from the beginning of the second millennium B.C. a hound with long thin legs and an excessively long body is depicted in a sitting position (Fig. 104); the unnatural position of the tail is probably due to the lack of space on the stone, preventing a more correct representation. A mechanical toy, dated to the late XVIIIth Dynasty (c. 1375 B.C.), represents a hound whose lower jaw can be moved by manipulating a lever (Fig. 106).

Hounds from Ancient Ethiopia

There are numerous records of packs. In a coloured fresco a pack of hounds is represented among other animals which a subject Ethiopian tribe sent to the court of Thotmes III as tribute. A similar pack is depicted in a tomb at Gurneh, near Thebes, and another one is known from Karnak. Twelve centuries after Thotmes III had obtained a pack of hounds from Abyssinia, Alexander the Great received, among other presents, a similar pack of ninety hounds from Queen Candace of Ethiopia. These were especially trained to hunt down fugitives, and the letter which accompanied the gift mentioned that they were an indigenous product of Ethiopia (Siber, 1899).

Brachymely in Egyptian Hounds

Brachymely in Egyptian dogs is recorded from the XIIth Dynasty on, at Beni Hasan, El Bercheh and several other localities. It first appeared among hounds (Hauck, 1941).

Apparently, brachymely is a common mutation in dogs; for short-legged dogs are also known from Mohenjo-daro in the Indus valley, where they seem to have been evolved independently of the Egyptian type, from the Inca period of Peru, and the period of the Roman conquest of Germany. In China a short-legged type of dog, called "pai", is reported from the end of the 1st century A.D., but may have existed considerably earlier (Collier, 1921).

Crania of Ancient Egyptian Hounds

Crania of ancient Egyptian hounds are known in considerable numbers (Figs. 108 and 109). Hilzheimer (1908) has described two mummified skulls from Asyut, which are short and broad, but with relatively small and narrow neurocrania. The parietals are only slightly vaulted and the temporal region is only moderately constricted, the skulls broadening out orally into very wide and slightly dished foreheads. The moderately curved supratemporal arches diverge to such a degree that the sloping postorbital processes are vertical with the cranium. Owing to the relatively high splanchnocranium, the stop is rather shallow, the profile sloping evenly from the centre of the forehead to the tips of the nasals. The inferior orbital edge is wide, and the oral part of the splanchnocranium remarkably broad.

The "Bloodhounds of Katsina"

At the present time hounds no longer exist in native Africa. They were linked with a particular social environment obtaining in dynastic times, and disappeared when the ancient civilisation of the Nile valley fell into decay. Nor are any remnants of them

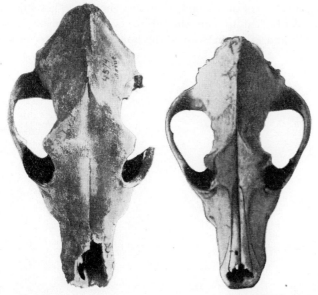

108. (left) Skull of a mummified hound from Asyut, Egypt (after Hilzheimer)

109. (right) Skull of an Ancient Egyptian hound from Abydos (after Antonius)

110. The "Bloodhounds of Katsina" (after a drawing by Harvey)

left in the deserts that fringe the valley of the Nile, for the climate and flora of the steppe and desert which favour a dog hunting by sight, operate against a hound working only on the scent. Arab traders on their slave expeditions are said to have been accompanied by ferocious bloodhounds; but no traces of such hounds have survived in Africa. In the twenties of the 19th century A.D., Col. Denham brought three hounds to the Tower of London from his expedition through the country of the Tibu and Tuareg to Bornu and Lake Chad. They became known as the "Bloodhounds of Katsina"; but they were certainly not scent hounds, but rather greyhounds, possibly with a strain of pariah blood. According to Denham, who had used them on antelopes in Africa, they followed the scent only occasionally, although they were capable of taking up a spoor two hours old (Bennett, 1829). Most of the time they hunted by sight. Harvey's drawing does not convey a very clear picture of the racial type of these hounds. They show the high setting of the ears, slender tucked-up body and rat-like tail of the greyhound, while the ears appear too broad and long, the stop in the hound in front too pronounced, and the neck too heavy for purebred salukis. They were short-coated and reddish yellow in colour.

IV. Origin and Descent of the Dogs of Africa

i. The Influence of Domestication on the Canine Cranium

The wild Canidae exhibit a remarkable degree of variability in body size, coat colour and cranial conformation, a variability characteristic of generalised types on the threshold of an outburst of variation. In the course of domestication of the dog, this great variability has tremendously increased. There is hardly any feature, any part of the canine body that has not been influenced to some extent; but the greatest and, from an evolutionary point of view, most important changes concern the conformation of the cranium. The latter has been affected by the domestication process more profoundly than in any other domesticated species. This, however, should not veil the fact that the body, in Weidenreich's (1941) words, constitutes a totality, i.e. "a unique construction in which all parts harmonize from the beginning of its organization and in which every essential alteration must be accounted as a consequence of a change in the entire construction".

The Influence of Captivity on Wolf Crania

Captivity profoundly affects the crania of wild Canidae. In crania of wolves born in captivity, Wolfgramm (1894) described a change of all dimensions, such skulls being shorter, broader and higher than those of wolves matured in wild surroundings; but the restriction in length is limited to the splanchnocranium, the brain case remaining relatively unaffected by the general reduction in size. The relative positions of the neurocranium and splanchnocranium also differ in wolves born wild from those born in captivity. In the former the neurocranial and splanchnocranial axes form one continuous slope in the same plane or two parallel slopes, whereas in the latter the splanchnocranial axis is higher in front than at the back, while the anterior part of the neurocranium is slightly elevated, accompanied by an increase in width and a considerable expansion of the frontal sinuses. This produces the large vaulted frontal surface so typical of the skull of the domesticated dog. At the same time the facial musculature and the crests and ridges to which the muscles are attached develop considerably less in wolves reared in captivity than in the wild beast, although the degree of diversity largely depends on the diet. These changes in cranial conformation, which are not restricted to wolves bred in captivity but apply also to jackals and foxes, are accompanied by changes in the size and position of the teeth. The canines and carnassials grow smaller, and the gaps disappear.

While Wolfgramm's investigation shows the trend in cranial changes in wild Canidae reared in captivity, it should be borne in mind that he dealt with cage animals whose cranial metamorphosis may fundamentally differ from that of the dog's ancestors at the earliest stage of their symbiosis with man.

Strebel (1905) has produced a sketch of the crania of four wolves, one shot on the free range, the second captured as a cub and reared in captivity, the third born in captivity, and the fourth from parents already born in captivity (Fig. 112).

In comparison with the skull of the wild beast, the cranium of the wolf born wild but reared in captivity is distinguished by the depression of the profile at the posterior end of the nasals; but the concavity is so shallow that it can hardly be called a stop. The skull does not reach the full length of the skull of the wild beast; the neurocranium is moderately vaulted above the orbits, the occiput slightly slopes backwards, and changes are noticeable in the position of the teeth, the third upper premolar in particular showing a slight inward turn.

In the skull of the wolf born in captivity the splanchnocranium is still shorter, the nasals are higher aborally, and the neurocranium, considerably elevated behind the orbits, markedly slopes backwards, approaching the characteristic conformation in the domestic dog. The brain, Strebel (1905) remarked, had apparently moved forwards, giving the brain case an almost spherical shape. But this is a fallacy, as the elevation of the anterior part of the neurocranium is due mainly to the increase in cranial width and the greater development of the frontal sinuses. The third upper premolar has turned nearly vertical to the normal position of this tooth in the wild wolf, and the dentition is much closer.

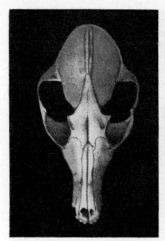 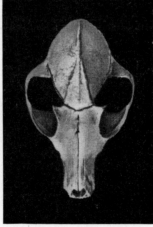

111. Crania of a fox born wild (left)—basilar length 123 mm and a fox born in captivity (right)—basilar length 107 mm (after Klatt)

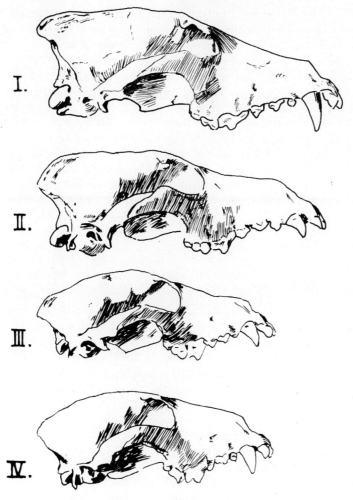

112. Sketches of four crania of wolves (after Strebel)
  I. Cranium of wolf shot in Poland
  II. Cranium of wolf born wild, but reared in captivity
  III. Cranium of wolf born in captivity
  IV. Cranium of wolf from parents born in captivity

The skull of the wolf bred from parents already born in captivity hardly resembles the skull of a proper wolf. The evolution initiated in the second and the third skull is continued. The cranium is so short that the mandible is already undershot. Not only has the third upper premolar turned completely vertical to its normal position in the wild beast, but the upper carnassial has also begun to turn inwards (Strebel, 1905).

The changes in the crania of wolves bred in captivity have been ascribed to the different diet and mode of life (Wolfgramm, 1894). In cage animals the maxillae and masticatory muscles are developed far less than in animals growing up in their natural environment. This is reflected not only in the development of the zygomatic arches and cranial superstructure but in the conformation of the entire skull.

The far-reaching effects of diet and restriction of movement in captivity on the development of facial musculature and cranial conformation are obvious. But in addition to these effects profound cranial changes occur in domesticated Canidae owing to variations in the hereditary constitution. Indeed, the majority of the different cranial types known in domestic dogs cannot be explained in any other way than as the results of selection of inheritable variations. It is significant that the changes encountered in the cranial conformation of cage wolves are not constant. Thus Antonius (1922) found that in one of two wolves from the same litter the splanchnocranium was a little shorter than the splanchnocrania of its parents, while in the other one the splanchnocranium exceeded those of the parents in length, its long narrow head resembling a greyhound's. This illustrates the great individual variability characteristic of the wolf.

The Effect of Domestication on Body Size

Environmental conditions obtaining during the early stages of domestication are generally less favourable to the physical development of the animals than is the normal life of the wild beast. The beginnings of domestication are therefore frequently accompanied by a reduction in the body size of the domesticated stock, owing to selection in relation to unfavourable conditions, accelerated by inbreeding, somewhat similar to the decrease in size observed in island fauna or near the limits of the range of wild species. To this rule the dog forms no exception.

The decrease in the size of the body reacts on the development of the various organs. A reduction in the size of an organ the form of which remains unchanged increases the proportion of surface to volume, while the degree of certain body functions depends on surface development rather than on volume (Klatt, 1927). This is of consequence to the conformation of the cranium. The cranium of the small domesticated animal is not an exact miniature replica of the cranium of the larger wild or domesticated beast. The reduction in the size of the various cranial bones does not occur proportionally, nor does every individual bone decrease in proportion to the general reduction in body size. These differences in the relative decrease in size are governed by physiological laws which are an important guide in the quest for the ancestors of the domesticated animals.

The Effect of Body Size on Cranial Proportions

Several authors have made a study of these laws (Dubois, 1897; Brandt, 1898; Weidenreich, 1925; 1941). Klatt (1913) has analysed the influence of differences in the basilar length of the cranium, reflecting body size, on the cranial conformation of various types of domestic dog. In a small dog the neurocranium is relatively larger and the splanchnocranium shorter than in a large dog of a similar racial type. The cranium of

ORIGIN AND DESCENT

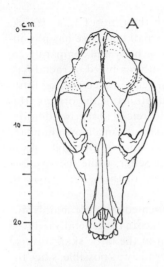
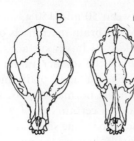

113. A—Cranium of a large Poodle;
 shoulder height 60.5 cm
 weight c. 21 kg
 B—Cranium of a Miniature Poodle;
 shoulder height 27 cm
 weight c. 3 kg
 C—Cranium A reduced to the basilar length of cranium B
 (after Roehrs)

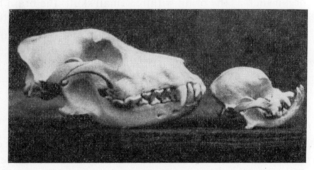

114. Dolichocephalic skull of a large dog, and brachycephalic skull of a dwarf dog, frontal and lateral views

an adult dog of a small breed resembles to some extent that of a pup of a larger breed of similar type. This is due chiefly to the small body size common to them. As the functions of various organs do not depend solely on body size, but on many additional factors, such as age for example, the resemblance between the crania of adult dwarfs and immature normal or large dogs of a similar racial type is not complete.

The relatively large size of the neurocranium in small breeds of dog is due to the size of the brain which is relatively larger in animals of a small body size, be this a racial feature or due to immaturity, than in larger specimens. The relative brain size is in inverse ratio to the body size (Weidenreich, 1941). In small dogs with a basilar length of the cranium ranging from 7 to 8 cm the brain has a volume of approximately 60 cc; in large dogs with a basilar length of 20–24 cm the brain volume is 125 cc, that is only about twice the brain volume of small dogs. While domestic dogs may vary between $1^1/_2$ and 75 kg in body weight, that is as 1 : 50, the brain weight of these extreme types varies merely between 50 and 150 g, i.e. in the ratio of 1 : 3 (Starck, 1962). The relatively large size of the brain in animals of small body size has physiological reasons. It is due to the relatively larger surface of the body and comparatively more intensive metabolism by which the smaller animal within a breed or racial type is distinguished from the larger. Even a small body must rely on a certain quantity of nervous substance to maintain the normal somatic functions of the organs; this quantity cannot be reduced beyond a certain limit (Weidenreich, 1941). In addition, the relatively larger surface of the body of the small animal results in a more abundant supply of nerve fibres. In the dwarf Bolognese the number of fibres of the sciatic nerve per 100 g body weight is almost five times that in a large Setter, while the absolute brain weight of the Bolognese amounts to only 63 per cent of the Setter's (Brandt, 1898).

The cranial differences between small and large domesticated dogs are mainly due to the fact that the brain occupies a relatively much larger space in the generally reduced skull of small dogs than the proportionately small brain in the large skull of large types. The cranial cavity in small dogs tends to expand as widely as possible, since the sphere is geometrically the most economical form to accommodate a relatively large organ such as the brain within a small skull: the brain becomes brachencephalic and the skull brachycephalic. At large body size there is no difficulty in the accommodation of the relatively small brain in the skull, and there is no necessity for the approximation to spherical form: the brain therefore is dolichencephalic and the skull dolichocephalic (De Beer, 1949).

Difference between Domesticated and Wild Canidae

However, domesticated and wild Canidae differ in this respect. Small wild Canidae, in contrast with the domestic dog, are diminutive forms of the large types. The volume of their brain follows a decrease in body size more closely than in domesticated dogs. Brain size is not a uniform factor; it reacts intraspecifically differently from interspecifically. In spite of the close zoological relationship between Canis aureus and Canis lupus, there exists no simple allometry in the capacity of their brain cases in relation to body weight, but an abrupt difference. As Roehrs (1959) has pointed out, in canine species of different body size intraspecific allometric values for brain weight in relation to body weight differ markedly from the interspecific values.

The wolf has more brain substance in relation to body size than have domestic dogs of the same size—1 : 270 in the wolf as against 1 : 355 in dogs. The difference in relative brain size between wolves and domestic dogs is attributed to the reduction of particular brain regions under domestication (see also p.92). On the other hand, domesticated dogs have approximately 20 per cent more brain weight than jackals of similar body size. A female jackal with a body weight of 7.7 kg has a brain weight of 59.4 g; a female dog, weighing 7.6 kg, a brain weight of 73.0 g. In the fox the ratio of brain to body weight is 1 : 118, as against 1 : 88 in dogs of a similar size. The brain weight of a fox of 6 kg body weight amounts to 50 g, that of domestic dogs of the same size to 69.5 g, i.e. 39 per cent more (Klatt, 1921).

Therefore, the cranial cavities of small wild Canidae are only slightly larger than

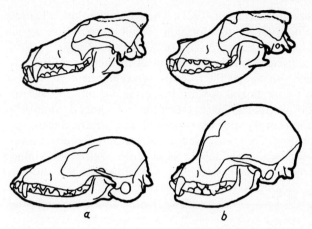

115. Sketches of 4 canine crania, reduced to the same basilar length (after Klatt, 1949)
a) above—wolf
   below—fennec
b) above—mastiff
   below—dwarf dog
(Note: wolf and mastiff skulls are of similar size; the same applies to fennec and dwarf dog crania)

implied by the size of their bodies and the length of their skulls. Consequently the relative difference in size between the neurocranium and splanchnocranium in small and large domestic dogs considerably exceeds that in wild Canidae. As the comparatively broad neurocranium in the small domestic dog influences the entire skull, including the anterior part of the splanchnocranium, the latter appears heavier in dwarf domestic dogs than in wild Canidae of a similar body size. This difference in the slenderness of the splanchnocranium between dwarf domestic and wild Canidae is enhanced by the relatively heavy dentition by which the domestic dwarf is distinguished from the wild (Klatt, 1927).

The Brain Case

There is a distinct negative correlation between the relative size of the cranial cavity and the total length of the skull. The greater the length of the cavity, the shorter the total length of the skull and conversely. In the wolf and the Irish Wolfhound the cranial cavity amounts to less than 50 per cent of the total skull length, whereas in dwarf domestic dogs, such as the King Charles Spaniel, it may exceed 90 per cent (Weidenreich, 1941).

Owing to the large volume of the brain, the anterior and posterior cranial widths, in relation to the basilar length of the cranium, are considerably greater in small than in large domesticated dogs; in dwarf breeds with a basilar length of 7–8 cm the posterior cranial width is 4.9 cm and the anterior 4.3 cm; in large breeds with a basilar length of 20–24 cm—6.2 and 4.8 cm respectively. With increasing cranial size a very marked decrease in skull width relative to skull length is common (Lumer, 1940). The length of the brain case, on the other hand, does not depend solely on the volume of the brain but also on the development of the muscles of mastication. This development is governed partly by the absolute size of the dog, and partly by its age, diet and mode of life. The shape of the brain and brain case is modified also by racial type. The tendency in small dogs to a shorter, higher and broader brain is enhanced by the peculiar general conformational tendency in brachycephalic dwarfs, and reduced by the opposite tendency in dwarf greyhounds (Klatt, 1921). These extreme types illustrate that cranial characteristics are not independent formations, but symptoms of certain plans of organisation.

In large types the frontals participate in the formation of the neurocranium only with their posterior ends, whereas in dwarfs the frontals partake in their entirety, so that the cranial cavity extends forwards and downwards to the nasion. As the brain case simultaneously increases in width and height, the bony substance available for the formation of the neurocranial walls appears exhausted and insufficient to meet the requirements. Hence the walls of the brain case are considerably thinner in dwarf dogs than in large types (the thickness of the tuber region of the parietal amounts to only about 1 mm in the King Charles Spaniel as against 3 mm in the Irish Wolfhound), and the cranial sutures remain patent (Weidenreich, 1941).

Cranial Superstructure

The muscles of mastication, especially M. temporalis, considerably influence the development of the frontal sinuses and cranial superstructure. In dogs M. temporalis is less developed than in wild Canidae of a similar body size; but the brain, more especially its posterior portion, is also less voluminous in the dog than in the wolf, the cranial capacity of the dog being 15.6 per cent smaller than in a wolf of about the same size. The smaller neurocranium of the dog, as compared with the wolf's, offers less surface for muscular attachment, increasing the necessity of superstructural development. This outweighs the consequences of the weaker muscles of mastication characteristic of the dog, so that the development of the dog's cranial superstructure generally exceeds that in wolves of a similar body size (Klatt, 1927).

In small dogs M. temporalis is relatively less developed than in the larger breeds. At the same time, the large neurocranium of the small dog offers an extensive surface for attachment. While in large dogs the splanchnocranium and temporal muscles are comparatively powerfully developed, the small brain case offers only a restricted surface for muscular attachment. This necessitates the development of large frontal sinuses and the ascent of the temporal muscles to the median line of the neurocranium where the parieto-frontal ridges join into a prominent sagittal crest offering a strong gripping surface to the temporal muscles. In small Canidae, immature or adult, where the temporal muscles do not require the sagittal crest for attachment, the parieto-frontal crests remain separate throughout, or unite only far back at the occipital protuberance.

The reduction of the masticatory apparatus fundamentally alters its mechanical correlation with the neurocranium. Weidenreich (1941) has pointed out that if the brain case is small but the masticatory apparatus large, the cranial surface is not sufficiently expanded to provide the necessary space for the attachment of the massive musculature and not strong enough to withstand the chewing force. In consequence of this disproportion, the cranial surface becomes enlarged and strengthened by superstructures. If the brain case is large but the masticatory apparatus small, the cranial surface is large enough to accommodate the reduced muscles and at the same time strong enough to serve as a buttress for the diminished chewing force. The latter, in addition, attacks at a more posterior level, owing to the changed topographical relation between brain case and jaws. Since from the architectonical standpoint air sinuses are regions left vacant between the mechanically essential pillars of the constructing parts of the splanchnocranium and neurocranium, they become smaller or disappear completely with the decrease in congruity between the individual constituents of the skull.

These conditions can be observed in nearly every cranium of domesticated as well as wild Canidae. For although in dwarf wild Canidae, in contrast to the domesticated dwarf, the brain and cranial cavity are only slightly larger than implied by the size of their skulls and bodies, the formation of their cranial superstructure approaches the corresponding types of the domestic forms. The sagittal and nuchal crests of large wild canine types are well developed, and the protuberance between the postorbital process and the coronal suture which lodges the frontal sinus is quite pronounced. In the small types of wild Canidae there is no sagittal crest, while the nuchal crest is weak and the temporal ridges reach the middle of the cranium only within the lambda region. In both the fennec and common fox the frontal sinus is lacking. These peculiarities are due to the fact that the entire masticatory apparatus is relatively less developed in small than in large wild Canidae (Weidenreich, 1941).

There is a close relation between the frontal superstructures and frontal sinuses. In large dogs the space between nasion and anterior pole of the cranial cavity is occupied by a large sinus on either side. In dwarf dogs the cranial cavity extends forward up to the nasion, the sinuses having disappeared completely. Weidenreich (1941) regards the formation of the supraorbitals as a consequence of the construction of the skull and of the special arrangement of its two constituents, the neurocranium and splanchnocranium. If the splanchnocranium is situated more or less in front of the neurocranium, there is a static and dynamic impulse to strengthen the superior splanchnocranial parts by projecting structures serving as transmitters or buttresses, especially in animals in which the dentition and masticatory apparatus are strongly developed. If the splanchnocranium is reduced and more or less underlies the neurocranium, such protruding formations are not required, the neurocranium itself being strong enough to serve as an adequate recipient.

Orbital Size and Shape

Conditions similar to those of the brain obtain in the size of the eyes. Small animals have relatively larger eyes than animals of larger size. In the cranium this is reflected in the size, shape and position of the orbits. But here again wild Canidae differ from domestic dogs in that the eye follows a decrease in body size more closely (Klatt, 1927). The difference in relative eye size affects also the size of the brain; in man, for example, just as many sensory nerve fibres proceed to the brain from the optic nerves as from the entire spinal cord (Dubois, 1898).

According to Weidenreich (1941), the size of the orbit is dependent on the size of the eyeball in a most general way only. Being determined by different factors, the two structures vary to a considerable degree. While the eye develops as a part of the brain in a very early stage, at a time when the central nervous system exceeds all other organs, the orbit is constructed by skeletal elements belonging partly to the neuro-

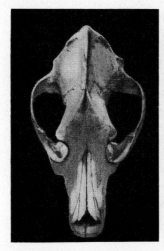 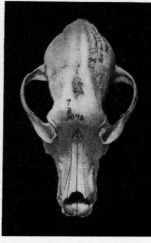

116. Mastiff (left) and terrier (right) skulls (reduced to the same basilar length) showing different development of the sagittal crest (after Klatt)

cranium and partly to the splanchnocranium. In the construction of the neurocranial portion the supraorbitals take an active part. In foetal and juvenile stages the disproportion between eye and orbit is therefore less pronounced than in adults where the supraorbitals and the splanchnocranium have attained their definite size.

A change in the shape and position of the orbits occurs also owing to the modification of the habits of the dog under domestication; this is reflected in the difference in the position of the eyes between dogs and wolves (Dahr, 1942). The eyes of the wolf are somewhat lateral and slit-like in relation to the roundish and more forwardly directed eyes of the dog.

Development of the Stop

The cranial profile, especially the development of the stop, is likewise influenced by body size. The height of the splanchnocranium is proportionate to the basilar length of the skull. But the height of the neurocranium depends in small crania entirely, and in large skulls at least partly, on the volume of the brain. In small skulls the relative neurocranial height therefore exceeds that in large skulls. This results in a steeper ascent of the profile at the posterior end of the nasals and a more clearly defined stop in the smaller than in the larger specimens and breeds of similar racial type. In extreme cases of dwarfism the maxillae are enormously reduced in length and height (Weidenreich, 1941); this reduction, together with the strongly bulging forehead, produces a deep incurvation of the nasal bridge.

Klatt (1921) has shown that the anterior (frontal) section of the brain is larger in dogs than in wolves of a similar body size, the parietal section is about equal, whereas the aboral part shows a reduction in the superior, inferior and posterior regions. The process of domestication has brought about an increase in the centres of association, and a reduction in those of projection. These changes are mainly responsible for the steeper ascent of the frontal, and the marked descent of the occipital region, as well as for the more vertical position of the orbits by which the crania of dogs are distinguished from those of wolves of a similar body size.

The Splanchnocranium

There is a sharp increase in the relative length of the facial portion of the skull with increasing total cranial length (Lumer, 1940). On the other hand, the greater the space required by the brain, the more other parts of the skull must suffer and yield to that need. The enlargement of the neurocranium in domestic dogs of small body size results in an extensive reduction in splanchnocranial length, with a consequent shortening of the palate and dental arch. In large types the entire set of teeth is situated in front of the anterior part of the cranial cavity; in dwarf types the whole set of teeth is shifted considerably towards the rear, with the molar row underlying the entire orbital and frontal portion of the cranial cavity. The shortening of the palate in small domestic dogs restricts the space occupied by the teeth. The gaps between them become smaller and the premolars and molars shift their direction from a longitudinal to a transverse position. The size of crowns and roots is greatly diminished and, in extreme cases of dwarfism, premolars and molars may be reduced in numbers and their patterns become remarkably simplified through the loss of the cingulum and even certain cusps, with the remaining cusps and crests lowered and rounded.

The lower jaw is also adapted to the different conditions of body and neurocranial size. In large canine types the mandible represents a straight bone, harmonising in length with the upper jaw; in dwarfs the mandible is shortened and curved and, in

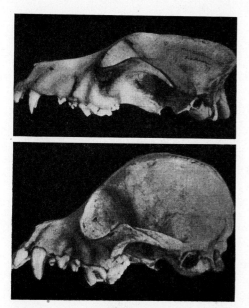

117. St Bernard (above) and Papillon (below) skulls (reduced to the same basilar length) showing different development of the stop (after Klatt)

cases where this transformation is insufficient to compensate for the shortening of the maxilla, it is undershot (Weidenreich, 1941).

However, these remarks on splanchnocranial development in dogs of different body size, more especially Lumer's (1940) assertion that there is a sharp increase in the relative length of the facial portion of the skull with increasing total cranial length, apply only to dogs of a similar growth type, and not to breeds of a fundamentally different cranial conformation, e.g. the Boxer and Dachshund (see pp. 96–98).

Craniological Comparison between Dogs of Different Sizes

Weidenreich (1941) has compared the cranial structure of adult dogs of three different types; a dwarf (King Charles Spaniel), a large (Irish Wolfhound), and an intermediate (English Bulldog) type, the last-mentioned somewhat closer in conformation to the dwarf than to the large breed.

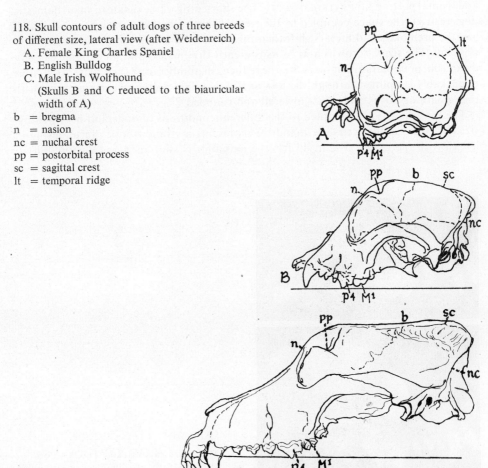

118. Skull contours of adult dogs of three breeds of different size, lateral view (after Weidenreich)
  A. Female King Charles Spaniel
  B. English Bulldog
  C. Male Irish Wolfhound
    (Skulls B and C reduced to the biauricular width of A)
b = bregma
n = nasion
nc = nuchal crest
pp = postorbital process
sc = sagittal crest
lt = temporal ridge

The Cranium of the Large Breed

In the skull of the large breed (Fig. 118-C) the neurocranium is long and relatively low. A very pronounced triangular postorbital process (pp) projects considerably over the surface of the cranial wall, continuing backwards with a heavy crest (sc) and increasing in height towards the occipital region where it communicates with a corresponding nuchal crest (nc). The lateral surface of the neurocranium is formed by two rounded elevations, an anterior and a posterior one. These elevations are separated from each other by a distinct depression which coincides with the course of a rough line, i.e. the completely obliterated coronal suture. In the frontal view of the skull the anterior elevation is rhomboid-shaped, its angle being represented by the postorbital processes on either side. The posterior elevation is much smaller and of ovoid form.

The Cranium of the Dwarf Breed

The skull of the dwarf dog (A) differs from that of the large breed not only in size but also in form and proportions of neurocranium and splanchnocranium. Only the skull of the large type shows those features that are considered characteristic of the genus Canis, i.e. a long narrow form with the brain case relatively small but carrying heavy superstructures and the jaws far projecting. In the skull of the dwarf dog the neurocranium is enormously enlarged, short, high and bulging. The tuber-like frontal part projects outwards and the same holds of the occipital part. The postorbital process (pp) is reduced to a faint triangular pattern on the lateral surface. There are no traces of

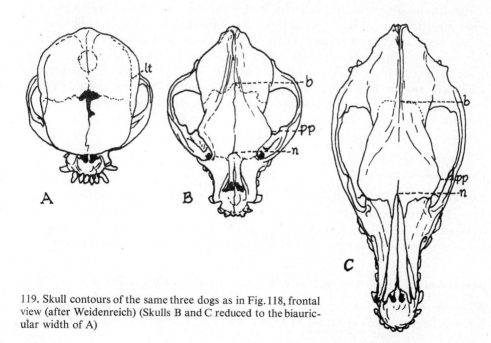

119. Skull contours of the same three dogs as in Fig. 118, frontal view (after Weidenreich) (Skulls B and C reduced to the biauricular width of A)

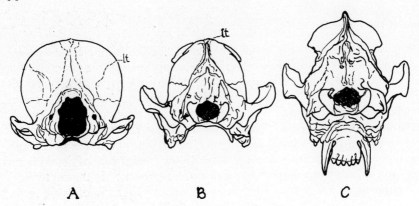

120. Skull contours of the same three dogs as in Fig. 118, Nuchal View (After Weidenreich) (Skulls B and C reduced to the biauricular width of A)

crests, instead of which there is a fine temporal line (lt) running about halfway between the squamosal and sagittal sutures. Seen from above, the neurocranium presents a unitary globule-like body with well marked frontal and parietal eminences. The coronal, sagittal and metopic sutures, and occasionally even the fontanelles remain patent until the end of life. The bregmatic region is occupied by a wide fontanelle of irregular form.

The fundamental differences between the large and the dwarf skulls are most marked when viewed from behind. The dwarf skull strikingly exhibits its globular form, with the greatest width far above the biauricular plane. There are no crests or other superstructures, the temporal ridge (lt) being represented by a faint line. All sutures are patent. The Foramen magnum is very wide, and developed particularly in a vertical direction. In the large skull the neurocranium appears in a narrowed triangular form, its greatest breadth coinciding with the biauricular plane. The cranium is crowned by a high thick crest. The latter ends at the front on either side in a large superstructure corresponding to the anterior elevation and the postorbital process. The sutures are obliterated, and the Foramen magnum is relatively small and especially developed in a transverse direction.

The Cranium of the Bulldog

The skull of the English Bulldog (B) stands intermediate in regard to these peculiarities. The cranium is considerably shorter and relatively broader than the large skull, but there are still the two elevations and the superstructures, although much less pronounced. The neurocranium is somewhat wider and the sutures are still patent.

Independent Effects of Growth Type and Body Size on Cranial Conformation

Weidenreich's theory of constant proportions between cranial and brain lengths in domestic dogs has been criticised by Starck (1962) because it fails to take into account the independence of the two phenomena, dwarfism and brachygnathia, which may occur either alone or in combination. In dogs of any body size relatively broad skulls combined with short splanchnocrania, and narrow skulls combined with long splanchnocrania are encountered. These very different growth types, which are connected by intergrading forms, may occur even in the same breed (Fig. 121). Brains of similar size may be associated with totally different splanchnocranial lengths (Stockhaus, 1962). A King Charles Spaniel with a brain volume of 71 cc has a lateral facial length of 28 mm, and a Dachshund with a brain volume of 70 cc a facial length of 70 mm; in a Bulldog a brain volume of 95 cc is associated with a lateral face length of 64 mm, in a German guard dog a brain volume of 91 cc with a facial length of 94 mm. Again, the Dachshund and Boxer have similar cranial lengths. The discrepancy between these two breeds in the ratio of cranial length to body size is due to the difference in the splanchnocranial length which is independent of body size. The Boxer has an absolutely much shorter splanchnocranium than the dachshund. In the Boxer the same cranial length is composed of a long neurocranium and short splanchnocranium, and in the Dachshund of a short neurocranium and long splanchnocranium (Bohlken, 1962). In Weidenreich's examples the effects on cranial conformation of body size and growth type are superimposed, invalidating his conclusions with regard to the influence of body size. In addition, his representation of the topographical relations between brain and cranium is faulty, for in large long-headed dogs the frontal part of the brain and the neurocranial cavity extend much farther rostrally (the difference being as much as 30 per cent), and in medium-sized dogs (English Bulldog), too, Weidenreich's drawing of the cerebrum does not correspond with the factual condi-

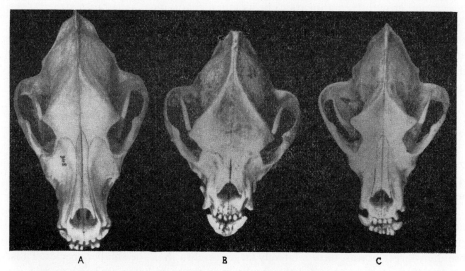

121. Different growth types in Boxer skulls (after Stockhaus)
A and B of similar brain volume, but different splanchnocranial size;
C of smaller brain volume than A or B, but of similar splanchnocranial conformation to A

tions (Starck, 1962). Similarly, Rosenberg (1965) has pointed out that the splanchnocranial and neurocranial conformations in dogs are not as closely connected as Weidenreich assumed. Even the shapes of the maxillae and mandibles may be inherited independently. Weidenreich's assertion that dwarfing is accompanied by a shortening and elevation of the splanchnocranium stems from his mistake in choosing the King Charles Spaniel for comparison with larger dogs. The King Charles belongs to the short-headed breeds of dog. Its cranial conformation is a peculiarity of its growth type, and not only a function of its body size. Shortened and elevated splanchnocrania may be found also in large breeds of dog. Different growth types are inherited independently of body size. The existing effects of the latter on cranial conformation can only be correctly judged by comparing breeds of different size but of similar growth type.

Weidenreich (1941) used his findings of the effects of body size on cranial conformation in dogs as basis for his autogenetic theory of human evolution, which has been criticised by Nesturkh (1959) on account of its idealistic conception, its over-emphasis of the effects of changes in body size and its failure to take sufficient account of changes in diet and feeding or of selection of "indefinite accidental variations".

Genotype in Cranial Conformation

In discussing the problem of whether the structural peculiarities of canine crania of different sizes "belong in the category of genotypical or phenotypical phenomena", Weidenreich (1941) asserted that neither term is applicable since cranial conformation is a reaction to the size of the brain "which alone appears genotypical for the dog". However, the validity of this dictum is questionable in view of the above-mentioned fundamental differences in cranial growth type existing in domestic dogs of any body size; cranial growth type appears to be as much genotypical as is brain size.

The genotypical differences in the effect of brain size on cranial conformation between wild Canidae and domestic dogs of different body sizes emphasise the importance of growth type. Cranial conformation is as much influenced by the specific difference between wolf and jackal as it is by brain size and growth type in the domestic dog. In the latter the size of the brain, determining the presence or absence of certain structural peculiarities of the skull, including the dentition, is conditioned by the size of the whole body. In wild Canidae this also applies, but in a totally different degree (see pp. 87–89 and Fig. 115). It would appear that the artificial selection practised by man in intensifying intraspecific variation in body size has produced less change in brain size than is apparently brought about by natural selection working on wild forms (Huxley, 1932). Klatt and Vorsteher (1923) have attributed the fact that in large dogs the brain is smaller than in large wolves, but in small domestic dogs larger than in small wild Canidae, to fundamental differences in the linear proportions of their bodies and, to some extent, to the different proportions of the association and projection centres in their brains. Generally, the larger species of a genus, characterised by absolutely larger brains, have relatively larger cerebra than the smaller species. Thus, in the wolf the average share of the cerebrum in the total brain amounts to 81.5 per cent, but in the jackal to only 77 per cent. In the domestic dog, on the other hand, the share of the cerebrum in the entire brain averages 78.5 per cent, irrespective of the absolute size of the brain (Klatt, 1921; Herre, 1955). As the brain and eyes, i.e. organs which are larger in domesticated than in wild Canidae of small body size, ontogenetically develop from the ectoderm, the genotypical differences between the crania of large and small wild and domesticated Canidae may derive, apart from brain size and growth type, from fundamental physiological and psychological differences between wild Canidae and the domestic dog.

Comparison between C. F. palustris and C. F. matris optimae

Under the aspect of the relation between body size and cranial conformation, Klatt (1913) has attempted an analysis of the differences between C. f. matris optimae and C. f. palustris. The C. f. matris optimae skull, as above mentioned, is distinguished from that of C. f. palustris by its greater absolute size, the longer, narrower and more pointed splanchnocranium, shallower stop and lower neurocranium. In C. f. matris optimae the sagittal crest is long and well developed; in the turbary dog it is weak and short, commencing only in the vicinity of the occipital protuberance. Klatt observed that these cranial differences were due solely to the difference in body size between the two types, and not to racial difference. No other conclusion could be drawn from them than that domestic dogs of larger and smaller size occurred in prehistoric Europe. Dahr (1942) arrived at a similar conclusion in his critique of the classification and phylogeny of recent dogs on the basis of their cranial resemblance to fossil types.

But the limitation of this approach becomes apparent if a large representative of the palustris type is compared with a small dog of matris optimae type, or if Studer's other basic types of prehistoric dogs are compared with one another. For example, in body size C. f. intermedius stands about midway between C. f. palustris and C. f. matris optimae. The slightly greater height and width of the intermedius skull as compared with the cranium of C. f. matris optimae, as well as the stronger development of its teeth, are still explainable by the difference in body size. But already the broader and shorter splanchnocranium cannot be fully explained in this way; and if the more extreme cranial types, such as the mastiff or the greyhound, of whatever body size, are taken into consideration, the overwhelming role of selection of inheritable variations not only as a breed- but also as a type-forming element becomes manifest.

Three Types of Cranial Characters

Weidenreich (1941), consequently, distinguishes between three different types of morphological characters:—(1) Fundamental characters of the first order, specific to the dog, not subject to alteration, and common to all variations of dog independent of the size of the body or of the brain. (2) Special characters of the second order, specific to the type of dog and dependent on the size of the brain in relation to body size. (3) Relatively unimportant characters of the third order; with no bearing on either the fundamental or the special characters which are determined only by the individual belonging first to the family Canidae and then, within this group, to a certain order of size.

Relative Growth Rate

Hilzheimer (1926) attributed the cranial differences in the various types of domestic dogs partly to the premature arrest in the growth of different bones at different age stages of development, and partly to the continuance of growth beyond the age limits of their development in wild and normal-sized domesticated Canidae.

In the course of the ontogenetic development the nervous system is first differentiated and its growth rate exceeds that of all other organs. As long as this rate remains unchanged, the brain determines the predominance of the neurocranium over the splanchnocranium. Subsequently the growth of the central nervous system diminishes until it ceases entirely. Then the hitherto restrictive organs become propulsive and the proportions of brain case and face reverse (Weidenreich, 1941). In the newborn sheepdog, as Becker (1923) has pointed out, the neurocranium considerably exceeds the splanchnocranium in size, forming a laterally slightly compressed and orally narrowing hemispherical case which sharply slopes backwards to the occiput and forwards to the nasals. The latter, markedly sloping orally, are convex above the anterior edge of the orbits, and considerably restricted in the region of the Foramina infraorbitalia. Owing to the steep descent of the frontals, the posterior end of the nasals is situated considerably below the surface of the frontals. This results in the two breaches in the descending slope of the profile. The premaxillae are very short, and only slightly exceed the nasals in height. The zygomatic arches extend in the direction of the posterior part of the markedly protruding maxilla, almost parallel with the cranial axis. From below, the maxilla appears therefore as a line nearly rectangularly broken be-

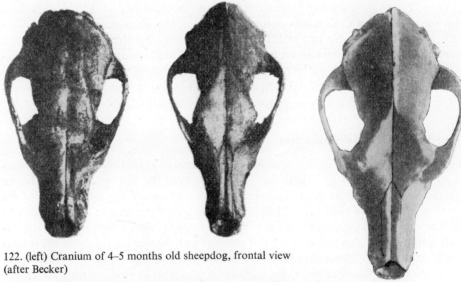

122. (left) Cranium of 4–5 months old sheepdog, frontal view (after Becker)

123. (centre) Cranium of 5–6 months old sheepdog, frontal view (after Becker)

124. (right) Cranium of an adult sheepdog, frontal view (after Becker)

tween the alveoli of the first and second temporary molars. The anterior edge of the premaxilla forms a short high arch, and the short mandible is nearly bow-shaped. With the exception of the supraorbital processes, the cranial superstructure is completely undeveloped.

Cranial Changes in Early Life

Until the shedding of the temporary teeth, the principal cranial changes consist of the expansion of the neurocranium into a pear-shaped case, and the extension of the splanchnocranium in length and height. The increase in height reduces the sharp demarcation of the stop, so that the profile descends in an unbroken slope from the highest point of the forehead to the constriction of the nasals in the region of the Foramina infraorbitalia. The extension of the splanchnocranium concerns chiefly the region in front of the Foramina infraorbitalia, more especially the premaxillae, the anterior ends of which are now situated considerably in front of the tips of the nasals. The crests and ridges for muscular attachment appear at the occipital part of the skull.

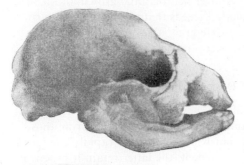

125. Cranium of a newborn sheepdog, lateral view (after Becker)

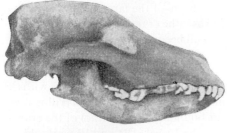

126. Cranium of 4–5 months old sheepdog, lateral view (after Becker)

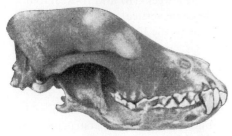

127. Cranium of an adult sheepdog, lateral view (after Becker)

In the centre of the line where the roof of the brain case is joined to the posterior wall of the skull the occipital protuberance begins to extend backwards, accompanied by the elevation of the originally oblique posterior wall to a more vertical position. The bullae are nearly fully developed, and the mandibles have grown in length; but the splanchnocranium still remains shorter than the neurocranium which has not yet fully lost its spherical shape.

Later Cranial Changes

From the time of the shedding of the milk teeth the skull develops its final shape. The parietal ridges appear, first as separate crests connected only at the occipital protuberance, but later joined into the fully developed sagittal crest. The frontal region becomes distended by the development of the frontal sinuses. The superior and inferior orbital processes and the zygomatic arches receive their final shape. The roof of the brain case is flattened out, and the constriction of the skull in the temporal region is completed. The proportion between the neurocranium and splanchnocranium becomes final, with the latter nearly reaching the size of the former.

Parallel with the excess of splanchnocranial over neurocranial growth in the postnatal skull, the relative growth in cranial length exceeds that in width. From birth to the termination of growth the width of the palate gradually decreases in relation to length, accompanied by a relative reduction in cranial height.

Cranial Comparison between Mature Dwarf Dogs and Puppies of Large Breeds

Schmitt (1903) has demonstrated that skulls of newborn puppies of large breeds, such as the St Bernard or the Newfoundland, do not differ essentially from those of dwarf dogs, like the Black-and-Tan Terrier; but while the former rapidly change the original form in the course of postnatal development, the latter retain it practically unchanged to the most advanced age. In the newborn of the large breed the frontal bone forms the entire anterior part of the brain case, the cranial cavity extending forward up to the nasion. The more elongated the skull becomes by pushing the face forward, the more the brain case falls back, so that after 10 weeks only the posterior portion of the frontal bone participates in the neurocranial formation, while its major portion now serves to construct the face, that is to say, the postorbital process with the enclosed air sinus. The extraordinary elongation of the splanchnocranium is illustrated by a comparison of the position of the anterior boundary (dotted line) of the cranial cavity in the newborn (Fig. 128 A) with those in the 4-week- (B) and 10-week-old (C) puppies. In the new-born this boundary lies on the same vertical plane as the anterior site of the junction of the zygomatic arch, whereas in the 10-weeks-old pup (C) it is situated on the posterior site.

Median Cranial Sections

Weidenreich (1941) has demonstrated the elongation of the splanchnocranium and the corresponding development in the height and bulkiness of the superstructures by a comparison of median sections through the crania of newborn and adult specimens of large breeds (Fig. 129). The dwarf type, retaining the foetal conditions to a considerable degree, markedly contrasts with the cranial evolution of large breeds. This is illustrated by the median section through the cranium of a newborn of a large breed (the same specimen as in Fig. 129) superimposed on that of an adult dwarf (Fig. 130). There is no difference in the proportions of the entire skull and the relations of neurocranium and splanchnocranium between the newborn of a large, and the adult of a dwarf breed.

Arrest of Cranial Growth in Dwarfs

Hilzheimer (1926) attributed the spherical neurocranium characteristic of dwarf breeds of dog to the arrest in cranial growth approximately at the stage reached by large breeds shortly before the shedding of the temporary teeth. After birth the crania in the

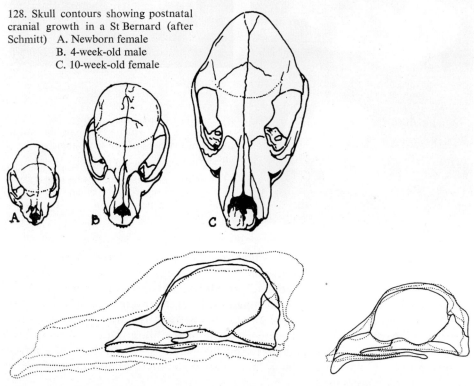

128. Skull contours showing postnatal cranial growth in a St Bernard (after Schmitt) A. Newborn female
B. 4-week-old male
C. 10-week-old female

129. (left) Diagrams of median sections through the skulls of an adult Newfoundland (....) and a newborn Wolfhound (——) superimposed and reduced to the same size of cranial cavities (after Weidenreich)

130. (right) Diagrams of median sections through the skulls of an adult dwarf Pinscher (——) and a newborn Wolfhound (....) superimposed and reduced to the same size of cranial cavities (after Weidenreich)

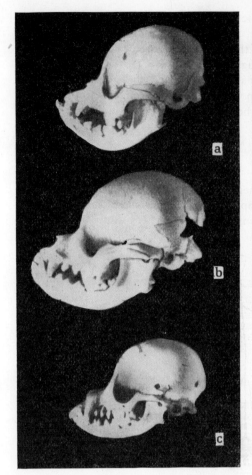
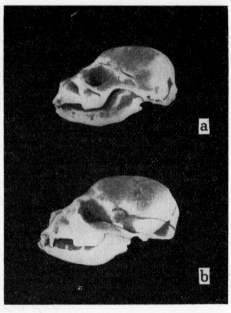

131. (left) (a) Japanese Spaniel
　　　　　 (b) Pug
　　　　　 (c) Pomeranian
Note: breed differentiation in conformation at the age of 3–5 months still slight (after Baumann and Huber)

132. (right) (a) St. Bernard—8 days old
　　　　　　(b) Borzoi　　—10 days old
Note: conformation at this age still similar (after Baumann and Huber)

various types of domestic dogs show a high degree of similarity, since the characteristic racial differences begin to appear only with the shedding of the temporary teeth (Figs. 131 and 132). The crania of adult toy breeds which terminate their growth at this early stage are therefore difficult to distinguish. Hilzheimer admitted, however, that the arrest of cranial growth in toy breeds was not absolute, as additional factors, such as hereditary hydrocephalus and the greater use of the brain in the adult dwarf as compared with the pup of a large breed, might operate in the direction of a larger neurocranium, while the weak facial musculature in the toy dog left the juvenile spherical neurocranium unaffected by superstructures.

Cranial Development in Medium-sized and Large Dogs

The crania of medium-sized terriers, according to the same author, represent approximately the stage at which larger dogs shed their teeth. At this stage both the neurocranium and splanchnocranium have already grown in length, and the temporal constriction and frontal sinuses have begun to develop, while the surface of the neurocranium is still convex and, owing to the early stage in superstructural development, fairly smooth.

A later age stage in cranial development is reached by the large mastiffs in which the neurocranium is furnished with powerful crests and ridges for muscular attachment, whereas the maxillae generally remain short. In some breeds, however, the mandibles are undershot and very powerful, requiring strong muscles of mastication which in turn react on the development of neurocranial superstructure. Hilzheimer (1926) remarked in respect of such undershot mandibles that they had evidently grown beyond the stage at which the maxillae ceased to develop.

The above examples of arrest in cranial growth at different age stages can be increased by the multitude of intermediate stages at which the various parts of the cranium in the different types and breeds of domestic dogs terminate their growth.

Heretofore only such types have been discussed which Hilzheimer believed to be evolved from the normal type by the retention of juvenile features. He held, however, that in some domestic dogs the crania reached age stages beyond the normal termination of growth in large domestic and wild Canidae. In large greyhounds, for instance, the splanchnocranium attains a length exceeding that in any other domesticated dog or in the wolf.

On the "Retention of Juvenile Characters", Foetalisation and Retardation

Klatt (1927), while acknowledging Hilzheimer's consistency in extending the principle of the retention of juvenile features to the acquisition of post-adult characters in certain domestic dogs, has pointed out that this merely amounts to expressing size in different terms. While the introduction of the principle of the retention of juvenile characters may have some justification from the purely morphological point of view, it is completely misleading in an examination of the physiological causes of the cranial differences encountered in various types and breeds of domestic dog. Its application is justified only in obvious cases of infantilism where not only the development of the cranium but also the differentiation of tissues is arrested at a premature stage. This does not apply to the crania of toy dogs which represent merely the extreme limits of complete series leading through a multitude of intergrading forms to the largest crania within their respective types.

Actually the dwarf type of skull is due, not to the retention of visible foetal features, but to the preservation and persistence into later stages of the early high relative growth rate of the central nervous system which causes the predominance of the brain over the other organs during the foetal and juvenile stages of life. In dwarf dogs the foetal condition of the growth relation between neurocranium and splanchnocranium persists, while in large dogs this relation becomes reversed with the definite arrest of the brain growth. As compared with large dogs, the growth of dwarfs is not retarded but rather accelerated: dwarf dogs reach sexual maturity earlier than large breeds (Baumann and Huber, 1946). At the time of birth the skull of a dwarf Black-and-Tan Terrier has already attained 34 per cent of its final length, and at the age of

5–6 months 87.5 per cent; in the large St Bernard, on the other hand, only 28.5 and 82.3 per cent respectively (Schmitt, 1903). To speak of a retention of foetal features confuses the real conditions. The skull of an adult dwarf, though resembling in form and proportions the skull of a newborn, cannot be considered as merely a foetal form arrested in its development; morphologically and physiologically it represents the skull of an adult animal. Adult dwarf and juvenile large forms show a similarity in appearance because "the same smallness in size in both cases maintains certain functions at approximately the same level" (Klatt, 1913).

Bolk (1926) regarded foetalisation as a consequence of retardation, and attributed it to changes in the internal secretory complex. But Slijper (1936) has denied any causal connection between foetalisation and retardation. The term "foetalisation" refers to a phylogenetic process by which descendants acquire a certain specialised feature through cessation of its ontogenetic development at an earlier stage than occurs in their (in this respect) more generalised ancestors. The term "retardation" denotes merely a temporary delay of certain phases of the developmental course. In cases of foetalisation, final phases of the life cycle extant in ancestral forms are wanting; in cases of retardation the end phases are eventually arrived at. With regard to the attempts at ascribing retardation and foetalisation, or intraspecific form differences in domestic animals to the effects of endocrine factors, Starck (1962) has pointed out that so far these have remained entirely hypothetical.

Schaeme's Classification of Canine Cranial Types

Schaeme (1922), on the basis of measurements of the maxillary and basal widths in relation to the basilar length as well as the lengths of the maxilla and zygomatic arch, classified all wolves and dogs, including Studer's prehistoric forms, into two basic hereditary types:—(1) the brachygnathic platocephalic decumanides, characterised by a broad skull, short frontals and maxillae and long parietals, as represented by the Great Dane; (2) the dolichognathic leptocephalic veltrides, distinguished by a narrow skull, long frontals and maxillae and short parietals, as represented by the German Shepherd dog. Both forms occur in four main variations: dwarf, small, intermediate ("normal") and giant.

Goetze and Dornheim (1926) believed that the decumanides and veltrides types represented ancestral forms of which the majority of modern breeds were hybrids. However, their interpretation of statistical frequency distribution of certain size indices has been rejected by Lumer (1940) who found no valid basis for Schaeme's system of classification. Similarly, Dahr (1937) has pointed out that the decumanides and veltrides types are two arbitrarily chosen stages from a continuous range of variation, and as such without phylogenetic significance.

Dahr's sentence may well be applied also to Baumann and Huber's (1946) construction of two separate phylogenetic series: mastiff and greyhound. The mastiff line of descent is drawn from the heavy forest and mountain wolf through the mastiff, Boxer and French Bulldog to the Pug; the greyhound line from the leggy steppe wolf (Fig. 133; see also Fig. 12) through the greyhound, Borzoi and Whippet to the Italian Greyhound. The mastiff line shows an increasing tendency to the shortening of cranial length

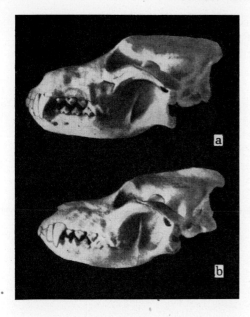

133. (a) Skull of a Russian forest wolf
(b) Skull of a Labrador steppe wolf
(after Baumann and Huber)

coupled with a relatively constant width, while the greyhound line tends to an ever narrower width of skull combined with a relatively constant length.

Marchlewski's Classification of Canine Cranial Types

Marchlewski (1930) distinguished between three principal canine types: C. f. leineri, C. f. matris optimae and C. f. decumanus, on the basis of an analysis of cranial proportions by the method of least differences, in which he compared the sum of a number of measurements, each expressed as a percentage of basilar length, in different breeds, regarding those that differed least in this respect as the most closely related.

Lumer (1940) has questioned the validity of this method, because crania differing in many proportions may nevertheless give similar values of the sum used as a criterion, even despite possible correlations between the measurements. Also this approach suffers from the defect that only similarities in adult proportions are considered, while the possibility that differences in these may be merely a consequence of differences in absolute body size is neglected.

Wagner's Classification of Canine Cranial Types

In consideration of the well-known phenomenon that within a group of related types of organism the adult proportions are frequently correlated with total body size, Wagner (1930) has plotted various size indices of different canine breeds against the length of the cranial cavity, selected as a measurement representative of body size. For some indices the points for the various breeds form a single band, but for most they form two distinct bands, the breeds occupying each being the same in all cases. From these results Wagner concludes that Canis familiaris may be subdivided into two major groups, including the following breeds;—

Group (a)	*Group (b)*
German Herders' guard dog	Great Dane
Dachshund	St Bernard
Setter	Newfoundland
Dobermann Pinscher	Bulldog
Pointer	Old-type Bulldog
Dingo	French Bulldog
Poodle	Boxer
Dunker	Pug
Lapland dog	Pekingese
Iceland Sheepdog	Japanese Spaniel
Fox Terrier	
Schnauzer	
Borzoi	
English Greyhound	
Large Whippet	
Whippet	
Rat Terrier	
Bolognese	
Toy Terrier	
Monkey Terrier	

Wagner thus places three (and if we consider the dingo as a representative of the C. f. poutiatini type, four) of Studer's basic types of dogs, i.e. C. f. palustris, C. f. intermedius and C. f. matris optimae, into a single group, on the grounds that the differences between them are due entirely to differences in absolute body size, which are not an adequate basis for racial distinctions. The wolf holds a position intermediate between Wagner's two basic groups; for a number of relationships the two bands representing groups (a) and (b) intersect at approximately the point representing the wolf. Wagner regards this as supporting the theory that the wild ancestral type of the dog corresponds to Canis lupus in size and form.

Lumer's Classification of Canine Cranial Types

Lumer (1940) has applied allometric analysis to the classification of the different breeds of domestic dog. This enabled him in the first place to rule out the great majority of earlier classifications as being based solely on adult proportions (percentage ratios of various measurements). By plotting various absolute adult measurements of different-sized breeds on a double logarithmic scale, he obtained evolutionary growth-constants, and was then able to group the various breeds into six allometric tribes, in accordance with the law of evolutionary allometry or relative growth which has been found in a large number of cases to be a valid empirical representation of ontogenetic growth relations in a single type (Huxley, 1943).

I. Terrier Tribe
German Herders' guard dog
Dachshund
Setter
Pointer
Dobermann Pinscher
Dingo
Poodle
Dunker
Lapland dog
Iceland Sheepdog
Fox Terrier
Schnauzer
C. f. palustris
C. f. palustris ladogensis
C. f. matris optimae
C. f. inostranzevi
C. f. intermedius
C. f. poutiatini

II. Toy Terrier Tribe
Rat Terrier
Bolognese
Toy Terrier and Monkey Terrier

III. Pug Tribe
Pug
Pekingese and Japanese Spaniel
C. f. inostranzevi
  (Bulldog-like variety)
Inca Bulldog

IV. Bulldog Tribe
Bulldog
Old-type Bulldog
French Bulldog
Boxer

V. Great Dane Tribe
Great Dane
St Bernard
Newfoundland

VI. Greyhound Tribe
Borzoi
English Greyhound
Whippet
Large Whippet
Primitive greyhound

Each allometric tribe is characterised by the possession of a particular set of growth-coefficients and by the fact that for any given relation all its members form a single curve conforming to the law of allometry expressed by the simple power function $y = bx^a$, where x and y are the variants corresponding to two parts, and b and a are constants. Different tribes may show the same growth-coefficients for certain organs. Thus, for example, the terrier and the greyhound tribes share the length relations of the snout, and presumably diverged later in respect of their width relations, while the Bulldog and Great Dane tribes became more extreme in both relations, in opposite directions. The conformity of the breeds within a tribe is interpreted as indicating that they are relatively homogeneous genetically with respect to ontogenetic relative growth and differing genetically chiefly in adult body size. In other terms, the process of evolution within a tribe involves primarily hereditary variations affecting only absolute body size, followed by changes in proportions, whereas the evolution of a new tribe involves also hereditary changes affecting the growth-coefficients of particular regions and producing marked changes in relative growth. Lumer (1940) thus includes in his terrier tribe, in addition to C. f. palustris, C. f. intermedius, C. f. matris optimae and the dingo, already combined by Wagner (1930) in a single group, also one form of C. f. inostranzevi, following Wagner in the contention that the differences between these prehistoric canine types can be ascribed entirely to differences in absolute body size.

In stressing the tentative nature of certain parts of his classification, Lumer (1940) remarks that the distinction between the Bulldog and Great Dane tribes may prove to be invalid on the inclusion of breeds of intermediate size, and that where a tribe (as the pug tribe) is represented by only a few breeds it is doubtful whether much significance can be attached to the particular values obtained.

The wolf generally conforms either to the terrier or the Great Dane tribe, except in the case of the snout length-cranial length relation where it stands at the intersection of the two main curves on to which all the forms (except the Toy Terrier) fall, and in the molar-premolar relation, in respect of which the wolf deviates markedly from the curve characterising the various breeds of domestic dog; not only are the two series of teeth in the wolf greater in absolute length than in any of the dogs, but there is also a difference of proportions, the wolf having a relatively shorter molar series than any type of dog. From the general conformity of the wolf in other respects either to the terrier or the Great Dane tribe, Lumer (1940) concludes that the wolf may be regarded as corresponding to the ancestral type of these two tribes, and the remaining tribes as having been subsequently derived from them.

ii. The Origin of the Pariah Dogs of Africa

Two different theories have been advanced on the origin of the pariah dogs, namely, 1) that they represent a transitional stage in evolution from wild ancestors to domestic dogs proper (Studer, 1901); 2) that they are the descendants of domesticated dogs which have turned semi-feral.

The Dingo

The problem of the origin of the pariahs is closely connected with the question of the descent of the dingo (Canis dingo Blbch. = C. familiaris dingo) of Australia and of the extinct dingo race believed to have formerly existed in New Zealand. Dingo is the Australian aborigines' name of contempt for the white man's dog; they called the wild dog the Warrigal (Le Souef and Burnell, 1926). The latter is about the size of an average pariah dog. Its head is distinguished by relatively small prick ears and the powerful development of the masticatory muscles. The body is moderately long, the tail bushy and commonly carried low. The coat shows a reddish cast produced by yellow guard hairs covering the grey underfur clear to the black muzzle. Occasionally the nose, feet and tip of tail are white.

Physiologically, the dingo displays the comparatively inoffensive smell of the wolf, and the habit of silent hunting; and in both these characters it differs widely from the jackals (Jones, 1921).

The Dingo Skull

The basilar length of the dingo skull ranges from 165—174 mm (Studer, 1901). In 22 genuine dingo skulls Jones (1921) found an average basi-condylar length of

134. Dingo pair

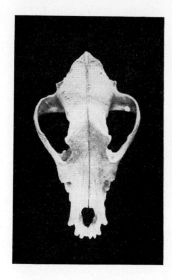

135. Dingo skull, frontal and lateral views (after Studer)

177.3 mm. The neurocranium is long, the parietal area vaulted, occasionally exceeding the width between the auricular points; the temporal region is markedly constricted, and the area behind the supraorbital processes bulges out. The high sagittal crest is extended into a strong occipital protuberance. The zygomatic arches are wide, the narrow forehead is dished along the median line, and vaulted above the orbits. The constriction in front of the Foramina infraorbitalia is negligible, and the stop is well defined, at any rate in dingos born in captivity. For in dingos shot in the wilds of Australia Tichota (1937) noted a nearly straight cranial profile, while Wagner (1930) found the degree of facial curvature in the dingo to bee appreciably less than in other dogs of a similar cranial size. Allen (1920) thought that in palatal length the dingo closely resembled the larger breeds of European dog, but showed a wide departure from wolves and less modified breeds of dog, such as the aboriginal dogs of the American Indians. Actually this character is of no specific value (Young and Goldman, 1944; Haag, 1948). The teeth of the dingo are strong, with P_4 as long as the two molars combined; but dingos reared in captivity develop relatively smaller teeth than is normal in the race. In 22 dingo skulls Jones (1921) recorded an average upper carnassial length of 20 mm. In the dingo the antero-posterior length of the carnassial is greater than 10 per cent of the basi-condylar length of the skull, whereas in other types of domesticated dog investigated it is less than 10 per cent (Longman, 1928). The dingo may have a basi-condylar skull length of only 165 mm, and yet possess an upper carnassial as large as that of a St Bernard with a basi-condylar skull length as great as 248 mm. In the relatively large size of the carnassials the genuine dingo approximates to the wolf more closely than do most other breeds of dog. Jones (1921) has pointed out that of the series of dogs' skulls he examined the Chow shows the greatest likeness to the dingo in the form and proportions of the upper carnassial teeth. But the kind of diet is doubtless a major factor in carnassial development. Thus, Greenland and Alaska Eskimo dogs, which in cranial measurements very closely approximate to the dingo, have larger carnassials (Haag, 1948).

The similarity in upper carnassial proportions, as Tichota (1937) has shown, is grossly misleading with regard to the proximity between the dingo and the Chow. The latter is a typical representative of the cranial C. f. palustris type, whereas the skull of

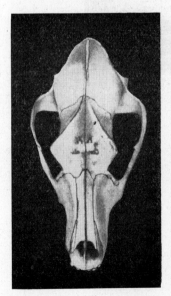 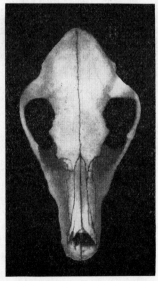

136. Crania of a dingo caught wild in Australia (left) and of a prehistoric dog of C.f.matris optimae type (right) (after Tichota)

the dingo, with its relatively long splanchnocranium, nearly straight profile, high sagittal crest, flat zygomatic arches and powerful dentition, differs fundamentally from C. f. palustris, and resembles the crania of small lupine races, such as C. l. pallipes and C. l. hodophylax. Of Studer's prehistoric cranial types, it shows the closest approximation, in basilar length and cranial indices, to C. f. matris optimae.

Comparison between Dingo and Pariah Skulls

Generally the cranium of the dingo is stronger and heavier than the pariah skull, with a broader and blunter splanchnocranium and greater development of cranial super-

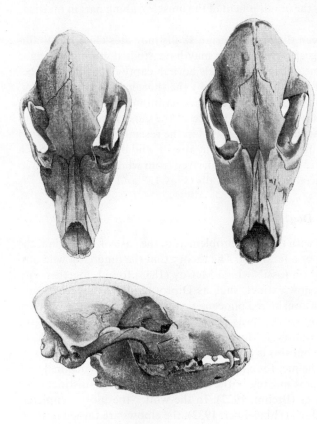

137. Crania of young dingo (left) and young pariah (right), frontal view (after Studer)

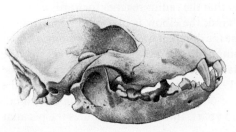

138. Crania of young dingo (above) and young pariah (below,) lateral view (after Studer)

8 Epstein I

structure, such as the sagittal crest and supraorbital ridges. But some pariah skulls hardly differ in these respects from the dingo skull. Young pariah and dingo skulls are very similar in the development of the maxillae and superstructure.

It would appear that the differences between mature dingo and pariah skulls are due principally to different diets and modes of life, the dingo being mainly a carnivorous beast hunting its prey, while the pariah hardly ever hunts game, but feeds on anything it may chance upon on the garbage heaps it frequents. The dingo's mode of life causes a more powerful growth of facial musculature with a correspondingly strong development of the maxillae and the crests and ridges to which the muscles are attached. Anthony (1903) has shown by experiment that the height of the sagittal crest in the canine cranium is influenced by the development of the muscles taking part in mastication.

While the resemblance between dingo and pariah skulls indicates close racial alliance, Antonius (1922) has suggested that Studer may have studied crania of dingos kept in menageries for some time or even of animals born in captivity, such skulls being distinguished from those of wild dingos proper by the shorter splanchnocranium, more clearly defined stop, and stronger slope of the neurocranium. But this conjecture, should it even be correct, cannot shake the evidence of the close racial relationship between the pariah dogs and the dingo, as expressed in the resemblance of the pariah skull to the crania, not only of menagerie dingos but also of wild dingos proper. Such a cranium, which Antonius (1922) admitted to be derived from wild stock, is shown in Studer's (1905) drawings of pariah and dingo skulls (Figs. 137 and 138).

Is the Dingo a Wild or a Feral Dog?

In the dingo we are confronted with the same problem as in the case of the pariah, the problem of whether it is a wild or a feral dog. The theory that the dingo is a wild animal proper is based principally on fossils which McCoy (1883) found, together with Australian tertiary marsupials now extinct, such as Diprotodon, Thylacoleo, Nototherium, etc., in what were then considered pliocene strata at Lake Timboon, western Victoria. "The Dingo," McCoy wrote, "was really one of the most ancient of the indigenous mammals of the country." More recent investigators regard the Lake Timboon strata as pleistocene; but this is of no fundamental importance to the issue.

In two albinotic dingos, Hilzheimer found the position of the forelegs identical with that in wild Canidae, these dingos moving in the manner of wild dogs, distinct from the movements of domestic dogs (Brehm, 1922). In the wolf, "the most completely developed type of the trotting dog" (Hilzheimer, 1932), the elbows are turned inwards owing to the narrowness of the shoulders, so that the radius reaches forwards and outwards. In the domestic dog, on the other hand, the elbow is placed more outwards, and the forearm inwards. This seems to cause the different manner of running observed in wolves and dogs, the wolf always keeping with the hindfeet in the tracks of the corresponding forefeet, whereas the dog places the hindfeet to the left or right of the tracks made by the forefeet (Hilzheimer, 1926). In galloping the movements of the wolf and dog are similar; both animals leave double footprints, thrusting their hindlegs in front of the forelegs which are placed apart. Iljin (1941) attributes the peculiar gait of the wolf partly to certain external characters which may occur in domestic dogs as the result of unfavourable conditions of development. He refers particularly to the legs of the wolf being pressed to the chest and its paws turned outwards, a character with which the hollow back of the wolf is believed to be connected, as this is encountered also in domestic dogs with a similar position of the paws. Working dogs may show an almost typical wolf gait as a consequence of an unusual development of the leg set and of the relations of the angles between the extremities and the body. The different gaits of the wolf, domestic dog and dingo are thus due to slight differences in their anatomy.

Although Nehring (1888) concluded from his examination of skeletal material that the dingo was not a domesticated dog returned to the wild but a true wild race, opposing evidence, showing that the dingo reached the Australian continent not as a wild but as a domesticated dog, is overwhelming. Were the dingo a wild dog proper, it would be the only larger placental mammal of Australia, the smaller placentalia being represented by a few Chiroptera and Muridae. "The evolution of a modified northern wolf in an isolated portion of the Southern Hemisphere, tenanted solely by Ornithodelphians, Didelphians, and a few stray Monodelphian rodents and bats, is a thing which is zoologically inconceivable," Jones (1921) has pointed out. Obviously their power of flight carried the bats to Australia, while the Muridae probably reached the Continent on drift wood or in human vessels. A wild dog, on the other hand, could hardly have immigrated from the oriental region into Australia in any other way than over a land bridge; and had such a land bridge existed, it would be strange that none of the Suidae, Cervidae or other oriental placentalia availed themselves of that passage. Further, there are no wild dogs, allied closely enough anatomically to the domestic dogs or the dingo to be considered among their ancestors, in the part of the oriental zoological region nearest to Australia, wolves and jackals being absent in the Malay Archipelago where Cyon is the sole representative of the family.

Dogs on Pacific Islands

Domestic dogs, on the other hand, occur widely on the islands of the Pacific. Cook encountered them on Tahiti. Fitzinger (1876) pointed out that they had doubtless been brought to the Society and Sandwich Islands from the Indian continent. The dog was found in New Zealand in association with moa hunters; it had been introduced by the Maori on their second migration, about A.D. 1150, along with the rat (Francis Smith, 1952). It was also present in New Zealand at the time of first European contact, but did not get to the Chathams, east of New Zealand (Sharp, 1957). The early Maori of New Zealand distinguished between two different types of dog, namely, ruarangi and mohorangi; now the Maori dog is extinct. In historical times there were large numbers of feral dogs, distinguished by prick ears and a bushy tail, in the interior (Lang, 1955). Fitzinger (1876) described them as closely resembling large pariahs, but with slightly wider muzzles and longer bodies; most of them were russet red, some black, white or variegated.

Jones (1929; 1931) has described crania of native dogs of C. f. palustris type, exhumed from graves of the pre-European period of New Guinea and Hawaii. In New

Guinea similar dogs have been observed in a feral state. They are of a small size, standing 29 cm at the shoulder, the head and body measuring 65 cm in length. The head is distinguished by a deep and narrow muzzle, slightly oblique eyes, and short erect ears; the neck is short and thick; the bushy tail reaches to the middle of the lower leg; dewclaws are absent. The coat is black-and-white in colour, with short hair closely adpressed, without under-fur, forming a ruff on the neck. The basi-condylar length of the skull measures 150 mm at most; the length of the carnassial in two specimens has been recorded as 15.5 and 16.0 mm respectively (Longman, 1928). A similar dog, standing about 35 cm at the shoulder and weighing approximately $11^1/_2$ kg, with a flat skull, small and erect ears, a bushy, gaily carried tail, and a short and smooth, black coat with white markings on the chest, feet and tail, was found on Goodenough Islands, north of the eastern extension (Papua) of New Guinea, in 1896 (Hubbard, 1946). No white man had previously visited the islands (Le Souef and Burnell, 1926). Tichota's (1937) generalisation that the dog did not occur on any of the islands situated between Asia and Australia, including New Guinea, prior to the arrival of Europeans is therefore erroneous. Indeed, on some islands dogs were eaten, and on others they were sacrificed or otherwise ceremonially used.

The dog was well adapted to survive accidental voyages. Canoes swept away with dogs aboard probably arrived at distant islands more than once in the several thousand years of Polynesian dispersal. Owing to its high fertility, only one or two individual dogs were necessary to establish the species in a whole island group; as Mayr (1963) has pointed out:—"The establishment of highly successful colonies by single founders is not only feasible; it seems to be the normal method of spreading in many species of animals and plants." But on islands where there was no wild game in the interior the dog, when living with man, had not a high survival value, being his competitor for animal food. In times of famine the dog was likely to be eaten by man, except in the bigger islands where it ran wild in the interior. Feral dogs occurred, in addition to Australia, New Guinea and New Zealand, also in West Samoa and several other islands (Williams, 1837). On Tuamotu Islands they lived on fish (Lang, 1955). Sometimes the dog was considered as a nuisance; Will Mariner, who lived among the Tongans from 1806–1810 after the wreck of his ship, told of how the chief of one of the Tongan Islands had all the dogs destroyed for this reason (Martin, 1817). The dog's absence in the Marquesas, Chatham, Cook, Tubuai, Easter and other islands of the Pacific may therefore in some instances be due to the lack of system and design in the dispersal of the animal, and in others to its subsequent extinction (Sharp, 1957).

The Tengger Dog

In the Tengger mountains of eastern Java existed formerly a dog which Kohlbrugge (1896) believed to be a true wild animal and called Canis tenggeranus. Its body measured 98 cm in length, the tail 30 cm, and the prick ears 10.5 cm (Jentink, 1896–97). The head showed a well defined stop. The fleecy coat was light brown in colour with a reddish hue and a number of blackish brown stripes, one extending from the posterior end of the nose along the ridge of the back to the tip of the bushy tail, a second forming a collar, and others leading from the back down to the belly and thighs. The ears, nose and eyebrows were a very dark brown, the extremities light brown, and the underline was white. In skeletons of both the Tengger dog and the dingo the humerus exceeds the radius, and the femur the tibia in length. Studer (1901) stressed the marked resemblance between the dingo and Tengger dog in cranial and general conformation, ascribing the fleecy coat which distinguished the Tengger dog from the dingo to adaptation to the cool mountain climate of its range. From this resemblance he drew the conclusion that the Tengger dog represented the last remnant of the wild dingo group in the oriental region.

139. Head of Tengger dog (after Jentink)

Antonius (1922), who examined the skull of a Tengger dog in the Leiden Museum, has pointed out that it shows the characteristics of a domesticated dog in such a pronounced manner that it could not possibly have belonged to a wild dog proper. The splanchnocranium is short and wide, the stop well marked; the neurocranium with its very high frontal sinuses rises steeply from the posterior end of the nasals; the superstructure is well developed, and the skull deeply dished between the orbits. The tympanic bullae and the dentition likewise show the characteristics of a domestic dog. Finally, the type and coloration of the coat, as described by Jentink (1896–97), point to a domesticated or feral dog.

Strebel (1905) considered the Tengger dog as an intermediate form between the common pariah of India and the dingo, while Antonius (1922), on craniological grounds, placed it between the Chow, a representative of the C. f. palustris group, and the dingo.

The Tengger dog therefore provides a link, not between the dingo of Australia and the true wild dogs of the oriental region, as Studer assumed, but between the dingo and the pariahs.

Significance of the Dingo's Diversity of Coat Colour

In addition to the above zoo-geographical considerations, there is other evidence to show that the dingo is not a true wild dog. Among individual wolves and jackals variability in coat colour is considerable, but in the dingo this is far surpassed. Even before the introduction of domestic dogs from Europe into Australia, there occurred melanistic and leucistic animals, while in some of the common reddish brown dingos the tip of the tail, the feet and muzzle were white, in others of the same colour as the rest of the coat. The first white men who came in contact with the dingo remarked that black dogs and red dogs were common (Mivart, 1890).

Barking in Dogs

It may be mentioned that in captivity the dingo soon learns to bark. But this carries no proof of the relationship of dingos to dogs since sometimes dogs cease barking and sometimes captive wolves learn to bark like dogs. The dogs of the Guinea coast and of several central African tribes, in addition to a few aboriginal Mexican breeds, are silent, while the domestic dogs which became feral on the island of Juan Fernandez lost the power of barking within thirtythree years, but gradually reacquired it on their removal from the island. The Hare Indian or Mackenzie River dogs make an attempt at barking which usually ends in a howl; but the young of this breed born in zoological gardens possess this faculty to the full extent (Encyclopaedia Britannica, 1902).

Feral Dogs

Although Siber (1899) claimed that domestic dogs never went wild, "at least not in Africa", there are many records of the contrary. At Brava, Somaliland, Brenner (1868) encountered a feral dog resembling a guard dog in size and conformation, with long black hair broken by large yellow patches on the hindquarters. Drake–Brockman (1910) has recorded that a pariah dog which had formerly belonged to some Midgan hunters in Somaliland, but had been beaten und driven from the karia, turned wild and disappeared into the bush where it hunted with a black-backed jackal as a companion. Nor may Siber's assertion be applied to the dingo or the extinct Tengger dog. Entering a continent where they found a favourable environment and no competition from other carnivora, the domestic ancestors of the dingo seemingly multiplied to such an extent that large numbers no longer found enough sustenance about the camps and in the wake of the small bands of nomad hunters. They were forced to seek their prey among the game of the country. This would accord with the fact that some of the dingos continued to live in a state of symbiosis with the aborigines. "When an Australian native wants a hunting dog, he goes out and catches a puppy which, as it grows up, is taught to help in tracking and finding wounded animals. When the dog becomes fully adult, it usually runs away and returns to its wild life. However, there are occasional dogs, especially bitches, who become so strongly attached to the human family that they will not run away" (Linton, 1956). The loose connection between the dingo and the Australian aborigines explains Brehm's (1915) remark that it was difficult to find out whether the Australians embarked in taming the dingo before or after they had become acquainted with domestic dogs introduced by Europeans.

The Antiquity of the Dingo

Schoetensack (1901) tried to explain the occurrence of dingo fossils in pleistocene Victorian strata by suggesting that the ancestors of the Australian aborigines accompanied by dingos might have immigrated at the end of the tertiary or the beginning of the pleistocene period. This view is untenable as dogs were not yet domesticated at the beginning of the pleistocene epoch. The occurrence of fossilised dingo remains together with those of extinct marsupials may be ascribed to the canine habit of burrowing in suitable ground, which Antonius (1922) repeatedly observed among pariahs. The proximity of the dingo bones to the marsupial fossils deposited at a much earlier period may therefore be mere coincidence. The fossilised state of the dingo bones could be due to their great age, since the introduction of the dingo into Australia probably dates back several millennia. Zeuner (1963a) suggests that the association of dingo bones with those of some extinct marsupials need not imply a very great antiquity, as some such extinctions have occurred at no great distance of time, and these animals were probably the contemporaries of the existing aborigines.

The earliest inhabitants of the Australian region were apparently not in possession of dogs. No canid remains occur in either Tasmania or Kangaroo Island (Howchin, 1925–30). The absence of dogs among the earliest inhabitants refers in particular to Melanesia which comprises those islands of the western Pacific which are inhabited by Oceanic negroids. These islands stretch from New Guinea south-eastwards to the Fiji Islands and New Caledonia, the other main groups being the Bismarck Archipelago north of New Guinea, the Solomon Islands and the New Hebrides. The extinct aborigines of Tasmania seem also to have been of negroid stock. The early Melanesians comprise three main physical types: 1) the pygmies, who are believed to be either the aboriginal inhabitants or to be derived from Papuan stock through local specialisation in remote and inhospitable mountain areas of New Guinea and formerly also in some of the other islands; 2) the Papuans who inhabit the major part of New Guinea; 3) the Melanesians in a racially and linguistically restricted sense, who inhabit eastern New Guinea, the Bismarck Archipelago and the islands eastwards to Fiji and who seem to be the result of the mixture of immigrant strains with a basically Papuan population (Cranstone, 1961).

The dingo seems to have reached Australia with Malayo-Polynesian immigrants who constitute the main body of the aboriginal population of the Australian mainland. Jones (1921) held that the domestic ancestors of the dingo (C. familiaris dingo) arrived in seaworthy boats with the progenitors of Talgai man, who carried a knowledge of the boomerang, of the basis of a totem system, and various other cultural features, all bearing a strange suggestion of very distinctly western origin. Zeuner (1963a) suggests that for geological reasons this event is likely to be at least as remote as the last phase of the last glaciation, when the sea level was low, so that the dingo may be older than the mesolithic dogs of Asia and Europe. But this is contradicted not only by several cultural traits of the ancestors of Talgai man, but also by the complete lack of evidence for the domestication of the dog during the Pleistocene.

The reversion in the conformation of the dingo's skull and forelegs to a condition closely approximating to that in wild Canidae proper may be attributed to natural selection of adapted heritable variations in a feral state so thorough as to be unknown in any other recent dog, and to the doubtless very primitive generalised type of its domestic ancestors.

If the dingo, prototype of the pariah, is to be regarded, then, not as a wild but a feral dog, it may reasonably be assumed that the pariahs themselves, far nearer than the dingo to a state of proper domestication, do not represent a transitional phase in evolution from wild ancestors to domestic dogs, but a stage of transition from domesticated to feral dogs. Thus, Soman (1963), referring to the pariah dogs of India, points out that these are not wild dogs attracted to the dwellings of man by an easy means of obtaining food, but the descendants of domesticated scavenger and guard dogs which have degenerated into a semi-feral condition, yet remain, as though by an inherited habit, in association with mankind.

Pariah Cats

The pariah dogs are not an isolated case. In the towns and villages of the Near East large numbers of cats may be observed that have similarly turned pariah. A group of dustbins which they defend against all newcomers usually forms the centre of a family group of such pariah cats. Like pariah dogs they have no master and no home; they are not fed nor bred or protected by anyone. They do not get accustomed to a home, remaining wild and ferocious unless taken in at a very tender age. Although resembling pariah dogs in their mode of life and state of domestication, these cats have never been claimed to constitute a transitional stage from the wild *Felis libyca* to the properly domesticated cat, so obvious is their derivation from domestic stock.

Variability in Pariah Dogs

There are several indications that in body size, colour and character of coat, and cranial as well as general conformation a group of domestic dogs of a variety of breeds revert to the pariah type if left to find their own food and shelter and to interbreed freely for a number of generations. Dahr (1942) has drawn attention to the peculiarity that the pariah dogs which live outside the breeding control of man have approximately the same body size everywhere throughout their tremendous range of distribution.

However, the pariah is not a type that would fit readily into the narrow limits of a breed standard. Nearly a century ago, Fitzinger (1876) already distinguished between a large, a small, and a short-legged pariah type. The Indian pariah dogs are larger in size than those of Egypt, and the latter larger than the pariahs of Sumatra. The pariahs of India are commonly distinguished by elongated splanchnocrania, while those of Sumatra generally have well defined stops (Studer, 1901). A random group of pariahs usually exhibits a considerable range of individual variation; thus, in four paria skulls from Istanbul basilar lengths ranging from 143—176 mm have been recorded (Werth, 1944). Moreover, the variational range in the pariahs of Japan differs from that of Indian pariahs and the latter from that of the African. Nor are the pariahs in the different parts of India or Africa of the same type.

Among the pariah dogs of south-western Asia several cranial and conformational types can be distinguished: a heavy guard dog type, a heavy medium dingo-like, light medium collie-like, light greyhound-like and small toydog-like type; in the north the guard dog type predominates, in the south the greyhound-like type (Menzel and Menzel, 1948). The pariah of guard dog type has a compact, rather square-shaped body, broad skull with a moderate stop, short wide muzzle, button or pendulous ears, and a bushy tail curled across the back; the coat is either wavy or woolly, or short and smooth, white or yellow in colour, occasionally black or spotted. The dingo-like pariah varies in appearance between the dingo and the Eskimo dog; its head is pear-shaped, the neurocranium considerably broader than the muzzle, with prick or semi-prick ears; the legs are rather short, the hair is rough, long, straight and stiff, the tail curved over the back. The collie-like type stands about midway between the Eskimo dog and the Collie; it is distinguished by a fairly square body with tucked-up belly; the head appears more elongated than in the guard dog and dingo-like pariah types; the bushy tail is carried across the back, and a profuse mane is common in this type; the coat may be self-coloured white, fawn, reddish brown or black, or variegated with white, or grey with lighter-coloured legs or black-and-tan. The greyhound-like pariah is always short-haired, sometimes feathered; the head is elongated, the skull narrow and the muzzle relatively long; dogs of this type have either pendulous or rose ears, occasionally prick ears; the chest is deep and narrow, the belly tucked up greyhound-like; the tail is long, set on low and carried in a curve; the prevalent colours are sandy or yellow, rarely with white markings. The toydog-like pariah, while considerably smaller in size than any of the above types, ranges in build from heavy to light; it often resembles a Basenji in colour and conformation. These five principal types are connected by numerous intergrading forms. It should be possible, Menzel and Menzel add, to breed from them, in a comparatively short time, nearly every Northern type of dog, save the Mastiff and Dachshund, and even these might be produced given the necessary time to wait for adequate mutations.

Even in so small a country as Palestine the pariah dogs show a definite tendency to turn out smaller in size and more slender in conformation as one proceeds from north to south. In a random collection of five pariah skulls from Palestine (Figs. 140 and 141), furnished by Mrs. R. Menzel, only two specimens—a (juvenile) and c (adult)— approach the poutiatini type, Studer's description of the type specimen (see p.29) applying particularly well to skull c. Skull d holds an intermediate position between C. f. poutiatini and C. f. matris optimae, while both b and e, though greatly differing in basilar length, display the slender forms of greyhound skulls. In these circumstances, Zeuner (1963a) is hardly justified in speaking of the "curiously constant characters" of the pariah group.

While in most countries where pariah dogs occur the poutiatini type predominates in cranial conformation, in some areas the range of variation tends to the palustris, in others to the matris optimae type; in some districts slender, in others coarser forms

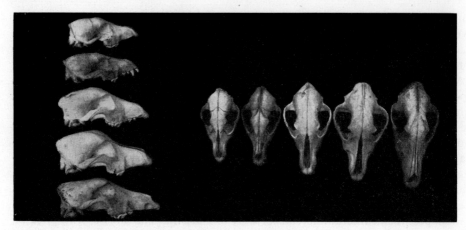

140. (left) Five pariah skulls from palestine, lateral view
141. (right) Five pariah skulls from palestine, frontal view

prevail. And this phenomenon, as indicated by the large number of different African pariah breeds—if breeds they can be called— may not be due only to natural selection under varying environmental conditions, but possibly also to differences in the ancestral domestic stocks. This would imply that a population of pariahs derived from domestic stock with a preponderance of greyhound blood would show a somewhat different range of variation from a pariah population descended from dogs among which hound or guard dog types predominated. Differences in the ancestral domestic stocks may thus be partly responsible for both regional and individual variations in the pariah type.

Petters' (1934) craniological analysis of the Bantu dogs of South Africa (see pp. 46– 49) indicates that among a group of recent African pariahs definite basic cranial types are discernible. Among the pariahs of Ethiopia, Somaliland and other parts of Africa, pronounced racial types, such as the palustris, are frequently encountered, notwithstanding the absence of any influence of European dogs. Further, the dogs in the earliest rock drawings in the Eastern and Western deserts of Egypt do not all show the generalised pariah type—as far as the primitive drawings permit judgement on this point—but rather specialised types, similar to the recent Berber dogs, prick-eared greyhounds and Pomeranians. Again, the majority of the dogs painted or sculptured by the ancient Egyptians belong to definite racial groups other than the generalised pariah, and the same applies to crania of mummified dogs from the Nile valley (see pp. 55–58).

Early Uses of the Dog

Werth (1944) has ventured the opinion that originally the Shenzi (Werth's proposed term for pariah) was kept by African, Asiatic and American hoe-cultivators solely for its meat; in fact, that it was domesticated for this purpose. Its use in the hunt was secondary and not general; hunting peoples, such as the early Bushmen, Ituri pygmies, Melanesians and American Indian hunters, he writes, did not employ the dog in the pursuit of game. The influence of Islam and European penetration restricted the eating of dog meat; hence the animal became degraded from a useful meat supplier to a useless pariah.

But this theory, however plausible it may appear, is not in agreement with the factual situation. Curwen and Hatt (1953) stress that the dog was not domesticated as a source of food, although, like every other edible thing within reach, he was eaten when he was too old to serve as an assistant in hunting. The earliest animals domesticated for their food value were those we still use for the same purpose, namely, cattle, sheep, goats and pigs.

Again, in archaic shell heaps in Kentucky and Alabama, eastern United States, the canine skulls showed no holes struck in the side to remove the brain. Dogs were carefully buried, and in several instances dog teeth were used for necklaces; but there is no evidence that the animal was used for food. It is in the hunter-fisher-gatherer horizons all over the world that the dog is prominent in their cultural context. With the advent of agriculture the dog lost much of its importance. It is true that the dog was used as a burial offering for some time after the introduction of agriculture both in Egypt and America; but the retention of this practice is paralleled by many similar survivals in culture throughout the world (Haag, 1948).

Forde (1934) has pointed out that apart from some specialised uses of the dog, as for drawing sledges in north-eastern Asia and in Arctic America and the travois on the American plains, for carrying small packs by a few people, for food by several hoe-cultivators in the Congo region, Oceania and America, and for its hair used as wool for blankets by the Araucanians of central Chile, the Coast Salish of British Columbia and formerly also by the Maori, the animal is over a wide area the companion and assistant of the hunter, and less often the guardian of the camp or flock. Not infrequently, however, it is put to little use and is, as in Australia and among most African forest peoples, mainly a companion of man. This does not imply that the dog was domesticated first for hunting or companionship. Its earliest and still widespread relation with man is that of an uncared-for scavenger and parasite on the camp group or household. Indeed, it is in this parasitism that Hahn (1896) sought the beginnings of its domestication.

The earliest pictorial evidence of the dog invariably shows it as helper in the hunt (Figs.148, 149), which contradicts Werth's (1944) generalisation that the hunter peoples did not originally know or employ the dog. Again, the same author's assertion that the Bushmen did not use the dog for hunting, even after they received it from the Hottentots, is contradicted by the unanimous reports of all, including the early, authors (Barrow, 1801–04; Lichtenstein, 1811–12; Schultze, 1907; Herbst, 1908; Passarge, 1908; Range, 1914); Dornan, 1925; and Bleek, 1928) on the Bushmen's use of the dog in the hunt (see also Fig. 57). To quote only one of these:—"It is really wonderful with what courage these lean mongrels will tackle a leopard, a courage no doubt begotten of hunger and an absolute confidence in their masters, who, with their knobkerries, are never far behind as soon as the dog has gripped his prey" (Herbst, 1908). The Naron employ dogs in hunting jackals, leopards, lynxes and other animals whose skins are valued for making karosses (Schapera, 1930).

Sauer (1952) has criticised the story that as a camp follower of hunters the dog gradually joined the camp and became their hunting companion, as a myth projecting modern European romantic views into the past. Hahn (1896) long ago pointed out that hunting dogs were late specialisations among certain peoples of advanced culture; really primitive peoples do not hunt with dogs, though these may trot along with the hunting party. Dogs appear archaeologically first with mesolithic folk, who were not much engaged in hunting, at least not on the land. Domestication presupposes permanent dwellings and a comfortable surplus of food, such as existed in the early planting cultures. Its initial purpose is not economic, but a cultural bent giving varied aesthetic satisfaction; the economic benefit man will gain from the domestication of a particular beast is as yet unknown. Only very young and helpless animals, wholly dependent on foster parents, form a strong attachment to the household in which they are raised. They share food and roof with the family, are fed and trained by the women, become playmates of the children and thus part of the household. By such means the dog was first adopted into the human family.

This is a sober view of the domestication of the dog, although Sauer's postulate that the initial purpose of domestication is to provide aesthetic satisfaction may be disputed.

In hoofed animals the critical period of attachment to dam or foster-mother is within a few hours of birth; in the dog, which is born more immature physiologically than the ungulates, after about three weeks (Reed, 1959). It would appear that among the animals fostered by men some are psychologically prepared to remain henceforth with the foster-mother (for the dingo see p. 118). That is to say, the fostered wild animal, more especially of a species normally living in herds or packs, is ever ready to enter into domestication; this applies to the palaeolithic period no less than to the mesolithic or neolithic. It did enter domestication the moment this became environmentally possible with the qualitative change of human society from the food-gathering to the food-producing (cultivating) stage when some of the animals that had become part of the human household remained alive long enough to produce offspring. At this stage domestication is apparently a spontaneous process. For several reasons it would be easier for the dog than the ruminants to pass into permanent domestication. The dog has a far higher fertility rate and can subsist on the offal of human meals (in these respects it is equalled by the pig); moreover, it does not require protection from predators.

Initially the domesticated dog was of no economic use to man, save for its doubtful role as scavenger. Only gradually and in the course of its diffusion it became a utility animal—an object of sacrifice, of ceremonial or profane consumption, or employed as guardian of the home and flocks, in the hunt, for draught or for its wool, or merely tolerated as a scavenging pariah. It may be assumed that it was only after realisation of the usefulness of the first domesticated species that the idea and practice of domestication were transferred to other species.

Probable Cause of the Pariah's Feral Existence

It is probable that the pariah dogs came into existence in ancient times, possibly soon after domestication of the species, in those places where, owing to their high rate of propagation or owing to political, social or other upheavals, not all domesticated dogs reared found masters, so that some were left to their own devices in obtaining food and shelter. It has been observed in Palestine that wherever a new human settlement is established, a new border zone for the pariah is automatically called into existence; "the pariah dogs arrive at once, as if it were out of nothingness" (Menzel and Menzel, 1948). In the same country pariah cats and wild jackals have shown a rapid increase in number with an increase in human habitations. For ages the roving bands of pariah dogs have interbred with domestic dogs and been joined by those abandoned. In other instances, where they were required for work, food, the hunt or as guardians, pariahs have been returned to a state of proper domestication. From the absence of pariah dogs in the northern parts of Asia and Europe it would further appear that domestic dogs do not generally turn feral in the range of the larger wolves and in a colder environment where they remain dependent on the warmth and shelter or on the food offered by human habitations. In this connection, the author, travelling for several weeks from yurt to yurt in Inner Mongolia (People's Republic of China) and the Mongolian People's Republic (Outer Mongolia) in 1963, asked numerous old herdsmen if they had ever encountered or heard of a homeless dog (there were a number of large guard dogs at every yurt for the protection of herds and flocks from wolves) that had turned wild. None had ever heard of such a case, not even during the period of political and social disturbances, though some had knowledge of "mad (hydrophobic) dogs" that had been killed by the herdsmen.

The Status of C. F. poutiatini

Antonius (1922) classed the pariahs and the dingo with C. f. poutiatini which he believed to represent the earliest stage in the domestication of the dog, the beginning of cranial evolution under domestication. From the generalised poutiatini type, he assumed with Studer (1906), the more specialised basic types known in domestic dogs were derived. Hilzheimer (1926) held that the skull of a dog derived in equal shares from the crossing of hounds, greyhounds and primitive collies should resemble that of C. f. poutiatini, adding:—"In Canis f. poutiatini we therefore find one of those stages of domestication in the dog at which the separation into different lines such as hounds, collies, etc. has not yet taken place."

We are thus confronted with the problem of whether C. f. poutiatini stands at the root of the pedigree of the various basic lines of domestic dogs, or whether it is a form generalised either through cross-breeding of different more specialised strains or the evolution of a feral type in dogs, or through both cross-breeding and a feral existence; or, finally, whether C. f. poutiatini represents the root of the family tree of the domestic dog as well as its feral and semi-feral forms.

Were C. f. poutiatini only the earliest stage in the domestication of the dog, the dingo, the most primitive recent representative of the poutiatini type (according to Tichota [1937], of the C. f. matris optimae type, which stands very close to C. f. poutiatini), would have to be regarded either as a wild dog proper or as a feral descendant of dogs introduced into Australia in the earliest stage of their domestication. The pariah dogs of poutiatini type would then represent a further stage in evolution from wild or semi-wild ancestors to domestic dogs proper. Actually, however, such an alternative does not exist, as both the dingo and the pariahs are feral or semi-feral dogs descended from domestic stock. If, on the other hand, the poutiatini type represents merely the feral form in dogs, the generalised link between the wild ancestor and the more specialised basic cranial types of domestic dogs would still be missing. While the type specimen of C. f. poutiatini may or may not represent this generalised link, it may be assumed that the dog at an early stage of domestication, possibly as a result of interbreeding of various strains, develops, amongst others, a generalised cranial type approximating to C. f. poutiatini. However, it should be noted in this connection that Werth (1944), on craniological grounds, sees no justification for the insertion of a poutiatini type between intermedius and matris optimae; and Hauck (1950) says:—"Canis poutiatini as a special type is doubtful and cannot therefore be considered as parent stock."

Absence of the Dog in Palaeolithic Sites

We have seen that among the African pariah dogs not only the poutiatini type, buc several cranial types, some of them similar to Studer's type specimens of prehistorie palaearctic dogs, can be distinguished. During the seven decades that have passed sinct the publication of Studer's work, a considerable amount of new cranial material has been recovered, which has thrown new light on the prehistoric dogs of Europe and western Asia.

Up till now no skeletal remains of the domestic dog have been found in any of the palaeolithic deposits of central and western Europe, and no dog remains are present in palaeolithic sites in central Asia (Nelson, 1927). Canine coprolites have been found in late palaeolithic levels in the Crimea and Siberia. Reports of the occurrence of the dog in late palaeolithic sites in the Crimea are based on the presence of canine coprolites at Tash-Air I (Hančar, 1958)—rather weak evidence in view of the possibility that these may be derived from wolves. At Afontova Gora II on the Yenisei river, south-western Krasnoyarsk, and at Verkholenskaya Gora, near Irkutsk, Siberia, canine bones are associated with late Magdalenian (late palaeolithic) implements (Von Merhart, 1923; Hančar, 1955), while at Timonovka on the Desna river canine coprolites and fragments of jaws occur in a Gravettian context (Zeuner, 1963a). But, as Haag (1948) contends, it is impossible at present to determine whether these remains belong to a domestic dog or some other canine form. Wherever the dog is present in a palaeolithic context, as in North America, it has been introduced from outside.

Presence of the Dog in Mesolithic Sites

In the Old World domesticated dogs first appear with certainty during the epipalaeolithic (= mesolithic) period within the range of both the larger northern and the smaller southern wolves. In Great Britain, Denmark, the Crimea and Transcaucasia remains.

of dogs have been found in Azilian stations; in Denmark they occur at Maglemose, coeval with, and culturally a northern phase of, the Azilian–Tardenoisian of France and Spain, the earliest phase of which synchronises with the epipalaeolithic (early mesolithic) Capsian and almost immediately succeds the upper palaeolithic Magdalenian.

The Capsian culture, named after the ancient Tunisian town of Gafsa, is traceable from Algeria, Tunisia and Morocco through the Iberian peninsula, southern France and Italy to an as yet undefined south-eastern European zone, and through Egypt to Palestine and Syria. In the main, the Capsian is therefore a Mediterranean culture. Davison (1944) suggested that it may have developed in Spain from the Chatelperron, and have spread much later to Tunis and penetrated into the western Sahara. Leakey, on the other hand, held that the Kenya Capsian reached East Africa from Palestine, via Arabia and Bab el Mandeb, and that from East Africa it spread north-westwards across the desert to North Africa, where it appears far later than in Kenya (Cole, 1954). While Coon (1968) concurs that Capsian affinities are broadly Palestinian and that the Capsian culture arrived in North Africa from south-west Asia in post-pleistocene time, he disagrees that it spread from East Africa north-westwards, holding that it was carried across the Sahara from North into East Africa as far south as Olduvai and beyond where Capsian skeletons bear a family likeness to those north of the Sahara.

Early Records of Dogs in the Iberian Peninsula and Brittany

Three primitive dogs, still with the straight tail characteristic of wild Canidae, are depicted on a cave wall at Alpera, eastern Spain (Fig. 142). In another section of the same cave a hunter is shown encouraging his dog to follow the trail (Breuil, 1912). In the shell mounds of Mugem near the mouth of the Tagus river in Portugal, classed as epipalaeolithic of the final Capsian phase, the fauna includes the dog which, according to Obermaier (1912), "is apparently indigenous to Spain". "It is probable," Forde (1934) writes, "that the Capsian hunters, who occupied North Africa and Spain during the last phases of the Ice Age, already had domestic dogs, for their bones have been found in shell middens of this period in Algeria (for refutal of this view see p. 52), and they are represented in hunting scenes painted on rock faces in south-eastern Spain. Shortly after this the dog appears to have spread farther into Europe with the northward advance of these peoples in Epipalaeolithic times."

The domestic dog occurs, along with sheep or goats, in a late Tardenoisian context at Téviec-Hoëdic on the Atlantic coast of Brittany. Téviec-Hoëdic and Mugem appear to have been contemporaneous with neolithic inland sites. At any rate, both stations are younger than the mesolithic sites in northern Europe and south-west Asia for which the domestic dog is attested (Narr, 1959).

Early Records of Dogs in Northern Europe

The dog occurs in mesolithic sites in the Crimea, Germany (Senckenberg, Naeselov, Ellerbeck, Husum), Denmark (Mullerup culture), East Baltic (Kunda) and Norway. In midden materials from the late mesolithic Oban culture of Scotland no dog remains have been found; but the dog was apparently present since bones thought to have been gnawed by dogs have been reported from a cave at Oban (Movius, 1942; Lacaille, 1954). Canid remains from the Maglemose site of Star Carr in Yorkshire, England, including the maxilla of a young animal with an estimated basilar skull length of 150 mm, have been assigned to the wolf (Fraser and King, 1954); they are radiocarbon-dated to 7538 ± 350 B.C. But since the cranial bones, in addition to being much too small for a wolf, closely resemble the respective bones of domestic dogs from the Maglemose of Denmark, nearly 1000 years later than Star Carr, Degerbøl (1962b) refers them to the domestic dog.

Another early mesolithic dog, represented by an entire skeleton from the Sencken-

142. Wall painting of three dogs, hunters and wild animals in a cave at Alpera, Eastern Spain (part of Fig. 258) (after Obermaier)

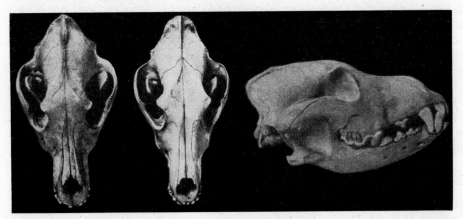

143. Skulls of the Senckenberg dog, frontal (left) and lateral (right) views, and of a dingo (centre) (after Mertens)

berg moor in western Germany, has been dated by pollen analysis to the same (late Preboreal or early Boreal) period as the Star Carr remains (Baas, 1938); according to the Danish chronology of peat bogs, this period is dated to 8000–7500 B.C., but Swiss peat chronology would place the Senckenberg dog later (Welten, 1944; Zeuner, 1963a). Mertens (1936) compared the animal to the prehistoric C. f. poutiatini and the recent dingo. The basilar length of the Senckenberg skull is 169 mm, that is the exact basilar length of the type specimen of C. f. poutiatini, but considerably below the basilar skull length of the European wolf. The length of the frontal cavity in relation to the length of the splanchnocranium is 1:0.9, as compared with 1:1 or 1:1.1 in the wolf. The forehead is only moderately high, the posterior section of the profile line slopes but slightly, and the sagittal crest is strongly developed (Mertens, 1936). The teeth are fairly large, although not quite as large as those of the Star Carr dog (Degerbøl, 1962b).

The mesolithic crania already show a considerable range of morphological types. The dog population from mesolithic levels of Naeselov, Ruegen, comprises nearly all forms of prehistoric dogs accumulated in the course of time, except the specialised and chronologically much later C. f. leineri and C. f. decumanus. Some of the Naeselov skulls in conformation and measurements resemble C. f. palustris ladogensis, others C. f. matris optimae or C. f. intermedius; among the mandibles of Naeselov dogs the same types occur in addition to C. f. palustris and, possibly, C. f. poutiatini. A skull from the mesolithic (Ertebølle) site of Ellerbeck has been classed with C. f.

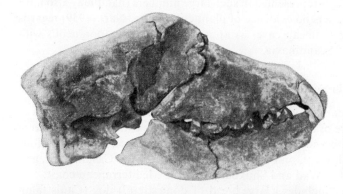

144. Skull of a mesolithic Naeselov dog (C.f.palustris ladogensis type with tendency to C.f.intermedius) (after Werth)

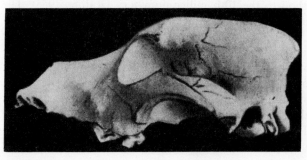

145. Skull of a mesolithic Ellerbeck dog (C.f.palustris ladogensis type) (after Gehl)

palustris ladogensis, and another one from mesolithic strata of Husum with C. f. intermedius (Gehl, 1930).

Among the cranial remains of dogs from the mesolithic Mullerup culture of Sweden and Denmark (Maglemose, Svaerdborgmose, Holmegaardmose), dated to 6800–5000 B.C., C. f. inostranzevi, C. f. palustris ladogensis and C. f. palustris have been identified. Dog bones have been found in all larger sites from the (Boreal) Maglemose period of Denmark, but they are much rarer than those of other animals. The flesh of the dog was apparently eaten, for some neurocrania were broken to extract the brain, and several bones show knife marks. The size of the Maglemose dogs of Denmark was very variable; the basilar skull lengths range from 143–190 mm. But these dogs were relatively much shorter in the leg than the Star Carr dog (Degerbøl, 1962b). The canid cranial fragments from the middens (shell heaps) of the following Ertebølle culture of Denmark have basilar lengths ranging from 126–170 mm (Winge, 1900; Brinkmann, 1921–24). The Ertebølle culture is still mesolithic and an adjustment of the classical Maglemosian; but its pottery is now suspected by many authorities to be due to culture contact with neolithic cultures farther to the south (McBurney, 1960).

Early Records of Dogs in the Crimea

The bones of a larger and a smaller type of dog, found at Shan-Koba, Sjuren II and Fatma-Koba in the Crimea, are distinct from those of both the wolf and the jackal (Hančar, 1937). To judge from the condition of the bones, the dog occasionally served as food. In these sites the pig is represented in such large numbers that it may also have been domesticated; but there is no evidence of plant cultivation. Narr (1959) regards the mesolithic dogs of the Crimean caves as approximately contemporaneous with those of the Maglemose of Scandinavia.

Early Records of the Dog in South-West Asia

In the quest for the early canine types introduced into Africa, the ancient records of dogs in south-west Asia are of considerable importance. The earliest sites in this area from which the possible presence of the domesticated dog has been reported include the mesolithic levels of Belt Cave, on the southern foreshore of the Caspian Sea, and the mesolithic Natufian of el Wad and Shukbah, near the Mediterranean coast.

The fauna of Belt Cave "includes a large breed of domestic (?) dog (Canis familiaris)" dated by the C^{14} method to the 10th millennium B.C. (Ralph, 1955). The same level had previously been referred by the same method to the 7th or 6th millennium B.C. (Arnold and Libby, 1951).

A similar uncertainty shrouds the canid remains from the Natufian levels of the caves of Mount Carmel, Kebarah, Zuttiyeh and Shukbah in Palestine (Bate, 1927; Garrod and Bate, 1937; Garrod, 1943). Bate inferred from an almost complete canine cranium from the Natufian of Mount Carmel (Fig. 146) that the Natufians were in possession of a domesticated dog of large size (in another place of her description she says, of a medium-sized dog). This skull is distinguished by a short and wide splanchnocranium and weak teeth, especially the carnassial the length of which, taken along the external border, fails to equal that of the two molars. The muzzle is short, the forehead low, and the interorbital portion low and flat. The flatness of the interorbital region is due to the small extent of the frontal air sinuses. Viewed from the front, the muzzle is seen to be stout in correlation with the wide flat palate, and its hinder border is deeply emarginate. The basi-occipital is wide, measuring about 21 mm between the bullae, while the length of the skull taken in a straight line is 205 mm. No skulls of recent dogs

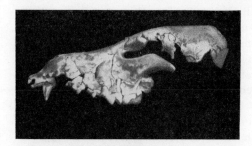

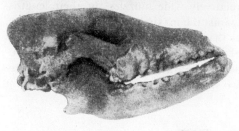

146. (above) Cranial fragment of a Natufian dog from Palestine (after Clutton–Brock)
147. (below) Cranium of C.f.matris optimae type from Anau, Turkmenia (after Duerst)

can be said to resemble at all closely that of the Natufian dog. The type to which the Palestine form seems to show the most resemblance in size and cranial conformation is C. f. matris optimae (Garrod and Bate, 1937).

While Bate thought that the Natufian dog might be a domesticated descendant of Canis aureus lupaster, Reed (1959; 1960) suggested that the skull could be that of a large wild jackal; he now regards it as the skull of a small southern wolf within the range of variation of Canis lupus pallipes (Reed, 1961). This may well be the case, judging from the similarity between the cranial fragment from Mount Carmel (Figure 146) and the skull of Canis lupus arabs (Fig. 13) and especially of C. l. pallipes (Fig. 16). On the other hand, the likeness may be due, not to racial identity, but to the similar body size of the Natufian dog and these small wolves, and the similar size and type of their skulls; for the Natufian skull also nearly resembles that of a domestic dog of matris optimae type from Anau (Fig. 147). Again, the presence of a cranium of a small wolf among artifacts in an incipient planting culture in this area and at this early period (recent determination of the radiocarbon age of the Natufian layers in the tell of prehistoric Jericho gave 8840 B.C.—Zeuner, 1963a) may indicate an early stage of commensalism in which the wolf was still a wolf, albeit a tame one.

Clutton–Brock (1962) points out that in view of the small size of the Indian Canis lupus pallipes and the even smaller size of the southern Arabian Canis lupus arabs it is extremely difficult and sometimes impossible to diagnose the specific status of canids from fragmentary remains of teeth and jaws from early prehistoric sites in south-west Asia. The teeth, especially the canines and carnassials, of Canis lupus pallipes are usually larger than those of the pariah dog and dingo, but there is not the same sharp distinction in tooth size and shape that exists between European wolves and recent domestic dogs. The relative length of the upper carnassial in comparison with the combined lengths of the two upper molars is no reliable diagnostic character, for other than in European wolves the relative carnassial length in the Indian race is as short as in the domestic dog (Gaudry and Boule, 1892). Thus the carnassial lengths of the specimens from Natufian levels of Mount Carmel and Kebarah, identified as domestic dog (Bate, 1927; Garrod and Bate, 1937), overlap the range for the small Asiatic wolves and are outside the size range of the present day pariah dogs. Again, a comparison of cusp shape and overall pattern of the teeth of Indian and Arabian wolves and pariah dogs shows a wide overlap in the variability of these characters. While, then, the Natufian canid remains, both of upper and lower jaws, cannot be specifically identified on the basis of measurements, the upper jaw fragments from the Natufian of Mount Carmel and Kebarah show one structural feature that is suggestive of domestication: both specimens have relatively wide muzzles for their lengths. Measurements of the length of the palate against the maximum width of the maxillae show that the facial region is relatively wider than in the skulls of Canis lupus pallipes and C. l. arabs. Shortening of the muzzle, relative to its length, combined with a reduction in size and crowding of the teeth is a definite character of domestic dogs. It may therefore be justifiable to suggest that the two specimens from the Natufian of Mount

148. Rock drawings of dogs and ibex from the Natufian of Kilwe, Jordan (after Rhotert)

Carmel and Kebarah represent domestic or perhaps tamed animals (Clutton–Brock, 1962).

While the zoological evidence that the Natufians were in possession of a domesticated dog is inconclusive, the cultural evidence, represented by the presence of the animal in several rock drawings of hunting scenes in a naturalistic style from the Natufian of Kilwe in Jordan (Fig. 148) suggests that they were (Rhotert, 1938).

At Çayönü near Ergani in the Turkish province of Diyarbakir, west Kurdistan, two canine rami, probably a pair, have been found in strata dated around 7000 B.C. These are remarkably thickened latero-medially and the teeth are large and plumb, with the anterior premolars crowded. The tooth row is shortened and bends steeply up posterior to M_1; the right ramus lacked a third molar. Additional characters indicative of domestication are a pathological condition of the margin of the right jaw involving externally the alveolus of the anterior root of P_4 and the posterior of P_3. Referring to it as "the oldest positively identifiable dog from the Near East", Lawrence (1967) added that this small heavy-boned dog occurred over rather a wide area, for a jaw of very similar proportions was collected from a neolithic layer at Mugharet el 'Aliya in Morocco.

In southern Anatolia, hunting scenes depicted on temple walls at Çatal Hüyük show the use of dogs by stag hunters who are armed with bows and arrows (Fig. 149) (Mellaart, 1963; 1965). These paintings, which have been dated to the 7th–6th millennia B.C., represent the earliest pictorial evidence for the occurrence of the domestic dog in Asia Minor. The presence of the dog at Çatal Hüyük is also confirmed by skeletal remains recovered from the temple ruins (Brentjes, 1967).

In pottery neolithic strata at Munhatta, in the Mediterranean zone of Palestine, dated to the 5th millennium B.C., Perrot (1963) found a mandible, several teeth and skeletal parts of a domestic dog, which, in view of an upper carnassial length of 16.8 mm and a total length of two molars of 17.8 mm, have been assigned to a small pariah.

A nearly complete cranium, a mandible and skeleton, excavated in a Ghassulian (chalcolithic) site in the Wadi Ghazzeh, in the semiarid zone of Palestine, dated to the 4th millennium B.C., have been referred to the domestic dog. Ducos (1968) has described the cranium as follows:—Small height, short, narrow, convex and anteriorly rounded facial part, and moderately large and antero-posteriorly dished brain case;

149. Stag hunt with dog, Çatal Hüyük (after Mellaart)

narrow, flat, well set off forehead; large, anteriorly rounded orbits; low occipital triangle with prominent posterior protuberance; well marked crests and ridges for attachment of muscles; length of superior carnassial equal to the combined lengths of two molars. With a basilar (basion-posthion) length of 155.2 mm, the cranium approximates to the lower limit of the typical Canis familiaris matris optimae skull, which it also resembles in the narrowness of the forehead. It is, however, distinct from the matris optimae cranium in the shortness of the facial part and the low occipital triangle. In some respects the skull resembles the C. f. inostranzevi type, and in others the C. f. palustris, C. f. intermedius and C. f. leineri types respectively. In short, the skull is of a generalised pariah type, such as is characteristic of the majority of recent pariah skulls.

The dog was found associated with the Yarmukian culture of Palestine which Stekelis (1951) considered to be mid-neolithic, but which actually corresponds with the al-Ubaid period of Mesopotamia, dated to the middle of the 4th millennium B.C. (Reed, 1960).

The records from Jericho in the lower Jordan valley and Jarmo in north-eastern Iraq indicate the occurrence of the domestic dog in south-west Asia in the 7th–6th millennia B.C. At Jericho two different types of dog are reported from a lower plaster-floor level (pre-pottery neolithic B—about 6710 B.C. ± 150): one nearly as small as a Fox Terrier, and the other almost the size of a wolf (Zeuner, 1958; 1963a). While the presence of two varieties, differing in size and possibly also in appearance, would suggest a preceding period of evolution that may go back to the mesolithic Natufian, Clutton-Brock (1969) recently declared that, contrary to Zeuner's statement, careful measurement of all the Jericho specimens does not show any great diversity in size of bones and teeth; many are the size of pariah dogs.

At Jarmo the evidence is mainly cultural; a number of clay statuettes of dogs, the domestic state of which is manifest from the tail curled over the back, have been found among several thousand figurines, many of them identifiable as mammals native to the area. Canine bones from the midden remains of Jarmo are so badly fragmented that distinction between a wolf and a large dog is rendered difficult (Reed, 1959). In the larger canids from Jarmo a tendency can be detected towards a shorter mandibular tooth-row, more pronounced in the premolar series but indicated also for the molars, than is characteristic of recent wolves from Palestine and the Zagros mountains; the same tendency appears also in the upper carnassial (P_4). But owing to the great individual variability of C. l. pallipes and the paucity of the archaeological material of large canids from Jarmo, differentiation by proper statistical techniques was considered by Reed (1961) to be still impracticable. However, further examination of the remains of the large canids from Jarmo was revealed these undoubtedly represent dogs and not wolves (Reed, 1969).

Boessneck (1961), discussing the presence of dogs of large and small palustris size in the pre-ceramic settlement of Argissa–Magula near Larissa in Thessaly, rightly cautions against the recent tendency to deny the dog the rank of the earliest domestic animal or even a place among the oldest ones. During the early period of domestication, the dog did not belong to the customary meat animals, although its flesh was not disdained. For this reason it usually occurs in prehistoric sites in smaller numbers than the common farm animals; and in small settlements it may be missing altogether. Hence, with regard to the period of its domestication, no significance may be attached to the paucity or absence of its remains.

The Anau Dog

The skull of a dog of medium size, similar in conformation to the skull of the Natufian canid, is represented at Anau, Turkmenia (Fig. 147). The Anau skull has a basilar length of 164 mm, and is relatively low, approximating in this respect to both C. f. poutiatini and the dingo, comparatively narrow, and only slightly concave in the temporal region. The sagittal crest is strongly developed, while the tympanic bullae are relatively small. The splanchnocranium is characterised by the relatively broad, short palate. The length of the carnassial, measuring 17 mm, comes to 25 per cent of the whole row of cheek-teeth, in which the three premolars occupy a space of 42.5 per cent. Except in its more pointed upper jaw and the slightly wider muzzle, the skull bears a strong resemblance to that of the C. f. matris optimae type. Duerst's (1908a) description ends with the words:—"The domesticated dog of Anau belongs accordingly to the subspecies of Canis familiaris matris optimae in a form which stands craniologically very near to the dingo and to Canis poutiatini Studer, but which is distinguished by a rather broad muzzle." Duerst believed that this dog reached Anau together with the camel and the goat. The camel and the goat were probably introduced into Turkmenia from the south.

The Indus Valley Dog

The wide range of this canine type in ancient Asia is illustrated by its presence at Mohenjo-daro in the Indus valley, where the mutilated skull and jaws of a domestic dog have been dug up. Cranial remains of a similar dog, with an estimated shoulder height of 50 cm, are reported from Harappa on the Ravi (Prashad, 1936). Marshall (1931–32) pointed out that the Mohenjo-daro dog is nearly related to both C. f. poutiatini and the Australian dingo, and that the Anau and Mohenjo-daro dogs, the dingo and the recent pariah dogs of India may possess a common ancestry.

Early Dogs of Indian America

The dog of Indian America was introduced from Asia across Bering Strait. In its new home it developed in the course of time into three major types and sixteen lesser varieties (Allen, 1920). It had long been thought that the animal arrived in the New World only with the last non-pottery, non-agricultural immigrants from north-east Asia not much earlier than 1000 B.C. (Haag, 1948). But recent excavations have put the date of its arrival in America very much further back, further indeed than the earliest certain records of the occurrence of the dog in either Asia or Europe.

A small and slender jaw from a postglacial, late Wisconsin alluvial gravel bed in Coles County, Illinois, dated not later than 6000 B.C., has been referred to a domesticated dog (Galbreath, 1938). Rami and a maxillary fragment from Jaguar Cave, a hunting camp in Lemhi County, Idaho, dated to about 8400 B.C., also belong to domestic dogs. The jaws are massive, both deep dorso-centrally and thick latero-medially, and the tooth rows are short compared with the size of the individual teeth. This shortening is accomplished partly by crowding, particularly of the anterior premolars, so that their sockets do not lie in a straight line, and partly by tilting upward of the level of the tooth row. In addition to being shortened, the jaws are widespread in typical dog fashion. The maxillary fragment provides further evidence that these dogs had rather short noses and broad palates, and there is also evidence that they were broad-headed (Lawrence, 1967). There cannot therefore be any doubt about the domesticated state of these dogs, although the dating of their remains may seem exceedingly early for American dogs.

Similar Variability in Populations of Mesolithic and Recent Pariah Dogs

The essential difference between the various cranial types of mesolithic dogs, both palaearctic and southern, is the difference in size; thus the smaller skulls fall into the palustris group, then follow the morphologically closely allied intermedius and poutiatini, with matris optimae and inostranzevi at the other extreme. In spite of their variability in size and cranial conformation, the mesolithic dogs represent a population of a single racial type, a biological continuum, that does not justify even subspecific taxonomic assignment. Werth (1944) has therefore proposed for the mesolithic dogs the use of the oldest name, i.e. C. f. palustris Rüt. Showing a remarkable intergradation in size and conformation from C. f. palustris to C. f. inostranzevi, the prehistoric dogs strangely resemble some of the recent regional pariah populations which for all the cranial and conformational variability encountered among them cannot possibly be separated into distinct breeds or racial types. This illustrates the unsoundness inherent in taxonomic conclusions based on single specimens or small samples.

Causes of Cranial Variability in Primitive Dogs

Cranial variability in regional populations of primitive dogs, mesolithic or recent pariah, must be ascribed to changes in cranial proportions due to variations in body size as well as to hereditary variations in the growth-coefficients of different cranial regions. In the African pariah dogs cranial variability doubtless derives from the widely differing conditions of life which these dogs encounter in the various parts of Africa and the racial mixture of the parent stocks. The pariah bitch in heat is served by one male after another, from the same or neighbouring territories, in the order of precedence decided by the preliminary fighting; this is an additional factor accounting for the diversity of types (Menzel and Menzel, 1948). It is therefore futile to seek in the case of every individual pariah skull with some distinctive features a special wild ancestor, as Hilzheimer has done for the crania of Egyptian dogs in his Canis pallipes domesticus, C. doederleini domesticus, C. hadramauticus? (sacer?) domesticus, C. lupaster domesticus, and C. studeri domesticus. Studer (1901; 1906), in his description of the cranial types of prehistoric dogs, Klatt (1913), in his study of the influence of body size on cranial conformation, and Lumer (1940), in his application of allometric analysis to the classification of the breeds of domestic dogs, have shown that there is no difficulty in tracing the various basic cranial types known in domestic dogs to a single wild ancestor. But while there is no difficulty from a theoretical evolutionary point of view, we cannot be certain that actually all domestic dogs are of monophyletic origin.

There are many references in the literature to interbreeding of wolves and dogs. Although such interbreeding may happen occasionally, and some authenticated cases of wolf-dog crosses are known, there are few factual data to support these claims. Most of them stem from the remarkable similarity of appearance of certain dogs and wild Canidae. However, in analogy with other domesticated animals, it is likely that during the early stages of domestication and diffusion of the dog, tamed wolf cubs were incorporated with the domesticated stock throughout the wolf's range and that the infusion of wild strains affected the genetic pattern and variability of the as yet numerically small dog populations.

Early Opinions on the Descent of Domestic Dogs

Darwin (1868) has enumerated a whole series of instances of domestic canine breeds exhibiting similar features to those in the wolves of the same region. And while noting that the observations to this effect were not always exact, he tended to support the theory of the polyphyletic origin of the domestic dogs.

Studer (1901) made interesting survey of the differences of opinion among earlier authors on the wild ancestral stock of the domestic dogs:—"According to several writers, such as De Blainville, Pictet, Bourgignat and Woldřich, the domestic dogs are descended from wild diluvial Canidae which were neither wolves nor jackals, but true Canes familiares. Pictet calls the diluvial dog Canis familiaris ferus, and Bourgignat refers to it as Canis ferus, both assuming therefore only a single ancestral form, while Woldřich believes in several ancestors, the direct progenitors of Canis familiaris palustris, C. f. intermedius, and C. f. matris optimae respectively. Other scientists, opposing this view, regard the recent wolves and jackals as the progenitors of the domestic dogs. Gueldenstaedt already considered the jackal as the ancestral form, writing in an analysis of the differences between the wolf, fox and jackal: '... inter canem et schacalam nullam differentiam specificam existere...'. Isidore Geoffroy Saint-Hilaire likewise regards the jackal as the ancestors of the dogs save the greyhound which he traces to Canis simensis. According to Jeitteles, Canis familiaris palustris is a domesticated descendant of the jackal, Canis familiaris matris optimae of the Indian Canis pallipes, the Egyptian pariah of Canis lupaster Ehbg. et Hmpr., and the African greyhound of Canis anthus Cuv. The European wolf is said to have shared in the evolution of the various canine breeds only to the extent of its having been crossed with domestic dogs in several localities. Nehring on various occasions has stressed the extraordinary variability of the wolf, and the changes to which its skull and dentition are subject in captivity. He has shown that all the different lupine races, diluvial and recent, fall under a single species, a view supported with regard to the diluvial wolves by Hagmann on the basis of exact comparisons. The discovery of crania of a large prehistoric canine breed, Canis decumanus Nehrg., and their affinity to the cranium of the wolf, have convinced Nehring that the latter is the ancestor of the larger breeds of dog, while the smaller ones are descended from the jackal. Von Pelzeln endeavours to trace the genealogy of the wolf-like dogs to the wolf, the Pomeranian-like to an extinct quaternary species, the jackal-like to the jackal, the greyhounds to Canis simensis, and the Indian and Oceanian dogs to Canis pallipes."

Studer (1901) himself held that during the diluvial (pleistocene) epoch a small Canis ferus occurred throughout the range of the wolves, going beyond it only in the south. This species, resembling the wolf in the high degree of its variability, comprised two main races: a northern—Canis ferus Bourg.—and a southern-oriental—the dingo. Both groups entered at first into pariah-like relations with man, and were then properly domesticated, the larger canine breeds being subsequently evolved by crossing the domesticated descendants of Canis ferus with wolves.

Studer's theory has proved unacceptable, chiefly because fossil or subfossil remains of a Canis ferus are unknown not only in the area of distribution of the pariah dogs but also in the northern parts of Europe and Asia. Antonius (1922) has pointed out that most of the remains believed to belong to Canis ferus are actually either those of jackals, whose range formerly extended farther north than at the present time, or of small wolves, while in others the age has been overrated. Zeuner (1963a), who favours Studer's theory of the descent of the domesticated dog from a wild dingo, adds, though, that the difficulty remains of explaining why a form so closely related to the wolf should have been able to maintain itself within the area inhabited by the Indian wolf.

Exclusion of the Jackal from the Dog's Ancestry

If, then, there is no evidence for the existence of Canis ferus in Studer's meaning of the term, the wild ancestors of the pariah dogs of Africa have to be sought among the true wolves and jackals, as the remaining members of the family of Canidae are eliminated from the genealogy of the African dogs for zoological or geographical reasons.

From his study of a very strong male Schnauzer-jackal hybrid, which had very small testes without any sign of spermatogenesis, Matthey (1969—personal communication) inferred that jackal-dog hybrids are sterile. But his specimen must have been exceptional, for Gray (1954) has pointed out that reciprocal crosses between jackal and dog are possible and that the hybrids are fertile in both sexes.

The wolf and the dog have the same chromosome number of 78 (Ahmed, 1941). Borgaonkar et al. (1968) found no differences in the chromosome count of $2n = 78$ (74–80, with 86 per cent 78) or in chromosome morphology between the European Beagle, Shetland Sheepdog, African Basenji and Malayan Telomian, and their findings have been confirmed by other studies of European, American, Indian and Japanese dogs.

Matthey (1954) counted 74 chromosomes in the African jackal, in contrast with the 78 chromosomes of the wolf and the dog. Ranjini (1966) found 78 chromosomes in the Indian jackal, i.e. the same number as in the wolf and the dog. Rather than considering polymorphism in the African and Indian jackals, Matthey (1967) regards it as possible that owing to technical difficulties an error may have occurred in his own work and that Canis aureus, Canis lupus and the domestic dog have the same chromosome number of 78. No doubt, this question will soon be solved by further studies.

It has been shown that already in the first generation of wolves and jackals born in captivity a reduction in cranial size is noticeable. The basilar length of the dingo skull measures 165–174 mm, and of a recent African pariah 158 mm, reaching 171 mm in larger pariah dogs from Upper Egypt and the northern Sudan (Nubia) (Studer, 1901; Hilzheimer, 1908). In C. f. matris optimae the basilar cranial length is 171–189 mm, in C. f. poutiatini 169 mm, in C. f. intermedius 164 mm, and in the early (mesolithic) C. f. palustris, i.e. C. f. palustris ladogensis, 145 mm. The basilar length of the Canis aureus cranium, on the other hand, ranges from 139–143 mm. "It is therefore no longer possible to believe in the descent of Canis palustris from the jackal, as I formerly did," wrote Hilzheimer (1926). And what holds of C. f. palustris, obviously applies even more to the larger cranial types of mesolithic dogs, such as C. f. poutiatini, C. f. intermedius and C. f. matris optimae, as well as to the recent dingo and pariah dogs.

In addition to the fact that Canis aureus is too small to be considered as the ancestor of the mesolithic dogs (which, no doubt, were already smaller than the wild ancestral stock), the jackals in general differ from domestic dogs in that their smallest forms have the most convex, the smallest dogs the most concave, profiles; further, jackals are distinct from dogs in the flatter forehead, smaller frontal sinus, more oblique position of the eyes, slenderer mandible and more pointed splanchnocranium (Werth, 1944). In all domestic animals with a body size similar to that of their wild ancestors the brain has become smaller under domestication; this applies also to the dog in relation to the wolf (see p. 88). The fact that the jackal has a smaller brain than a dog of its size excludes it, as Roehrs (quoted by Herre, 1955) has pointed out, from the ancestry of the domestic dogs. Schaeme (1922) excluded the jackals from the ancestry of domestic dogs also because the two basic cranial trends in dogs, i.e. decumanides and veltrides (see also p. 106), occur in wolves but not in jackals. Again, Miller (1920) has pointed out that in all specimens of domestic dogs, representing very diversified breeds, the cranium and teeth remain fundamentally true to the type which of all wild Canidae is peculiar to the wolves. This type, particularly as regards the cheek-teeth, does not represent a primitive condition which might be expected to occur in various members of the family without having any special significance. On the contrary, in respect of the development of a combined cutting and crushing type of carnassial and molar teeth it is the most highly specialised type now in existence. In every type of domestic dog, including the dingo and pariah, the specialised first upper molar wolf tooth with the reduced cingulum, and not the primitive jackal tooth, is present.

While Klatt (1927) agreed that Miller's rule did apply to the vast majority of dogs, whether large or small, he noticed a very few skulls that differed in upper molar formation from the general wolf type and approximated to that of (Eritrean) grey jackals. Tichota (1937) thought that the jackal could be excluded from the ancestry of the dog, more especially of the C. f. palustris type, only if in the cross-breeding of jackals and wolves, dingos and dogs the specialised upper molar wolf tooth with the reduced cingulum were to prove a recessive character. But he did not give any reasons for this opinion.

While Canis aureus of Asia and the grey jackals of North Africa, save Canis aureus soudanicus, are eliminated from the quest for the ancestors of the African pariah dogs owing to their small body size, C. a. soudanicus (C. doederleini), with the basilar length of the cranium ranging up to 187 mm, is excluded because of its peculiar cranial conformation and its jackal-like and disproportionately weak dentition. There remains, therefore, only the wolf sufficiently large in size, strong in dentition, and with the reduced or absent part of the cingulum in the first upper molar to have provided the ancestral stock of the domestic nd pariah dogs as well as of the dingo.

The monophyletic descent of all breeds of dog throughout the world is illustrated by the sense of unity retained by the extreme representatives of the dog tribe, such as giants and dwarfs, bulldogs and greyhounds (Hauck, 1950). Finally, while craniologically and anatomically the widely differing types of dog can be interpreted as variations of a single ancestral form, genetically and cytogenetically the wolf and the dog are closely allied.

Wolf-Dog Hybrids

In the hybrids derived from crossing wolf and dog, Iljin (1941) has demonstrated typical segregation for many different characters, such as colour and pattern of hair, eye colour, shape of ear, size, and various cranial features. The segregation of the colour of wolf-dog hybrids proceeds in the same way as in the progeny of dogs mated to dogs. Characters recessive in dogs are also recessive in wolf-dog hybrids; characters dominant in dogs are dominant in wolf-dogs. The genes of dogs, introduced into crosses with the wolf, behave in the same way as when they meet with their normal allelomorphs. The wolf and the domestic dog possess certain identical genes in the same linkage groups, and identical loci. With regard to conformation, Iljin holds that "there seem to be very few or even no real external differences between the wolf and the working dog. There exist but modificatory changes under unfavourable external conditions of development." The differences between the dog and the wolf in the tail set are believed to be partly modificatory and partly genetical. Schenkel (1947) observed in wolves practically all caudal positions characteristic of domestic dogs and dingos. The wolf can learn to wag its tail, a trait connected with emotional moments in wolf life. Jackals and foxes display the same faculty. The semi-erect ear in domestic dogs is completely dominant, and the lop-ear incompletely dominant, to the erect ear of the wolf.

Regarding the crania of wolves, Iljin (1941) lists the following six principal characters in which these differ from the skulls of dogs:—

(1) The orbital angle, formed by the orbital plane, drawn through the upper and lower marginal edges of the eye, and the horizontal plane, drawn through the upper margins of the Ossa frontalia, the orbital angle in wolves being more acute than in domestic dogs.
(2) The shape and length of the Processus zygomaticus maxillae, more especially the angle at which it projects from the zygomatic arch, the angle being typically more acute in wolves than in dogs.
(3) The shape and size of the Bullae tympani, the bullae in wolves being large, spherical and devoid of ribs, and in dogs small, flat and ribbed.

(4) The basilar length of the skull.
(5) The volume of the brain case.
(6) The border line of the palate.

The three first-mentioned characters, considered to be the most marked and reliable points of distinction, are analysed genetically. In the crossings of wolf and dog there is uniformity in the orbital angle in the first hybrid generation, and evident segregation in the second, the type of inheritance being conditioned by polymeric factors, with the number of genes concerned not more than two. The shape of the Processus zygomaticus maxillae is believed to be genotypical, the inheritance of the acute wolf angle being governed by at least two independent, completely or almost completely dominant genes, and that of the dog type by two recessive genes. The wolf type of the angle found in certain Laika dogs is ascribed to frequent crossings of this breed with wolves and its consequent "saturation" with wolf genes. As to the tympanic bullae, the large size of the wolf bullae is almost completely dominant to the small size of the dog bullae; in the shape and presence of ribs the bullae are intermediate in the first hybrid generation, while in the second generation there is segregation in all these characters. The genetical analysis of other cranial features shows greater maxillary breadth dominant to smaller, the shortened parietals dominant to the elongated ones, the shortened maxilla dominant (or intermediate) to the elongated, and the elongated zygomatic arch dominant to the shortened. Iljin suggests that his investigations prove independence in the manifestations of different hereditary characters of the skull, but he admits that certain of these characters may be reflections of the same hereditary character. This remark is significant in view of the probability that several important differences in cranial conformation are due to differences in body size. Iljin concludes that his investigations serve to emphasize the very close genetic similarity of the wolf and the dog, and suggest the possibility of the origin of the various breeds of Canis familiaris from a single wild species, viz. Canis lupus.

Theories of the Dog's Descent from Small Southern Canidae

Dahr (1942) and Hauck (1950), on the other hand, infer from a comparison of dentition that the dog could not be descended from the wolf. The wolf is the most carnivorous of the genus Canis, and the descendant of less carnivorous ancestors. The dog in its dentition approximates to the more primitive ancestral type rather than to the wolf. The improbability of reversion to a more primitive dentition is regarded as sufficient reason for excluding the wolf from the ancestry of the dog. This argument disregards the influence of the changed diet and mode of life under domestication. Zeuner (1963a), indeed, points out that similar dental changes have occurred in the domestication of the cat.

Haag (1948), again, concludes from the small size characteristic of the aboriginal dogs from the oldest archaeological horizons of America that no living circumpolar wolf can be considered as the source of the dog; this must have been an Asiatic pleistocene "small wolf-like canid greatly differing in size from any of the living wolves." This theory ignores not only the absence in the pleistocene Siwalik strata of North India of any Canidae that could have been ancestral to the dog (Colbert, 1935), but also the more important consideration that the small size of the earliest aboriginal dogs of America was obviously due to the low environmental level at which they existed among peoples of a relatively low degree of cultural attainment.

A similar notion led Antonius (1922) to regard the tendency in domesticated dogs living outside the breeding control of man to revert to pariah size as evidence that their ancestors had to be sought, not among the northern wolves but among one of the smaller southern races, such as Canis lupus pallipes Syk. of southern India, Canis aureus soudanicus (C. doederleini) (which Antonius erroneously classed not with the grey jackals but with the wolves), and possibly Canis aureus escedensis Kretzoi, 1947 (erroneously described by Mojsisovics as a wolf with the name Canis lupus minor). However, the present vulpine races and their ranges may differ from those at the time of the domestication of the dog; for, as Gehl (1930) has pointed out, the range of the wolf has been markedly restricted by man. This has led to geographical and genetic isolation, resulting in racial differentiation.

Werth (1944) argues from his theory that the pariah (Shenzi) dog was domesticated by early hoe-cultivators for the sake of its meat (see also p. 122) that the Indian Canis lupus pallipes was the wild ancestor, for it was only in India that the range of the wolf and the sphere of hoe cultivation overlap.

While neither Antonius' nor Werth's arguments are convincing, the archaeological evidence renders it probable that the first domesticated dog was derived from a southern wolf. For, as Narr (1959) has pointed out, the earliest records of the dog in south-west Asia antedate the first appearance of the animal in the Baltic–Scandinavian region; and Zeuner (1963a) writes that in mesolithic times the domesticated dog came into Europe ready-made.

On the Cranial Conformation of the Earliest Domestic Dog

We are still ignorant of the cranial type of the earliest domesticated dog. All mesolithic groups had acquired, or themselves domesticated, dogs "of a wolfish type". "From the Crimea to Portugal the bones of wolf-like dogs are found first in mesolithic settlements" (Childe, 1944; 1958). But these mesolithic dogs already exhibit a wide range of variation in size and cranial conformation from the large C. f. inostranzevi to the small C. f. palustris; and they were probably just as variable in colour, coat and appearance. Theoretically we may surmise that the earliest type of dog on the threshold of domestication was closer to the cranial type or types of the wild ancestor than were later evolutionary stages of the domesticated dog. This we may presume notwithstanding our knowledge of the immediate influence of captivity on the crania of wolves, established by Wolfgramm with regard to cage animals, notwithstanding Haag's (1948) conclusion from his osteometric analysis of aboriginal American dogs that the greater the antiquity of a site and the lower the cultural level of its occupants, the smaller the size of the dog, and notwithstanding the superficial resemblance in size, head shape and general appearance of the recent Eskimo dog to a modern wolf.

Of Studer's basic cranial types of prehistoric dogs, the type closest to the cranium of the wolf is C. f. inostranzevi, which Brinkmann (1925) has traced to the early mesolithic Azilian of Denmark. It is so close to the wolf that Studer (1901) and Antonius (1922) considered it to be derived from the crossing of C. f. palustris and C. f. poutiatini with the wolf. Hilzheimer (1926) objected to this view on the questionable evidence that C. f. palustris and C. f. poutiatini did not appear prior to the late Mesolithic, whereas C. f. inostranzevi occurred already during the early Mesolithic. Hence he concluded that the Azilian C. f. inostranzevi was a domesticated wolf. Since every dog is a domesticated wolf, Hilzheimer's assertion can only mean that the nordic C. f. inostranzevi was a *recently* domesticated wolf, and that it was domesticated in the Baltic area. But for this there is no evidence, although Brinkmann (1925) referred one of the cranial fragments from the Maglemose period of Denmark to a tamed or domesticated wolf, because no domestic dog had such strong premolars, great palatine width, or short molar series. As Hauck (1950) has pointed out, none of these features is a valid criterion of distinction between the wolf and C. f. inostranzevi. Prehistoric dogs in general have larger teeth than recent dogs of comparable body size, and crania similar to the Maglemose (Sværdborg) skull in question are found in many recent Eskimo dogs (Degerbøl, 1927).

That C. f. inostranzevi had retained the cranial conformation of the wolf more completely than had the smaller mesolithic dogs is obvious. But this does not imply that the European mesolithic dogs of inostranzevi type stand also chronologically nearer to the wolf. Some recent dogs are remarkably wolf-like. "I have seen Eskimo dogs that correspond hair for hair with the Arctic wolf," Kumlein (1879) wrote. Yet, as Haag (1948) has pointed out, the Eskimo dog of Arctic America is probably as far removed from the original wild form that gave rise to the domestic dogs as any living breed of dog of the American aborigines.

The Probable Domestication Centre of the Dog

Several authors (Obermaier, 1912; Forde, 1934) have suggested that the wolf was first domesticated by Capsian hunters, fishers and gatherers in the Iberian peninsula, whence it was carried to northern Africa by way of the Straits of Gibraltar, as well as to northern and south-eastern Europe (see also p. 127). But this is not the generally accepted view. Menghin (1931) inferred from the chronological precedence of the mesolithic canine remains in the Baltic area to the single specimen from the late hunting community at Mugem in Portugal, and from other ethnological evidence, a northern origin of the dog; and a similar view has been expressed by Herre (1958b).

However, the evidence of a (north- or south-)western European domestication centre of the dog is probably prejudiced by the fact that so far only this area has been thoroughly explored. With a fuller knowledge of early mesolithic sites in Asia the weight of evidence regarding the domestication centre of the dog may shift to the east, supporting the assumption that Canis lupus pallipes or one of the allied lupine races of south-west Asia was the ancestor. Thus Miles (1934) assumes that the dog originated in Iran, east of the Caspian Sea, whence it was diffused north-west by way of southern Russia to the Carpathians, west to Asia Minor, south-west to North Africa, north-east to central Asia, and south-east to India. The assumption that this or a not distant, perhaps more southern, region with a planting culture was the domestication centre of the dog finds support in the very early occurrence of the animal in the diffusion areas mentioned. But Flor's (1930) hypothesis, arrived at on culture-historical grounds, that the dog was domesticated in Siberia by the bearers of the proto-Eskimo-Samoyed culture, although finding some support in the recent discoveries of early dog remains in Idaho and Illinois, U.S.A., remains doubtful in view of the peripheral situation of this culture and the difficulty in accepting that domestication of the wolf was accomplished by a nomad hunter people with a palaeolithic culture.

Nor is Sauer's (1952) theory that the dog, like other household animals, such as the pig, fowl, duck and goose, and unlike the herd animals, was domesticated in the forested monsoon lands of south-east Asia, and that it originated from a wild dingo-like dog, perhaps resembling the fox in food and social habits, convincing. For this region is inhabited by the genus Cyon and lies outside the range of the wolf. The belief, shared by Sauer, that the recently extinct Tengger dog of Java was wild has been shown to be erroneous (see pp. 116–117), while the early or middle pleistocene Chou-k'ou-tien caves near Peking, from which remains of different wild Canidae have been reported, are far away from the area in south-east Asia where Sauer believes the dog to have been domesticated. Nor is there any valid evidence that the Canidae at Chou-k'ou-tien included a wild dingo, although Zdansky (1928) referred a single mandibular fragment to "Canis sp. cfr. dingo". Pei Wen-chung (1934) recorded the following Canidae at Chou-k'ou-tien:—1) Canis lupus L.; 2) Canis lupus variabilis Pei; 3) Canis cyonides Pei 1934; 4) Nyctereutes sinensis Schlosser; 5) Cuon cf. alpinus Pallas. Since the three last-mentioned forms have no bearing on the ancestry of the domestic dog, there remains only the undisputed fact that wolves of different sizes occurred in the Pleistocene of north-east China. While Sauer's theory has been wholly or partly accepted by Dittmer (1954), von Wissmann (1957) and Smolla (1960), La Baume (1962) has convincingly pointed out that there is not a shred of evidence for the origin of the mesolithic dogs of south-west Asia and north-western Europe from the littoral of the Bay of Bengal, southern China or any other part of south-east Asia.

The Antiquity of the Dog's Domestication

There is considerable disagreement among authors on the antiquity of the domestication of the dog. Lowie (1934) refers its inception to the latter part of the palaeolithic age in Asia, and Werth (1944) similarly assumes that it began 15,000 years ago in India. Zeuner (1963a) suggests that the dog was probably present in the south at the height of the last pluvial period, phase 3, and possibly present in the north at the height of the last glaciation 3, both periods being dated to c. 18000 B.C. But, as Haag (1948) has pointed out, the domestication of the dog may be not nearly so ancient a cultural trait as formerly thought. The wide distribution of the dog, including its presence in prehistoric America and Australia, cannot be interpreted as a substantiation of the claim for the antiquity of its domestication, but is more likely the result of the ease with which dogs may be taken along by immigrants. Yet, in North America the presence of the dog is attested for 8400 B.C. In the Old World the mesolithic Capsian and Azilian sites, where the earliest records of dogs have been found, are dated to about 7000 B.C. In south-west Asia the earliest certain remains of the dog are of a similar age, while the mesolithic Natufian, in which the dog probably occurred, lasted from about 9000–6000 B.C., preceding the earliest Neolithic in this area which has been carbon-dated to the 8th or the end of the 9th millennium B.C. (McBurney, 1960). It may be taken for granted that the dog was then known also throughout the major part of the vast territory that lies between south-west Asia, North America and western Europe. However speedily the diffusion of the animal may have proceeded after its domestication, the latter must certainly be referred to a date several hundred or thousand years before its earliest certain occurrence. The great cranial variability encountered in the mesolithic dogs of the Old World indicates that these had passed already a considerable period of symbiosis with man under varying conditions of environment and nutrition. In the light of the early dates of the presence of the dog in North America and of domesticated cattle and sheep at Çatal Hüyük in Asia Minor (see pp. 245 and II, 75), the reported dog from Belt Cave on the shore of the Caspian Sea from the 10th millennium B.C. takes on additional importance (Lawrence, 1967). It would appear that domestication of the dog took place in south-west Asia on the eve or the inception of the Mesolithic as early as 10,000 B.C.

Ignorance of the Racial Type of the Earliest African Dogs

In view of the cranial variability characteristic of the mesolithic dogs, we cannot be certain of the type of the earliest dogs introduced into Africa. Except in the case of the greyhound, the ancient pictorial records and skeletal remains do not offer sufficient evidence to distinguish definite canine types in predunastic Egypt.

Should future excavations vindicate Bate's assumption that the Natufians were in possession of a domestic dog resembling cranially the dogs of Anau and Mohenjodaro and approximating to the type specimens of C. f. matris optimae and C. f. poutiatini, this type of dog may have been introduced by them into Egypt. The Natufians belonged to the ancestral Semito-amitic stock. Although the source of their culture has not yet been determined, it is probable that they entered Palestine from the north or north-east. At Jabrud (Damascus) a succession of eight separable layers of blade industry, the lowest of which may date from about 12,000–14,000 B.C., ends with a well known local culture, the Natufian (Rust, 1950); this last is believed to precede immediately the earliest Neolithic in the area, for which a date in the 8th millennium B.C. or slightly earlier is suggested by carbon readings (McBurney, 1960). A few sites of Natufian culture have been discovered also in Iraq. At a later phase the Natufians penetrated into the Nile valley. On the route from Palestine to Egypt a Natufian station has been located in the desert, 70 miles east of Ismailia, and in Egypt there is an important site with a closely related culture at Helwan south of Cairo (Albright, 1949; 1960).

Although the Natufians, to our present knowledge, did not apparently penetrate farther west, it is significant that the racial affinities (Mediterranean) of the late hunting communities at Mugem in Portugal and Téviec-Hoëdic in Brittany (for whom the domestic dog is attested—see p. 27) show a very close likeness to the Natufian type of Palestine, while the Mugem people are associated with a culture showing obvious points of resemblance to the Upper Capsian (Vallois, 1952) (earliest carbon reading of capsian variously given as 6800 B.C. by McBurney, 1960, or 6450 B.C. by Coon, 1968).

The original dogs of the Capsians, as depicted in some of the ancient rock drawings in the Western and Eastern deserts of Egypt (Figs. 59 and 60) and in the Alpera cave of eastern Spain, may have been of the generalised type known from Anau and Mohenjo-daro and, possibly, from Bate's Natufian type specimen. The same generalised type of dog is represented on the ivory knife handle from Gebel el 'Araq (Fig. 178) and

150. (left) White limestone statuette of an Egyptian dog
151. (right) Black marble statuette of an Egyptian dog

152. Pomeranian-like dogs following onager-drawn chariots (ancient Babylonian seal cylinder, c. 3000 B.C.)

probably also by two later statuettes of well-bred Egyptian dogs in the Louvre (Figs. 150 and 151); one, in white limestone, with a slender body, long orally pointed head, and long prick ears; the other, in black marble, with a rather short head, deep oral part, and shorter ears.

The Pomeranian-like dogs which subsequently entered Africa from the east may have approximated to the prototype of the ubiquitous C. f. palustris. A dog of this type, following a two-wheeled chariot drawn by a domesticated onager, is depicted on an ancient Babylonian seal cylinder (Fig. 152). The palustris type apparently represents a general phase of canine evolution under certain conditions of domestication, and as such seems to have developed independently in different parts of the world. For example, of the various forms of neolithic dogs found in Japan nearly all are referred to C. f. palustris (Hasebe, 1924); and the dogs of pre-European Hawaii and New Guinea were of the same type (Jones, 1929; 1931).

Conclusion

The pariah dogs of Africa, then, trace their genealogy to several types of dog introduced into Africa in prehistoric times, in addition to the greyhound.

iii. The Origin of the Greyhounds of Africa

In the quest for the origin of the African greyhounds a discussion of their relationship with the greyhounds of Asia and Europe is essential.

1. Recent Greyhounds in Asia and Europe

In western Asia greyhounds occur over a wide area extending from Sinai and Arabia through Palestine to Syria, Iraq, Turkmenia, Iran, Afghanistan and north-western India. They are commonly lop-eared. In crossbred dogs of saluki-pariah descent prick ears are occasionally encountered. Hamilton Smith's drawing of a greyhound from Aqaba (Fig. 154) seems to be that of a saluki with faultily depicted ear tips. The purest salukis, outside of Africa, are owned by the Arabs near the southern frontier of Palestine (Murray, 1935). Antonius (1922) observed a strain of purebred greyhounds among the Turkmen. The variation encountered among the Asiatic salukis is similar to that in the greyhounds of North Africa; it is restricted mainly to body size and coat character.

The Persian Greyhound

In the Persian Greyhound or Tasi the coat is generally longer than in the saluki proper; but as there is much cross-breeding between the saluki and the Tasi, many transitional

153. Short-haired pariah dog of greyhound type from Palestine (after Antonius)

154. Greyhound from Aqaba (after a drawing by Hamilton Smith)

forms occur from the smooth- to the long-coated variety. Cross-breeding between greyhounds and pariah dogs or segregation within pariah populations may be responsible for the primitive dogs of greyhound type sporadically met with all over south-western Asia (Fig. 153).

The Afghan Greyhound

The Afghan Greyhound is native to Balkh, north-eastern Afghanistan. Rock engravings in the caverns of Balkh date it back at least as far as 2200 B.C. (Hubbard, 1948). At the present time these greyhounds are bred mainly by the Sirdars of the Barukhzy family. They hunt in couples—male and female, an interesting survival of a practice known in Egypt several thousand years ago (see Figs. 85 and 86). Afghan Greyhounds stand about 68–72 cm at the shoulder and weigh 27 kg or more. They may be of any colour, but the majority are reddish fawn, almost wheat-coloured. The body is covered with profuse soft hair extending over the ears, shoulders and half-way down the legs, the lower half of the latter being bare of long hair, although the toes are heavily feathered. The tail is not bushy. But for the abundant fine silky hair on the arched toes, and the top-knot of long fine hair on the head, the Afghan resembles an oversized Tasi or saluki. A local variety of the Afghan, called Zadin, is distinguished by an even greater profusion of hair extending down the legs and the creamy rather than golden or reddish fawn colour of the coat. It has also a tuft of hair on the head. The head of the Afghan is narrow, with an oval skull and prominent occiput, long splanchnocranium, and little or no stop. While resembling the head of the Persian Tasi in general conformation, it is more powerful. The ears are long and pendulous, the neck is long, strong and arched, running in a curve to the long sloping shoulders which are well laid back. The loin is strong, slightly arched, and falling away towards the stern, with the underloins well ribbed and tucked up. The hip bones are high and wide apart, the tail is set low, but the tail carriage is high with a curve at the end. The forelegs are straight and strong, of great length between elbow and ankle; the hindquarters are well muscled and very long between hip and hock. With its heavy coat the Afghan hound is well fitted for the very cold climate of mountainous Afghanistan. Its large feet are better suited for travelling over rocks than are the smaller cat-like feet of other greyhounds used in less mountainous countries than Afghanistan (Lloyd, 1937). The Afghan hunts by sight; but on the flat it is not so fast as some of the other greyhound breeds.

155. Persian Greyhound

156. Afghan Greyhound

The Rampur and other Greyhounds of India and Pakistan

In India, Afghan hounds were used on a large scale during the period of the Mogul Empire. At the present time several greyhound breeds are encountered in India and Pakistan.

In northern India greyhounds are represented by the Rampur Hound, named from Rampur where it is used for coursing hare, jackal and sometimes deer. The peculiarity about the Rampur is its coat which is like that of a freshly clipped horse; if longer, it is considered as a sign of impure blood. The colour is black, mouse-grey or black-and-tan. The Rampur is a powerfully built greyhound, standing 70 to 75 cm at the shoulder and weighing 30–36 kg. It has an elongated strong skull, flat between the ears, and powerful jaws. The eyes are light yellow in colour, the ears filbert-shaped and of a fairly large size, and the body is somewhat coarse, but of fair length (Lloyd, 1937).

The Banjara or Vanjari Greyhound is bred by the nomadic Banjara (Vanjari) tribe of northern India. It resembles the Persian Greyhound, but is slightly stouter and has a squarer muzzle. This is attributed to the fact that it is not commonly found pure-bred, but usually crossed with pariahs. The true breed is little smaller than the Rampur, reaching a shoulder height of 62–70 cm and a weight of 20–30 kg. The coat is either rough or short and silky, and usually black mottled with grey or blue, sometimes sandy, wheaten, fawn or brindle in colour. The ears, legs and tail are feathered. Banjara hounds hunt by sight as well as by scent. They are said to be as fast as Scottish Deerhounds, with great staying power. But whereas the Deerhound seizes the deer by the throat, the Banjara always goes for the hindquarters (Soman, 1963).

The Vaghari Greyhound is bred by the nomadic Vaidava Vaghari tribe of Kathiawar. It stands 50–60 cm at the shoulder, and has a thin, short and shiny coat which may be tan, brown, black, or a mixture of colours (Soman, 1963).

The Mudhol Greyhound, called also Kathiawar, Mahratta or Pashmi, is found scattered all over Maharashtra. It is a graceful hound, intermediate in size between an English Greyhound and a Whippet. It has a long and slender head with finely textured ears which are turned to the back, a long muscular neck, very deep chest, long body with well muscled shoulders and thighs, and long thin legs. The tail is carried over the back when the hound is not hunting. The coat is usually black-and-tan (Soman, 1963).

In the districts around Madras the Poligar, a dog of greyhound type, similar to and crossbred with the Rampur, is extensively used for hunting (Hubbard, 1946).

The Chippiparai is the greyhound of southern India. It has a racy body and long legs, and is commonly white in colour, more rarely variegated. It is employed in hunting hares.

The Alunk It or Sonangi is a nearly naked miniature greyhound with a mottled skin. It is found chiefly in Tanjore, south-west India (Soman, 1963).

The Greyhounds of China

Greyhounds have been introduced into China from India and Persia. The earliest certain records of their presence in Shantung province date from the Eastern Han

Dynasty (A.D. 25–221); they are represented in hunting scenes on Han vases and funerary stones. Subsequently they were depicted in dog books and scrolls in which the court painters illustrated the Chinese emperors' favourite dogs. Two greyhounds in a painting of the Ming Dynasty (A.D. 1368–1644) closely resemble the recent Persian Greyhound or tasi (Fig. 157). They give the impression of grace and speed, being characterised by a very long and narrow head with a flat skull, pointed nose and practically no stop, large eyes, long and slender ears covered with long silky hair, a long supple neck, deep chest, sloping shoulders, slightly arched back, long slender legs with well arched toes, and a very long tail feathered on the underside and carried in a curve. The coat is smooth and devoid of feathering on the legs or at the back of the thighs. In paintings of hunting scenes from the reign of Ch'ien-lung (Kao-tsung, 1736–95) the greyhounds have erect ears, strong tails and stockier bodies than those characteristic of either the recent saluki, Tasi or Afghan Hound (Epstein, 1969).

Both rough- and smooth-coated varieties of greyhound existed. Dr. Lockhart in 1867 referred to the rough-coated type as the "shock-greyhound" and described it as similar in all respects to the European greyhound, except for its shaggy coat (Gray, 1867). In the Mohammedan districts of Kansu and Shensi a short-coated greyhound is used for hunting hares and foxes. It has large erect or fly ears, a short and smooth coat of

157. Painting of two greyhounds of the Ming Dynasty (A.D. 1368–1644) attributed to Hsüan-Te (c. 1426–1435)

yellow, fawn, cream or chestnut colour with a white front, and a slightly feathered tail. It is bred in two sizes. The smaller type stand sbetween the Italian Greyhound and the Whippet; the larger variety is only slightly smaller than the English Greyhound. Coursing dogs were formerly found in the encampments of Mongolian princes, many of whom possessed 40–50 couple which were used for hunting wolves and foxes.

In Tibet, Marco Polo observed several kinds of dog including sporting dogs (Yule, 1875). Bogle reported in 1774 that the Pyn Cuchos at Rinjaitzay Castle, an estate and country seat of the ruling family situated north-west of Shigatse, kept a parcel of all kinds of dogs, among which were also greyhounds (Markham, 1876). Das (1885) probably also referred to greyhounds in his note that after the death of the Panchen Lama of Tibet there were found at his seat, the great lamasery of Trashilhünpo, "large packs of hounds and mastiffs which the Grand Lama had kept for sporting purposes, though the sacerdotal functions precluded him from shooting animals". Hounds, hunting mainly by scent, such as Bloodhounds, Foxhounds or Harriers, seem to have been absent in China and Tibet.

The Borzoi

The greyhounds of Persia and Afghanistan probably represent very nearly the ancestral type from which the Borzoi of Russia is descended. The early history of the borzoi is not well known. The breed seems to be derived from greyhounds brought to the north from one of the desert regions of south-west Asia, crossed with large powerful domestic dogs distinguished by a primitive cranial conformation. The latest research into the origin of the Borzoi has brought out that it goes back to salukis imported into Russia in the very early part of the 17th century. As these dogs with their thin coats were unable to stand the severe Russian winters, they were crossed "with a native Russian breed somewhat similar to the Collie of to-day, but slightly different in build, having longer legs, longer gracefully curved tail, neck slightly longer and more powerful, very heavy furred ears and a carriage more like the wolfhound of to-day. This dog's coat was very heavy, wavy or curly and with tendency to be woolly. In colour he was sable red or grey". The result of this crossing was the Borzoi (C.D.B., 1943).

The head of the Borzoi is cleanly cut so that the shape and position of the cranial bones as well as the course of the principal veins and arteries are clearly discernible. The neurocranium is narrow and slightly domed, and the splanchnocranium long and slender; the transition between them is marked by a stop so slight as to be hardly perceptible. The dentition is very strong, the somewhat obliquely set eyes are almond-shaped, dark in colour, lively and expressive, and the ears are small and thin, with a high setting and slightly rounded tips, lying back on the neck when in repose, but raised when at attention. The neck is muscular, arched, and somewhat shorter than in the greyhound, the chest very deep but not wide, the ribs are flat, and the back rises at the loins in a graceful curve. The loins are broad and muscular, but rather tucked up owing to the great depth of the chest and comparative shortness of back and ribs. The rump droops towards the long sickle-shaped tail. The forelegs are narrow in front, straight, and placed close to each other; from the side they appear almost

158. Borzoi bitch

wedge-shaped, broadening out towards the shoulder. The shoulder and thigh muscles are long, but not prominent. The hindquarters are long, very muscular and powerful, with well bent stifles and strong second thighs; hocks broad, clean and well let down. The feet are hare-shaped, long and with close toes and well arched knuckles. In the male the shoulder height is from 72 cm, in the female from 67 cm upwards. Generally the coat is long, soft and silky, on some parts of the body curled but not woolly, forming a profuse frill or collar at the throat and nape of the neck, and with plentiful feathering over the hindquarters and lower part of the body, less so on the chest and back of the forelegs. On the head, ears, anterior part of the legs, and the toes the coat is short and smooth. The colour is usually white with blue, grey or fawn markings of different shades, the latter sometimes deep orange, approaching red. Whole colours are considered unsatisfactory.

The Ibiza Dog

In southern Europe a primitive greyhound with large erect ears, a smooth-, coarse- or long-haired coat, and racy body, 55–65 cm in height and 16–24 kg in weight, occurs in Crete and the Balearic Islands where Ibiza is the chief breeding centre. These dogs, which are generally either white-and-red, white-and-yellow or red-and-yellow in colour, are similar in appearance to the prick-eared greyhounds represented in ancient Egyptian frescoes. On the Balearic Islands the breed is known as Perro ibizenco or Cà Eivissenc, that is, Ibiza dog; and tradition is that these greyhounds were originally bred exclusively in the Island of Ibiza, or Pityusa, whence they were exported to Mallorca. At the present day they have disappeared from the smaller island, and are bred only in Mallorca. In former times the breed was probably introduced also into the south of France, and of late years it has been imported into the neighbourhood of Montpellier, southern France, also to Tenerife, Canary Islands. It is now found in fairly large numbers in Catalonia, Valencia, Roussillon and Provence, and is known by the names of Mallorquín, Xarnelo, Mayorquais, Charnègue, Charnigue and Chien de Baléares (Pinto, 1948). The Cirneco of Sicily, the Spanish Podenco and the Portuguese Podengo are of a similar type (Menzel and Menzel, 1960).

Siber (1899) believed that the Ibiza dog was known also in Morocco, where a slender, leggy, pariah-like dog, red in colour, with large prick ears and a straight tail, was employed in hunting wild boar, antelope and jackal. But as there is no further evidence to show that the prick eared greyhound has survived in North Africa or been introduced thither from the Balearic Islands, the Moroccan hunting dogs referred to were probably of pariah descent or pariah-saluki crossbreds.

Hilzheimer (1932) has pointed out that in several respects the Ibiza dog is unlike the original Egyptian greyhound in build. In the latter the back seems to have had a much more pronounced curve, and the proportion of the upper part of the hindleg to the lower, and similarly of the upper and lower parts of the foreleg seems to have been different. But in particular the cranial conformation differs completely. "The head of the Balearic greyhound appears remarkably hollow in front of the eyes. In this way, and not only because of the peculiar square-shaped erect ears, the head is entirely different from that of the Eurasiatic greyhound. This is best seen from the profile: in the latter type the part before the eyes is more or less convex, in the African dog unmistakably concave. Of course, this conformation has nothing to do with a higher development of the brain, but depends upon the hollows of the brow which extend from the nose to the frontal bones. I have proved the same peculiar skull-formation to exist in early Egyptian mummified dogs, which I should like to think belong to the Tesem breed."

159. Ibiza Greyhound
(after Hilzheimer)

The Italian Greyhound

The Italian Greyhound is bred as a toy rather than a sporting dog. Commonly weighing between $2^1/_2$ and $3^1/_2$ kg, it is a very small miniature of the coursing or racing greyhound, but with a higher action of the forelegs, resembling the step of a hackney. There is also a heavier class. The eyes are large and prominent, and the neurocranium is wide and vaulted. The ears are either pendant at the tips or large and erect. The skin is very fine, on some parts of the body of a transparent thinness, and the hair is short and silky. All colours and colour combinations are permitted.

Vesey-FitzGerald (1957) has pointed out that miniature greyhounds appear first in several pictures by early Flemish and German painters, e.g. in Jan Van Eyck's (1390?–1440) 'Hours of Turin' and Hans Memlinc's (1430–1495) 'Martyrdom of St. Ursula'. More than one hundred years were to elapse before an Italian artist portrayed the breed, which acquired the name 'Italian Greyhound' from its representation in Veronese's (Paolo Cagliari, 1528–1588) famous painting. From this Vesey-FitzGerald concludes that the diminutive greyhound originated not in Italy but in northern Europe and probably in the Low Countries. In view of the delicate constitution and intense craving for warmth displayed by the breed, it is questionable whether this conclusion is justified.

The English Greyhound

The English Greyhound has a long, narrow head, fairly wide between the ears, long and pointed muzzle, little or no development of nasal sinuses, scarcely perceptible stop,

160. (left) Spanish Podenco (after Krichler)
161. (right) Italian Greyhound

thin and slightly arched neck, and long and fine tail tapering with a slight upward curve. The eyes are large and very keen, the ears small and fine in texture, thrown back and folded, the teeth are very strong, and both jaws may be of equal length or the upper jaw slightly overshot. The shoulders are very long and oblique; the chest is deep but not so wide as to impede speed. The thighs are long and muscular, and the loins broad, of good depth of muscle, well cut up in the flanks, and in common with the back slightly vaulted. The forelegs are long in the upper, and short in the lower part, and the feet are hard and close, rather more hare than cat feet. The shoulder height in the male is approximately 68 cm, and the weight 30 kg. The Greyhound now accepted as of pure blood has a short smooth coat, but half a century ago wire-haired or rough-coated mountain greyhounds, somewhat similar in coat to the Deerhound, were not considered as of impure blood (Lloyd, 1937). A Greyhound may be any colour, but those preferred are black, red, brindle, fawn, and blue, whereas white Greyhounds are unpopular.

The old English Greyhound was employed in coursing the red and fallow deer. It was a much more powerful animal than the modern English breed in which everything has been sacrificed to speed. It seems to have been bred in two different sizes: a heavier type, the Greyhound proper, and the lighter gaze hound used in rabbit coursing (Strebel, 1905). These two types appear to have merged into the modern Greyhound.

The Whippet

The whippet is a small dog of greyhound type, its shoulder height approximating 40 cm, and the weight varying between 5 and 10 kg. It was first produced in the north and north-west of England from crosses made between Greyhounds proper, Terriers and Italian Greyhounds. Over a distance of 200 yards the Whippet is the fastest of all dogs of its height and weight.

The Irish Wolfhound

The Scottish Deerhound and Irish Wolfhound, while not of pure greyhound stock, carry a share of greyhound blood only slightly less than in the Russian Wolfhound

162. Common Greyhound (after an etching by A. Bell, 1726–1809)

(Borzoi). In former times the Irish greyhound or wolfhound was employed in Ireland in hunting the wolf and the stag; but the extinction of these beasts of chase led to the neglect and consequent degeneracy of the breed, and its becoming almost extinct in that country. A few decades ago the breed was resuscitated from crossbreds derived from Scottish Deerhound and German Boarhound or Great Dane stock to which borzoi blood was added later. The Irish Wolfhound is the largest of hounds, the tallest of all hunting dogs, and the most active for its size. The minimum height of males is 78 cm, and of females 70 cm, the desired height being between 80 and 85 cm. The average weight of a full-grown male is 60 kg, and of a female 50 kg. The recognised colours are grey, brindle, red, black, pure white or fawn.

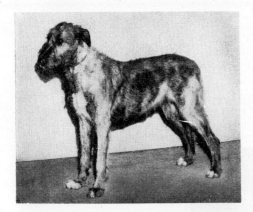

163. Irish Wolfhound bitch

164. Scottish Deerhound bitch

The Scottish Deerhound

The old Irish Wolfhound, which was a heavily built, smooth-coated greyhound, is believed to have been introduced into Scotland, where its modified descendant, the Scottish Deerhound, in hunting the stag bears testimony to the great strength and agility of its progenitor. In general outline the Deerhound is like a Greyhound proper, but heavier. Caius (John Keys or Kays) (1576) wrote with regard to greyhounds:—"Some are of a greater sorte, some of a lesser; some are smoothe skynned and some

165. Crania of English and dwarf Italian Greyhounds, frontal view (after Studer)

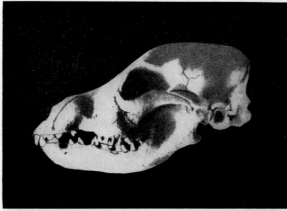

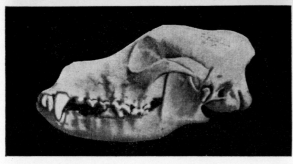

166. Crania of Arabian Greyhound (above) (after Baumann and Huber)
and English Greyhound (below) (after Studer)

curled, the bigger therefore are appointed to hunt the bigger beastes, the buck, the hart, the doe." The Deerhound is a hard-coated, shaggy dog of various colours: dark blue-grey, brindle, yellow and sandy, red or red-fawn, often with black points. The male stands 75–80 cm at the shoulder, and the female 70–75 cm. The weight of the male is 40–50 kg, and of the female 34–43 kg.

Greyhounds have an old history in Scotland. The fragment of a sculptured shrine at St Andrews Cathedral, dated to soon after A.D. 800, shows two greyhounds with the curled-up tail characteristic of the ancient Egyptian Tesem. This may be either a case of convergence or a cultural trait introduced by the neolithic ancestors of the early Scots from North Africa and the Iberian peninsula (see also p. 237).

Cranial Conformation of Recent European Greyhounds

In the greyhounds of Europe, not employed for hunting, the neurocranium is relatively larger than in the saluki or the Borzoi, the cranial superstructure is less developed, the splanchnocranium high and narrow, but at the same time considerably shorter. In the diminutive Italian Greyhound the skull shows the characteristic features of the dwarf: the neurocranium is large, smooth, and devoid of ridges, while the splanchnocranium, although high and narrow, is relatively short and marked off from the neurocranium by a fairly well defined stop (Studer, 1901) (Figs. 165–167).

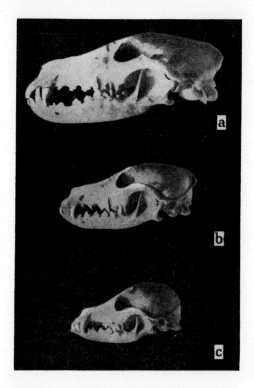

167. Crania of (a) Borzoi, (b) Whippet and (c) dwarf Italian Greyhounds, lateral view (after Baumann and Huber)

2. Greyhounds in Ancient Asia and Europe

Greyhounds in Ancient Mesopotamia

In ancient Mesopotamia greyhounds are depicted on vessels and seal cylinders from Tell Halaf and Uruk levels at Nineveh, Tell Arpachiyah and Tepe Gawra (Christian, 1940), also at Uruk and Nimrud. A greyhound skull was excavated at Tepe Gawra (Hauck, 1947). The ancient Mesopotamian greyhounds were prick eared, but in contrast with the curled-up tail of the Egyptian Tesem they had rat or sickle tails (Figs. 168–170).

Greyhound-like Crania at Harappa

The greyhound has not been reported from Mohenjo-daro in the Indus valley. But from Harappa on the Ravi river the cranial remains of dogs, greyhound-like in their elongated splanchnocrania, have been described by Prashad (1936), who suggested that this breed was the ancestral form of the recent Indian greyhounds and that in cranial conformation it was allied to Canis lupus pallipes. Hauck (1950) has criticised this theory mainly on the ground that there is no evidence for a specifically Indian prehistoric greyhound, whereas such evidence does exist with regard to Egypt and Mesopotamia.

Greyhounds in Ancient Crete, Sicily and Italy

In Crete the most ancient record of a greyhound is represented by a dog's head in pottery of neolithic date at Knossos (Hutchinson, 1962). Another early representation

168. (left top) Greyhounds on pottery from Tepe Gawra XI (after Christian)

169. (right) Greyhound on a vessel from Nimrud (after Rawlinson)

170. (left bottom) Greyhound on a seal cylinder from Uruk (after Christian)

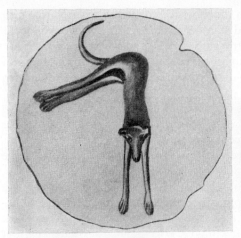

171. (left) Prick eared greyhound on a black steatite vessel from the Early Minoan period of Crete
172. (right) Prick eared greyhound on a Greek coin from Panormos (Palermo), 4th century B.C.

of a greyhound dates from the Early Minoan bronze age (c. 2500–2000 B.C.), i.e. a period corresponding roughly with the IVth to XIth Egyptian dynasties. It shows a prick-eared dog in recumbent posture, painted on the cover of a black steatite vessel (Fig. 171) (Eastman, 1916). On Cretan coins from the 6th century B.C. and earlier, greyhounds with erect ears, slender flanks, and long and light limbs are frequently depicted. They show a remarkable resemblance to the primitive greyhounds still encountered in the island.

Formerly the range of this greyhound extended also to Sicily and the southern part of the Italian mainland. On numerous coins from the Greek colonies in Sicily and Italy prick-eared greyhounds are shown in a form closely resembling the recent Ibiza dog.

Ancient Greek and Roman Greyhounds

Greyhounds are represented also in many ancient Greek and Roman sculptures and paintings. Greyhounds of saluki type are depicted on a wine jug of the 6th century B.C. (British Museum), and on many Greek vases, bowls, jars, coins and gems of later periods. A variety of greyhound with a feathered tail is shown on a 6th century B.C. hydria of Athenian fabric entitled 'Departure of Warriors' (British Museum). This variety is also represented on an amphora of the same period, and many times thereafter (Vesey-FitzGerald, 1957). Greyhounds seem to have been common at the time of Xenophon (434?–355? B.C.), and they are frequently depicted on monuments as companions of Meleager in the hunt of the Calydonian boar.

Greyhound Skulls in Ancient Europe

Few skeletal remains of prehistoric greyhounds have been found in Europe. A skull of greyhound type (C. f. leineri Studer) from the turbary station of Bodman at Lake

Überlinger (Constance) has been dated to the most recent neolithic period (see also p. 23). As this specimen is quite isolated and no other remains of C. f. leineri type occur in other neolithic stations, C. Keller (1909) regarded it as a later accidental insertion into the turbary strata of Bodman. It is significant that the type specimen does not belong to a greyhound proper, but rather to a crossbred dog, similar to the Scottish Deerhound or the Irish Wolfhound.

Crania of C. f. leineri type become more frequent in Europe in bronze and iron age strata; Gehl (1930) believed that they all date from the iron age, although slender crania, resembling C. f. matris optimae, occur earlier. In Austria the cranial remains of a fairly powerful dog of this type were recovered from a pile dwelling at Lake Starnberg, dated to the late bronze age (MacCurdy, 1924). In Ireland two large skulls, one of them similar to the type specimen of C. f. leineri, the other closely resembling the cranium of an Irish Wolfhound, were found in the iron age crannog of Dunshaughlin (Studer, 1901). C. f. leineri is known also from iron age deposits in Denmark (Dosenmoor) and Schleswig-Holstein. Brinkmann (1921) has pointed out that the earliest true greyhound of Europe is represented by a skeleton from a Viking interment at Errindlev, Denmark. The skull, with a basilar length of 198 mm, resembles the cra-

173. Greyhounds hunting hare (Greek painting)

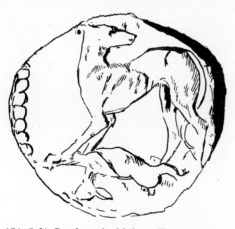

174. (left) Greyhound with hare (Roman coin)
175. (right) Greyhound from the Roman period, Stuttgart, Wuerttemberg

nium of the Borzoi from which it is distinguished merely by the larger relative width of the palate and zygomatic arches, due, probably, to the weaker masticatory musculature and elongated nasal part in the Borzoi. The extremities of the Errindlev dog are slightly shorter and thicker than in the modern greyhound.

The introduction of the greyhound into southern Russia is dated to the middle of the 2nd or, at the latest, the middle of the 1st millennium B.C. (Gehl, 1930).

176. Greyhounds at play, Roman marble (Vatican)

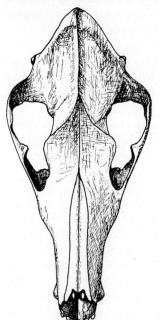

177. Skull of Errindlev greyhound (restored) (after Brinkmann)

3. Origin and Descent of the African Greyhounds

Chronology of Diffusion

The area of distribution of the greyhounds in ancient and modern times, and the successive stages of their appearance permit certain conclusions with regard to the centre of their evolution and routes of diffusion. Moreover, the conformational and physiological peculiarities of the greyhound offer an indication of the geographical environment in which the type was first evolved.

In ancient Egypt greyhounds are recorded already from a period prior to the dynastic conquest. In Sumer and Elam they also occur at a very early time; in Afghan Turkestan they can be traced back to approximately 2200 B.C., but at Mohenjo-daro the breed was apparently unknown, while the assignment of the Harappa skulls to greyhounds is open to doubt.

In Crete greyhounds appear during the Early Minoan age, but on the European continent they cannot be traced earlier than the Graeco-Roman period. Their introduction into southern Europe may have been initiated by Phoenician colonists settled on the coast of North Africa, and by Greek traders. Their diffusion in Europe reached its height at the time of the expansion of the Roman empire. In central and western Europe the greyhound appears rather suddenly during the iron age (earlier records being doubtful), apparently as a result of trade with the south.

The very early and nearly simultaneous occurrence of greyhounds in Egypt and Mesopotamia renders it difficult to decide whether the type was first developed in Africa or in Asia. The peculiar conformation of the greyhounds and their relative lack of scent point to their evolution in an open steppe or desert country where hunting by sight is an advantage. But this indication is of little help in the quest for the country of origin of the greyhounds, for deserts abound both in western Asia and in Africa east and west of the Nile valley. However, the anatomical aspect is not entirely devoid of significance. The Mesopotamian greyhounds had rat-like tails, but in all early African records of the greyhound the animal is depicted with a tightly curled-up tail. The latter, as Hilzheimer (1932) has pointed out, is anomalous in a racing dog, and should be regarded as a specialised character, evolved by breeders from the straight tail through the sickle and ring stages. In this respect the generalised tail conformation of the ancient Mesopotamian greyhounds appears to have been nearer the original greyhound type.

Introduction into Ancient Egypt

During the dynastic period Tesems were repeatedly imported into Egypt from Nubia and the Land of Punt, a country between Suakin and Assab, or further east along the Somaliland coast (Clark, 1954). This may have been the early breeding centre of the Tesem. But whether the original parent stock from which the Egyptian breed was evolved was autochthonous in that part of East Africa or thither introduced from western Asia via Arabia is still uncertain, although the scanty evidence available tends to favour the latter alternative. An independent evolution of the Egyptian and Mesopotamian breeds is unlikely; for the steppe and desert country intervening between the Nile valley and Mesopotamia forms no barrier to the greyhound which is especially adapted to such an environment.

A primitive greyhound may have been brought into East Africa by the Gerzeans or by the ancestors of the dynastic people, neither of whom were autochthonous in Africa; the Gerzeans are believed to have come from Syria or the Red Sea region, and the dynastic people from Elam. Culturally these immigrants were not primitive hunters, but seem to have formed societies characterised by the accumulation of large herds and flocks, sea-faring vessels, and slaves in the hands of their chiefs. They were in possession of dogs and other domestic animals already prior to their arrival in Africa. Their skill at handling beasts and birds is indicated by their domestication in Africa of the ass, the cat and Nile goose, and the taming of several other African animals which never became properly domesticated. Their ruling class used to hunt with falcons and dogs. In the steppe and desert surrounding their habitations they employed a fast dog hunting by sight. The evolution of the greyhound, then, should be regarded as an event accruing from a peculiar social pattern in a peculiar environment. On an ivory knife handle from Gebel el 'Araq (Fig. 178), a national monument of conquest which is to Egyptian history what the Bayeux tapestry is to English history, are two figures of dogs belonging to the Babylonian myth of Etana on the flying eagle. Above them is a figure of a hero or divinity subduing two lions, and below there are exquisitely spirited figures of both wild and domestic animals, including another dog (Petrie, 1939). The dogs represented on the knife handle are probably of the type the ancestors of the dynastic people brought from Elam to East Africa. They can hardly be designated as greyhounds; rather their sturdy bodies, prick ears and shallow stop suggest a generalised form within the type range of C. f. matris optimae or C. f. pou-

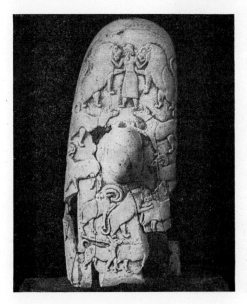

178. Knife handle from Gebel el 'Araq

tiatini. But the erect ears and negligible stop already foreshadow those of the Tesem, while the sickle tail, verging on the ring tail, may have been the forerunner of the Tesem's coiled-up tail.

Theory of the Greyhound's Descent from Simenia simensis

Several theories have been advanced on the ancestry of the greyhounds. Geoffroy Saint-Hilaire believed that the kaberu or Abyssinian wolf (Simenian fox), which takes its Latin name, Simenia simensis, from the district of Simen in Ethiopia, was the ancestral form. This theory was upheld by von Pelzeln (1886), C. Keller (1900)

179. Simenian fox (Simenia simensis) (after Rueppel)

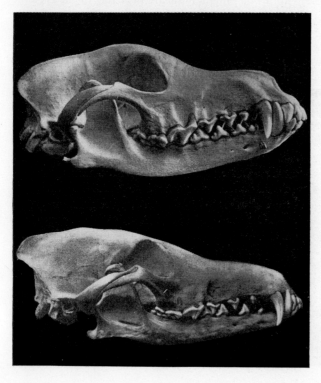

180. Crania of an Ibiza dog (above) and Simenia simensis (below) (after C. Keller)

and Eastman (1916). Keller concluded from the cranial likeness between Simenia simensis and the Borzoi that the greyhounds were derived from Simenia simensis, "without, however, supporting this view by exact measurements", as Studer (1901) commented. Eastman, without producing any evidence, noted that the Abyssinian wolf "persists even at the present day in a feral (sic!) state in the Ethiopian region, and appears to have been domesticated at an extremely remote period among the primitive inhabitants of northern Africa".

The kaberu is a slender high-legged animal. Its very long and narrow cranium has indeed a superficial resemblance to that of a greyhound; but a closer examination of the skull reveals that this is fundamentally a fox skull drawn out in length, rather than that of a wolf, with which it merely shares the presence of a frontal sinus albeit exceedingly small, and the absence of a hollow on the supraorbital processes. The teeth of Simenia simensis are set far apart and are very small in relation to the length of the skull, far smaller than in wolves and greyhounds of comparable body size. For this reason, Nehring (1888f) excluded it from the ancestry of the greyhounds and, for that matter, of any other domestic dog. Studer (1903), Hilzheimer (1908) and Antonius (1922) concluded from independent craniological investigations that Simenia simensis had a vulpine skull, especially in respect of the small size of the premolars and superior sectorius, the very slight development of the frontal sinuses, and the slender arched canines; therefore, it should be classed with the foxes, which it resembled also in its mode of life, rather than with the wolves or jackals.

Theory of the Greyhound's Descent from Canis Aureus Lupaster

Among the grey jackals of Africa, Canis aureus lupaster (see pp. 7, 8) resembles the greyhounds in the slenderness of the skull. Studer (1901) remarked that the lupaster skull "in its drawn-out shape and long narrow splanchnocranium nearly approximates to Canis simensis". From the slenderness of the Canis aureus lupaster skull Hilzheimer (1908) inferred that this North African jackal was the wild ancestor of the Egyptian greyhound.

A mummified skull of greyhound type from Asyut, northern Upper Egypt, distinguished by a large neurocranium with strongly vaulted parietals, a considerably constricted temporal region, very narrow splanchnocranium, and similar relative measurements to those of the Canis aureus lupaster skull, served Hilzheimer (1908) as proof of his theory. The resemblance between the Asyut skull and the skull of Canis aureus lupaster is so striking that it could be doubted whether the mummified head belonged to a dog rather than to a jackal, were it not for the relatively slight cranial superstructure and the osteolytic destruction wrought along the maxillary edges. Even so, the Asyut skull could be that of a tamed jackal kept in captivity, for jackals were frequently tamed in ancient Egypt.

There is no further evidence in support of the theory that the ancient Egyptian greyhounds were derived from North African jackals. The rather late appearance of the greyhounds, in relation to the domestication of the dog in general, renders their independent domestication from wild Canidae improbable. Nor is it likely, in view of the dental and cranial differences between dogs, including greyhounds, and wolves on the

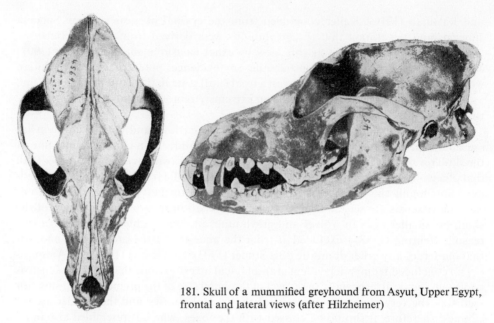

181. Skull of a mummified greyhound from Asyut, Upper Egypt, frontal and lateral views (after Hilzheimer)

one hand and jackals on the other hand, that a jackal strain shared in the evolution of the greyhounds.

As to the recent greyhounds of Asia and Europe, and the closely related saluki which he considered to have been introduced into Africa at a comparatively recent period, Hilzheimer (1908) held that they had nothing in common with the early Egyptian greyhound, but were descendants of slender wolves. He based this theory on a comparison of the crania of European greyhounds with the Asyut skull. In the former the neurocranium is relatively long and the parietal region narrow, whereas in the latter the parietal portion is particularly strongly developed; in European greyhounds the frontal region is broad and strong, never vaulted, and owing to the shorter and more convex supratemporal arches different in shape from the frontal region of the Asyut skull. Differences are noticeable also in the splanchnocranium and the upper dentition, the Asyut skull being distinguished from the crania of European greyhounds by the larger P_1 in relation to both the basilar length of the skull and the size of the molars and remaining premolars.

As Hilzheimer's theory rests solely on the skull from Asyut, of doubtful provenance, it has little enough to stand on. Its author offered no evidence for his suggestion that the saluki was imported into Africa from elsewhere. Indeed, the chronology of the ancient records indicates the early introduction of the greyhound from Africa into southern Europe. In a later publication (1932) Hilzheimer says:—"I believe that all breeds of dog, with the single possible exception of the Egyptian and its derivative the greyhound, have been evolved ultimately from the wolf." And while in his former work (1908) he stressed the difference in cranial conformation between the Asyut skull and European greyhounds, in the later publication (1932), in discussing the difference in build between the Ibiza dog and the early Egyptian greyhound, he likens the latter to the Borzoi type.

Birkner (1900–02) regarded the recent European greyhounds as phylogenetically distinct from the southern dogs, but on quite invalid craniological evidence, as Hauck (1950) has pointed out.

We should not conclude from the historical evidence of the southern origin of the European greyhounds that there was a lack of suitable parent stock in Europe for the development of greyhounds. The greyhound-like cranial conformation encountered in many modern sheepdogs classed with C. f. matris optimae indicates that greyhounds could readily have been evolved from autochthonous European stock. But historically and phylogenetically the African and European greyhounds cannot be separated. The differences in cranial conformation, within the compass of the general type, between the ancient Tesem and the recent greyhounds of Africa, Asia and Europe may be due to continued selection of a highly specialised dog hunting by sight either from among the original material, as probably in the case of the saluki, or from among crossings of greyhounds with other types of domestic dog, as in the case of the Borzoi.

Theory of the Greyhound's Descent from Canis Lupus Pallipes

Brinkmann (1921) suggested that the large greyhounds were descended from the small and slender Canis lupus pallipes of India, which their primitive European representative, the Errindlev greyhound of Denmark, resembled in several cranial features, such as neurocranial conformation, frontal width, the lateral walls of the face, length and width of palate, and the position of the Processus postglenoidales. However, it is doubtful whether such evidence may be regarded as sufficient;—not so much because there are also important differences between the pallipes and Errindlev skulls in the outline of the facial profile and the development of the sagittal crest, zygomatic arches and dentition, but because it would be surprising if two Canidae of the same species, similar in slenderness of conformation, and nearly identical in basilar skull length did not show considerable cranial resemblance. Brinkmann's view that the greyhound is a direct descendant of Canis lupus pallipes cannot be rejected on zoological grounds, since the ultimate descent of the greyhound from the wolf is not disputed. It is the lack of evidence of the evolution and early presence of the greyhound in India, in addition to the fact that the domestication of the dog and its diffusion throughout the Old World considerably antedate the greyhound's first appearance in Egypt, Mesopotamia and India, that refute Brinkmann's theory. For the ubiquity of the dog must have rendered repeated domestication of the wolf superfluous.

Theories of the Greyhound's Descent from Pariah and Other Domestic Dogs

Nehring (1888) suggested that the prehistoric greyhounds might have been obtained through selection from heavier breeds of dog in steppe regions where wolves and jackals as well as domesticated dogs display a tendency to slender forms. This view has been elaborated by Baumann and Huber (1946) who have pointed out that the evolution of the greyhounds shows a progressive reduction in cranial width. The most primitive forms are traceable to a greyhound-like C. f. inostranzevi type, regarded as a domesticated steppe wolf. In the English Greyhound the reduction in cranial width is still moderate, while in the Borzoi it is so marked that the skull appears long, narrow and high. The Whippet and especially the Italian Greyhound display, along with the narrow splanchnocranium characteristic of all greyhounds, the vaulted neurocranium typical of small dogs.

While Baumann and Huber have shown that morphologically canine skulls can be arranged in a progressive series of decreasing width from the slender steppe wolf through the English Greyhound, Borzoi and Whippet to the Italian Greyhound, phylogenetically such a series is quite meaningless. More especially, there is no evidence whatsoever of the descent of the primitive greyhounds from strains of C. f. inostranzevi greyhound-like in cranial and body types. Nor is there any evidence to indicate that slender C. f. inostranzevi strains are descended from steppe wolves, and heavier C. f. inostranzevi strains from forest or mountain wolves.

For the soberest view on the origin of the greyhounds we are endebted to Studer (1901) who considered them to be derived from pariah stock through a marked elongation of the splanchnocranium in relation to the neurocranium. Hilzheimer (1908) regarded the slenderness of the greyhound skull as due, not so much to a greater development in length as to a restriction in width, although conceding that in several highly specialised greyhound breeds the part in front of the Foramina infraorbitalia may also be lengthened.

In support of his view of the pariah descent of the greyhounds, Studer (1901) referred to the skull of a European greyhound, with a basilar length of 140 mm, which closely resembled the skull of a pariah from Sumatra. Antonius (1922) has confirmed that no well defined line of demarcation can be drawn between slender greyhound-like pariahs and coarse greyhounds proper; it would be possible to range such breeds as the Bengal dog, the Shilluk greyhound and the Podenco with either the pariah or the greyhound. This supports Studer's view that no separate wild ancestor is responsible for the evolution of the greyhounds, but that these are of pariah descent, selected and bred with a view to slender forms. Additional support for Studer's theory is provided by the great anatomical and physiological variability of the pariah group; this ranges from the extreme respiratory type, as represented by the Bengal dog, to the extreme digestive type, characteristic of some of the breeds of black Africa bred for meat. Considering however that the pariahs, in the course of their history, have absorbed many different canine strains, the possibility cannot be dismissed that in some of the greyhound-like pariahs, particularly in North Africa, the slender conformation may not be primary, but due to the more recent influence of greyhound blood.

iv. The Origin of the Mastiffs of Ancient Egypt

Mastiffs in Ancient Mesopotamia

The mastiffs occasionally portrayed in the art of ancient Egypt were not autochthonous in Africa, but came from western Asia. In Mesopotamia large mastiffs, standing about 80 cm at the shoulder, are recorded from the end of the 3rd millennium on (Fig. 182). They were employed there as war dogs, a practice that subsequently spread all over western Asia. Herodotus recorded that when the Perinthi were attacked by the Paeoni on the Propontis 'man was matched against man, horse against horse, dog against dog'; and Aelian described the employment of war dogs in action among the Magnesians and Hyrcanians (Hilzheimer, 1932). Mastiffs were used in Mesopotamia also as watch dogs. Their statuettes were regarded as apotropaia, i.e. charms averting evil. They seem to have contributed to the evolution of the powerful hounds used for hunting purposes in Assyrian times.

The occurrence of mastiffs in western Asia during the 14th century B.C. is shown by two scenes of combat between a lion and a large mastiff on a basalt orthostat excavated at Beth-shan, an Egyptian fortress and garrison town in Palestine. The slab, of Syrian origin, seems to have been part of the Egyptian spoils of war (Albright, 1949).

Mastiffs in India and Turkestan

The mastiff type was known not only in ancient western Asia but also in India, as indicated by the steatite figurine of a mastiff-like hound found at Mohenjo-daro. Alexander the Great encountered similar dogs in the Punjab in 327–325 B.C. In Sind

182. Mastiff. Babylonian relief, c. 2200 B.C.

183. (left) Mastiff from Tello, c. 2000 B.C.
184. (right) Statuette of a prophylactic dog

they are used to this day for killing wild boars tied to poles (Marshall, 1931–32; Mackay, 1935).

Hilzheimer (1932) reported the survival of a similar type in Turkestan:—"In 1930, at the Industrial Fur-Trade Exhibition, I saw an animal among the Turkestan shepherd-dogs which reminded me at once of the early Babylonian mastiff. Of course there were other examples of these dogs, smaller and slighter, showing that the breed is not so entirely stabilised as those systematically bred in countries with a high degree of culture."

Theory of Descent of the Mastiffs from Tibetan Dogs

Albrecht (1903) believed that the ancient Mesopotamian mastiffs originated in Tibet and entered Mesopotamia by way of the highland of Elam; hence the Assyrians called them Ilamti. According to C. Keller (1902), mastiffs first spread from Tibet into Nepal and India whence they reached China and Persia; from Persia they were introduced into Mesopotamia, and from there Xerxes sent some to Epirus. Alexander the Great is credited with having brought mastiffs from India to Macedonia, whence the Romans received the breed which they called Canis molossus. At the beginning of the Christian era, mastiffs reached the Roman colonies north of the Alps, and these formed the foundation stock of the mastiffs of Switzerland (from which the St Bernard was subsequently evolved) and of Britain and central Europe.

But the validity of this theory is doubtful; for dogs of mastiff type were known north of the Alps nearly one thousand years before the Christian era. Also, it is an unproved conjecture that Canis molossus was derived from Tibetan dogs; the Molosser originated in southern Albania (Antonius, 1922). There is therefore no evidence of any direct connection between the Tibetan dog and the mastiffs of Europe or, for that matter, between the latter and the mastiffs of ancient Mesopotamia and Egypt, although Vesey-FitzGerald (1957) asserts that the mastiff of ancient Greece, as portrayed in sculpteurcs, arvings and paintings, "is the Mastiff of Egypt and Assyria, such differences as there are being attributable to the different technique of the artist rather than to any actual difference in the dog."

The Tibetan Mastiff is the largest of all the many breeds of mastiff. It so closely resembles the Mongolian Mastiff that it is not feasible to distinguish between them (Vesey-FitzGerald, 1948). The earliest Chinese record of the Tibetan Mastiff, dating from 1121 B.C., refers to a dog that had been sent by the Liu, a people in the west of China, to the Chinese emperor Wu Wang; this dog is reported to have stood 4 feet at the shoulder. Marco Polo referred to their size "as large as asses", and in the description furnished by Samuel Turner (who visited Tibet by order of the East India Company in 1800), their large size is stressed in similar terms. According to Count Béla Szechenyi, who studied these dogs in their home country, they resemble the New-

185. (left) Tibetan dog—pariah type (after Strebel)
186. (right) Tibetan dog—mastiff type (after Brehm)

foundland but are broader and heavier, with fiercer eyes, a shorter head, wider forehead, more powerful dentition, and heavy flews covering the teeth (Studer, 1901; Strebel, 1905). Siber (1897) described the Tibetan Mastiff as a large well proportioned dog, reaching a shoulder height of 70–90 cm in the male, and 65–80 cm in the female. The head is large and heavy, the nasal part very deep but only moderately long and broad, the forehead wide and vaulted, the stop well defined, the eyes are dark brown in colour and of medium size; the skin above the eyes, on the forehead and at the side of the muzzle is folded. The ears, set on high and broad, are triangular in shape, pendulous, and of medium size. The neck is short and strong, its powerful appearance being enhanced by a long mane. The body is deep and heavy, with a broad chest, straight back, rather weak and flat thighs, and short straight legs. The feet are small, with well closed toes, and dewclaws on the hindlegs (see p. 176). The tail is of moderate length, bushy and straight or curled on to the back. The coat consists of long hair, either straight, rough, curly or silky in texture, with a soft woolly underfur which is shed in summer. On the back, neck, thighs and tail the hair reaches its greatest length, while the head, ears and feet are covered with short hair. The common colour of the coat is black-and-tan, with white marks on the chest and feet. The underfur is grey. However, it should be noted that Tibetan dogs, as Antonius (1922) has pointed out, are highly variable in size and conformation, the only general features being the length, thickness and dark coloration of the coat and the moderate size of the pendulous ears.

Theory of Descent of the Tibetan Mastiff from the Tibetan Wolf

C. Keller (1902) held that the Tibetan Mastiff, and hence the entire mastiff group, was descended from the Tibetan wolf (Canis lupus chanco), a large, powerful, relatively short-legged animal, with a long shaggy coat, extending into a mane on the neck and chest, and frequently black in colour. The Tibetan wolf resembles the Tibetan Mastiff in the shagginess and dark colour of the coat.

Theory of Descent of the Tibetan Mastiff from Dingo-like Ancestors

Studer (1901) has pointed out that the skull of the Tibetan dog is characteristic of a breed domesticated for a long time. This is apparent in the obliquity of the orbital plane and the weakness of dentition. The skull shows little resemblance to the crania of northern dogs but a great deal to the dingo's; this resemblance is so striking that the skull of the Tibetan dog may be described as a greatly enlarged dingo skull. Accordingly, Studer classed the Tibetan dog not with the mastiffs of the northern hemisphere but with the dingo group of the southern. He considered that it was evolved from a dingo-like ancestral form, and not from pariah stock, and that the marked constriction of the neurocranium in the temporal region and the development of the sagittal crest indicated an early crossing with the wolf.

Antonius (1922), following Studer, suggested that the Tibetan Mastiff was evolved from dingo-like pariahs through the direct or indirect introduction of a lupine strain, passing

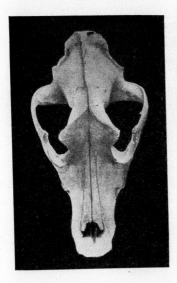
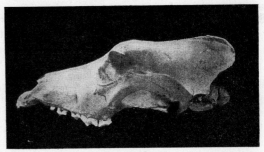

187. Skull of Tibetan dog, frontal and lateral views (after Studer)

ORIGIN AND DESCENT

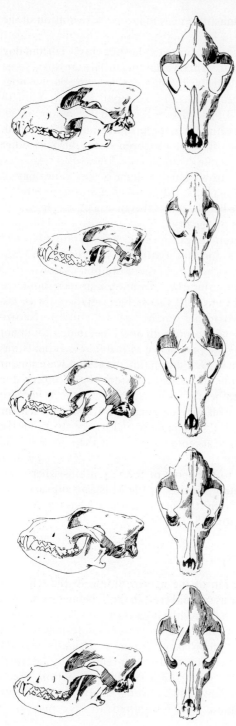

188. Cranial outlines (after Strebel)
  I.   Tibetan wolf
  II. Dingo
  III. Tibetan dog
  IV. Tibetan dog
  V.   German Mastiff

through the stage of the guard dog to the mastiff type, similar to the evolution of the St Bernard from Swiss Mountain dogs.

Strebel (1905), while admitting the resemblance between the dingo and Tibetan dog skulls examined by Studer, stressed the marked cranial variability in Tibetan dogs; some show pure mastiff type, while others are hardly distinguishable from pariah dogs with which they apparently often interbreed, especially in Nepal, Bhutan, Yunnan and Lahore. Among Studer's material skulls of pariah type seem to have predominated. The cranial variability of Tibetan dogs, ranging from the primitive, albeit oversized, pariah through the guard dog to the mastiff type, has been noted also by other writers (Antonius, 1922).

Cranial measurements therefore do not throw sufficient light on the genealogy of the Tibetan dog, nor on its relation to the mastiffs of Europe. According to Strebel, such measurements indicate a close alliance between the Tibetan dog on the one hand, and the dingo, wolf and European mastiffs on the other hand (see also Fig. 188); and also between the mastiffs of Europe and arctic dogs. In these canine cranial types of the southern and northern hemispheres (as, for that matter, in all other types) Studer's distinction between palaearctic and southern dogs therefore breaks down.

Although Strebel arrived at the conclusion that the Tibetan dog proper should be classed with the mastiff group, he opposed C. Keller's theory of its share in the evolution of the mastiffs of Europe, and also of its descent from the Tibetan wolf; the relatively frequent occurrence of melanism in both the Tibetan wolf and Tibetan dog could not be regarded as evidence of such descent, as the prevalence of melanistic animals was observed in many southern wolves. Antonius (1922) has pointed out that were similar coat colour a criterion of descent, it would be as well to consider the Leonberg dog (a breed fairly recently developed in Württemberg from crosses between the St Bernard and the Newfoundland) with its coat of wolf-grey colour as the link between the wolves and the larger and heavier breeds of domestic dogs of Europe. Nor can the occurrence of dewclaws in Tibetan Mastiffs be accepted as proof of their descent from the Tibetan wolf, since dewclaws may occur in any canine breed. Keller's suggestion that they were developed in Tibet owing to the snow of the mountains where they provide a broader pad is unfounded; Tibetan Mastiffs of the Mustang subvariety as well as arctic dogs are devoid of them; only the black-and-tan Lhasa variety of Tibetan Mastiffs has the fifth digit (Hodgson, 1832).

On the Origin of the Inostranzevi Cranial Type

Studer's theory that the marked constriction of the neurocranium in the temporal region and the development of a conspicuous sagittal crest in the Tibetan dog point to the latter's descent from wolves crossed with dingo-like dogs is obviously fallacious in view of our knowledge of the influence of body size on cranial conformation. The cranial peculiarities referred to are normal functions of the relatively small size of the brain and brain case in the large Tibetan dog as compared to their relatively larger size in the smaller dingo and pariah. Hilzheimer's doubt, in a similar connection, that a domestic breed with an average shoulder height of 40 cm would freely interbreed with wolves standing at least 70 cm at the shoulder, may be ignored; this is a poor argument, as outcrossing of domesticated dogs with wolves reared in captivity, whether immature, stunted or of full normal growth, has probably been practised at some time or other throughout the entire range of the wolf. Hence it cannot be disputed that different geographical races of Canis lupus furnished the material for the evolution of the numerous breeds of domestic dog. Disputed is the theory that the Tibetan dog owes its origin to such a cross, or that the different inostranzevi types, as Antonius (1922) held, could be traced to clearly definable outcrosses of domestic dogs with wolves. Thus Antonius asserted that the type specimen of C. f. inostranzevi showed a combination of the cranial characteristics of the wolf with those of C. f. poutiatini; the recent arctic and subarctic dogs of inostranzevi type a combination of the cranial features of the wolf with those of C. f. palustris ladogensis; and the guard dogs, likewise of inostranzevi type, the cranial features of the wolf combined with the characteristics of C. f. matris optimae. Antonius's observation merely indicates that the C. f. inostranzevi and C. f. matris optimae cranial types cannot easily be separated and that both may have contributed to the evolution of the guard dogs (Hauck, 1950). It also confirms Woldřich's (1882), Brinkmann's (1923–24) and Wagner's (1930) opinion that the cranial forms included in the inostranzevi group constitute a very variable material pointing in different evolutionary directions.

That outcrossing of dogs with various geographical lupine races, as well as interbreeding of different types of domestic dog, has contributed to the genetic complexity and variability of the progeny, of this there can be no doubt. And that this increased variability under domestication has facilitated the task of breeders in the selection of a multitude of types in accordance with the needs of human society under the most different conditions, is equally obvious. While the earliest dogs of inostranzevi type may owe their close cranial resemblance to the wolf to their recent derivation from the latter, some of the later inostranzevi breeds may have been evolved independently from variable stocks of domestic dogs. Lumer (1940) has pointed out that the prehistoric and ancient bulldog-like forms have most probably arisen independently of the modern bulldog from an entirely different stock. The genetic peculiarities of some of the recent breeds classed with the inostranzevi group demonstrate their evolution through artificial selection of inheritable variations. In the St Bernard, for example, large size in conjunction with a heavy jowl and other effects of one-sided pituitary action has amassed combinations of modifiers which protect the organism against the harmful effects of its exaggerated glandular development. Matings between St Bernards produce normal litters; but when the breed is crossed with the Great Dane, a simple giant type with no hyperpituitary characters, the protective genes are diluted and a considerable proportion of the progeny show hydrocephalus, paralysis of the hindlegs and other serious disturbances clearly of genetic origin (Huxley, 1943).

The similarity in cranial conformation of the inostranzevi type to the wolf, on which Studer and Antonius based their theory of the descent of the inostranzevi group, merely shows that the latter has undergone less far-reaching changes in the course of domestication than have other cranial types of domestic dogs. The same phenomenon is observed in all domesticated animals, some breeds displaying a closer, and others a remoter resemblance to the wild ancestral stocks.

Herders' Guard Dogs

From the history of the St Bernard we learn that the mastiffs can be traced to the herders' guard dogs from which they were evolved. Herders' guaro dogs are large dogs with a shoulder height of over 65 cm, a long, often flat, neurocranium with a markedly developed sagittal crest and occipital protuberance, a moderately short, aborally broad splanchnocranium, pronounced stop, long pendulous ears (only occasionally prick ears, as in a guard dog on the Pergamon altar frieze), a straight, rarely sickle, tail, and a curly or wavy coat.

189. Pyrenean Mountain Dog

Guard dogs are distributed over a vast area in Europe and Asia. Although for the greater part they are now restricted to the mountain zones extending from the Pyrenees where the Pyrenean Mountain dog (Fig. 189) has its home, through the Alps, with their various breeds of mountain dogs, Apennines and Balkans to the Caucasus, Turkestan, the highland of Tibet and Mongolia (Altai), a few breeds, such as the Old English sheepdog, the Russian Ovcharka, and the Hungarian Komondor and Kuvasz, are still found in the plains. Formerly their range seems to have been even wider, including such famous breeds as the Molosser and Epirote.

Occasionally the guard dog type may be observed among the Berber dogs and Armentis of North Africa, in which it may have been prevalent before these breeds, under the influence of pariah and other blood, acquired their present cranial variability.

With this wide distribution of the guard dog type it seems feasible that mastiffs could have been evolved independently in different regions and at various times. This appears to be definitely the case with the Tibet dog and the St Bernard; and it seemingly also holds true of the ancestors of the mastiffs of ancient Egypt, i.e. the Mesopotamian mastiffs, which were probably imported from Elam or Turkestan.

Theory of the Mastiffs' Descent from Swedish Wolves

Hilzheimer (1909a) held that guard dogs and mastiffs were evolved independently, their anatomical resemblance being due to their common descent from the wolf. He believed all mastiffs, including those from Mesopotamia, to be derived from the powerful short- and heavy-headed wolf of central Sweden, whence the type dispersed fan-like to the south (Brehm, 1922). However, in view of the evolution of the St Bernard from Swiss guard dogs, and the probable descent of the mastiffs of Tibet from local stock of guard dog type, the theory of an independent evolution of the mastiffs and guard dogs cannot be supported. This is not in contradiction to the results of Lumer's (1940) investigation, indicating that the prehistoric and ancient bulldog-like forms may not be regarded as direct ancestors of the modern bulldog. Even the small autochthonous Inca or Chincha bulldogs of Peru, which, according to Hilzheimer (1936a), are not found anywhere else on the American continent, were probably evolved from a South American guard dog type (Nehring, 1884b).

Chronological reasons also invalidate Hilzheimer's theory of the descent of all mastiff breeds from Swedish wolves. Several skulls of mastiff type have been found in prehistoric European stations: Nehring (1884c) has described two mastiff skulls of prehistoric or early historical (bronze?) age from the vicinity of Berlin with the name C. f. decumanus. A mastiff skull from the early Hallstatt period (earliest iron age, about 900 B.C.) has been described by Studer (1907), and a primitive bulldog skull from the iron age turbary station of Walthamstow, near London, by Poetting (1909). The mastiff was not present in Britain during the bronze age proper; it was probably introduced by the Celts in the course of the Hallstatt invasions (Vesey-FitzGerald, 1957). Van Giffen (1929) found a predominance of mastiffs and other large dogs in late iron age tumuli in Friesland and Groningen. During the same period mastiffs were also common in Denmark. Brinkmann (1921) has described a mastiff skull from a Viking grave near Naestved, and two broken skulls of a similar type from an iron age settlement at Vejleby on Lolland, Denmark. The Naestved skull is strong and massive, with a broad forehead, short wide splanchnocranium, and concave facial profile but no distinct stop. It resembles the medium-sized, comparatively slender mastiff type characteristic of smaller specimens of the recent Great Dane breed. These seem to be the earliest representatives of the mastiff group north of the Alps. In Italy the mastiff is represented on a pre-or proto-Etruscan late bronze age situla from the Certosa at Bologna (MacCurdy, 1924).

In Europe mastiffs are therefore traceable to the iron age, at most to the late bronze age. Their occurrence in the oriental region preceded their appearance in Europe by a long period of time. This makes it impossible to trace the Mesopotamian and Egyptian mastiffs to a hypothetical mastiff breed evolved from Swedish wolves.

Theory of the Mastiffs' Partial Descent from Northern Wolves

Antonius, who regarded the tendency of domestic dogs living outside the breeding control of man to grow to pariah size as evidence of their descent from small southern wolves, concluded from the fact that most dogs of inostranzevi type are larger than the average pariah that a strain of large northern wolves had shared in their evolution.

However, there are several factors that weaken the force of this argument. There is no proof that every domesticated species tends to revert to the average size of its wild ancestral stock, if left to breed in a haphazard manner. This is illustrated by the aboriginal dogs of America associated with various archaeological horizons: the older the horizon and the lower the cultural development of the human inhabitants, the smaller the dog (Haag, 1948). We may therefore conclude that a random breeding population of domesticated animals need not conform to the size of the wild ancestral stock; it exceeds it in some populations and in some environmental conditions, and falling short in others. The fact that the average size level of the domestic dog approaches that of the pariah is therefore no proof that the wild ancestral stock is to be found among the smaller lupine races, although actually this may be the case.

Nor is large body size in dogs an indication of the influence of northern wolves, although these may have been occasionally crossed with domesticated dogs. There is no reason to doubt that the large inostranzevi breeds could be derived from the same wild stock from which the smaller breeds are descended. Fossil wolf skeletons are racially indistinguishable, their variability being merely individual (Hauck, 1950). The recent smaller southern and larger northern wolves, as all authors on this subject agree, cannot be specifically differentiated, the southern races passing imperceptibly into the northern. But within the different races variability in size is very considerable; while the southern races are commonly small and the northern large, individual specimens among southern wolves may be of large, and among northern of small size. To a considerable extent size in the various lupine races apparently depends on environmental conditions which seem to favour a smaller type in the south and a larger in the north, but not in the extreme north, the pressure of selection being responsible for the corresponding differences in the genetic composition of the lupine populations of various regions.

Large Size in Dogs due to Selective Breeding

Since guard dogs are employed as guardians of homes, herds and flocks, they are required to be large. Selection would therefore aim at bringing to the fore the genetic factors making for large size. A similar condition is observed in Arctic American and Siberian dogs which have long been an important means of transportation. As early as 6000 B.C. there is evidenced use of the dog-drawn sledge in the area east of the Baltic (Childe, 1939). Selection of sledge dogs for strength, in combination with the high vitamin D content of their fish (liver) diet, would ensure the prevalence of the larger forms. Initially, breeding for size might be more difficult and take longer to accomplish in a canine population evolved from southern wolves than in one derived from the larger northern races. But once accomplished, the genetic results would be similar.

Possibly climatic conditions influence size in wolves and dogs more than other environmental factors. For the largest breeds of dog are found in the central and northern regions of Europe and Asia and in the cool mountain zones of the southern, where the wolves also reach their largest size. In the far north, again, where the wolves grow smaller, the arctic and subarctic dogs of inostranzevi type do not reach the size of the cranially nearly allied herders' guard dogs farther south.

v. The Origin of the Hounds of Ancient Egypt

Origin of Hounds in Western Asia

Although hounds appear in Egypt not much later than greyhounds, there is reason to believe that they were not evolved in the Nile valley but were introduced from Asia. Environmental conditions in subtropical regions do not favour a dog hunting by scent; and while hounds were bred in Egypt during the dynastic period for various forms of the hunt, they disappeared towards the end of the dynastic era.

In Mesopotamia powerful hounds, approaching the guard dog type, are represented on numerous monuments from Assyrian times. They can be traced as early as the Sumerian period, and were employed in hunting onagers and other game. However, it is doubtful if these hounds were actually evolved in the sultry valleys of the Tigris and Euphrates; more likely they originated from the cooler mountain regions in the north (Kurdistan). In Anatolia (Sinjerli) hounds are represented in rock-reliefs from the 14th and 13th centuries B.C. In the Caucasus they are recorded from the beginning of the first pre-Christian millennium; hunting scenes in open-work bronze plaques show hounds with long ears and erect sterns, giving tongue while attacking deer.

Theory of Descent of the Egyptian Hounds from a North African Jackal

In cranial conformation hounds show a general resemblance to the type specimen of C. f. intermedius. Hilzheimer (1926) has pointed out that the crania of ancient Egyptian hounds, while resembling C. f. intermedius in a general way, differ from the latter in the length of the sagittal crest, the shape of the neurocranium and tympanic bullae, and the inclination of the jugals. Owing to these differences, he doubted whether the prehistoric hounds of Europe and the hounds of ancient Egypt were racially connected. The latter, he suggested, might be derived from the North African jackal, Canis doederleini (= C. a. soudanicus—see pp. 8-10), which is distinguished by a similar neurocranial conformation and an equal width of the temporal constriction to those in the Egyptian hounds; moreover, the forehead and supratemporal arches of this jackal are of a similar shape to those in the skull of the Asyut hound (see p. 81), and the inferior orbital edge is equally broad (Hilzheimer, 1908).

However, in view of the striking changes to which the canine cranium is subject under domestication, the very resemblance between Canis doederleini and the crania of ancient Egyptian hounds negatives any close relationship between them, and has to be attributed to convergence. In this connection a remark passed by Antonius (1922) with reference to Woldřich's attempt at tracing the genealogy of C. f. intermedius to a special quaternary wild canine is significant:—"A wild Canis intermedius is just as unthinkable as a wild Canis palustris, since the characteristics distinguishing it from true wild Canidae are nothing else but the results of domestication." Hilzheimer himself does not seem to have been fully convinced of the existence of a Canis doederleini domesticus, judging from his remark that "the only point opposing its (i.e. C. doederleini's) parentage is the weak stop in the domestic hound, while experience has taught us that in all domestic animals the profile is angulated far more than in their wild ancestors".

190. Assyrian hounds in a relief from the palace of Ashurbanipal at Kouyunjik (Nineveh) c. 668 B.C.

191. Hound from Kouyunjik (after Hauck)

192. Onager hunt, Assyrian relief, c. 650 B.C.

Origin of the C. F. Intermedius Cranial Type

Klatt (1913), in opposing Woldřich's (1882a) attempt to trace C. f. intermedius to a wild ancestor of its own, rather one-sidedly ascribed the peculiar intermedius features, such as the great height and width of the neurocranium and shortness of the splanchnocranium, to the size of the intermedius skull in relation to other cranial types of dogs. He conceded, however, that artificial selection might be a contributory factor to the shortness of the splanchnocranium, as pariahs and wolves also display considerable individual variability in splanchnocranial length. In a later work (1927), in discussing the differences in cranial conformation between C. f. intermedius and C. f. matris optimae (which stand very close to each other in cranial size), Klatt emphasised the factor of domestication as the cause of brachycephaly in C. f. intermedius, admitting size and selective breeding to be equally important factors in the evolution of the intermedius skull from a less specialised cranial type.

Studer (1901) pointed out that the intermedius type did not appear all of a sudden, but gradually emerged from C. f. palustris in the most recent neolithic strata of the Swiss lake dwellings. At this level, skulls were found which are larger than the palustris type, with a wider and flatter forehead, broader maxilla, and shorter spherical oral part. Finally, crania with a basilar length ranging from 150–160 mm appeared, which can hardly be distinguished from the typical intermedius skull, except by the slightly shorter neurocranium and relatively less elongated frontals. This gradual development, Studer held, warranted the conclusion that C. f. intermedius had been evolved from C. f. palustris.

More recent investigations, however, have shown that crania resembling C. f. intermedius occur already in mesolithic strata, as at Naeselov and Husum (see p. 130); and as the intermedius type occurs also among recent pariahs in different parts of their range, we may conclude that it is not necessarily a developmental stage of the palustris, but one among several cranial types characteristic of primitive dog populations. Again, if the intermedius type lies within the natural range of variation of primitive dogs, hounds might have been evolved repeatedly and independently in different parts of the world from different parent stocks of this ubiquitous cranial type. More especially this may apply to the hounds of Europe and ancient Egypt. But of this we cannot be certain; for in view of the relatively speedy rate of diffusion of new creations and acquisitions even in primitive society, if economic, social and environmental conditions favour diffusion, it is possible that the southern and northern hounds, or the majority of them, derive from a single ancestral stock of C. f. intermedius type. This finds support in the late appearance of hounds (though not of the intermedius cranial type) in Europe as compared with Egypt. Into the economy of the lake dwellers of Switzerland they were probably introduced from the south-east (Anatolia). It is significant that C. Keller (1911) regarded the intergrading forms from the Swiss lake dwellings, referred to by Studer (1901), as crossbreds of the autochthonous C. f. palustris and the newly introduced bronze age hound.

Chapter II

CATTLE

I. On the Classification of Cattle

The Classification of the Bovini

The Bovini are classed into five genera; two of these, namely Bubalus and Syncerus, have horns angular in cross-section, the other three, including Bibos, Bos and Bison, horns of an oval or round section (Lydekker, 1913; Duerst, 1931; Pilgrem, 1939; 1947; Sokolov, 1954; Bohlken, 1958).

Bubalus and Syncerus

In Bubalus (Asiatic buffalo) and Syncerus (African buffalo) the parietals are situated in a wide zone on the roof of the cranium, which is thereby separated from the frontals. The forehead is short and narrow. The distinction between the two genera is based on the formation of the vomer and palatinum; in Bubalus these are fused, but in Syncerus they remain separate.

Genus Bibos

Of the Bovini with horns of an oval or round cross-section, the genus Bibos, including two species, i.e. the banteng and the gaur, is distinguished by long frontals, broad

193. Kouprey bull (Cambodian postage stamp)

parietals flanking the interparietal to its base, posteriorly projecting horns, and the presence of a dorsal ridge due to the elongation of the spinous processes of the 3rd–11th thoracic vertebrae.

The position of Bos (Bibos) sauveli, i.e. the North Cambodian forest ox or kouprey, is still uncertain. The animal was discovered only recently, and its total population is very small. Bulls attain a shoulder height of 190 cm. The horns are cylindrical and widely separated; in old bulls they are recurved and occasionally frayed at the tips, producing a shaggy appearance; the cows have lyre-shaped horns. The coat is of a blackish colour, with whitish stockings (Walker, 1964). The kouprey has been variously regarded as a species of Bibos (Urbain, 1937; Sauvel, 1949a; 1949b; Lekagul, 1952; Sokolov, 1954); or as the sole representative of the genus Novibos (Coolidge, 1940; Harper, 1945) or of the subgenus Novibos (along with the subgenus Bibos) of the genus Bos (Haltenorth, 1961a); or as a primitive species closely related to Bos primigenius (Braestrup, 1960); or as a hybrid of Balinese (domesticated banteng) and zebu cattle with the name 'Boeuf des Stiengs' (Duerst, 1931) or of banteng and either zebu, gaur or buffalo parentage (Edmond-Blanc, 1947). The possibility that the kouprey is a crossbred animal is also mentioned by Ellerman and Morrison-Scott (1951) and Haltenorth and Trense (1956). Bohlken (1958) regarded the kouprey as a hybrid of banteng and zebu; but in a subsequent study (1961), comparing the kouprey's skull, horns, dewlap, tail, hooves, dorsal stripe and coat colour with those of other Bovini, he could not decide whether it was a true wild species or a feral type derived from interbreeding of zebu and banteng. However, since the animal could not be separated from either Bos or Bibos and since, in view of the kouprey's position, Bos and Bibos themselves could not be clearly separated from one another, he tentatively classed the kouprey as a species of the subgenus Bibos of the genus Bos.

Genus Bos

The genus Bos, which comprises two subgenera, namely the extinct urus and the yak, has also long frontals which extend laterally on to the occiput. But the parietals are narrow and do not share in the formation of the forehead, being situated in the plain of the occipital area. Only a small portion of the interparietal reaches the upper edge of the frontal region. The horns project upwards, with a forward slope.

The urus has a much larger skull than the yak, with a relatively longer forehead; the occipital conformation is also different. Sokolov (1954), following Gray (1843), therefore separates the yak from Bos, recognising a separate genus Poëphagus. But the majority of recent taxonomists (Simpson, 1945; Ellerman and Morrison-Scott, 1951; Bohlken, 1958) class Poëphagus as a subgenus of Bos L.

Genus Bison

The genus Bison, comprising the European and American bisons, is characterised by short and broad frontals and small size of the parietals which are separated by a large interparietal. The latter occupies nearly the entire parietal region which is situated behind the horns on the roof of the cranium. The neural spines of the anterior thoracic vertebrae are markedly elongated. The European and American species are alike in cranial conformation and completely fertile in cross-breeding.

References to the Urus Races, Banteng and Arnee

The three geographical races of the extinct urus, as well as the banteng (Bibos javanicus) and the arnee (Bubalus arnee), are discussed in connection with the domestic types derived from them: Bos primigenius opisthonomus Pomel on pp. 231–235, Bos primigenius namadicus Falconer et Cautley on pp. 253–258, and Bos primigenius primigenius Bojanus on pp. 227–231; the banteng on p. 526, and the arnee on pp. 562–563.

The Classification of Domesticated Cattle

The domesticated cattle of the taurine group (Bos) have been divided by Rütimeyer (1861; 1865; 1867; 1877) into three basic types according to differences in cranial conformation: 1) Bos taurus primigenius, 2) Bos taurus brachyceros, 3) Bos taurus frontosus.

Bos taurus primigenius

The skull of Bos taurus primigenius is distinguished by the great length of both the neurocranium and splanchnocranium, straight cranial outlines, short molar dentition, and the size, shape and direction of the horns which resemble those of the wild Bos primigenius Boj. The length of the forehead, amounting to 47 per cent of total cranial length, exceeds the width. The distance between the temporal fossae equals the lateral forehead length; the interorbital width is less than the median forehead length. The forehead is flat, the intercornual ridge nearly straight; the orbits are oblique with a forward inclination, not prominent, relatively small, and square in outline. The horn cores rise from the skull without intervening necks. Their direction is first backward,

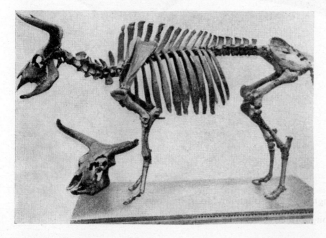

194. Skeleton of Bos primigenius primigenius Boj.

frequently extending behind the plane of the Torus frontalis, then sideways, forward and slightly upward, finally straight upward. The cross-section appears compressed and devoid of edges; the ratio of the horizontal to the vertical diameter is 5:4 or 4:3. The substance of the cores is compact, the surface smooth, with fine but distinct vascular lines, a wreath of bone pearls at the base, and deeply marked longitudinal furrows at the posterior and inferior surfaces. The occiput forms a right angle with the forehead, and the Torus frontalis rises little or not at all above the occiput. Below the horn bases the occiput is markedly constricted by the temporal fossae, so that the smallest occipital width is one-third to one-half less than the greatest. The Condyli occipitis and Processus occipitales converge markedly towards the median line. The lateral surface of the cranium is characterised by the long, straight and horizontal temporal fossa which penetrates deeply underneath the forehead, although its height is only moderate. Both the temporal ridge and jugal arch are nearly straight; behind the jugal arch the temporal fossa shows no expansions; the parietal point is separated from the sphenoid wing. The splanchnocranium is long and devoid of cavities. The nasals are long, little shorter than the forehead, and markedly curved, occasionally with a slight elevation at the bases; the tips are only slightly bifid. The maxillae are elongated, and the long premaxillaries, distinguished by broad alveolar branches, extend to the nasals. The molar set is short, and the toothless portion of the maxilla long in consequence; the length of the molar dentition amounts to 25–27 per cent, and that of the toothless part to 30–32 per cent of the total skull length. The mandible is also characterised by a short molar set, which measures about one-third of the entire mandibular length. The posterior ramus is oblique and the anterior branch strong but not high, ascending from the middle of the dentition in a straight line. The symphysis is very long, and the incisive part fairly broad. The teeth are strong, the molars and premolars wide and

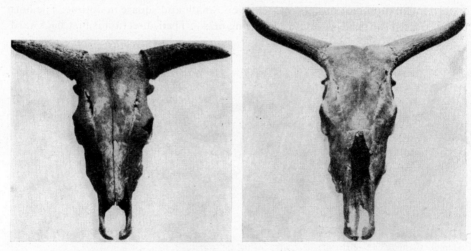

195. (left) Skull of a domestic longhorn bull, Longhour Moor, Ireland
196. (right) Skull of a domestic longhorn cow, Stanway Moor, Essex

short, with well marked dental prisms. The incisors have square crowns, the incisive set is only slightly curved, and the palate is flat (Rütimeyer, 1861).

In Europe the primigenius cranial type is characteristic of cattle from archaeological sites of the neolithic age. It has been reported from neolithic levels at Vila di San Pedro in Portugal, neolithic stations in the Netherlands, Bundsø and Dyrholmen in Denmark (Degerbøl, 1936; 1942), Maiden Castle, Windmill Hill and other neolithic settlements in England (Howard, 1962), St. Aubin and Egolzwil in Switzerland (Hescheler and Rueger, 1939; Dottrens, 1947), Alsonemedi and Tószeg in Hungary (Bököny, 1951; 1952), Cmielow and Grodek in Poland (Krysiack, 1950; 1951; 1956), and Weissenfels in the G.D.R. (Behrens, 1953). The remains from all these widely separated places show a remarkable similarity in general dimensions and the size and curvature of the horn cores. Associated with the bones of these neolithic cattle are often found those of the much larger wild Bos p. primigenius. While Howard (1962) and Zeuner (1963a) suggested that interbreeding of wild and domesticated animals might account for the very primigenius-like appearance of some neolithic domestic cattle, the validity of this explanation is by no means certain in view of the survival to this day of "very primigenius-like" breeds in countries where the urus has long been extinct.

Variability of the Primigenius Cranial Type

Rütimeyer's description of the primigenius type in domesticated cattle is based entirely on cranial conformation, which he erroneously regarded as the firm and unchangeable basis of the head on which the soft parts were modelled. He had no knowledge of the causes of cranial variations; and, when describing the primigenius type, he had only three skulls for comparison with fossils from Swiss lake dwellings (Duerst, 1931). Subsequent authors have found considerable cranial variability among the different breeds classed with the primigenius group, only two important features retaining a certain measure of constancy: 1) the relatively large size of the primigenius skull, and its similarity in this respect to the cranium of the wild Bos primigenius, a point stressed particularly by Klatt (1913); 2) the large size of the horns in domestic breeds of primigenius type, resembling the horns of the wild Bos p. primigenius. Duerst (1926a, 1931) has drawn attention to the marked effect of the size, shape and direction of the horns on neurocranial conformation, pointing out that what appears to be constant in neurocrania of primigenius type is to a great extent attributable to the large size of the horns, and what is variable in the neurocrania of different primigenius breeds and individual specimens, to the variable shape and direction of the horns, i.e. their moment of rotation. However, this seems to be applicable in a general sense only. As Requate (1957) has pointed out with reference to the wild Bos p. primigenius:— "A strict correlation in Duerst's ... sense between horn shape and size and the conformation of forehead, frontal edges and occiput cannot be substantiated in the urus." This applies, Bohlken (1958) adds, as far as the same sex is concerned. If crania of different sex but of similar size are compared, significant differences, especially in occipital conformation, are observed, confirming the influence of horn development on cranial conformation.

As the long horns are the principal characteristic of the primigenius type of cattle, the author has chosen the term 'longhorn' for the humpless native cattle of Africa classed with this type.

Bos Taurus Brachyceros

The second basic type of domesticated cattle postulated by Rütimeyer (1861), i.e. the brachyceros type (Bos taurus brachyceros), is characteristic of archaeological sites from the iron age of Europe (Howard, 1962). In Britain it first appeared during the bronze age—in the beaker settlement of Eastern Down in Wiltshire and in beaker levels in the Skendlebury long barrow in Lincolnshire (Jewell, 1963). It also occurs exclusively in several early neolithic sites of Switzerland, e.g. in the lowest stratum of

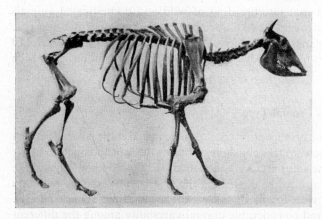

197. Shorthorn skeleton from Schussenried Moor (after Duerst)

Saint-Aubin, on Lake Bienne (Dottrens, 1946a; 1946b; 1947). Recent radiocarbon dates indicate that some brachyceros cattle arrived in Europe as early as 3000 B.C. (Zeuner, 1963b). In some of the most ancient Swiss lake dwellings the brachyceros type is found in association with the remains of Bos primigenius and of domesticated cattle larger in size than the typical brachyceros type (Hescheler and Rueger, 1942).

The holotype was first figured and described by Owen in 1843 in a museum catalogue with the name Bos brachyceros. Subsequently (1846) he withdrew this name because it had already been used by Sundevall for Bubalus brachyceros Gray, and replaced it by the term Bos longifrons. Rütimeyer (1861; 1876), however, rejected the name Bos longifrons as unsuited and incorrect, and revived the original name Bos taurus brachyceros, furnishing the following craniological description of the type:—

The forehead is very uneven, the orbits are large, prominent, and lateral in direction, the horns short, markedly bent, and set close to the head. The splanchnocranium is short and wide, with a long molar series and slender incisive portion. Owing to the shortness of the splanchnocranium, the forehead is longer in relation to the whole skull, at the same time broader and squarer, than in the longhorn (primigenius) type, its length ranging from 50 to 52 per cent of total cranial length. The smallest forehead width between the temporal ridges equals the distance between the posterior part of the horn base and the centre of the orbit. The greatest forehead width between the orbits frequently equals the median forehead length, which is not the case in the longhorn type, not even in male specimens. In the brachyceros skull the forehead is uneven in outline; the large orbits are considerably elevated above the frontal surface, and markedly project beyond the lateral plane of the skull. The supraorbital fossae are short, broad and deep, converging orally; the frontal area situated between them is concave. Above the dished part, the forehead rises to a high narrow Torus frontalis which slopes steeply down to the horn bases, but extends considerably backwards beyond the intercornual line. As a consequence thereof, the horn cores are placed at some distance in front of the posterior frontal edge. Duerst (1931), however, has pointed out that the reverse of this statement of Rütimeyer's is true, the situation and direction of the horn cores being the primary condition and the shape of the torus consequential. The cores are attached to the head without necks, so that the forehead appears as though constricted by the cores (Rütimeyer, 1861). (This generalisation is incorrect; many brachyceros cattle have moderately developed frontal necks, whose presence or absence depends solely on the moment of rotation of the horns—see also p. 320). The cores grow first outwards, then forwards and upwards, and are slightly twisted around their axes, so that the tips of the horns may be directed upwards, backwards or forwards. The cores are short, thick and conical, smooth, without edges, and devoid of bony tubercles at the bases. The vascular part, furnished with short wide vascular openings but usually lacking in longitudinal furrows, is distinctly marked off

198. (left) Skull of Celtic Shorthorn bull, Longhour Moor, Ireland

199. (right) Skull of Celtic Shorthorn cow, Walthamstow Crannog, Essex

from the forehead. The ratio of the horizontal to the vertical diameter of the cross-section at the base ranges from 7:6 to 1:1. The oval-shaped occiput, considerably exceeded in height by the Torus frontalis, forms an acute angle with the forehead. The height of the occiput above the Foramen magnum equals the distance between the horn bases. The ratio of smallest to greatest frontal width varies between 1:1·48 and 1:1·70. The Condyli and Processus occipitalis are less oblique than in longhorn skulls. The temporal fossa is short and rather shallow, its superior edge strongly curved and pressed down by the low-set horns. The jugal extends backwards, the temporal fossa behind it being wide open. The parietal point nearly touches the sphenoid wing. The maxillary part of the splanchnocranium is long, and the oral part short and marked by several cavities. There is a large triangular gap at the anterior point of the frontals, and a smaller one at the anterior point of the lacrimals. The maxillae and the short slender premaxillae barely touch the nasals, which are narrow, with parallel edges, slightly vaulted, deeply indented at the anterior part, and from 22 to 34 per cent shorter than the forehead. The maxillae rapidly come to a point in front of the molar series; the length of the latter amounts to 29–31 per cent of total cranial length, being about equal to the anterior toothless portion. The mandible is slender, the posterior ramus nearly vertical, the anterior low and ascending only slightly. The toothless part and the symphysis are short; the incisive portion is narrow and slender. The length of the inferior molar series exceeds one-third of the total mandibular length, being longer than either of the toothless mandibular sections situated in front and behind; the anterior toothless section is slightly longer than the posterior. In a later work (1867) Rütimeyer described the teeth of Bos taurus brachyceros as follows:—The dentition is very close. The upper molars are nearly square, in later stages often broader than long. The dentinal pillars in both the upper and lower cheek-teeth are strongly developed, cylindrical or extended laterally. The valleys are horseshoe-shaped, with weak plications; all accessory parts, lateral plications and supporting pillars are also weak. The lower molars are short, strong and thick, with well developed pillars. The incisor set is narrow and compressed. The body of the tooth is remarkable in shape and direction; the upper and lower molars are not quite square, and oddly arranged, the upper molars forming backwardly deflected, and the lower molars forwardly deflected, squares. The position of the teeth in the jaw bones conforms with this situation, the upper molars projecting backwards, the lower molars obliquely forwards.

There are two outstanding differences between the brachyceros and the primigenius skull. One concerns the small total size of the former, to which Klatt (1913) attributed most of the differences between the two types; the other the small size of its horns. The effect of the small horns on neurocranial conformation in Bos taurus brachyceros has been stressed particularly by Duerst (1926a; 1931). Rütimeyer himself paid tribute to the importance of the short horns by calling this taurine type 'brachyceros' in preference to Owen's term Bos longifrons.

Short horns being the principal characteristic of the brachyceros cattle, the term 'shorthorn' (brachyceros) is retained for the humpless native cattle of Africa classed with this type.

Additional Cranial Types

The third type of domesticated cattle, i.e. Bos taurus frontosus (Fig. 200), was established by Nilsson (1849). Rütimeyer (1861) originally accepted this type with the name Bos (taurus) trochoceros, but later (1867) discarded it because he considered it merely as a variation of Bos taurus primigenius. Dawkins (1867) also disputed Nilsson's identification, since a complete series of intergrading forms could be arranged between B. t. frontosus and the small B. t. brachyceros (longifrons); he remarked that all basic types founded on the different development of the frontal sinuses should be rejected as this depended on sex, age and environmental effects (Reynolds, 1939). Antonius (1922) attributed the 'frontosus' character in part to the influence of shorthorn cattle on the longhorn type. The difference in size between the remains of B. t. frontosus and B. t. brachyceros is about the same as that between B. t. primigenius and B. t. frontosus (Howard, 1962).

To these basic types Wilckens (1878) added Bos taurus brachycephalus, characterised by considerably larger size than B. t. brachyceros and by a particularly short splanchnocranium. Wilckens' type specimen originated from deposits of the Laibach (Ljubljana) moor. The cranial type is common, along with the brachyceros type, in cattle from the Roman period of Europe (Revilliod, 1926; Hescheler and Rueger, 1939; Whitehead, 1953). Apparently it was introduced by the Romans in the provinces of the Empire in an attempt to improve the native breeds.

Bos taurus brachycephalus as a special type has neither been accepted by Rütimeyer nor by subsequent authors, some regarding it as the result of crossbreeding of primigenius, frontosus and brachyceros cattle, as Antonius (1922) did with reference to its European representatives, and others ascribing it to a variation in hypophysial secretion (Adametz, 1923). Duerst (1931), on the other hand, whose examination of the hypophysis of brachycephalus cattle showed no aberration from the normal, attributed the bend of the skull axis in front of the orbits mainly to the moment of rotation of the horns. Since brachycephaly occurs among both longhorn (Niata, Devon) and shorthorn cattle (Jersey, Tuxer, Eringer) it cannot be regarded as a basic racial characteristic.

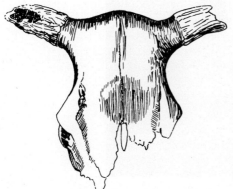

200. Holotype of Bos taurus frontosus Nilsson

Polled Cattle

Among all types of African cattle polled animals occur sporadically. In the following they are not discussed separately nor assigned to a racial group of their own, although Duerst (1931) stressed the high degree of similarity displayed by hornless crania of whatever original racial type. Polled cattle from predynastic or early dynastic times of Egypt are classed with the Hamitic type, commonly long-horned, occasionally gianthorned; those from the time of the New Kingdom with the shorthorn type. The hornless cattle encountered among recent longhorn cattle and shorthorn cattle and among various humped breeds are classed with their respective relatives. As in nearly every instance their origin is obvious, they do not require special consideration.

The Orthoceros Type

Stegmann von Pritzwald (1912) has added the orthoceros to the list of basic cattle types. According to C. Keller (1913), the orthoceros type, represented by the Kalmyk cattle of the steppes of the Caspian Depression, western Siberia and Turkmenia, is derived from a cross of west Siberian cattle with zebus. Stegmann von Pritzwald (1924) at one time regarded this type as the hybrid progeny of zebu, banteng and European cattle; subsequently, as descended from European and Bali cattle, a theory untenable

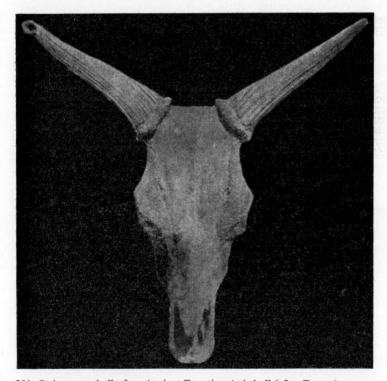

201. Orthoceros skull of an Ancient Egyptian Apis bull (after Duerst)

202. Orthoceros Kalmyk skull (after Kulagin)

for anatomical and geographical reasons (Hilzheimer, 1926). The humped cattle encountered among the orthoceros (Turano-Mongolian) breeds of the Kalmyk and Kirgiz are derived from relatively recent introductions of zebus from the south. Adametz (1925) has pointed out that the orthoceros type is merely a variation of the ordinary longhorn. The difference in cranial conformation between orthoceros cattle and other longhorn breeds is due to the effect of the peculiar upright direction of the horns on the skull. Crania of cattle, believed to be the ancestral type of the Kalmyk, from prehistoric Mongol graves in Transbaikalia, resemble the skulls of ancient Egyptian longhorn cattle far more closely than do recent crania of Kalmyk cattle (Pia, 1941). As orthoceros specimens occur among Apis (see pp. 223–226) skulls from ancient Egypt as well as among recent humped African cattle, they are dealt with in connection with the respective basic types to which they belong.

The Zebu Type

In addition to the humpless longhorn and shorthorn types, another basic type of domesticated cattle is generally accepted—the humped zebu (Bos indicus). The term zebu was introduced by Buffon who heard it at the Paris fair of 1752. In India, as Yule (1875) wrote, this name was unknown. It has been suggested that the term is connected with the Portuguese and Spanish giba (gebo) = hump or hunchback (Latin—gibbus or gibba = hunch or hump). But according to Webster's International Dictionary, 1919, p. 2368, it is probably derived from the Tibetan zeu, zeba = the hump of a zebu or camel.

The hump is the most important and persistent zebu feature, although it is not equally well developed in all zebu breeds, and not all cattle in which it occurs are true zebus.

The vast majority of zebus are distinguished by bifid spinous processes of the dorsal vertebrae from the seventh vertebra on. Again, from the eighth dorsal vertebra the spinous processes appear as though compressed antero-posteriorly in the upper third.

Their direction also is remarkable, the processes being bent backwards considerably more than in humpless cattle (Epstein, 1955b).

Zebu cattle have the same number of chromosomes, i.e. 2n = 60, as humpless cattle. The karyotype is identical except for the morphology of the Y chromosome; in humpless cattle this is a small metacentric, but in zebu cattle the Y chromosome is a small acrocentric (Hsu and Benirschke, 1968; Kieffer and Cartwright, 1968).

In size the zebus vary greatly; some breeds reach 180 cm at the withers, others measure less than 90 cm. In general, zebu cattle are distinguished by a narrow body, long legs, drooping rump and fine bone. According to Darwin (1868), they differ from other cattle in general conformation, the shape of their ears, the point where the dewlap commences, the typical curvature of the horns, the manner of carrying their heads when at rest, and in their ordinary variations of colour. Also, they are supposed to have different habits, and a voice different from that of other cattle; but this the present author, who has handled many Africander cattle, humped sangas, thousands of Iraqi, Iranian and south-east Asian zebus, and American Brahman cattle, cannot confirm: breed and individual differences exceed those of type. Zebus have a large, loose, often pigmented skin, and are able to regulate their body heat better than European cattle. In some breeds, as in the Kankrej zebu of Gujarat, the skin is thrown into numerous folds extending on to the face, and the ears are large and pendulous. Zebus are supposed to be more resistant to and tolerant of ectoparasites and to possess a considerable degree of resistance to disease (Du Toit, 1936). Owing to their hard feet and speedy movement, they are more suitable to take the place of horses than other cattle. However, the majority of these characteristics seem to depend on breed and environment rather than to apply to the zebu group in general.

The Zebu Skull

The typical zebu skull is remarkable for its lack of elevations and cavities. The forehead is commonly convex; this is more marked in bulls than in females in which the

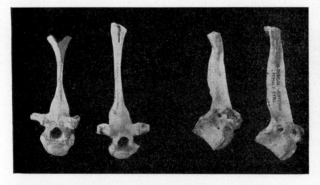

203. Frontal and lateral views of 9th dorsal vertebrae of a Bengal zebu bull and cow (after Curson)

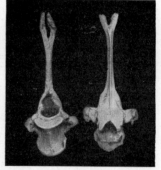

204. Ninth dorsal vertebra of an Indian zebu (after Gans)

forehead is often flat or dished. The highest point of a zebu bull's forehead usually lies immediately behind the eye sockets. In general, the zebu skull is distinguished by great length and narrowness, and also by the lack of prominence of the eye sockets which are situated more outwards and less forwards than in humpless longhorn cattle. The skull somewhat resembles that of a horse. Its narrowness is apparent especially in the interorbital distance; in relation to basilar skull length, this amounts to 45 per

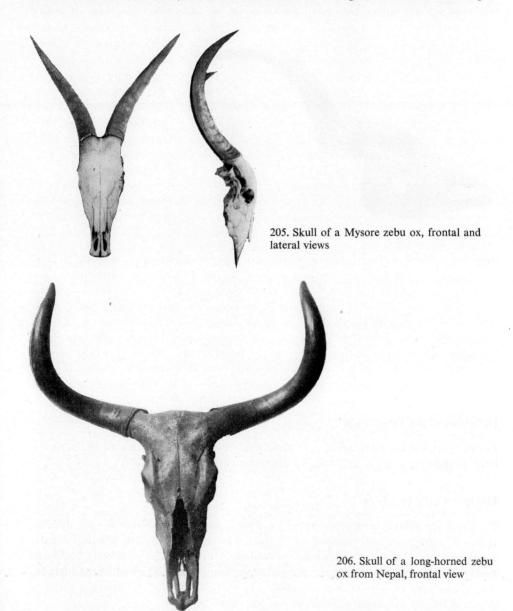

205. Skull of a Mysore zebu ox, frontal and lateral views

206. Skull of a long-horned zebu ox from Nepal, frontal view

cent in Bos indicus (zebu), 48 per cent in Bos taurus primigenius (longhorn), and 49.3 per cent in Bos taurus brachyceros (shorthorn) (Lydekker, 1912a). The frontals become narrower aborally (in this respect the zebu cranium resembles the longhorn skull); the angle formed by the occipital and frontal bones is commonly acute. The horns vary in size; in some breeds they are remarkably large, in others small or absent. They are generally slender, and often yellow in colour. Typically their direction is lateral or upright with a backward tendency, i.e. of auchenokeratos type; in profile

207. Skull of a long-horned zebu ox from Nepal, lateral view

they are set well behind the face. Rütimeyer (1861) has given the following cranial description:—"The zebu skull is characterised by the decided backward direction of the horns, the vaulting of the forehead in all directions and its marked contraction towards the back, the slight prominence of the orbits, the convexity and considerable upward extension of the nasals, the small height but great lateral extension of the parietals, at the expense of the frontals, in the occipital and lateral regions (where the parietals approach the sphenoid wings in the temporal fossae, almost crowding out the frontals), and by the length of the jugal process of the temporal which extends to the frontal process of the jugal."

Definition of the Term 'Zebu'

For our purpose the term 'zebu' may be defined as humped cattle devoid of characteristics suggesting a major influence of humpless cattle.

Humped Cattle in Africa

In Africa the zebu type is represented by a great variety of different forms. In addition, there are numerous humped cattle south of the Sahara desert, which, while including a strain of zebu, are not zebus proper. All these humped breeds, zebu and partly zebu, are classed under the general term 'humped cattle' which constitutes the third basic type of African native cattle. The classification of the humped cattle of Africa is set out in a separate section (pp. 327–339).

II. The Humpless Longhorn Cattle of Africa

i. Recent Humpless Longhorn Cattle in Africa

Humpless longhorn cattle in Africa are now restricted to three relatively small areas in the west. These are not contiguous, but form enclaves in the midst of the general breeding areas of humpless shorthorn cattle and of humped cattle.

The N'Dama

The most typical of the humpless longhorn breeds encountered in West Africa, and the morphologically nearest to Rütimeyer's Bos taurus primigenius, is the N'Dama which originates from the Fouta Djallon plateau in Guinea, and is now found also in Gambia, Sierra Leone, Liberia and western Mali. Into Ghana and Liberia it has been imported during recent years for experimental purposes. In addition to N'Dama, the names Gambia Longhorn, Futa, Futa Longhorn, Malinke and Mandingo are used for this breed (Mason, 1951a).

The N'Dama has a short and broad head, straight in profile, with prominent zygomatic arches and a broad muzzle. The horns vary in size. Typically they are about

208. N'Dama bull, Sierra Leone

209. N'Dama heifer, Liberia (after Johnston)

210. N'Dama cow (after Johnston)

60 cm long, curving upward and outward or lyre-shaped. Some animals have upward curving horns only 30 cm long, or still smaller downward and inward curving horns, or small buds only 2–5 cm long. Finally, loose-horned and polled individuals occur in the N'Dama breed, especially in Sierra Leone and Portuguese Guinea, but also in adjacent territories.

The compact body is set on short legs of fine bone. The neck is thick and deep; the back straight from withers to tail head, of good width and well fleshed. The hindquarters are fairly deep and well muscled. The dewlap and umbilical fold are poorly developed.

The typical coat colour is some shade of fawn with darker extremities and lighter underside. Most common are whole colours, including light to dark fawn, grey, dun, light red, chestnut, red with a black head. In dun animals a dark dorsal stripe, dark circles around the eyes, and a light-coloured ring around the black muzzle are common (Stewart, 1937; Ross, 1944). White may be found on the belly and the lower part of the tail. The skin pigment is either red or black. The horns are white with dark-coloured tips; the hooves red or black.

The N'Dama is distinguished by excellent beef conformation. The oxen make fairly good work animals; but the cows are poor milkers, averaging only about 450 kg milk per lactation. Mason (1951a) quotes the average live weight of N'Dama cattle from sources in Mali as 250–350 kg, and the average withers height as 95–120 cm. For adult N'Dama bulls in Nigeria Faulkner gives an average weight of 320–360 kg, and for adult cows 250–270 kg (Faulkner and Epstein, 1957). The animals reach full maturity at the age of 4–5 years. In Gambia the N'Dama is somewhat bigger than in Fouta Djallon, possibly owing to interbreeding with zebu cattle. In Guinea there are two distinct types, namely a larger plains type and a stockier hill type, both being pure N'Damas (Mason, 1951a).

One of the most valuable characteristics of the N'Dama is its high degree of natural non-specific resistance to trypanosomiasis. It is also markedly resistant or immune to tick-borne infections; but the breed is said to be less resistant to rinderpest than are

West African zebu cattle. Yet, during the outbreak of rinderpest in 1890–91, central Fouta Djallon was the only region in West Africa where cattle proved resistant. Johnston (1906) recorded that the N'Dama cattle seemed to stand the moist forest climate of the tropics fairly well, but had not become so completely acclimatised as the smaller parti-coloured short-horned (brachyceros) cattle which are the prevalent breed on the west coast of northern Africa. This is significant as it may indicate one of the reasons for the extensive replacement of the longhorn by the shorthorn type in North Africa in the course of the last three or four thousand years. However, opinions on the relative immunity of the N'Dama and west coast shorthorn are not unanimous; many veterinarians now believe that the N'Dama is superior in this respect. Payne (1964), for example, writes with regard to trypanosomiasis that the resistance of the West African shorthorn cattle is not as marked as that of the N'Dama cattle, possibly because less time has elapsed since they arrived in the forest environment, and perhaps because they have lived in isolated open forest glades and have therefore not been so continuously subjected to the attentions of the tsetse-fly. The resistance to trypanosomes, Mason (1951a) says, is to local strains; and tolerance diminishes if the cattle are removed from their own locality.

On the Origin of the N'Dama

Several authors have commented on the origin of the N'Dama cattle. It is most unlikely that their ancestors were introduced from Portugal into the Mandingo country by early Portuguese seafarers, for European cattle, imported in relatively small numbers, would hardly have survived in the unhealthy environment of West Africa, much less have given rise to a well established, hardy and disease-resistant breed. Occasional importations of Portuguese humpless longhorn cattle would not be discernible in the racial character of the N'Dama breed.

Adametz (1920) has described the N'Dama cattle he saw at the Paris World Exhibition in 1889 as entirely humpless, long-and-slender-horned, of small size, sand-coloured, and unquestionably of pure primigenius type, closely resembling the longhorn cattle represented in ancient Egyptian paintings. Pierre (1906) has classed the N'Dama with the group of humpless (taurine) cattle, adding that the horns are long and slender, while Mason (1951a) says:—"The fairly long aurochs-type horns suggest that it may have been derived from the Hamitic Longhorn..." Similarly, Johnston (1906) wrote that the N'Dama (Mandingo) ox "in some respects suggests a dwarf variety of Egyptian longhorned, straight-backed, uniform-coloured cattle", but added that it differed from the long-horn cattle of ancient Egypt in horn shape. While this is true of the smaller-horned N'Dama type, the illustrations in Johnston's standard work on Liberia show typical longhorned N'Damas with sickle- or lyre-shaped horns which resemble those of ancient Egyptian longhorn cattle. Regarding their origin, Johnston suggested that, although the cattle of the Mandingo fundamentally belong to the Egyptian type, they probably include a strain of Mauritanian cattle. The latter he believed to be descended from European (Iberian) animals of Bos taurus (i.e. Bos taurus brachyceros) stock, an assertion for which, however, he furnished no evidence.

Curson and Thornton (1936) agree with Johnston that the N'Dama (Mandingo) cattle may be a cross between longhorn and brachyceros stock. Johnston, they write, "believes the Mandingo to be a hybrid between Bos taurus (brachyceros) and Bos aegyptiacus which view appears very probable, the Hamitic characteristics predominating except possibly for size".

However, the small size of the N'Dama cattle, taken as the only criterion for their partial brachyceros descent, is not conclusive. Longhorn cattle of primigenius type do not retain a large frame under every environmental condition, just as brachyceros cattle are not always small in stature. The Brown Swiss, essentially of brachyceros stock, are large animals, while some of the longhorn cattle from the Balkans and Asia Minor are dwarfed (see Fig. 370). Since many N'Dama cattle to all outward appearances except size suggest the longhorn type, their small size may not necessarily be due to interbreeding with dwarfed shorthorn cattle in West Africa, but to the unfavourable environmental conditions in this region where also brachyceros cattle and other domesticated animals have become dwarfed.

Horn length in the N'Dama seems to be a more significant criterion in the study of their descent than body size. In this respect the N'Dama displays considerable variability, short-horned animals being encountered along with the typical long-horned. The smaller horns point to cross-breeding of the original long-horned type with humpless shorthorn cattle; some may be due to decrease in horn size in herds where selection for long horns is no longer practised.

The Boenca and N'Gabou

The N'Dama is found also in Portuguese Guinea where it is called Boenca. The Boenca has horns of medium size, or may be loose-horned or polled. The common coat colour is golden-yellow. In addition to the N'Dama or Boenca, Tendeiro (1954) records the N'Gabou. The latter is white with a black muzzle and occasionally with black spots at the lower extremities. Tendeiro suggests that the N'Gabou arose from the crossing of pure N'Dama and Niger or Senegal zebu (Fulani) cattle; but the moderate horn size and the absence of even a vestige of a hump contradict this assumption.

The Jakore (Djakoré)

In the provinces of Sérères, Cayor and N'Diambour, at the Gulf of Saloum, Senegal, humpless N'Dama longhorn cattle have been crossed with Senegal Fulani (Gobra) cattle, the crossbred progeny being called Jakore (Djakoré). A similar crossbred type is becoming increasingly popular in Gambia (Mason, 1951a). Pierre (1906) pointed out that the Jakore retained all zootechnical characters of the Gobra parent stock to the third or fourth generation; but, rather inconsistently, he asserted that "the (Gobra) hump has disappeared", whereas Curson and Thornton (1936) refer to the cervico-thoracic hump of the Jakore. Doutressoulle (1947) says that the hump is barely evident, and that the Jakore is not a fixed breed. From the N'Dama parent it has inherited the lightness of skeleton, and from the Gobra side its height and width. It has particularly well developed hindquarters, and produces a very good carcase, while the cows are fair milkers. The head is long and slightly heavy, with slender horns of variable size. The neck is short and fairly well muscled; its upper border is straight or slightly concave, its lower border is furnished with a thin dewlap extending from the lower jaw to slightly behind the forelegs. The prepuce and umbilical fold are little developed. The chest shows good development in depth and width; the heart girth is ampler than in the N'Dama. The hump, broad, low and scarcely prominent, is situated slightly in front of the withers, being separated from the latter by a slight indentation. In some

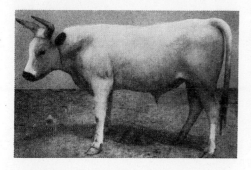

211. (left) N'Gabou bull, Portuguese Guinea (after Da Costa)

212. (right) Polled N'Gabou cow, Portuguese Guinea (after Da Costa)

213. (left) Loose-horned N'Gabou bull, Portuguese Guinea (after Da Costa)

214. Jakore ox (after Pierre)

215. Bambara bull (after Pierre)   216. Bambara cow

animals the hump has completely disappeared, but commonly it is more or less conspicuous. The back is straight, the loins are long, fairly wide and exceptionally thick. The rump is broad and well muscled, often a little short and sloping. The thighs are markedly developed and extend well down to the gaskins. The tail is well attached. The limbs are slender and fine of bone. The hide is thin and flexible (Doutressoulle, 1947).

The Bambara

The Bambara or Méré, of the southern and western parts of Mali, between the habitats of the Sudanese Fulani and the N'Dama, is a crossbred but well fixed variety, derived from long-horned Fulani and N'Dama parent stocks, similar in origin to the Jakore. It is fairly uniform in conformation, but varies in body size, the largest animals, up to 140 cm in withers height, being found in the north, and the smallest, standing only 100 cm at the withers, in the south. Cattle of the larger type weigh approximately 250–300 kg (Pierre, 1906; Doutressoulle, 1947). The head of the Bambara is short and generally straight in profile, with a broad, flat forehead which slightly recedes towards the occiput. The horns of the bull are about 25–30 cm long, with a basal circumference of 20–25 cm. In light-coloured Bambara cattle the horns are white in colour with black tips; in cattle of darker coat colours they are dark-coloured throughout. The neck is short and rather thick. The bull has a small muscular hump in front of the withers, indicating the humped Fulani cross and the change in hump situation from the thoracic in the Fulani to the cervico-thoracic in the crossbred type. Pierre (1906), noting that the withers are occasionally elevated and very heavy, apparently referred to the muscular hump of the Bambara bull. The dewlap is short and little developed, the chest is wide, the back generally straight, and the rump wide but slightly drooping. The loins are broad, the thighs full and well rounded, the legs rather short. In the northern part of their habitat Bambara cattle are commonly yellow or reddish brown in colour with darker legs; in the south darker shades are encountered. In light-coloured animals the mucosa is unpigmented, in animals of darker coat colours it is frequently slate-grey. Bambara cattle fatten readily and produce a good carcase; the cows are fair milkers (Doutressoulle, 1947; 1948).

The Namji

A dwarfed breed of longhorn cattle, numbering not more than one thousand head, has survived in a small isolated enclave in northern Cameroun, between the Fula settlement of Ngaundéré in the highland of Adamawa and the trading centre of Garoua on the Benoué river (Weidholz, 1939). This breed is owned by the Namji and Pape (Dupa), true African cultivators whose principal domestic animals are the goat, hen and pariah dog. The cattle are kept solely for sacral purpose; their hides are used to cover the corpses of the dead. They live practically wild in the hills, and the cows are never milked so that their udders and teats remain very small. Although small in body size—their withers height rarely exceeds 105 cm—these cattle are well proportioned. The back is long and straight, with slightly raised withers; but there is

217. Namji bull (after Weidholz)

218. Namji cow (after Weidholz)

no indication of a hump. The primigenius-like head, with a square, flat forehead and nearly straight intercornual line, is furnished with fairly long horns of medium thickness. The prevailing coat colours are reddish yellow, dark brown or black, occasionally with a lighter dorsal line; variegated animals are also common. The fact that Namji cattle show a high degree of immunity to trypanosomiasis, in contrast with the humped breeds in the surroundings of their range, indicates that the Namji has been kept in its present environment for a long time. Its classification with the longhorn (primigenius) group is confirmed by Antonius (1943–44) in the following words:—"In West Africa pure remnants of the ancient primigenius cattle have sporadically survived, as the large-horned cattle of the Buduma on the islands of Lake Chad and ... the degenerated Namji cattle of northern Cameroon." The photograph of a Namji bull and cow illustrating this passage is legended:—"Dwarfed primigenius type without zebu mixture."

A similar humpless longhorn breed was formerly bred by the Koma of the Alantika Mountains of northern Cameroun. It was wiped out by rinderpest, believed to have been introduced by Bororo (Abore) cattle nomads of Fula stock (Weidholz, 1939).

The Kuri

The peoples clustering round Lake Chad in the Bornu Province of Nigeria and in Chad keep a breed of longhorn cattle known as Kuri or Buduma. These cattle have become used to spending a considerable part of the day immersed in water with only the nostrils lifted above the surface. They are excellent swimmers and follow their herdsmen through the water as they travel from island to island in search of grass. While well adapted to the peculiar environmental conditions of their habitat, they fail to thrive outside the lake area.

The Kuri has a disproportionately large head, characterised by a convex forehead and great interorbital width. The horns are attached to large pedestal-like necks covered with skin and hair. They are 70–130 cm long, projecting outwards, upwards and slightly backwards (the backward sweep causes the convexity of the forehead), the tips sometimes curving inwards. In thickness of horn Kuri cattle are of two distinct types: one with long horns of moderate, normal, proportionate width and a basal circumference of 20–30 cm—this is the typical longhorn; the other with long bulbous horns, conical in shape, of large basal circumference that may exceed 60 cm. A pair of these horns, in the British Museum, measures 105 cm in length along the curve, with a basal girth of nearly 60 cm. These are exceeded by another pair in which the length along the curve is 117 cm and the maximum circumference 82 cm. This is a highly specialised form of horn, undoubtedly brought about by long deliberate selection, combined with isolation of the Kuri from other types of cattle. The horns are of porous material, fibrous, and not much thicker at the base than a human finger nail, so that they are very light, the pair in the British Museum weighing just over 4 lbs (H. Smith, 1827). Occasionally the thick bulbous horns are not longer than 20–30 cm, so that their basal width nearly equals their length. Loose horns are also encountered, and some Kuri cattle are polled (Doutressoulle, 1947).

The Kuri is a large-framed coarse-boned animal, one of the largest in West Africa. Mason (1951a) states:—"They are commonly 130–150 cm high at withers and may reach 180 cm. Their weight is 300–800 kg with an average of 480 kg." Faulkner gives the average weights, recorded at Maiduguri Livestock Centre, Nigeria, as 500 kg for bulls and 360 kg for cows; the average withers height of bulls 152 cm and of cows 134 cm (Faulkner and Epstein, 1957). The body is long and shallow, lacking in width, with raised withers. The dewlap is poorly developed and free from folds. Likewise the

219. Buduma cattle (after Adolf Friedrich, Duke of Mecklenburg)

220. Kuri bull, Northern Nigeria (after Faulkner and Epstein)

221. Giant-horned Kuri from Lake Chad (after Mason)

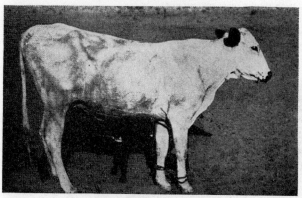

222. Polled Kuri heifer (after Doutressoulle)

223. Kuri bull, lateral view

224. Young Kuri, front view (after Faulkner and Epstein)

prepuce and umbilical fold are never marked. Typically Kuri cattle are humpless, those at any rate which are isolated from zebu stock on the islands of Lake Chad. The sacrum is high and the rump has a marked slope. The hindquarters are poorly developed, the thighs narrowing sharply as they descend to the gaskins. The legs are long with large spreading hooves. The gait of the Kuri when walking has appropriately been described as shambling.

The Kuri is generally white in colour, often with red or reddish brown or black splashes or speckled markings. Doutressoulle (1947) regards these as an indication of impure breeding. The muzzle and inside of the ears are black. The skin is also black.

The Kuri is not suitable for draught and not used for this purpose. Kuri cattle are also poor in beef qualities; nevertheless they form a high proportion of the slaughter cattle imported from the Chad region to the northern and central markets of Nigeria. The milk potentialities on the other hand are considerable. Records from the Maiduguri Livestock Centre give the average production over a number of years as 1260 kg per lactation, with a record of 2440 kg in 314 days (Faulkner and Epstein, 1957).

The Jot Koram and Other Kuri Crossbreds

At greater distances from Lake Chad the Kuri is found crossed in an increasing degree with short-horned zebu or long-horned Fulani cattle of the Rahaji (Red Bororo) breed. The crossbreds are distinguishable from the purebred Kuri by the presence of a small cervico-thoracic hump, less bulbous horns and a coloured coat. Crossbreds of Kuri

225. Jot Koram bull

226. Jot Koram cow (after Faulkner and Epstein)

and short-horned zebu parentage are encountered in Bornu Province, Nigeria, and in Kanem, Chad, where the cross is with the Shuwa zebu. Ross (1947) suggests that the Kilara, mentioned by Morton (1943) as a colour variant of the Wadara (=Shuwa) from northern Clad, is in faet one of these crossbred types. Another crossbred of Kuri and short-horned zebu or long-horned Fulani descent is the Jot Koram or Jotko, named after the tribe that owns this breed. The Jot Koram is smaller than the purebred Kuri, standing only 127 cm at the withers, and has longer and less bulbous horns (Mason, 1951a). The milk potentialities of the crossbreds are similar to those of purebred Kuri cows. Faulkner mentions a Jot Koram cow at Maiduguri, Nigeria, which gave 3152 kg milk in 356 days (Faulkner and Epstein, 1957).

ii. Ancient African Longhorn Cattle

Humpless longhorn cattle were once ubiquitous among cattle breeders throughout northern and eastern Africa. Early records in the Nile valley, and rock engravings in Ethiopia, the Sahara and the Eastern and Western deserts of Egypt indicate that this bovine type was bred in Africa for several thousand years before humpless shorthorn and humped cattle made their first appearance.

Badarian Cattle

In the Nile valley the earliest records of domesticated cattle are traceable to the neolithic encampments in Lower Egypt and the Badarian farming villages in Upper Egypt, whose cultures are derived from the same root. The Fayum encampments were situated on the edge of a lake, now dry, in the Fayum depression west of the Nile, dated by the radiocarbon method to 4440–4150 B.C. (Childe, 1958d). The Badarian is named after the village of Badari which is situated on the east bank of the Nile, 20 miles south of Asyut. Badarian objects have been found not only in Upper and Middle Egypt and down almost to the Delta, but also in the Northern Provinces of the Sudan, 400 miles west of the Nile in the latitude of Wadi Halfa (Frankfort, 1951). Petrie (1939) referred the Badarian age to the middle of the 8th millennium B.C.; but his dating of this, like that of the subsequent prehistoric periods of Egypt, is generally considered much too early, the Badarian period probably not being older than the second half of the 5th millennium. According to McBurney (1960), little more than a thousand years elapsed between the arrival of the first primitive hunter peasants in Lower Egypt (Fayum and Merimde) and the establishment of the Ist Dynasty.

The most usual amulet during the Badarian period was the bull's head, front face, with the horns curving downwards. It continued in use, conventionalised, till the late Gerzean age, and in a very rude form till the Ist Dynasty. The bull's head amulets are of bone, ivory, carnelian, sard, slate, black steatite, alabaster, and green noble, grey or brown serpentine (Petrie, 1917). Brunton and Caton-Thompson (1928) found Badarian cattle and other domestic animals wrapped in matting and linen. The Badarian animal burials resemble the much later burials of dogs, oxen, goats and sheep found in the course of the archaeological survey of Nubia. During the Badarian age the vegetation in Egypt was more luxuriant than now, and Badarian cattle found pasture in what is now desert. Remains of rough stone walls on the high desert near the great wadi may be parts of fences or cattle enclosures. Therefore, Reed's (1960) sweeping assertion that "Brunton's assumption of cattle domestication for the Badarian people must be completely discounted" is unfounded.

Amratian Cattle

During the early Amratian age of Egypt (called Naqada I or Early or First Predynastic in the older literature)—named after El Amrah—bovine figures occur on white-line pottery and painted vases. The Amratian graves at El Amrah yielded also several clay figurines of cattle, some of which were mounted four abreast on a single base and one of which showed a remarkably large udder (Randall-MacIver and Mace, 1902). But wild beasts, such as the hippopotamus, ibex, arui (Barbary sheep), chevrotain, giraffe and hartebeest(?), continued to be more frequently depicted than domesticated animals. A radiocarbon determination for this period gives about 3700 B.C., a date which is slightly later than the Arpachiya site in Mesopotamia (Zeuner, 1963a).

Gerzean Cattle

There are a few records of cattle from the early Gerzean phase (also called Naqada II or Middle Predynastic), in many ways a continuation of Amratian, but with the addition of new elements that point to fairly close relations with the East, with Sinai and with Palestine (Frankfort, 1951). The records include a bull's head amulet and several pairs of horns of slate used as magic pendants (Petrie, 1939). Although in possession of domesticated animals, the early Gerzeans do not seem to have had a pastoral culture; for their burnt offerings are entirely vegetable, containing not a single fragment of bone.

During the late Gerzean age figures incised or painted on pottery become more varied; but those of wild beasts and birds still predominate. An interesting specimen is a snuff horn of black pottery, closed at the wide end and shaped as an ox head at the tip—the prototype of the modern snuff horn of the Basuto. On one of the archaic statues of the god Min at Koptos, belonging to the end of the Gerzean period or to the Ist Dynasty, a hyena is depicted chasing a longhorn calf, both standing on hillocks.

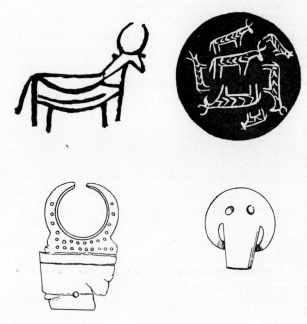

227. (left) Vase painting of bovine figure, early Amratian (after Petrie)

228. (right) Herd of cattle on early Amratian white-line pottery (after Petrie)

229. (left) Pair of bovine horns on magic slate, early Gerzean (after Petrie)

230. (right) Bull's head Amulet of ivory, early Gerzean (after Petrie)

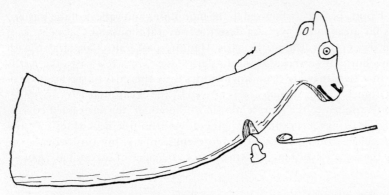

231. Snuff horn of black pottery shaped as ox head at tip, late Gerzean (after Petrie)

232. Hyena chasing a longhorn calf. From the Aunu figures of Min at Koptos (after Petrie)

In all these predynastic records of Egyptian cattle the horns are usually curving forwards, sometimes downwards, rarely upwards, and are never wide-spread. "The early type seems to have the incurving horn on a level, and this was somewhat varied both down and upward. The later type with wide-spread horns is unknown in the carvings" (Petrie, 1917).

Hamitic Longhorn Cattle in the Deserts of North Africa

No cattle are found in the rock drawings of the desert dating from pre-Hamitic times—the times of the early hunters; Hamites and cattle make their appearance simultaneously. In the early period, when giraffes and elephants still roamed the country which is now desert, the Hamitic cattle breeders had to share their territory with invaders from the east, likewise breeders of stock. Contemporaneous with both, but only at a later period, lived the Gerzeans (Early Nile Valley Dwellers), well known from the excavations of predynastic (Gerzean) sites in the Nile valley (Winkler, 1938–39). Cattle with long wide-spread horns are depicted by early Hamitic pastoralists (Autochthonous Hamitic Mountain Dwellers)—in all probability the ancestors of the Blemyans of the

Graeco-Roman-Coptic period—who lived in the mountains and valleys of the Eastern desert as well as far away in the Western desert for several thousand years.

In the Libyan desert the cattle of the early Hamitic pastoralists are drawn with somewhat shorter horns than those of the Eastern desert; in a few cases the horns point forwards, in others they are bent downwards along the sides of the head. That these cattle were domesticated and not wild is proved by the occasional deformation of the horns, the care applied to the representation of the udder, the variegated coloration and, in some drawings, little strips hanging down from the neck (Figs. 233 and 234). It has been suggested that these strips represent amulets; but this belief is unfounded. Rather they seem to be flaps of skin cut from the neck of the animals, a prac-

233. Rock drawing of longhorn cows from the pasture oasis of Uweinat, Libyan Desert (after Winkler)

234. Longhorn bull. Rock drawing from Goll Ajuz, Nubian Desert (after Frobenius)

235. Domestic cattle and man with spear. Rock drawing by Autochthonous Hamitic Mountain Dwellers, Eastern Desert (after Winkler)

236. Ox and driver (in the mask of an ass). Rock drawing from Habeter III, Fezzan (after Frobenius)

tice that has survived among several South African Bantu peoples. The Ama-Xosa cut strips of hide from the neck and chest of their cattle, leaving one end attached so that the flaps form a fringe (Lichtenstein, 1811–12; Fritsch, 1872). Among the Bantu of South Africa this custom was still prevalent at the end of the 18th and the beginning of the 19th century (Barrow, 1801–04; Alberti, 1815). Livingstone (1857) observed it among the Makololo. The Herero employ this practice to distinguish ownership of the cattle (Von François, 1896).

On the Tassili-N-Ajjer plateau in the southern part of the Algerian desert, there are numerous rock engravings and polychrome paintings of domesticated cattle from the 'Bovidian Period'. The cattle are represented either with long crescent horns or with rather thick horns projecting forwards (Fig. 237). The latter are referred by Lhote (1959) to the brachyceros type; actually they are longhorns, albeit distinguished by a peculiar horn direction which, however, is not unique nor restricted to Africa (see also p. 241). The Bovidian engravings are believed to be contemporaneous with the predynastic period of Egypt (before 3000 B.C.). But Cole (1966) contends that in spite of the presence of pottery and grindstones in the early Neolithic of the Sahara, which occurred during a period of increased rainfall, there is no certain evidence of food production until the time of the cattle paintings associated with the later Saharan

Neolithic, towards the end of the 3rd millennium B.C. The occasional presence of Nile boats in Bovidian drawings, of a type seen in ancient Upper Egyptian rock engravings, suggests that the Bovidian people had been in contact with Upper Egypt; indeed, it is probable that they originated from there. They were of Hamitic race, akin to, or identical with, the Autochthonous Hamitic Mountain Dwellers of the Western desert of Egypt.

Longhorn Cattle in Ancient West Africa

The earliest rock paintings of cattle in West Africa invariably depict the humpless longhorn type; for example, the latter is represented in an early painting in a rock shelter at Birnin Kudu in Northern Nigeria. In a cave at Geji, on the Bauchi plateau of Nigeria, are paintings of humpless longhorn cattle, in addition to small short-horned cattle without humps. The true African peoples of West Africa did not possess cattle until the latter half of the last millennium B.C. when they acquired these from Hamitic invaders or immigrants (Kennedy, 1958).

237. Polychrome painting of cattle from the "Bovidian Period" of Upper Jabbaren, Tassili Plateau (4th or 3rd millennium B.C.)

Humpless Longhorn Cattle in Ancient Ethiopia and the Horn of Africa

In Somalia and the Diredawa area of Ethiopia the earliest rock paintings of pastoral scenes depict cattle with long horns and no humps, for the most part carefully drawn in a naturalistic style. There is a close connection between these paintings and the neolithic paintings and engravings at Jebel Uweinat in the Libyan desert and at other places in the Sahara. In a rock shelter on the Gilf Kebir, a high sandstone plateau in the desert, 400 miles west of the Nile and 100 miles north-east of Jebel Uweinat, there are paintings of cattle with moderately long horns and large udders, believed to be the work of a pastoral people living about the date of the Egyptian predynastic period (Shaw, 1936). They are closely paralleled by rock paintings at Ein Dawa near Jebel Uweinat and those near Harar in Ethiopia, where the same prominence is given to udders and horns. But whereas sometimes two, sometimes four legs are represented in the Sahara group, only two are ever shown in the paintings from the Horn of Africa.

Ancient Cave Paintings of Humpless Longhorn Cattle on Mount Elgon

Mount Elgon on the Kenya–Uganda border is the southernmost point at which the former presence of humpless cattle with either long or gigantic horns is recorded (Fig. 240). Some of the longhorn cattle in the cave paintings of Mount Elgon show the opisthonomus curvature of the horns that is characteristic of many longhorn cattle in early rock drawings and paintings in the Sahara. The paintings on Mount Elgon, which have been tentatively dated to the 1st millennium B.C., are probably the work of Hamitic pastoralists who, some time before 1000 B.C., began to fan out from southern Ethiopia to the Azanian Coast (see p. 545), Kenya and Uganda, where they occupied elevated locations with substantial precipitation (Murdock, 1959).

Clark (1954) believes that the earliest painters of longhorn cattle in the Horn, to whom the camel, the horse and the zebu were still unknown, were of a similar race and practised a similar pastoral economy to those of the Hamitic painters and the later engravers in the Sahara, whose culture combines Egyptian elements of various dates. The Ethiopian paintings may be in time prehistoric, and datable to the 1st or 2nd millennium B.C. or earlier. "The apparent absence of Bos brachyceros from the Sourré paintings suggests that they may date prior to the Middle Kingdom in Egypt (2200 to 1780 B.C.) when short-horned cattle were introduced; this does not take into account

238. Cave painting of humpless Longhorn bull and cow from Birnin Kudu, Nigeria (after Kennedy)

239. Rock painting of herdsmen and Humpless Longhorn cattle at Genda-Biftou, Sourré, in Harar, Ethiopia (after Breuil)

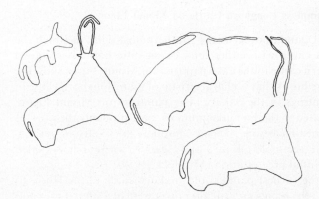

240. Outline of humpless longhorn (with opisthonomus curvature of the horns) and giant-horned cattle from cave paintings on Mount Elgon, Kenya (1st millennium B.C.) (after Wright)

any appreciable time-lag, however, before this breed was introduced into Ethiopia." Breuil (1934) dated the neolithic art groups of Capsian tradition in the Sahara to between the 5th or 4th and the 2nd or 1st millennium B.C., while considering the Sourré (= Sureh) paintings from Genda–Biftou in the Harar, which represent the earliest paintings of pastoral scenes in this region, as appreciably earlier than those at Jebel Uweinat. From this he inferred that the animals represented in the Sahara had been domesticated on the Ethiopian plateau, whence they had spread to the Nile valley and beyond. But this is refuted by the most various evidence on the domestication, source and diffusion of humpless longhorn cattle as well as by the gradient of decreasing skill in their representation as we pass south and eastward across the Sahara from the vicinity of the Maghreb (McBurney, 1960). As Clark (1954) writes:—"... the existing evidence seems to indicate that the spread (of the 'Neolithic of Capsian tradition') was from north-west to south-east."

The conventionalised and schematic paintings and engravings of pastoral scenes in Somalia and Ethiopia, which include the camel and zebu, possibly also the horse, in addition to humpless longhorn cattle, are of a much later date. They cannot be earlier than the 4th century A.D. and are probably considerably later (Clark, 1954). In these engravings the humpless cattle are commonly represented with wide-spread laterally projecting horns of gigantic size (Fig. 241).

Longhorn Cattle in Dynastic Egypt

In Egypt cattle with lyre- or sickle-shaped horns were common during the period of the dynastic conquest. Most of the colour patterns and markings known in recent domesticated cattle were found already in this ancient breed. In some animals the colour was uniformly red or black, or yellow and brown with black and dark red spots, occasionally dark with the topline and underline white. Others were spotted or brindled, or had the various colours distributed over the body in large zones.

The Egyptian longhorn cattle were large-framed beasts, their shoulder height reaching approximately 145 cm, and the length of body 170 cm. These measurements, calculated by Duerst (1899) from the anterior cranial length, correspond with the measurements of cattle in ancient Egyptian paintings. The head was short and broad; the eyes were large and prominent, the ears of moderate size, but the long hair at the tips made them appear longer than they were. The horns were light-coloured with dark tips, sometimes dark throughout. The crown was covered with a tuft of hair. The nostrils were large, and the muzzle was broad. The neck was rather short, with a moderate dewlap. The withers were high, but not as high as in the wild urus. There was no sign of a hump. The chest was deep, and in some animals of such dimensions as to recall some of the modern beef breeds. The shoulder bones were broad and well placed. Behind the withers the back was straight; in some beasts the loins were somewhat higher than the back, and the hindquarters drooping. The ribs were rather flat, judging from the numerous Apis statuettes unearthed, and the thighs rather thin, except in fattened oxen. The legs were short, and the hooves well shaped. The horns were of similar length in both sexes, a characteristic feature of many recent European longhorn breeds. Some animals were polled. Prince Me-henwet-Re (c. 2100 B.C.) kept 835 long-horned and 220 polled cattle on his estate (Boston, 1963). An early example of a hornless variant of the longhorn breed is represented by a relief of a fattened ox from the tomb of Achti-Hoteb, late Vth Dynasty (c. 2400 B.C.). Zeuner (1963a) writes that "from the sculptures and hieroglyphs of Egypt four different breeds of domesticated cattle can be deduced": a primigenius breed with small horns, a primigenius breed with lyre horns, a primigenius breed with double-lyre horns and hornless cattle. As the horned cattle are all referred to the primigenius cranial type, it seems unreasonable to speak of different breeds, instead of different horn types within a single breed.

The skeleton of an Egyptian longhorn (Cairo Museum) does not show specific differences from the skeletons of recent European longhorn breeds. It is notable that the spinous processes of the anterior dorsal vertebrae are not bifid (Fig. 244).

241. Humpless giant-horned cattle in conventionalised style from rock shelters in the Horn of Africa (after Clark)

242. Egyptian longhorn cattle

243. Fattened Egyptian longhorn oxen, Vth Dynasty

Apis Bulls

Several crania of Egyptian longhorn cattle are preserved in European and Egyptian museums. Among the more typical skulls is that of an Apis bull from Giza, in Berlin.

The custom of burying bulls and other animals in special graveyards developed since about 1000 B.C. from the animal cult practised throughout the whole period of ancient Egypt. The bull, in which the god Ptah was believed to be incarnated, was called Apis, probably the name of a bull-god (Barton, 1934). The earliest reference to an Apis bull in the reign of Khasekemui, last king of the IInd Dynasty (c. 2800 B.C.), is found on the Palermo Stele, while Manetho mentions an Apis bull in the reign of Kaiechôs, second king of the IInd Dynasty of Thinis. Apis was a black beast with white spots of special shapes and other body marks, in addition to a coloured spot on its tongue. The birth places of Apis bulls were dispersed throughout Egypt; but they were interred in a common underground rock tomb at Memphis (Saqqara), which was built at the time of Rameses II (1290–1224 B.C.). The last Apis bull was buried in the Serapeum (from the compound Ser-apis) at Memphis in late Ptolemaic time, but the Apis cult continued well into Roman times and the Christian era.

The Cranium of the Egyptian Longhorn

The Apis skull from Giza is distinguished by shortness and marked width, more especially of the forehead, and by a nearly flat frontal area. The length of the forehead exceeds 50 per cent of the total cranial length. The distance between the temporal fossae is shorter than the lateral length of the forehead between the opisthocranium and the frontal (Duerst, 1926b), and the width of the forehead exceeds its length. The orbital arches are only a little higher than the temporal ridges of the parietals, which is a characteristic feature in the African longhorn cattle as well as in recent European longhorn breeds, distinguishing these from animals influenced by shorthorn (brachyceros) or zebu blood, in which the difference in height between the orbital arches and the parieto-temporal ridges is considerable. The horns are slender and lyre-shaped, 62 cm long, with a basal circumference of 24 cm, and a core girth of 22 cm. The base of

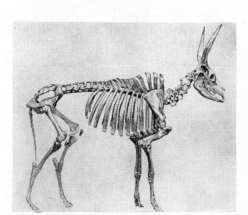

244. (left) Skeleton of an Egyptian longhorn
245. (right) Polled bulls or oxen from Ancient Egypt

the horn core is nearly circular. It has the appearance of a pearled wreath which considerably exceeds the neck of the core in height. The intercornual line, seen from the front, is nearly straight; but from the back of the skull the Torus frontalis, formed mainly by the parietals and supraoccipitals, is more conspicuous. There is a lap-like extension of the parietals into the forehead, noted because on this peculiarity several

246. Statue of an Apis bull from the Serapeum at Memphis

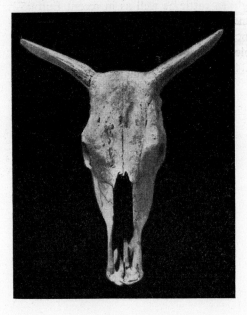

247. (left) Skull of a longhorn cow from Abadieh, Egypt (IVth Dynasty)

248. (right) Painted longhorn skull from the grave of a predynastic Egyptian chieftain

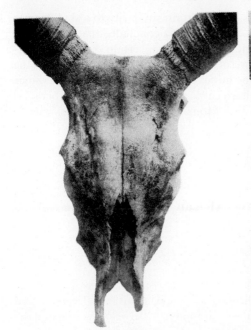
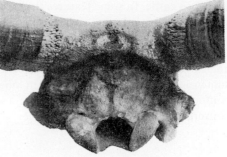

249. (left) Cranium of Apis bull from Giza in Berlin. Frontal surface (after Duerst)
250. (right) Cranium of Apis bull from Giza in Berlin. Nuchal surface (after Duerst)

251. Egyptian longhorn cow and calves in limestone. From the tomb of Ti at Saqqara. Vth Dynasty (c. 2500–2350 B.C.)

authors have based the theory of the racial alliance of several European longhorn breeds (Spain, Britain) to the Egyptian longhorn cattle, and of their common descent from the extinct North African urus.

Among 19 crania of ancient Egyptian (Apis) bulls dating from late predynastic to Roman times, which were examined by Pia (1941), only one showed a combination of primigenius and brachyceros characteristics. The remaining skulls were very similar to crania of Hungarian Steppe cattle, and only slightly different from skulls of Andalusian longhorn cattle. In some features there was also a certain similarity to a skull from a prehistoric Mongol grave in Transbaikalia. From this Pia concluded that the ancient Egyptian longhorn cattle were of primigenius type, and that certain cranial peculiarities might be due to a very early admixture of an Asiatic ox, similar in type to the ancient Mongolian breed. This suggestion, however, seems very far-fetched. Since similar horn direction (orthoceros) in longhorn cattle produces a similar neurocranial conformation, the resemblance to an ancient Mongolian skull is doubtless due to convergence.

iii. Origin and Descent of the African Longhorn Cattle

1. The Classification of the Extinct Urus Group

All domesticated true cattle throughout the world are of a single species.

There is nearly general agreement that the longhorn type of cattle has been evolved from the urus, a view based on the close anatomical similarity between the wild and the domesticated stock. The range of the urus formerly extended from the west coast of the Pacific through Asia and Europe to the eastern coastlands of the Atlantic Ocean, and from the northern tundras southwards into India and North Africa. As the urus is either very rare or absent in the lower Pleistocene of Europe, becoming abundant only after the end of the Ice Age, an Asiatic origin of the species is probable (Zeuner, 1963a). The animal has been extinct for more than 300 years; it died out first in Africa, then in Asia, finally in Europe where the last cow lived until 1627.

A species ranging over so vast an area, from the pleistocene epoch to the recent past, could be expected to show considerable variability. Yet the urus group is sub-divided into only three local races: the European Bos primigenius primigenius Bojanus, the Asiatic Bos primigenius namadicus Falconer et Cautley, and the North African Bos primigenius opisthonomus Pomel. These races are held to be distinguishable by differences in horn shape and body size. But in fact this division denotes little more than geographical range; for within each race variability in size and horn direction is not less than between the three geographical races.

Bos p. namadicus is believed to have been smaller than Bos p. primigenius. Yet in Europe a number of primigenius skulls have been found far below the normal size. Again, Bos p. opisthonomus is claimed to have been distinguished from Bos p. namadicus and Bos p. primigenius by its horns pointing forwards. But horns of opisthonomus shape are found also in the Spanish and French representatives of Bos p. primigenius, and in the British Isles large down-curved horns were comparatively common in the urus during the Great Interglacial (Zeuner, 1963a). On the other hand, a Bos p. opisthonomus skull from the Fayum has horns of a lateral direction, similar to those in a Bos p. namadicus skull from Mongolia, and a European Bos p. primigenius skull in the Paris Museum.

2. The European Urus

The European urus was a large animal, the male standing six feet (180 cm) at the shoulder, with high withers and large horns of variable shape. Skeletal remains indicate that the cow was smaller and more graceful than the bull (La Baume, 1959). In adult bulls the short coat was black in colour with a whitish stripe along the spine, white curly hair between the horns, and white around the muzzle; cows and calves were dark brown. The horns were yellowish white with black tips. The male was characterised by a strong musk (civet) smell (Duerst, 1931).

In palaeolithic wall paintings in French and Spanish caves, urus cows are frequently distinguished from bulls by smaller size, lighter build, a relatively longer and slenderer head, smaller horns, and red or reddish brown colour (La Baume, 1959). Windels (1949) and others mistook the sexual dimorphism of the urus, expressed in this manner, for specific differences, referring the paintings of urus bulls to Bos primigenius primigenius, and those of urus cows to Bos longifrons (= brachyceros), i.e. a wild bovid that did not exist (see pp. 312–317; also Koby, 1954).

The Bos Primigenius Primigenius Skull

Even in a large collection of urus skulls from a restricted geographical area hardly two specimens will be found to be alike in all particulars. They differ in size according to age and sex. The horn cores differ in length, thickness and direction (Gromova, 1931); in either sex they may be oval or nearly circular in cross-section at the base (Nobis, 1954). Horn necks may be present or absent. The Torus frontalis may or may not be conspicuous (Von Lengerken, 1953).

In spite of considerable variability in detail, the general morphological character of the Bos primigenius primigenius skull is fairly uniform: long and narrow, even in outline, with a flat forehead, only slightly prominent orbits, and deep temporal fossae. The horns usually curve first outwards and forwards, then upwards and inwards with the largest span just below the tips. In skulls with normally shaped horns the intercornual ridge is straight; deviations in the shape of the horns influence the intercornual ridge, rendering it more or less convex or concave. The dimensions of the horn cores in a number of specimens from the brick-earth of Ilford, now in the British Museum, are as follows:—Length along the outer curve 31–38 inches (77.5–95 cm), basal circumference 15–19 inches (37.5–47.5 cm), tip-to-tip distance 25–40 inches (62.5–100 cm) (Lydekker, 1898). In the largest Ilford skull the tip-to-tip distance is 42 inches (105 cm). On the basis of a core length of 100 cm, Duerst (1931) estimated the former horn length in this specimen at 145 cm. The horn core of the largest known Bos primigenius primigenius skull, from Monte Mario, Rome, has a basal circumference of 50.2 cm and a diameter of 17 cm. The large weight of the horns pulls the occipital crest outwards; in one of the Ilford skulls it is 10 cm wide. Commonly the horns are attached to the skull without Processus cornu ossis frontalis; only in crania of aged cows horn necks are occasionally encountered. In old bulls the bases of the horn cores and the adjoining parts of the frontals are frequently surrounded by a wreath of

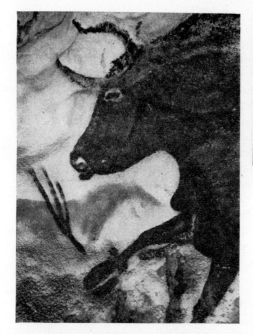

252. Head of an urus bull and urus cow from the Lascaux cave, France (late Palaeolithic—c. 13000 B.C.?)

253. (left) Bos primigenius primigenius Bojanus (after Hamilton Smith)
254. (right) Bos primigenius primigenius Bojanus (after Herberstain)

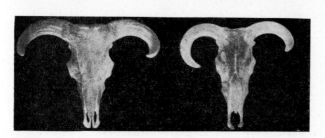

255. Crania of urus bull and cow (after Von Leithner)

horn pearls. The nasals are broad, markedly vaulted, and their lateral edges parallel. The premaxillae are strong, yet relatively slender. Normally their nasal branches touch the nasals; but occasionally they end a short distance below these. The mandible is very slender; the oblique position of the posterior mandibular ramus is due to the great length of the skull (Antonius, 1922).

Duerst (1931) estimated the weight of the head with the horns in the urus at 48 kg. In order to carry this enormous weight, the cervical muscles had to be very strongly developed and the neural spines to be erect, a position responsible for the great height of the withers. In the course of domestication the weight of the horns and of the cranial bones supporting these gradually decreased, the neural spines shortened and sloped backwards, and the withers became lower in consequence. Duerst regarded the subsequent effect of artificial selection on the slope and deflection of the spine in the region of the shoulder, with the purpose of levelling the withers, as one of the principal results of the domestication process, attributing the remaining differences between the domestic longhorn cattle and the urus to the normal reactions of the body to environmental factors encountered in domestication.

Dwarfed Bos Primigenius Primigenius Skulls

In central Europe Bos primigenius primigenius skulls have been found, which are about one-third less than the normal size, but do not differ from ordinary skulls in other respects. Hilzheimer (1926), pointing out that the majority of these skulls belonged not to young stock but to adult animals, considered it as possible that dwarfing in Bos p. primigenius was due to malnutrition; such animals might have been born either in autumn or in times of drought when cold weather or lack of feed during early calfhood left them stunted; or dwarfing might have occurred in the course of the first attempts at domestication. However, as the European subspecies of Bos primigenius was probably never domesticated independently of already existing domesticated cattle from western Asia, the last mentioned hypothesis is not acceptable. Von Leithner (1927), stressing the great sexual dimorphism in body size of the urus, referred the small crania to cows. Duerst (1931) believed the stunted specimens to have been the progeny of wild bulls and domesticated or half-bred cows, the size following the dams'. While this may apply to some dwarfed primigenius skulls, it is doubtful if the factor of occasional malnutrition of young wild stock can be excluded altogether.

On the Domestication of Bos Primigenius Primigenius

Nehring (1888e) and several subsequent authors have expressed the belief that Europe was one of the principal domestication centres of the urus, although in many prehistoric sites in Europe the earliest domesticated cattle are short-horned. In other neolithic stations longhorn cattle are found either exclusively or in association with shorthorn cattle (p. 191). On the coasts of the North Sea and Baltic the introduction of large cattle of primigenius (longhorn) type antedates the appearance of brachyceros cattle. Von Lengerken (1953) regards this as evidence of the domestication of the urus in northern Europe. However, it is most unlikely that cattle were domesticated in this culturally peripheral region. For in West Asia and North Africa they occur still earlier, and the chronological sequence of their appearance at intervening stations indicates the course of their migration routes from Asia and Africa to northern and north-western Europe. Duerst (1931) held that the domestication of the urus must have been such a

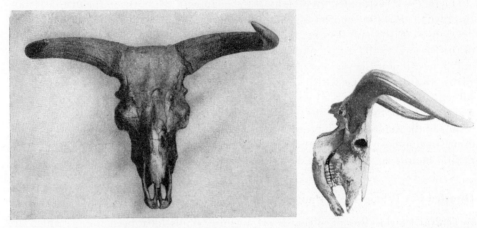

256. Skull of Bos p. primigenius Boj. from the Pleistocene of Ilford, Essex, frontal and lateral views

257. Prehistoric wall painting of Bos primigenius primigenius Boj. from Cogul, eastern Spain (after Obermaier)

258. Prehistoric wall painting of Bos primigenius primigenius Boj. and other animals from Alpera, eastern Spain (after Obermaier)

difficult task for primitive man that it was probably accomplished only in a single locality in Asia. Subsequently the domesticated cattle were brought to other regions where they interbred with the local races of the urus, receiving new features in colour and conformation. As Professor Duerst wrote to the author (22.3.1935):—"A completely purebred ox originating solely from Bos p. primigenius Boj. and having preserved its pure character through thousands of years does not exist. But the crossing of the domesticated descendants of Bos p. namadicus with the European urus, Bos p. primigenius Boj., or its African relative, is quite possible and must have occurred frequently."

3. The Role of the African Urus in the Origin of African and European Longhorn Cattle

Bos Primigenius Opisthonomus Pomel

In the quest for the wild ancestors of the Egyptian longhorn cattle we have first to deal with the theory that these are to be sought in the Nile valley. This view has been expounded by Adametz (1920) and others on the basis of a bovine skull fragment found by Markgraf in the Fayum in 1910, and described by Hilzheimer (1917). The fragment comprises the upper portion of the forehead with both horn cores, of which the left is almost intact and the right broken off in the middle, in addition to part of the occiput situated between the horn cores. The pleistocene (diluvial) stratum from which the piece was recovered renders it certain that the animal to which it belonged was not domesticated. The flat even forehead as well as the shape and direction of the horn cores preclude the possibility that the fragment might have belonged to a buffalo.

The frontal ridge of the fragment shows a remarkable curve. Seen from above, it runs forward from the direction of the cores, forming a round projection in the centre. The horn cores are turned slightly backwards. The tips appear to have been more slender than in the European urus, and to have lacked the upward curvature. A deep cavity below the high occipital crest indicates that the sinews at the back of the neck were strongly developed. The intercornual ridge is 27 cm (10.6 inches) long, the distance between the horns, taken at the lower edge of the cores, measures 36 cm (14.2 inches), and at the centre of the forehead 29 cm (11.4 inches). The distance from the tip of the left horn core to the centre of the Torus frontalis is 58 cm (22.8 inches), the height of the latter 6 cm (2.4 inches). The left horn core has a length of 85 cm (33.5 inches). Since its tip is broken off, the original length must have been about 89 cm (35 inches). Its basal circumference measures 38.5 cm (15.1 inches), that of the right horn core 36 cm (14.2 inches). The vertical diameter of the left core is 13.5 cm (5.3 inches), the horizontal diameter 10 cm (4 inches). The original distance between the tips of the cores is estimated at 72 cm (28.3 inches).

Hilzheimer (1917) named the wild ox of Egypt Bos primigenius hahni nova subspecies. But it is very doubtful whether the slightly different curvature of the horns, as compared with that in the urus of the Atlas countries, justifies the choice of a special term. Duerst (1900), as above mentioned, formerly identified the Egyptian urus as Bos

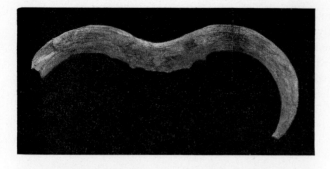

259. Cranial fragment of the Egyptian urus (after Hilzheimer)

260. Another view of the cranial fragment of the Egyptian urus (after Duerst)

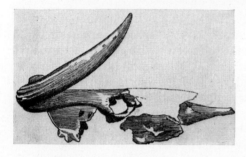

261. Cranial fragment of Bos primigenius opisthonomus Pomel (after Duerst)

262. (left) Rock engraving of male and female Bos primigenius opisthonomus from Thyout, Algeria (after Pomel)

263. (right) Male and female Bos primigenius primigenius with opisthonomus horn curvature, from the Cave de la Mairie (Teyjat), Dordogne, France (late Magdalenian)

(primigenius) opisthonomus, and there seems to be no necessity for subdividing the opisthonomus group.

The North African race of the urus, described by Pomel (1894) as Bos opisthonomus, and by Thomas (1881) as Bos primigenius mauritanicus, was nearly related to the European wild race, but its forehead was somewhat shorter, the horn cores curved less forwards and more downwards, and the limbs were relatively longer and slenderer (Lydekker, 1898). Herodotus, referring to Bos p. opisthonomus in his description of the wild cattle of the country of the Garamantes, the ancient inhabitants of Fezzan, stated that these animals were similar to domesticated cattle, but that they had thicker hides and were black in colour, while their horns were directed forwards and downwards so that the animals had to walk backwards when grazing as otherwise their horns would get stuck in the ground.

The most southerly finds of Bos primigenius opisthonomus occur at Tihodaïne in the Ahaggar massif of the central Sahara in late Acheulean (lower palaeolithic) times (McBurney, 1960).

The Urus in Egypt

Among the Sebilian cattle from the upper Palaeolithic of Kom Ombo in Egypt, Gaillard (1934) identified Bos brachyceros in addition to Bos primigenius (opisthonomus) —the former from a frontal with a slightly wavy outline of the crest. But since the Bovidae of Kom Ombo were certainly wild, and an irregularly shaped intercornual line is characteristic of many female urus skulls, this classification must be regarded as erroneous.

Among the records of the urus in Egypt there are two slate palettes used for eyelid paint: one from the time of the dynastic conquest, and a larger one of green slate, found at Hierakonpolis, which is dated to Nar-mer (Mena), first king of the Ist Dynasty. On an ebony tablet of Aha–Tetu, the successor of Mena, there is a scene depicting the netting of wild cattle. The shape and direction of the bull's horns closely resemble those of the Bos primigenius opisthonomus skull from the Fayum, while the high withers, repeated in several other representations of the urus from the periods of the dynastic conquest and the Ist Dynasty, recall the urus figures in the prehistoric wall paintings of eastern Spain.

C. Keller (1902) regarded the Bovidae on the slate palettes as domesticated bantengs, a theory without foundation, as the range of the banteng has never extended as far west as Egypt. Stegmann von Pritzwald (1924), mistaking the high withers for humps, considered these bovines as a kind of zebu, although zebu cattle were quite unknown in Egypt at that early period. Antonius (1922; 1933b), again, was inclined to regard them as buffaloes, basing his opinion on the shape of the horns and the absence of a dewlap. From a zoo-geographical point of view this belief is unobjectionable. But the absence of a dewlap does not disprove that the animals on the slate palettes represent the urus. Neither Herberstain's nor Hamilton Smith's reproduction of Bos p. primigenius Boj., nor the rock engravings and paintings in European and African sites, indicate that the urus had a conspicuous dewlap. As regards horn shape, Hilzheimer (1926) denied its similarity to that in the buffalo; were the African buffalo represented, the horn bases should be much larger and the horns joined in the centre. If, on the other hand, the artists meant to depict the Asiatic buffalo, they omitted to carve the transverse wrinkles typical of the arnee's horns. Moreover, the horns on the slate palettes give the impression of being round rather than flat at the bases. Hilzheimer therefore arrived at the conclusion that the animal represented on the ancient palettes was the urus.

In the tomb of King Sahurā (about 2700 B.C.), one of the most renowned rulers of the Old Memphis Kingdom, wall paintings show lions, hyenas, antelopes and a few

CATTLE

264. (left) Protodynastic slate palette used for eyelid paint (c. 3200 B.C.)
265. (right) Green slate palette of King Nar-mer (Mena), Hierakonpolis (Ist Dynasty)

266. Netting of wild cattle. Scene from the ebony tablet of Aha-Tetu, second king of the Ist Dynasty

267. Last Egyptian record of a wild bull hunt. Relief in the temple of Rameses III (1195–1164 B.C.) at Medinet Habu

wild oxen. The latter are reddish brown in colour with the top- and underline white, similar to the majority of the wild cattle depicted in the rock tombs at Beni Hasan (about 2000 B.C.). Some of these, however, are completely red, as are the bulls portrayed in hunting scenes in the tomb of Amenemhat and the tombs and mastabas at Saqqara. One of the hunted bulls in a wall painting from Beni Hasan has the following coloration: upper part of head black, face and neck light brown, chest and belly white; back, ribs, flanks and legs light brown with black spots, horns and hooves yellow (Duerst, 1899). In wall paintings from the 2nd and the end of the 3rd millennia the hunted oxen are frequently depicted with lyre-shaped horns, similar to those characteristic of the domesticated longhorn cattle of the Nile country. Such horns are seen in a bull hunting scene from the time of Rameses III (1195–1164 B.C.).

It is doubtful whether these animals may be regarded as representations of Bos p. opisthonomus. They probably depict feral stock or the progeny of wild bulls and feral cows strayed from the oases of the west across the ever expanding desert into the fertile plains of the Nile Delta. Rameses III's urus hunt (Fig. 267) took place, not in Egypt, but during one of the pharaoh's campaigns in Asia, as explained by the text referring to the relief (Boettger, 1958). In Egypt proper Bos p. opisthonomus became extinct probably in the 14th century B.C. Its latest occurrence in the Nile valley is recorded on a commemorative scarab of Amen-hotep III (1398–1361 B.C.). In the Atlas countries the animal may have survived longer. In Libya it is attested by Herodotus during the 5th century B.C. But it seems to have become extinct also in the Maghreb (the Atlas massif and its littoral plain, together with the steppe fringe of the desert south of the mountains from the Atlantic to the Gulf of Gabes) long before the Roman conquest, owing to the encroachment of the desert on the fertile savannas, the drying up of the springs and river beds, and the intrusion of domesticated cattle into its grazing grounds.

Theory of the Descent of African and Iberian Longhorn Cattle from Bos Primigenius Opisthonomus

Zeuner (1963a) holds that the occurrence of indigenous wild cattle in ancient Egypt explains the appearance of a pure and primitive primigenius breed there, from which the later long-and short-horned breeds are descended; but he does not produce any evidence in support of this theory. Adametz (1920), who likewise believed that the African longhorn cattle were domesticated from Bos p. opisthonomus in the Nile valley, based his view on the round projection in front of the Torus frontalis observed in the urus skull from the Fayum (see p.231). He considered this projection, which is caused by the extension of the parietals and supraoccipitals into the frontal region, as typical of the Egyptian longhorn cattle, but absent in the European and Asiatic uri and in all domesticated breeds derived from the latter. Conversely, he regarded several longhorn breeds of western Europe, the skulls of which show a similar projection, as descended from the African urus. Whereas Ulmansky (1918), in his description of Andalusian longhorn skulls, considered the extension of the parietals into the forehead, observed in one specimen, as exceptional and accidental, Adametz (1920) regarded this feature as characteristic of the Andalusian breed. "I can prove," he wrote, "that the Andalusian breed of cattle agrees with Bos primigenius hahni Hilzheimer in the latter's most important characteristics (i.e. the parietal extension into the frontal region, and the presence of necks to which the horn cores are attached). In other words, the Andalusian breed, extending over the entire southern third of Spain, belongs to the original unchanged Hamitic type and is identical with the ancient Egyptian longhorn cattle."

This theory was first suggested by Duerst (1899) who wrote with regard to the triangular lap formed by the extension of the parietals and the Occipitale superius into the forehead:—"I consider this parietal lap as a feature especially typical of this form of longhorn cattle." But Duerst had already encountered the parietal triangle in the upper frontal region not only in longhorn crania, but also in skulls of brachyceros cattle and of Asiatic zebus, as well as of the banteng.

Among European longhorn breeds, the extension of the parietals into the frontal region is not restricted to the Andalusian. It occurs also in several Portuguese longhorn breeds, such as the Barroso, Minho and Alentejo. Duerst (1899) found it also in the cranium of an ox of the giant-horned Franqueiro breed of Brazil, descended from Iberian longhorn stock. Representatives of these breeds frequently share with the ancient Egyptian longhorn cattle not only the parietal lap, but also several other cranial features, such as the presence of powerful horn necks, a wreath of horn pearls at the base of the horn cores, the porous substance of the latter, elevation of the frontal suture to a ridge, central depression in the nasal region, short nasal branches of the premaxillae, and right angle formed by the two mandibular rami.

British Longhorn Cattle

Adametz (1920) found the parietal extension into the forehead and the occurrence of long horn necks also in several British breeds of cattle (Devon, Highland, Welsh Black). From this he concluded that during the neolithic age Iberian tribes introduced longhorn cattle into Britain, where they are now encountered chiefly in the extreme

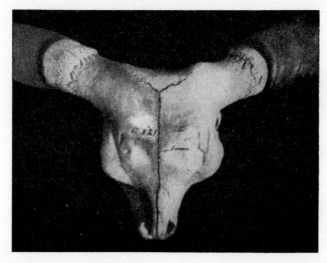

268. Skull of an Andalusian ox, fronto-parietal view (after Adametz)

west and south-west of England and in the mountains of north-west Scotland. During the neolithic and early bronze ages longhorn cattle were the only type found in Britain. They were larger than the cattle of late bronze age and iron age sites, and on the whole bigger than the cattle of medieval Britain; indeed, they were as large as Roman cattle and middle-sized modern breeds (Jewell, 1963).

In support of his argument that these longhorn cattle were introduced from the Iberian peninsula, Adametz emphasised the close racial relationship, evidenced by archaeology and physical anthropology, between the Spanish people and the Hamites of Africa on the one hand, and between the Spaniards and the prehistoric inhabitants of Britain on the other. This relationship has recently been confirmed by research into blood group frequencies which indicate that iron age and Anglo-Saxon immigrants into south and east England partly displaced to the north and west earlier inhabitants who were descendants of neolithic immigrants from North Africa and the Iberian peninsula (Carter, 1962).

The Parietal Extension into the Forehead of Longhorn Cattle

The main pillar on which Adametz's theory of the descent of the Iberian and British longhorn breeds from the longhorn cattle of Africa rests is the parietal extension into the frontal region of the skull. In reality this pillar is of no value for racial distinction and classification, and the theory of the descent of the African longhorn cattle from the Egyptian urus, based on this character, has no foundation (Epstein, 1958).

While preparing a number of longhorn skulls of Grey Steppe cattle from the Balkans, which had never been considered as of African origin, the author observed the parietal extension into the frontal area in several of these. Drawing Professor Duerst's attention to this fact, he asked for his present opinion on the value of this feature in the racial separation of African longhorn cattle and their supposed descendants from the longhorn (primigenius) breeds of central and eastern Europe. Professor Duerst was kind enough to give a detailed explanation of the subject (letters of 22.3.1935 and 15.5.1936).

Duerst's Evaluation of the Racial Significance of the Parietal Extension

In Bos p. primigenius Boj. (European urus), Bos p. namadicus Falc. et Caut. (Asiatic urus) and Bos p. opisthonomus Pomel (African urus) the triangular parietal lap is extremely rare, although it occasionally occurs. In these rare instances the intercornual ridge of the skull is not straight, but ascends to a high central torus. From this phenomenon Professor Duerst deduced that the projection of the parietal lap into the forehead in large long-horned skulls of Bos p. primigenius Boj. bulls was due to the direction of the horns and the resulting moment of rotation. The more clearly lyre-shaped the horns are, the more frequently the parietal wedge is visible in front. But lyre-shaped horns are rare in the urus.

As regards domesticated longhorn cattle, Professor Duerst stated:—"1) The stronger the lateral pull of the horn weight, the smaller the parietal lap becomes, until it disappears completely. 2) Conversely, the weaker the lateral pull, as in some brachyceros or polled cattle, the more the parietals share in the formation of the forehead. Therefore, I come to the conclusion that breed and race have no influence on this feature, as I still believed in 1899, but only the direction and weight of the horns, in other words, their moment of rotation; in as much as the intercornual ridge, not yet existing in calfhood, develops subsequently either more orally or more caudally, where it is formed by the growing bone as a mechanical result of static factors. The triangular lap is therefore no distinguishing genotypical factor, but one of phenotype, due to environmental influences."

"In Bos p. opisthonomus Pomel this parietal lap should really not appear at all; but the horns of this animal sometimes take such a downward direction that this again lessens the moment of rotation, thus increasing the parietal share in the intercornual crest."

Other Peculiarities in Longhorn Skulls

The presence of a frontal ridge, caused by the elevation of the frontal suture, occurs in individual instances in cattle as a consequence of muscular influences induced by the direction of the horns (Duerst, 1931).

The pearled wreath at the core base is found only in skulls of adult animals, where it is produced by the different growth rates in the length and width of the horn core. Particularly during sexual activity the damming up of the increased blood flow in the region of the horn base, and the consequent oversupply to the periosteum, cause the formation of horn pearls (Duerst, 1926a). Since this occurs in many different bovine types, it cannot be considered as a racial characteristic.

The characteristic conformation of the forehead in the African, Iberian and several British longhorn cattle, the central depression of the nasals, the shortness of the nasal branches of the premaxillaries, the broad temporals, and the peculiar angulation of the horizontal and vertical mandibular rami are all due, directly or indirectly, to the size and direction of the horns (hypsikeratos type), i.e. to their moment of rotation (Duerst, 1926a). Since these features occur in all cattle in which the static pressure of the horns on the skull resembles that in the ancient Egyptian longhorn, no specific value may be attributed to them.

The Processus Cornu Ossis Frontalis in Longhorn Skulls

Nilsson (1849) and Rütimeyer (1861), followed by Adametz (1920), regarded the frontal neck to which the epiphysis of the horn core is attached as of specific significance, indicating the racial genealogy of different types of domesticated cattle. But as this feature occurs among nearly all breeds of cattle, regardless of race and type, several authors have expressed doubts on the value of this character in the classification of cattle. Duerst (1926a) has pointed out that the variations in the length of the Processus cornu ossis frontalis are not due to racial peculiarities, but are purely mechanical. In his standard work on the horn of the Cavicornia he has explained that the length of the horn neck depends on the moment of rotation of the horns. The neck is elongated in skulls:

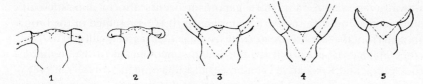

269. Outlines of various horn shapes combined with a long processus cornu ossis frontalis (after Duerst)

1) with large horns of a great effective moment of rotation, if such horns project ventrally at an angle of about 20° (negative angulation);
2) with small horns of a small moment of rotation, if such horns point sideways and forwards, their moment of rotation acting downwards at an angle of about 25° (negative angulation);
3) with large horns of a great effective moment of rotation, if such horns point upwards, forming an angle of 60° with the horizontal skull axis (positive angulation);
4) with very large horns of a great effective moment of rotation, if such horns point upwards at an angle of 60° with the horizontal skull axis (positive angulation);
5) with small horns of a small moment of rotation, if such horns point upwards at an angle of 50° (positive angulation).

Horn Gigantism in African Longhorn Cattle

In addition to the parietal lap, Adametz (1920) regarded the horn gigantism characteristic of many descendants of the African longhorn cattle as evidence of their descent from Bos primigenius opisthonomus. For the horns of the Fayum skull show that the African urus carried horns of gigantic dimensions.

Were megaloceraty restricted to Bos primigenius opisthonomus and African longhorn cattle, Adametz's theory might be valid. But this is not the case; Bos primigenius opisthonomus is not exceptional among uri in horn gigantism. Huge horns are found also in many pleistocene skulls of Bos primigenius primigenius and some of the Siwalik skulls of Bos primigenius namadicus, as well as in the Bos primigenius namadicus skull

270. Bronze statuette of a giant-horned bull from the late bronze or early iron age (Muraghian Culture) of Sassari, Sardinia (c. 800 B.C.) (after Zeuner)

from Süan-hwa-fu (see also p. 540). Horn gigantism occurs also in domestic cattle which have no connection with the African longhorn. It is represented in the bronze statuette of a bull from the late bronze or early iron age (Muraghian culture) of Sardinia, dated to c. 800 B.C. (Fig. 270). It is occasionally encountered in the commonly long-horned Kankrej and Kenwariya zebu breeds of India, and also in the kouprey of north Cambodia (Duerst, 1931). Therefore it is doubtful if the megaloceraty encountered among African cattle may be quoted as evidence of their descent from Bos primigenius opisthonomus.

Uncertainty regarding the Geographic Origin of West European Longhorn Cattle

None of the cranial features upon which Adametz based his theory of the descent of the African, Iberian and British longhorn cattle from the African urus is, therefore, of racial significance. It is possible that African longhorn cattle were at one time taken to southern Spain; we have no proof of this. Even if they had been, it would not necessarily follow that these were the ancestors of the Iberian long-horned breeds, and hence of the long-horned cattle of Britain, since, prior to the advent of the shorthorn, the longhorn type extended throughout the whole Mediterranean region.

Possible Influence of Bos Primigenius Opisthonomus on African Longhorn Cattle

But the weakness of Adametz's anatomical evidence does not preclude the possibility that Bos primigenius opisthonomus has influenced the racial character of the African longhorn cattle. Until the extinction of the urus in Africa, Asia and, finally, Europe, large numbers of cows and calves were caught in numerous localities to be incorporated into the existing domesticated herds. An analogy may be found in the domestication of four additional Bovini in Asia, i.e. gaur, banteng, yak and arnee. Cross-breeding between wild or tamed uri and domesticated or feral cattle doubtless occurred also in the Nile valley and the Atlas countries during the early period after the arrival of the first cattle breeders. Some of the domesticated cattle depicted in ancient rock drawings in Fezzan, Chad, Niger and Algeria (sites of Dao. Timni, Enneri Oudingueur, Gira Gira, Bardaï and Moritigui in the desert between Tassili-Oua-N-Ahaggar, Fezzan and

271. Group of domesticated cattle. Rock engraving at Habeter II, Fezzan (after Frobenius)

Tibesti—Huard, 1957) have the typical opisthonomus curvature of the horns (Figs. 237 and 271), a variant of the isopedokeratos type. Petrie's remark that in prehistoric records the horns of Egyptian cattle are usually curving forwards, sometimes downwards, only once upwards, and are never wide-spread is also significant. This remarkable curvature of the horns does not occur during the subsequent dynastic period. It may be due to interbreeding of domestic longhorn cattle, introduced by the earliest pastoral immigrants into Africa, and the African urus.

But as evidence of origin the occurrence of the opisthonomus type of horn in domesticated cattle is as unsafe as are the parietal lap, horn necks or gigantic horns. For horns of this peculiar curvature are not restricted to Bos primigenius opisthonomus; they are found in several palaeolithic paintings and engravings of the European urus in French, Spanish and Sicilian (Levanzo Island) caves (Fig. 263). In the British Isles large down-curved urus horns were comparatively common during the Great Interglacial (Zeuner, 1963a). Horns of opisthonomus type occur not only in prehistoric domestic longhorn cattle in North Africa, but also in ancient sculptures and drawings of longhorn cattle in the northern Caucasus (Maikop), and they are characteristic of the recent Mongolian cattle. There is therefore no valid evidence in support of Zeuner's (1963a) previously mentioned sweeping remark (see p. 235):—"That wild cattle were indigenous in Egypt explains the appearance of a pure and primitive primigenius breed there, from which the later long- and short-horned breeds are both descended."

4. *The Descent of the African Longhorn Cattle from the Asiatic Urus*

The question with regard to the origin of the earliest domesticated cattle of Africa is closely connected with the problem of the sources of the early neolithic cultures of Egypt and the origin of the dynastic people.

Origin of Neolithic Culture Traits of the Fayum and Merimde

In the neolithic Fayum encampments the evidence of food production is beyond dispute. Besides the remains of sheep or goats and of cattle in the debris, grains (emmer wheat, two-rowed and six-rowed barley) were found in considerable numbers in the interstices of large baskets dug into the ground to form storage silos. For the harvesting of the grain crop a primitive sickle was used, consisting of bifacial blades made from flint mounted in a groove in a straight wooden handle. This device offers an analogy to the most primitive sickles known, those of the Natufians of Palestine, who mounted flint blades in a grooved straight handle of bone. Also the burnished pottery of the Fayum culture has its nearest analogy in south-west Asia.

At Merimde beni Salama the presence of domestic animals, including sheep and goats, is highly probable. The general character of the equipment, particularly as regards sickles and pottery, is similar to that of the Fayum.

The analogy of the major cultural elements of the neolithic sites of the Fayum and Merimde in the early settled food-producing economy of south-west Asia suggests their derivation from the latter. For the comparable food-producing phases at Jericho, central Palestine, and in other sites of the Fertile Crescent antedate those of the Lower Nile valley. Moreover, while no food-producing culture had entered Lower Egypt prior to the period of the neolithic stations of the Fayum and Merimde (late 5th to early 4th millennium B.C.), the evolution of food production at Jericho can be traced, culturally and chronologically, to the initial Natufian layers, carbon-dated to the 8th or 9th millennium B.C. "It seems likely then that the earliest settled food-producing economy so far known belongs to this early date and was an indigenous invention of central Palestine" (McBurney, 1960).

Origin of Tasian Neolithic Culture Traits

The first phase in the succession of primitive farming cultures in Middle and Upper Egypt is termed Tasian after the site Deir Tasa, where a small semi-nomadic community possessed an economy that combined food-producing traits with a still basic element of food-gathering. Fishing was important and a barbed curved fish hook similar to the Natufian was used. Agriculture was based on the cultivation of barley and emmer, and domestic animals are believed to have included sheep or goats, but not yet cattle.

There is solid evidence of the source of the cultivated plants and domesticated animals of Deir Tasa in south-west Asia, where the wild prototype for the grain and livestock are best or exclusively attested (McBurney, 1960). Wild emmer has the most restricted distribution. It is found chiefly in Palestine and Syria, with a doubtful extension eastwards to the Zagros mountains. Wild barley occurs also in Syria and Palestine, in addition to Asia Minor, Transcaucasia, Iran, Turkmenia and northern Afghanistan, with an enclave in Tripolitania. Palestine–Syria is therefore the only area common to the two wild grasses which were first cultivated (Curwen and Hatt, 1953). The ancient Egyptian word for wheat (qm·hw), as that for vineyard (ka(r)mu), is of Semitic, more specifically Canaanite, etymology (Hitti, 1951).

As regards the two most important genera of domestic animals upon which the majority of early food-producing societies depended, namely sheep and goat, wild goats proper never occurred in Africa (see also vol. II, p. 221); the same probably holds true of sheep, since Ovis aries africana Pomel appears to have been not a pleistocene mouflon but a domesticated sheep bred by neolithic pastoralists (see vol. II, p. 6). Like wild emmer and barley, wild sheep and goats are native to south-west Asia; they still exist in the foothills of the Fertile Crescent, fringing the desert plains of south-west Asia.

Origin of Badarian Culture Traits

"The Badarians were not an isolated tribe, but were in contact with the cultures of countries on all sides of them" (Brunton and Caton–Thompson, 1929). Caton–Thompson and Gardener (1935) suggested that the ancestors of the Badarian had come from south of latitude 25°, i.e. from a region where the tabular flint available in the eocene cliffs from Upper Egypt to the Mediterranean does not occur; and Brunton

(1925), Junker (1828) and Scharff (1931) also insisted on Nubian analogies of the Badarian culture, claiming that its authors immigrated from the south. This does not necessarily contradict Petrie (1939) who considered Asia as the source of the Badarian people. For their migration route may have passed through Arabia and the Horn of Africa, and they may have been settled in Nubia before moving on to Middle Egypt. Badarian skulls are unlike the Abyssinian and Sardinian, the nearest in form to the Egyptian. They resemble the early Indian, but not the still earlier Dravidian or later Hindu. The physical type of the Badarians is related to that of the mesolithic Natufians of Mount Carmel as well as of skeletons from a chalcolithic (Ghassulian) cemetery at Byblos, excavated from beneath an Egyptian temple built before the end of the IInd Dynasty (Dunand, quoted by Albright, 1935). It is therefore to Asia, and not to Africa or Europe, that we should look for the cradle of the Badarian people. The use of corn by the Badarians, in the form of emmer wheat (Triticum dicoccum), points to the recognition of the corn god Osiris who brought Egypt out of savagery (Petrie, 1939). The Osiris myth is based on the Caucasus; "we must recognise that this land was the source of the Badarian people. Such source would well accord with some of them also migrating to India, after the primitive Dravidian and before the historic Hindu."

Amratian Invaders from Libya

The Amratians (Naqada I) of Middle Egypt form an exception among the prehistoric peoples of Egypt in that they did not come from the east, but seem to have been invaders from Libya. They may have obtained cattle from the Badarians or have brought their own stock along from north-east Libya whither a food-producing economy had by then been diffused from the Nile valley. At the cave of Haua Fteah, in the Derna district of Cyrenaica, there is evidence of food production (either domestic animals or agriculture or both) at a level carbon-dated to the second half of the 5th millennium B.C., while at Siwa oasis a bifacial sickle of early Fayum type and other elements in a neolithic context supply a valid link with the neolithic cultures of Egypt in the late 5th or early 4th millennium B.C. (McBurney, 1960).

Origin of the Gerzeans

In the course of the early Gerzean period an eastern folk with Magdalenean cultural affinities, living in rocky mountains, presumably in the Eastern desert, gradually entered and dominated Egypt. They were the first to introduce copper tools and pieces of glass into the Nile valley. Most likely they originated in Syria, and perhaps rather in the northern part, or possibly in the Red Sea mountains (Petrie, 1939).

Origin of the Dynastic People

The transition period from the Gerzean to the Protodynastic cannot be identified with an individual people or a distinctive culture. For some centuries large movements were going on, threats and influences from the south and east. An entirely new form of dagger appears during this period, "so closely like the later rapiers of Cyprus that it is certainly Asiatic". The slate palettes were influenced by the dynastic people coming in, with the falcon as their royal emblem. The dynastic invaders entered Egypt by the Quoceir-Koptos road, not far from the Fayum. They came from the land of Punt, up the littoral of the Red Sea. But the Horn of Africa was not their original home, although to the later Egyptians it was sacred as the source of their race. They probably originated in Elam whence they went down the Persian Gulf, entering Africa by way of Bab el Mandeb. On the ivory knife handle from Gebel el 'Araq (Fig. 178) two figures of dogs belong to the Babylonian myth of Etana on the flying eagle. The hero wears a thick coat and cap, and the lions have thick hair extending from the chest to the belly as a protection against the snow of a cold country. "It must be from mountainous Elam and not from the plains of Mesopotamia that the figures come," Petrie (1939) writes. But Brentjes (1962) points out that one of the two fleets represented on the knife handle consists of south Mesopotamian boats and the hero subduing the two lions is the south Mesopotamian priest-king. Any suggestion that the dynastic race derived from the earlier inhabitants of Egypt is refuted by the discovery in graves of the late predynastic period in the northern part of Upper Egypt of the anatomical remains of a people whose skulls were of greater size and whose bodies were larger than those of the natives (Derry, 1956; Emery, 1961).

Cultural Connections between Ancient Egypt and Mesopotamia

The connection between Asia and the predynastic and protodynastic cultures of Egypt is confirmed by the earliest documents bearing on the burials of the Egyptians, the spells in the Book of the Dead, in which the names of places closely conform to the geography of the Caucasus. At what point this mythology entered into Egyptian thought is not known. It cannot be before the use of corn in Egypt. It might however be connected with a later movement—Badarian or Gerzean, both Asiatic (Petrie, 1939). The evidence of Mesopotamian influence in pre- and protodynastic Egypt includes Mesopotamian objects found in the Nile valley, Mesopotamian usages temporarily adopted, and Mesopotamian objects, motifs and peculiarities of style represented on Egyptian monuments (Frankfort, 1951). Connections with Mesopotamia began on a small scale with the appearance of rare pottery types at the beginning of the Gerzean. Indications of intensified contacts with Mesopotamia occur in the later part of the Gerzean. The climax of connections was reached at the beginning of the Ist Dynasty with the introduction of niched brick architecture (Kantor, 1952). An invasion of Upper Egypt from southern Mesopotamia by sea around the Arabian peninsula at the beginning of the 3rd millennium B.C. is evidenced by rock drawings in Wadi-el-Hammamat, showing Mesopotamian boats with their crews (Brentjes, 1962).

Asiatic Origin of Early African Longhorn Cattle

It is to this early connection with Asia that we have to turn in our quest for the ancestors of the African longhorn cattle. It is significant that from their very first appearance in North Africa these cattle are encountered together with domesticated goats. There were no true wild goats in the restricted sense in Africa, from which the domesticated type could be descended, Capra nubiana and Capra walie being the only African representatives of the genus. The earliest domesticated goats came from Asia.

Domesticated cattle have been claimed to occur in Asia first in the Natufian of El Khiam and Mallahah near the Mediterranean coast (c. 8000 B.C.), later in neolithic strata of the Belt Cave at the Caspian foreshore and at Sialk I on the Iranian plateau (c. 6000 B.C.). They were possibly present at Amouq A, east of the Gulf of Iskenderun close to the Syrian–Turkish frontier (c. 6000–5500 B.C.). From the same period date the paintings on temple walls at Çatal Hüyük in southern Anatolia, which depict cult actions or games in which one person held a bull by the tongue while another one jumped over its back—a scene repeated at Knossos in Crete 4000 years later (Mellaart, 1966). The 'cattle-jumping ceremony' of the pastoral Banna of southern Ethiopia, in which young men, prior to marriage have to jump over a long row of up to 200–300 head of cattle placed alongside each other and held in line by head and tail (Jensen, 1959), may represent a reminiscence of this cult in recent times. Brentjes (1967) infers that domesticated cattle occurred in Anatolia as early as 7000 B.C.

Domestic cattle have been reported from Tepe Sabz at Deh Luran in Iranian Khuzestan, c. 5500 B.C. (Reed, 1969). At Qalat Jarmo remains of cattle have been found, but are extremely rare. Horn cores of bulls from the pre-pottery and pottery neolithic levels of Jericho are indistinguishable from urus horns; other skeletal evidence from bulls and cows does not indicate whether the animals were domesticated. But a crude and poorly preserved clay figurine from the pottery neolithic phase (5th millennium B.C.) may represent a domesticated bovid (Zeuner, 1963a).

Certain evidence of the occurrence of domesticated cattle is provided by bovine bones and teeth, considerably smaller than those of Bos p. namadicus and Bos p. primigenius, found in the Halafian site of Banahilk on the Diyala plain in northern Iraq (Braidwood and Howe, 1960); the Halaf period is dated to the 5th millennium B.C. (Frankfort, 1951). Bones and teeth of similar small cattle have been recovered from the same levels at Shanidar Cave, north-west of Banahilk (Perkins, 1960). This evidence is strengthened by a reproduction of a cow's head with forward curving horns excavated from a basal Halafian level at Arpachiya, just north of Hassunah, at the headwaters of the Tigris. The economy of Arpachiya was prosperous enough to afford the manufacture in quantity of luxury articles, so that it is very likely to have relied on a fully developed agricultural and stock-producing background (Zeuner, 1963a). Bulls were important in the emotional life of the Halafian people, as shown by their art and deduced from their religion (Mallowan and Rose, 1935; Reed, 1960). The presence of domestic cattle is established for the Ubaidian of Warka in Mesopotamia (early 4th millennium B.C.) and Anau II in Turkmenia (early 3rd millennium B.C.) (Reed, 1959; 1961).

Hence, in the Zagros hills and their grassy forelands cattle-breeding societies antedate similar cultures in the Nile valley by at least 1000 years. It is therefore reasonable to assume that the first domesticated cattle of Africa originated in Asia. But whether they came from the Caucasus or the Hilly Flanks area or another region in south-west Asia cannot be decided in the present state of our knowledge.

272. Bull Jumping scene, Çatal Hüyük V (after Mellaart)

Distribution of Longhorn Cattle in Ancient Times

There is an ancient stratum of longhorn cattle all along the shores of the Mediterranean—from Palestine through Egypt to North West Africa, and from Asia Minor through the Balkans to the Apennine and Iberian peninsulas, extending to the east, north and south as far as the Mediterranean race of man or his culture spread. Throughout this large area wall engravings and statuettes of longhorn cattle strikingly resembling each other have been found in large numbers in stone, bronze, silver and gold. There is little difference in conformation between the recent Highland cattle of Scotland and the sculptures of longhorn cattle made by the ancient Etruscans, between the modern longhorn cattle of Portugal and the breed of the ancient inhabitants of the Nile valley, between long-horned N'Dama cattle of West Africa and the clay, bronze and silver figurines of longhorn cattle from Mycenae.

Origin of Pastoral Culture

However, the cult of the domesticated ox did not originate in the Mediterranean basin (Frobenius, 1933). Whereas hunting culture and the cult of the wild ox are believed to have moved from west to east, pastoral culture passed from east to west. As we proceed

273. Bucranium design on pottery from Arpachiya (Tell Halaf period, c. 4500 B.C.) (after Mallowan and Rose)

274. (left) bronze head of a longhorn bull from Mallorca, Balearic Islands

275. (right) Head of a longhorn bull on a vase from Knossos, Crete

276. Ligurian longhorn oxen before the plough. Bronze Age (after Hoernes)

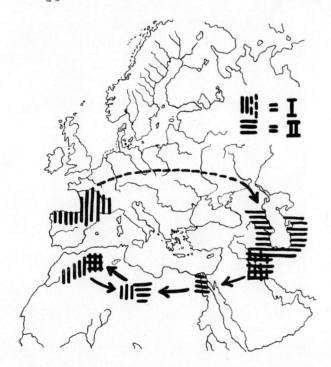

277. I. Wild ox culture
II. Domestic animal culture (after Frobenius)

CATTLE

278. Figurine of a longhorn bull from the tombs at Alaja Hüyük, Anatolia (second half of 3rd millennium B.C.)

279. Rock engravings of longhorn cattle in central Arabia, 3rd–2nd millennia B.C. (after Anati)

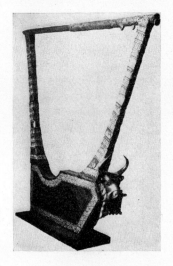 

280. (left) Gold and mosaic harp with a golden bull's head from the royal cemetery at Ur

(right) copper head of a longhorn bull from Tello

from the eastern shores of the Mediterranean eastwards, we find the same ancient stratum of longhorn cattle, encountered in the Mediterranean region, along our path. Humpless cattle with long lyre-shaped or crescentic horns, of a type similar to that represented in ancient Egypt and the Eastern and Western deserts of North Africa, are extensively depicted in the rock art of central Arabia, dated to the 3rd and 2nd millennia B.C. (Anati, 1968a). Before short-horned cattle swept over Anatolia, Syria and Palestine, longhorn cattle were bred there as far back as the domestication of cattle can be traced. The fauna of the neolithic settlement of Fikirtepe, near Kadiköy, northwest Asia Minor, dated to the end of the 4th millennium B.C., included large domestic cattle which had still retained the sexual dimorphism in size and the large teeth of the wild urus (Roehrs and Herre, 1961). Among the objects in gold, silver and copper from the copper-age tombs at Alaja Hüyük, north-east of Boghazköy in the Halys bend (second half of the 3rd millennium B.C.), are figurines of longhorn bulls (Fig. 278) totally unlike the shorthorn bulls represented in Asia Minor in Hittite times.

Longhorn Cattle in Mesopotamia

The earliest domesticated cattle of Mesopotamia belonged to the same longhorn type. A marble bowl from the Sumerian period (about 2700 B.C.) shows longhorn bulls carved in relief (Fig. 281). On the mosaic standard from Ur two bulls are led on ropes fastened to nose rings (Vol. II, fig. 212). The animals show the typical conformation of the longhorn cattle, their horns pointing sideways and forwards with the tips slightly upwards. A similar type is represented by the bull's head on the gold and mosaic harp from the great death pit at Ur and the copper head of a longhorn bull from Tello (Fig. 280). Longhorn cattle were bred also in Assyria at a later period, as indicated by the ivory figure of a longhorn cow, found in the course of the excavations of the Assyrian city of Nimrud (Kalah) (Fig. 283).

Longhorn Cattle in the Caucasus and Southern Russia

A few of the longhorn cattle represented on seal cylinders and other monuments from Sumer and Elam have horns projecting forwards and slightly downwards. Horns of a similar shape are characteristic of some of the longhorn cattle from the copper-bronze age kurgans of Maikop, in the northern Caucasus, which Stegmann von Pritzwald (1924) regarded as the descendants of a distinct local race of the urus—Bos balticus, a view supported by Friederichs (1933). However, there can be little justification for the assumption of a new local variety of the urus, together with a special type of cattle evolved from it, on the basis of so minor and variable a feature as the forward and slightly downward sweep of the horns; all the less as this particular curvature has been observed in urus skulls and in domesticated cattle in widely separated parts of the Old World. At Maikop another gold statuette shows a longhorn bull with the horns directed horizontally forwards and the tips upwards, with no downward tendency, illustrating how little importance may be attributed to this feature in the classification of the urus and its domesticated descendants.

281. Longhorn bulls in relief on a Sumerian marble bowl (c. 2700 B.C.)

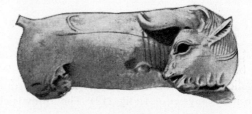

282. (left) Statuette of a longhorn bull from Babylonia

283. (right) Ivory figure of a longhorn cow from Nimrud (Kalah)

284. (left) Silver vase from Maikop, Northern Caucasus (c. 2000 B.C.) (after Friederichs)

285. (right) Gold statuette of a longhorn bull from Maikop (after Friederichs)

Crania of cattle from kurgans in the steppes of southern Russia, dated to the beginning of the Christian era, show similar proportions to urus skulls. Only the horns are a little shorter, thinner and more lateral in direction than the horns of the wild animal, the corresponding difference in the moment of rotation of the horns slightly affecting the shape of the frontals (Hancó, 1950).

Remains of Humpless Longhorn Cattle in Northern Baluchistan

East of Mesopotamia, in the Quetta–Pishin valley of northern Baluchistan (West Pakistan), bovine teeth, somewhat larger than those usually assigned to zebu cattle, have been found, associated with cultural debris, at early prehistoric levels, of which the lowest pre-ceramic may date from the end of the 5th millennium B.C. These teeth are nearly identical with the teeth of Egyptian cattle recovered at Deir el Bahari. "The latter are almost certainly teeth of the now extinct aurochs or urus" (Fairservis, 1956). Those from the Quetta valley most probably belonged to domesticated humpless longhorn cattle, since they occur together with the teeth and bones of sheep and goats slaughtered as yearlings, while the remains of wild fauna associated with them are extremely scarce.

Humpless Longhorn Cattle in the Indus Valley

In the ancient Indus valley sites of Mohenjo-daro and Harappa (3rd millennium B.C.) longhorn cattle are represented on a large number of seals and by a few statuettes. At Harappa the animals are shown with both horns, whereas at Mohenjo-daro only the right horn is visible on the seals (unicorn). Usually the horn is directed upwards and forwards, with the tip again upwards, and with numerous transverse wrinkles covering the main portion of the horn below the tip; but occasionally there is only an upward and forward sweep. A heart-shaped pattern, which Friederichs believes to indicate a cover of sacral importance, ornates the chest and withers. In most specimens the head

286. Longhorn bulls on seals from Mohenjo-daro, Indus Valley (after Marshall)

287. Clay figurine of a bull from Mohenjo-daro (after Marshall)

and anterior part of the neck are covered with wavy wrinkles; in some the head and neck are smooth except for a collar just behind the ears. The clay figurine of a bull with broken off horns, legs and tail shows that the wrinkles represent the natural folds of the skin on the neck of the animal, while the collar consists of a braided rope (Fig. 287).

Friederichs (1933) has assigned the seal engravings from Mohenjo-daro to two different bovine races: the wild or domesticated Bos p. primigenius, and Bos p. namadicus. There is, however, not a shred of evidence to suggest the occurrence of the wild or domesticated European urus in the Indus valley 4000 or 5000 years ago. The slight differences in the ornamentation of the head and neck, or even in the shape and direction of the horns, cannot be explained on racial grounds. The longhorn cattle of Mohenjo-daro and Harappa should be regarded as of Bos p. namadicus descent. It must be borne in mind that the ancient Indus valley stations do not represent the incipient stages of a pastoral civilisation; the wealth in the number and variety of their domestic animals suggests that the domestication of the majority of these was accomplished millennia before the period to which Harappa and Mohenjo-daro are assigned.

Theory of the Domestication of Cattle in Turkestan

In his discussion of the bovine remains from the excavations at Anau, southern Turkestan, Duerst (1908a) arrived at the conclusion that this site was to be regarded as the original domestication centre of Bos p. namadicus. In the lowest strata of Anau the bovine fossils consist only of those of Bos p. namadicus. At more recent levels these are successively replaced by the remains of domesticated longhorn cattle—"absolutely the same ox that was possessed by the ancient Egyptians".

The evidence produced by Duerst in support of his theory is not convincing. The Anau levels are now dated considerably later than they were in 1908 when Duerst described the animal remains. It is true that the Anau strata show a remarkably uninterrupted succession from wild oxen hunted by the earliest settlers to domesticated stock bred by later inhabitants. But this does not necessarily prove that the domesticated cattle were actually evolved at Anau; they may have been introduced by a pastoral people from elsewhere. For the neolithic culture of Anau is repeated at a number of widely scattered points in relation to which Anau occupies not a central but a peripheral position; and some of these sites show an equally uninterrupted succession of settlers from the hunting to the pastoral stage of culture to that found at Anau.

Uncertainty regarding the Domestication Centre of Bos Primigenius Namadicus

The original locality of domestication of Bos p. namadicus is, therefore, not yet known with any degree of certainty. It has probably to be sought at one of the ancient river systems of south-west Asia where hoe-cultivators then at a relatively high stage of economic development, probably already in possession of domesticated goats, sheep and dogs, were settled (see p. 258). Nor does the range of Bos primigenius namadicus permit any geographical restriction in this quest; for it extended over practically the whole realm of neolithic civilisations in Asia.

Bos Primigenius Namadicus in Palestine and the Lebanon

In Palestine and the Lebanon Bos primigenius namadicus was hunted by Tiglath Pileser I (about 1100 B.C.). At the time of the prophets the urus still occurred in the mountain ranges of the Hauran. It is the reem of the bible (called unicorn in King James' and wild ox in the Revised Version) and referred to in Job in the following passage:—"Will the reem be willing to serve thee, or abide by thy crib? Canst thou bind the reem with his band in the furrow or will he harrow the valleys after thee? Wilt thou trust him because his strength is great or wilt thou leave thy labour to him? Wilt thou believe him, that he will bring home thy seed and gather it unto thy barn?"

Bos Primigenius Namadicus in Mesopotamia

Sumerians symbolised domesticated cattle by $\bigvee$, adding the symbol for mountains $\bigvee$ when referring to the wild ox. In a temple palace of the Hittite settlement of Tell Halaf, on the upper Khabur river, von Oppenheim (1931) found basalt and limestone orthostats, covered with reliefs of numerous wild and domesticated animals. Some of these depict the hunting of the urus. On basalt reliefs only one horn of the animal is shown (unicorn), as on the seals of longhorn cattle from Mohenjodaro. But on orthostats of the softer limestone both horns are carved. The last report of a wild bull hunt in Assyria dates from the time of Ashurnasirpal (884–860 B.C.). The urus reliefs on buildings from the reigns of Nebuchadnezzar (605–562 B.C.) and Nebonidus (556 bis 539 B.C.) seem to be imitations of earlier works (Ebert, 1924–32).

288. Urus hunt on an orthostat from Tell Halaf (after v. Oppenheim)

289. Ashurnasirpal, king of Assyria (884–860 B.C.), hunting the urus

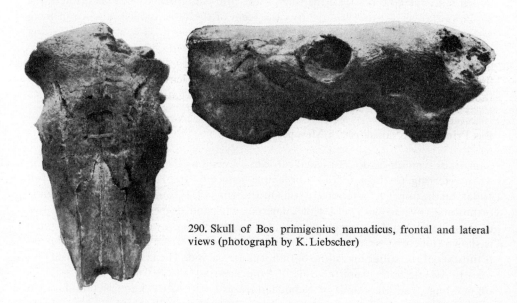

290. Skull of Bos primigenius namadicus, frontal and lateral views (photograph by K. Liebscher)

The Bos Primigenius Namadicus Skull

Fossil remains of Bos primigenius namadicus have been found throughout Asia, from the Lebanon to China, where urus remains are plentiful in loess (Boule et al., 1928). Nonetheless, our knowledge of the Asiatic urus is only scanty. The first references to a Bos primigenius from the Narbada valley of east India were made by Falconer (1859) and Blanford (1867). Rütimeyer (1878) classified the Bos p. namadicus skulls in the British Museum (Natural History) with the urus group; but Lydekker (1878) did not accept this because of the narrowness of the forehead and the high Torus frontalis as compared with Bos p. primigenius, classing them with the banteng. Against this Duerst (1931) has pointed out that the high Torus frontalis and narrow forehead in the Bos p. namadicus skull (Fig. 290) are due to the upward direction of the horns which exert less pressure on the cranium than the double-twisted horns common in Bos p. primigenius. Bos p. primigenius skulls which are exceptional in the upward direction of the horns, e.g. a specimen from Monte Mario in Rome and one of the Ilford skulls, show the high frontal torus and narrow forehead typical of Bos p. namadicus. On the other hand, the Bos p. namadicus skull from Süan-hwa-fu (Fig. 294), in which the cores point sideways, has the straight intercornual ridge and wide forehead typical of the male Bos p. primigenius.

Falconer (1859) described the cranium of the type specimen, which he called Bos namadicus, as follows:—"The forehead is flat and slightly concave above; it is square, taking the base between the orbits, its height is about equal to its breadth; the horns are attached to the extremity of the highest salient line of the head; the plane of the occiput forms an acute angle with the forehead (it is overarched) and the plane of the occiput is nearly quadrangular instead of semicircular." The longest horn cores in the Bos p. namadicus crania from the Siwalik hills are of the same length as those in the

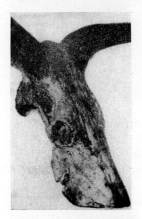
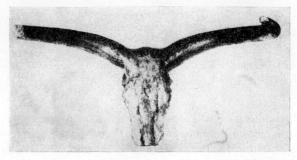
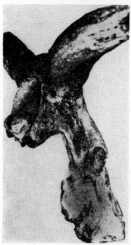

291. (right) Skull and horn cores of Bos primigenius namadicus, frontal surface. Calcutta

292. (left top and bottom) Skull of Bos primigenius namadicus, lateral views. Calcutta

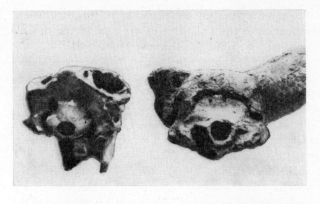

293. Two Bos primigenius namadicus skulls, nuchal surface (after Liebscher)

294. Cranial fragment of Bos primigenius namadicus from Süan-hwa-fu, frontal and nuchal views (After Duerst)

Bos p. primigenius skulls from Ilford. The core of the skull from Süan-hwa-fu is 90 cm long, with a basal circumference of 41.3 cm and a diameter of 13.8 cm (Duerst, 1931).

The most thorough craniological description and analysis of Bos p. namadicus skulls in the British Museum (Natural History) has been given by Liebscher (1926) in an unpublished thesis. Liebscher points out that the namadicus skulls display an even higher degree of variability than Bos p. primigenius skulls. In spite of this, there is a considerable general resemblance between these two geographical races, whereas a comparison between Bos p. namadicus and the crania of Bibos proves their specific separation, mainly owing to the completely different occipital conformation.

In Bos p. namadicus skulls the variability in the character of the forehead is far greater than in Bos p. primigenius Boj. In some specimens the forehead is convex, in others concave, with pronounced longitudinal furrows which Liebscher considers as typical of the Asiatic urus. The majority of the Bos p. namadicus skulls are distinguished by the markedly pronounced Torus frontalis; occasionally the frontal ridge is straight, but no specimens are known with the frontal ridge concave. Lydekker (1878) described the frontal torus as egg-shaped.

In the skulls in the British Museum the ratio of the lesser occipital height to the narrowest occipital width falls within the variational range of Bos p. primigenius skulls. The temporal fossa is deep and narrow, and the interorbital ridge convex with a forward slope. The massive nasals penetrate wedge-like into the forehead. Their greatest width lies at the height of the junctions between the frontals, nasals and temporals. Thence they gradually narrow down orally until the middle of their length whence they broaden again, reaching their greatest lower width in the region of the angle formed with the premaxillae. Nasals of this shape are not found in Bos primigenius primigenius skulls. In the latter the nasals are also relatively broad in the lower portion, but they taper gradually towards the tips, and no specimen is known in which the central portion is narrower than any part lower down.

The horn core preserved in one of the Bos p. namadicus skulls in the British Museum is attached to the skull without Processus cornu ossis frontalis, and is devoid of longitudinal furrows, of a basal wreath of horn pearls, and a twist on its axis. Liebscher considers the lack of these features as the most important difference between Bos p. namadicus and Bos p. primigenius skulls. But a close scrutiny of the photographs of the Bos p. namadicus skulls at Calcutta (Siwalik) and Paris (Süan-hwa-fu) reveals the presence of necks to which the horn cores are attached, of longitudinal furrows in the latter, and of a slight homonymous twist, while in the Calcutta skull a thin wreath of horn pearls surrounds the bases of the cores.

Distribution of Bos Primigenius Namadicus throughout Asia

Lydekker (1898) believed that the range of Bos primigenius namadicus had been limited to the Pleistocene of southern India. We now know that its distribution was far wider, embracing nearly the entire continent of Asia.

Evidence for the Domestication of Bos Primigenius Namadicus in South-West Asia

There are certain indications in support of our previous suggestion (see p. 253) that the search for the domestication centre of Bos primigenius namadicus may be restricted to the south-western part of its range. Dyson (1953) observes that an analysis of the fauna of a site over a period of time may at some point indicate a shift from reliance on wild game to reliance on 'prodomestic' game, a term referring to wild but potentially domesticable animals which are known as domesticated in later periods. Subsequently a second shift, this time in the age at which prodomestic animals are killed, may be indicated. When accompanied by a constant increase of the percentage of the prodomestic group in the total, these two shifts would seem to be reasonably good evidence for inferring cultural control over the animals in question.

Reed (1959) doubts the universal applicability of this statistical approach, for similar concentrations of submature individuals have been found in instances where the animals never became domestic. But, as Isaac (1962) has pointed out, such finds may mean that there had been an attempt at domestication which was later given up.

The statistical analysis of the fauna of widely separated sites indicates that a shift from a reliance on wild animals to a reliance on domestic animals took place in southwest Asia by the beginning of the 5th millennium B.C. In Europe similar shifts in faunal deposits occurred at least one millennium later, while in central, eastern and southeastern Asia the shift occurred closer to two millennia after that in south-west Asia (Isaac, 1962).

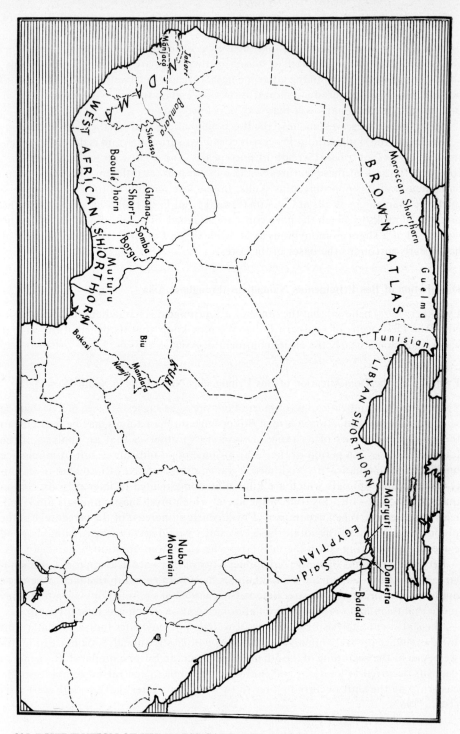

295. DISTRIBUTION OF HUMPLESS CATTLE IN AFRICA

III. The Humpless Shorthorn Cattle of Africa

i. Distribution and Characteristics of the Shorthorn Cattle of Africa

The term "shorthorn", as here used, is a literal translation of the Greek term "brachyceros" (more correctly brachykeratos); it has no connection with the British Shorthorn breed of cattle.

Humpless Shorthorn Cattle in the Nile Valley

Humpless shorthorn or brachyceros cattle are now the dominant type in Egypt and in the North African territories of Libya, Algeria and Morocco, whence they extend along the west coast of the continent as far south as Cameroun. In the Nile country all the native breeds of cattle are basically shorthorn, although more or less influenced by introductions of other types over many years. More than one hundred years ago, Hartmann (1864) described them as follows:—"The recent Egyptian cattle are very similar to the shorthorn breed of the ancient Egyptians. They have the same narrow forehead, short curved high-set horns, high withers, and long slender legs. They have long tails with a high setting, straight hocks, rounded hooves, and long slender dewclaws. They are of medium size, mostly red, coffee-brown, black, yellow or grey, occasionally white or black-and-white." Flower (1932) recorded that in his time the domestic cattle of Egypt were all of one short-horned type from Alexandria to Aswan. Then followed in Lower Nubia an entirely cattle-less country. In Upper Nubia from Dongola Province (Sudan) southwards cattle were again met with, but these were humped. When the importation of cattle from the Sudan to Egypt commenced in 1902, the large humped beasts with long upstanding horns were a matter of great interest to the peasants.

Owing to indiscriminate cross-breeding for numerous generations, all Egyptian cattle are now similar in conformation; yet geographically certain groups are recognised: the Damietta and Baladi in Lower Egypt, the Saidi in Upper Egypt, and the Maryuti or Arabian cattle of the desert (Curson and Thornton, 1936).

General Characteristics of Egyptian Shorthorn Cattle

All of these are distinguished by high withers. This does not always denote the influence of zebu blood. As the skeleton from Schussenried (Fig. 197) illustrates, high withers are a typical feature of primitive shorthorn breeds—a survival of the high withers of the wild urus. Resulting from the high spinous processes of the vertebrae they are not to be mistaken for a hump composed of muscle and fat tissue. In shorthorn breeds improved by selection the withers are more or less on a level with the back. This, as Duerst (1931) has pointed out, has not been achieved by the shortening of the spinous processes, but by the lowering of the spinal column in the region of the neck.

Among all types of Egyptian cattle, except the Saidi, red in lighter or darker shades is prominent. Red animals with black shoulders, necks and quarters are favoured. However, some cattle are black or spotted with white. Dark points, especially black muzzle, eye rings, eye lashes, hooves, and legs below knees and hocks, occur among all types. But animals with white hair and a black skin are not found.

The horns are nearly always cylindrical, not longer than 20 cm, and directed either horizontally outwards or downwards, with blunt tips. Only the Damietta cattle have finer and occasionally more pointed horns with an upward sweep.

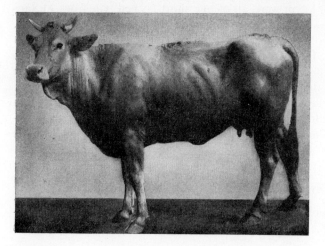

296. (left top) Damietta bull
297. (right) Head and neck of a Damietta bull
298. (left bottom) Damietta cow

Egyptian cattle are somewhat slow to mature. But they are hardy in conditions of drought and high ambient temperatures, and possess strong resistance to indigenous diseases.

The Damietta (Dumyati) Cattle

The Damietta (Dumyati) cattle predominate near the coast of the Mediterranean and north-east of the Delta. They are valued in their native district for dairy purposes (Curson and Thornton, 1936). Wahby (1938) describes them as slender in front and spacious behind, with fine legs, well developed udders, and prominent milk veins. The coat colour is usually red, with only the eyes and muzzle black; occasionally cream, black, or red with white markings. The horns sometimes resemble the Ayrshire's, but commonly they are very short and thin. Damietta bulls are often humped. However, the Director of Veterinary Services Cairo, replied to the author's enquiry about the photograph of a Damietta bull (Fig. 296):—"The bull in question is a typical Egyptian bull and has no zebu blood" (letter of September 27th, 1934) probably meaning that there has been no *recent* introduction of zebu blood (see also p. 263).

The Maryuti Cattle

The Maryuti is found in the desert near the coast of the Mediterranean to the north-west of the Delta, but is not numerous. It is very leggy, long and slender, poorly muscled, with a weak back and a high rump.

The Baladi (Beheri) Cattle

The Baladi or Beheri is found in Lower and Middle Egypt. Owing to its considerable size and strength, it is the most valuable of Egyptian cattle for work. The body is red in colour, with the muzzle, surrounded by a mealy-coloured ring, the eyes and switch of tail black. The long, shapely head tapers from the eyes to the muzzle. The black horns are very small. The mandibular muscles are particularly well developed. As compared with the Damietta, the Baladi has a shorter and thicker neck, with a prominent crest in the bull, a deeper body and more fleshy shoulder and chest. The withers are prominent and well muscled, and the muscular rump exceeds the back in height. The legs are relatively short and of strong bone. The tail root is thick, commonly with a high setting (Wahby, 1938).

The Saidi Cattle

The Saidi cattle are bred in Upper Egypt along the banks of the Nile. The head is relatively large, with a pyramid-shaped forehead and short, rather coarse horns. Owing to the lack of proper feeding and attention combined with the effects of the hot climate of Upper Egypt, the breed is smaller and leggier than the Baladi. The coat colour is commonly black or blue roan (Wahby, 1938).

CATTLE

299. Maryuti cow

300. Baladi bull

301. Baladi cow

In the Saidi the prominent cervico-thoracic hump is a remarkable feature for which, according to the Ministry of Agriculture, Cairo, the short-horned zebu is responsible. Curson and Thornton (1936), however, disagree:—"What has probably happened," they write, "is that formerly (up to some centuries ago) the cattle of Egypt were generally of the Sanga type; but that through the constant introduction of cattle of the Brachyceros type the conformation has become almost Brachyceros-like, the hump, however, retaining the characteristics of that of the Africander (or Sanga) type."

This explanation of the cervico-thoracic hump of the Saidi is not convincing, for 1) there is no evidence for the former existence of sanga cattle in this area; 2) constant introduction of humpless cattle into a sanga cross would reduce the hump much more, as evidenced by the Bonsmara cattle of South Africa; 3) a cervico-thoracic hump may result also from the crossing of humpless with thoracic-humped cattle; 4) the introduction of humped cattle from the Sudan, recorded by Flower (1932) (see p. 259), renders it likely that the cervico-thoracic Saidi hump is due to the influence of humped stock on the humpless shorthorn cattle of this area. Again, were it not for the proximity of zebu cattle to the breeding area of the Saidi, and the record of their importation into this area, the tendency to develop small cervico-thoracic humps in the absence of zebu admixture would also have to be considered in connection with the Saidi hump. In spite of the latter, the Saidi has been classed with the shorthorn type which it approximates in cranial and body conformation (straight rump, high tail head, small dewlap, tight skin of prepuce and umbilical fold).

Craniological Description of Recent Egyptian Shorthorn Cattle

Duerst (1899) has given the following description of the crania of recent Egyptian Shorthorn cattle:—The upper portion of the frontals rises to a prominent torus from which the frontal ridge extends to a distance of 75 mm. There is a depression along the side of this ridge. The forehead, devoid of a central elevation, is markedly dished, the depression starting below the frontal torus approximately at the level of the posterior

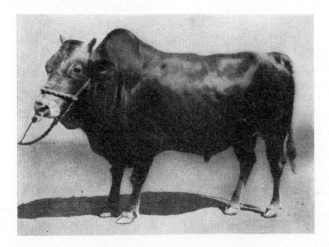

302. Saidi bull

orbital ridges, and ending at the root of the nasals. The orbital arches, which stand well out laterally, exceed the lateral elevations of the forehead in height. The supraorbital fossae, shallow but with clearly defined inner edges, converge in the direction of the lacrimals. The frontal edge of the lacrimals runs in a nearly straight line, interrupted only by a small projection of the frontal. The triangular aperture between the lacrimals, frontals and nasals is small. The outer edges of the nasals are considerably curved and the inner edges slightly indented. The nasal branch of the maxilla ends 7 mm below the lateral edge of the nasal. The maxillae are narrow; the malar point is not prominent; the palate is flat. The temporal fossa is shallow, but broadens considerably backwards. The Torus frontalis rises above the forehead and exceeds the bases of the horn cores in height. Its nuchal portion, hollowed in the centre by a triangular cavity, is only slightly higher than the squama from which it is divided by a smooth line. The squamosals are oval in shape and very little indented by the temporal fossae. In the male skull the occipital crest is flatter and the intercornual ridge wider than in the female. In the former the hollow in the centre of the crest is hardly noticeable. The squama forms an angle of 90° with the forehead. The frontal torus and the portion of the forehead situated between the horn cores are furrowed and scaly. The cores are very rough at the bases, growing smoother towards the tips. Tiny furrows proceed from the frontal ridge to the temporals. The horns are short and brightly coloured; their direction is upward, with the tips pointing inwards.

Shorthorn Cattle in the Sudan

Formerly humpless shorthorn cattle may have extended considerably to the south of Egypt. Hartmann (1864) and C. Keller (1896) recorded the sporadic occurrence of humpless shorthorn cattle in the Sudan in their time, among herds then already generally humped. The Annual Report of the Veterinary Department of the Anglo-Egyptian Sudan for 1925 groups the native cattle into a) short-horned zebus, b) sanga, c) a type along the Ethiopian border, apparently of brachyceros origin.

At one time shorthorn cattle were found over the whole Nuba mountain region of southern Kordofan (Sudan). But at the present time they are restricted to a small enclave in the isolation of an almost inaccessible habitat in the tsetse-infested Koalib hills around the village of Delami, where they number about 2000–3000 head.[1] In other parts of the Nuba mountains cross-breeding with Baggara cattle has taken place for a long time, with the result that the majority of the Nuba cattle have now humps varying greatly in size and shape (see p. 363 and Fig. 417). In the typical humpless Nuba, the head is of moderate length, with a prominent supraorbital region, strong and deep mandibles, broad muzzle and short horns of various shapes; polled cattle or animals with loose horns do not occur. Conspicuous are the high withers in Nuba cattle, due to the long spinous processes of the anterior dorsal vertebrae. In size the Nuba cattle are dwarfed, but in conformation deep-bodied and well proportioned. At the present day all colours are seen except roan, but the original type may have been black (Faulkner and Epstein, 1957).

[1] I. L. Mason has been told that it is now impossible to find humpless cattle even in the Kaolib hills.

303. Nuba Mountain cow (after Faulkner and Epstein)

Shorthorn Cattle in East Africa, Aden and the Islands of Socotra, Pemba, Mafia and Madagascar

There are a few doubtful records of the occurrence of shorthorn cattle south of the Sudan. In the Encyclopaedia Britannica (1929) humpless shorthorn cattle were still mentioned with reference to Ethiopia where now nearly all cattle are humped:—"There are also two breeds—one large, the other resembling the Jersey cattle—which are straight-backed." In subsequent editions this statement is not repeated. Yet, humpless shorthorn cattle may at one time have been introduced into Ethiopia, for in Shea Ghimirra, in the extreme south-west of Ethiopia, the small stocky cattle of the Chako tribe have poorly developed humps and are often entirely humpless (Straube, 1963).

Emin Pasha recorded that he encountered shorthorn cattle in Karagwe, west of Lake Victoria, amidst country generally occupied by humped cattle. On August 1st, 1889, he made the following note in his diary:—"Today we saw the first herd of cattle in Karagwe, slender, short-horned, humpless animals, white, brown, and black" (Stuhlmann, 1927). Ford (1964—personal communication) was informed by Mr. Mackintosh, Director of Veterinary Services, that there had been a short-horned humpless breed on the Sesse Islands in Lake Victoria before the sleeping sickness epidemic led to their evacuation between 1902 and 1909. According to native tradition, there was a short-horned humpless type of cattle in western Uganda before the Bahima arrived with their giant-horned sanga cattle; remnant herds of these cattle still existed in the nineteen-twenties (Ford and de Z. Hall, 1947). Ford saw humpless cattle in the Kigezi district, western Uganda, in 1934.

Shorthorn cattle are kept in small numbers by the Qarra of Aden, who speak not Arabic but a language closely related to the ancient Semitic language of the Minaeans (a Yemenite people and kingdom in the 2nd millennium B.C.), distinct from the speech of the tribes in their neighbourhood (Thesiger, 1959). Similar cattle of very small stature, without a trace of zebu blood and with short horns projecting slightly upwards, are found in the island of Socotra, off the Horn of Africa. The people owning these cattle in Socotra speak the same language as the Qarra and differ in physical type from the Arabs (Payne, 1964). It is however uncertain if these shorthorn cattle are derived from African stock; they may have been introduced into southern Arabia, and thence into Socotra, direct from southern Palestine prior to the arrival of short-horned zebu cattle in southern Arabia.

Tradition has it that similar cattle were formerly also found in the island of Pemba, and reports from the early nineteen-twenties suggest that they occurred at that time on the island of Mafia, off the coast of Tanzania (Payne, 1964). Dechambre (1951) mentions a non-zebu type of cattle, introduced into Madagascar by the Vazimbas in prehistoric times; but it is uncertain if these were short-horned or long-horned. From crossbreds of these cattle and zebus a small-humped feral type of Madagascar cattle, called Baria, is descended (Mason and Maule, 1960).

Shorthorn Cattle in North and West Africa

West of Egypt, shorthorn cattle are kept on the scrubland in the northern parts of the Libyan desert and the Barqa peninsula; they are found in Fezzan and Tripolitania, and extend through Tunisia and Algeria to the western parts of Morocco. The differences encountered among the shorthorn cattle of these countries are restricted mainly to colour and size, the latter depending foremost on environmental factors. Diffloth (1924) distinguished between two sub-types: one being bred in the fertile valleys and plains, and the other in the mountains and the fringe of the desert. In the lowland cattle the original racial character has been considerably modified by improved European breeds.

From Morocco shorthorn cattle have spread southwards, between the Sahara desert and the coast of the Atlantic, to the Guinea coast and the hinterland of Nigeria where their original racial type has been preserved relatively pure. Cameroun forms the extreme southern limit of their distribution in West Africa. They are found also in the westernmost extension of the Congo; but it is uncertain if they occurred there before imported by Europeans.

The Libyan Shorthorn

In Libya shorthorn cattle are found mainly in the coastal zone and in some parts of the semi-desert and low mountain regions where conditions for stock are exceedingly adverse. The Libyan Shorthorn is a small animal of fairly uniform type; adult bulls stand approximately 120 cm at the withers and cows 110 cm. The average weight of

304. (above) Libyan Shorthorn bull (after Faulkner and Epstein)

305. (below) Libyan Shorthorn cow

bulls is 380 kg and of cows 290 kg. The head is of moderate length, widest between the eyes, with a wide muzzle, and straight or concave in profile. The short round ears are carried horizontally. The horns, cream-coloured with dark tips, are 10–26 cm long, thin and round, projecting laterally, forwards and upwards. The neck is short and well attached to shoulders and brisket. In adult bulls the crest is prominent. The shoulders are compact but the withers rather high. The dewlap is fairly well developed, ending in two separate folds between and behind the front legs. The body tends to be long and often shallow. The topline frequently shows a slight depression in the middle.

The ribs are flat behind the shoulders, widening caudally. The hook bones are prominent, the pin bones narrow. The rump is long and narrow, with a tendency to be sloping from hooks to pins. The upper thighs are poorly developed and the lower thighs are very narrow. The limbs are fine and light of bone. The udder is small, well shaped and strongly attached; the teats are small but well placed. The tail setting is often higher than the top of the withers. The tail is long and slender. The skin is thin and pliable, pigmented either black, light brown or red. The hair is fine. The coat is generally self-coloured fawn, red or black. White patches are sometimes seen on the udder and belly. The breeders prefer dark colours which they believe to indicate constitutional strength. Libyan cattle are very docile and are used for ploughing. The cows give approximately 1250 kg milk per lactation (Faulkner and Epstein, 1957).

In 1946 the author handled several hundred young bulls from Benghazi, Cyrenaica (Fig. 306). They were of a remarkably uniform type, showing pure shorthorn character: slender, lightly built animals, with a fine nasal part of the head, a deeply dished forehead, short, thin, well-shaped horns commonly lateral, upward and slightly forward in direction, a moderately strong neck with a thin dewlap, narrow ribs, a narrow, slightly drooping rump, moderately long slender legs, and a thin tail reaching down to below the hocks. While from the fattening point of view they were among the worst cattle encountered by the author throughout the Near East, as they gained very little weight in fattening, the dressed weight in animals in good condition reached well over 50 per cent and the beef was of a good quality. The majority showed the following coloration: the forehead, poll, back, and the upper two-thirds of the tail brown, the rest of the head with the exception of a whitish ring around the dark muzzle, the neck, belly, legs and lower part of the tail dark brown, the scrotum light brown. Some were brown throughout, and several were brown with the inner sides of the thighs and the lower parts of the legs a light yellow.

Cranial Characteristics of the Libyan Shorthorn

The cranium of a young bull from Benghazi is distinguished by a deeply dished frontal region which ascends to the relatively flat upper portion of the skull and to a conspicuous bulge immediately below the prominent Torus frontalis which gradually slopes towards the short frontal necks to which the extremely porous horn cores are attached. The nuchal surface is situated well below the moderately wide intercornual ridge, and the interparietal does not extend into the frontal plane. The horns are fairly broad at the base where the cross-section is oval, growing thinner and round in cross-section beyond the lower two-fifths of their length. Their direction is mainly lateral and upward, with the tips growing forwards. The colour is greyish white near the base and dark olive towards the tips. The temporal fossae are very deep; narrow in front, and not particularly wide aborally. Laterally the central depression of the frontals gradually rises to the laterally prominent orbits situated completely on the lateral part of the skull. The orbits are quadrangular in outline with rounded corners. The supraorbital fossae are short and shallow and the fronto-lacrimal sutures nearly straight except for a small frontal process wedged into the lacrimal. The posterior ends of the nasals are situated at a relatively low level, owing to the marked forward extension of the frontal dish. There is a large central triangular aperture separating the frontal and nasal sutures. The Lacunae lacrymales are small. Below them the nasals slope down laterally and broaden out to the aboral end of the premaxillae whence they narrow slightly orally, each ending in two widely separated tips. The palate is flat, and the malar point little prominent. The mandible is strong and deep, the anterior ramus curving upwards orally from below the first molar. The cheek-teeth in both the upper and lower jaws are vertical.

Tunisian Shorthorn Cattle

All indigenous cattle of Tunisia, Algeria and Morocco are racially similar and of the Brown Atlas basic type.

The cattle of Tunisia, owing to selection of superior breeding stock, are conforma-

306. Young shorthorn bull from Benghazi, Cyrenaica

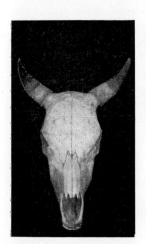
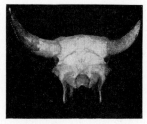

307–309. Skull of a young shorthorn bull from Benghazi, Cyrenaica, frontal surface (left), nuchal surface with right horn sheath removed (centre) and lateral surface (right)

310. Tunisian cow

tionally among the best cattle in North Africa (Saint-Hilaire, 1919). They are distinguished by a high degree of conformational uniformity, though even among this breed considerable variability in type and size is still encountered, partly owing to indiscriminate breeding in the past, partly owing to environmental differences. Some animals do not reach a withers height of 110 cm. The average live weight is approximately 200 kg. Generally the Brown Atlas cattle of Tunisia are characterised by a narrow, moderately long head, receding crest, and thin horns projecting forwards, with the dark-coloured tips upwards. The lower portion of the frontals is slightly concave, and there is a pronounced stop in the region where the frontals and nasals join. The supraorbitals are prominent. The body is well proportioned, the chest broad and fairly deep, the withers are narrow, and the flanks short. The bull has a sturdy body, with a well developed dewlap, short neck, and straight topline; the legs are short and strong, the buttocks well fleshed, and the long tail terminates in a well developed switch. The bone is strong in spite of its remarkable slenderness. The skin is supple, the colour of the coat usually dark brown, with the head and legs nearly black. The cows are remarkable for the slenderness and elegance of their forms; with proper feeding they milk fairly well.

In southern Tunisia the Djerba, a variety of the Brown Atlas type, evolved in the island of Djerba, off the south-east coast of Tunisia, is the most common variety. Though small, standing only 100–120 cm at the shoulder, it fattens readily. The head is long, the narrow forehead dished in the lower part, and the nasals are prominent. The back is straight, the rump rounded, and the legs are slender. The body is dark brown in colour with black legs.

In the northern parts of Tunisia various crossbred types are encountered, such as the Mateur and Cap Bon, derived from indigenous cattle and imported Charolais, Breton, Aubrac and Normandy stock, or Swiss and Italian cattle. These crossbreds are more susceptible to the unfavourable environmental conditions than are the native breeds, but their production is superior (Diffloth, 1924).

Cranial Characteristics of the Tunisian Shorthorn Cattle

The pure shorthorn character of typical Tunisian cattle is evident from Duerst's (1899) analysis of a male Tunisian skull, with two Algerian Shorthorn skulls, previously described by Rütimeyer (1867), used for comparison. In the Tunisian the forehead ascends to a slight torus whence a barely noticeable frontal ridge issues; in a male Algerian skull this ridge extends to as much as 65 mm in length. In the Tunisian the upper part of the forehead is relatively flat. There is a deep depression between the orbits. The lateral elevations are low, and considerably exceeded in height by the orbits. In the Algerian Shorthorn skulls (Figs. 314 and 315) these features are similar. The deep supraorbital fossae, distinguished by sharp inner edges, do not converge to any marked degree, ending at the lower part of the frontals at a distance of 20 mm from the nasals. The very prominent orbits are situated laterally. The upper edge of the lacrimals runs in an almost straight line, interrupted only by the projection of a small frontal process. There is a triangular aperture at the junction of the frontals, nasals and lacrimals. The nasals are broad and slightly bent in their upper portion; the lateral edges are curved and the outer processes well developed. The nasal branch of the premaxillae reaches the lateral edge of the nasals. The rough-edged malar point is very prominent, the palate flat. The temporal fossae are narrow in front and fairly deep, widening considerably backwards. The nuchal surface of the Tunisian skull closely resembles that of the Algerian. The horn cores, which are very short, deeply furrowed and devoid of necks, point horizontally outwards, with their tips slightly forwards. The narrow posterior mandibular ramus forms a right angle with the anterior ramus.

Algerian Shorthorn Cattle

Among the indigenous cattle of Algeria, which constitute a branch of the Brown Atlas type, two main groups are distinguished, i.e. Numidian and Mauritanian. The Numidian is sub-divided into the Guelma, Kabyle and Djerba breeds, and the Mauritanian into the Orano-Algerian and Moroccan (Cabannes and Serain, 1963). The Djerba has been discussed in connection with the shorthorn cattle of Tunisia (p. 270) where its main breeding centre is situated.

The breeding centre of the Guelma is on the high plateaux of eastern Algeria, around Constantine. The Guelma varies between 115 and 125 cm in height; cows weigh approximately 250 kg and bulls up to 400 kg. The head is small, with prominent eyes and slender, sharply pointed horns which curve first forwards and then upwards; they are white in colour with black tips. The neck is short and the dewlap well developed. The body is thick-set with well rounded ribs. The withers are heavy, the top line is straight, and the rump frequently narrow. The thighs are thin and the legs short and slender. The tail is long and the tail head set low between the pin bones. The skin is very fine. The colour of the coat varies from a deep iron-grey of the head, neck, shoulders, lower flank and legs to a pale grey elsewhere and a commonly white belly; many animals are light grey throughout. The muzzle is black, surrounded by a band of white hair; the eyelids are also black. The tail is light grey, ending in a long black tuft (Diffloth, 1924; Curson and Thornton, 1936).

311. Guelma cow

312. Cheurfa bull

313. Oran cow

The Cheurfa, which is bred in the Department of Constantine, was regarded by Saint-Hilaire (1919) as the original stock from which the Guelma was evolved, because its distribution, especially in the south of Constantine, is more extensive than the Guelma's. But Cabannes and Serain (1963) believe that it was evolved from Guelma cattle which had spread into the plains where a larger and heavier type was selected. The colour of the Cheurfa is similar to that of the Guelma; the bulls are generally darker than the cows.

The Kabyle, which is bred in the mountains between Constantine and Algiers has the same origin as the Guelma. But owing to grazing on poor pastures for many generations it has degenerated and become dwarfed, the withers height ranging from 80 to 115 cm. The Kabyle has a relatively larger skull and is more bony than the Guelma (Saint-Hilaire, 1919); a disproportionately large skull is frequent in animals that have become dwarfed in an unfavourable environment. Kabyle cattle are usually thin and poor and look neglected, although bulls may be strong and occasionally even fat. They are well acclimatised in their environment, hardy, resistant to disease, drought and acute climatic changes, alert, quick of pace and relatively easy to fatten. In the poor feeding conditions of the mountains the cows are indifferent milkers.

Another local variety of the Guelma, called Shawia (Chaouia), is bred by the inhabitants of the Aurès mountains in eastern Algeria. It closely resembles the Guelma in colour and conformation.

The Oran cattle, of the department of the same name in north-west Algeria, are a variety of the Moroccan type of Brown Atlas cattle. They are somewhat less regular in build than other Algerian cattle, but bulkier and larger in size, varying between 110 and 130 cm in withers height. The coat is of a reddish tawny colour. Several local varieties are distinguished, e.g. Beni Sliman, Cheliff, Oran, Tiaret (Cabannes and Serain, 1963).

Cranial Characteristics of the Algerian Shorthorn Cattle

The cranial conformation of Algerian shorthorn cattle has been mentioned in connection with the description of a Tunisian shorthorn skull (see p. 271). Rütimeyer (1867) described the skull of an Algerian shorthorn cow (Fig. 315) as follows:—"A narrow fine skull with small, markedly curved horns, placed at some distance behind the forehead. They are not attached to cranial necks, and compress the skull at their bases. The orbits are very prominent and lateral in direction. The splanchnocranium is short and slender."

The Shorthorn Cattle of Morocco

The cattle of Morocco resemble those of Algeria except for certain modifications due to environmental factors. In general, Moroccan cattle are of a more primitive and less uniform type. While strong of bone, their conformation is often faulty; the top line may be depressed in front of the sacrum, the flanks flat, the rump narrow, and the tail setting high. The dewlap is conspicuous in both sexes. Many cows have a deep umbilical fold extending to the udder. Commonly the coat is rather coarse; only the Meknès

314. Skull of an Algerian Shorthorn bull, frontal, lateral and nuchal views (after Duerst)

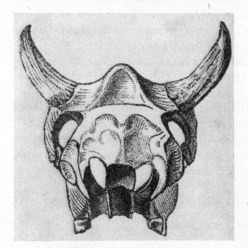

315. Skull of an Algerian Shorthorn cow, nuchal view (after Rütimeyer)

316. Moroccan Shorthorn cow

breed is distinguished by shorter and finer hair. In the eastern mountain region of Morocco the cattle are small and dark-coated, resembling the dwarf Kabyle cattle of Algeria; along the more fertile coast belt of the Atlantic they grow considerably larger (Saint-Hilaire, 1919).

In a publication issued by the 'Service de l'Élevage' in 1923 two types of Moroccan cattle are recognised: one dun with black extremities—Moroccan Brown; the other of a pale colour with the extremities slightly pigmented or unpigmented—Moroccan Blond.

The Moroccan Brown type, which represents the Brown Atlas breed in Morocco, predominates among the cattle of the mountainous parts of the country. The withers height varies from 115 to 135 cm, but in the mountains adult cows may not reach 115 cm. Bulls weigh 350–450 kg, and cows approximately 300 kg. The Moroccan Brown has a strong head with a straight profile, moderately developed orbital arches, and crescent-shaped horns which project laterally or slightly upwards. The brisket is prominent, the dewlap of moderate size, the top line is fairly straight, the flanks are flat, the rump is narrow, and the tail head not prominent. The legs are slender and well placed, and the hooves black. The skin is soft and oily. The coat is dark fawn, passing into a darker shade on the head and limbs, and a lighter hue on the top line. The muzzle is surrounded by a fringe of white hair. The skin of the scrotum, tail switch, muzzle, and mucosa of the palate, tongue, anus and vulva are black (Vaysse, 1952).

The Moroccan Blond type is prevalent among the cattle on the plateaux of Oulmès and Zair. Its withers height varies from 120 to 135 cm; the live weight averages 450 kg in the adult bull, and 300–325 kg in the cow. The head is rather long with a slightly convex profile, wide forehead, prominent poll, moderately developed orbital arches, broad ears fringed with long hair and directed backwards, and small horns projecting first horizontally, then upwards and forwards. The neck is short, dewlap large, chest deep, back straight, rump of medium width, thighs moderately well muscled, and tail head not prominent. The legs are fairly strong, with yellow or chestnut-coloured hooves. The coat of the bull is mahogany red, and in the cow of a dark wheat colour which becomes lighter with age. The mucosa is pink (Vaysse, 1952).

Vaysse (1952) writes that these two types have interbred for a long time, resulting in the appearance of numerous varieties which are called according to their origin, e.g. Zemour, Beni Ahsene, Branes, Demnat. Diffloth (1924) ascribed the variability encountered in Moroccan cattle to the haphazard crossing of three basic types. The first he refers to the red Mediterranean type or Sanson's Iberian. It occurs in the south of France, Spain and Portugal, also in Algeria and Tunisia. This is the Brown Atlas type. The second type, i.e. the blond Oulmès-Zair, is frequent in the mountain region of Zair; in Zemour it is commonly black. The third type is bred by dairy farmers in the surroundings of Meknès and to a lesser extent at Fez. It is of undetermined foreign origin, e.g. Breton, Bordeaux, Dutch; hence it resembles the cattle of northern Europe, i.e. Sanson's Netherlands type, being characterised by a long head, sickle-shaped horns, and a black-and-white coat, the white being concentrated mainly on the rump, underline and vicinity of the tail setting.

The validity of the theory that Moroccan cattle represent the crossbred progeny of different basic types is doubtful in view of the absence of any evidence that these assumed types ever existed in a pure form.

Shorthorn Cattle in West Africa

The shorthorn cattle of the west coast of Africa are closely allied to the Algerian and Moroccan shorthorn breeds, from which they are separated by a recent offshoot of humped cattle extending to the coast of the Atlantic north of latitude 14°. South of this line, trypanosomiasis and piroplasmosis are the factors limiting zebu extension (Curson and Thornton, 1936).

The number of shorthorn cattle in West Africa is relatively small; only in Dahomey and Nigeria, where they are kept farther inland, larger numbers are encountered. In Gambia, Portuguese Guinea, Guinea, Liberia, Ivory Coast, Ghana and Togo their range is commonly restricted to a coast belt not deeper than about 25 miles. In Cameroun they live in the mountains; this is due to the fact that their owners were forced to retreat from the fertile plains of the Diamaré and the Adamawa at the time of the Fula migration at the beginning of the present century.

Owing to their small size and the lack of any particular outstanding quality, the economic importance of the shorthorn cattle to the peoples of the west coast is negligible. Living in the neighbourhood of the villages in a half-wild state they do not constitute a commercial article, and the kirdi or pagans rarely concern themselves with them; for they do not drink milk and eat beef only in exceptional circumstances. On the death of a pagan one or more oxen are sacrificed and the corpse is rolled up in the hides of the slaughtered animals (Curson and Thornton, 1936). In hardiness the shorthorn cattle are superior to other breeds of cattle found in the coast regions of West Africa, being acclimatised to the tropical forest climate (Johnston, 1906). Leplae (1926) pointed out that in the forest regions of the Congo, where other breeds of cattle do not thrive, excellent results have been obtained with dwarf animals from Dahomey and Nigeria. In discussing resistance to disease, Stewart mentions that shorthorn cattle are more resistant to nagana, lung sickness and streptothricosis (contagious impetigo) than the zebu, while the latter is more resistant to rinderpest. The type "can exist and, what is more to the point, can thrive nearly everywhere" (Stewart, quoted by Curson, 1934).

Dwarfism in West African Shorthorn Cattle

In their pure form the Kirdi, Pagan, Somba, Muturu or Lagoon cattle of West Africa are of true shorthorn type, free from humpless longhorn or humped zebu admixture. Standing about 90–105 cm at the withers, they are among the smallest cattle in existence. The live weight of adult animals ranges from 155 to 300 lbs (70–135 kg). According to Henderson (1929), the Muturu is "not a special and distinct breed of small cattle", but the dwarf descendant of an at one time much larger and possibly humped type, dwarfing being due to the fact that these animals have been bred for generations under the combined adverse effects of trypanosomiasis and mineral deficiency. His evidence is based on the weight and size of the head and internal organs, which in most West African Shorthorn cattle are out of proportion to the general size of the body.

Stewart (1931) wrote that the specimens examined by Henderson "may have been dwarfed", but "the breed is not a dwarfed breed" (Curson, 1934). Yet the small body size characteristic of the majority of this type, and especially the disproportionately large head, support Henderson's statement. It is significant that in the region of the tropical rain forest dwarfism occurs among practically all domesticated animals, in addition to man. Moreover, the Pagan cattle of West Africa are not the only dwarfed shorthorn breed in Africa. The Kabyle and Nuba cattle are also of a very small size and distinguished by a disproportionately large skull. On the other hand, there is no foundation for Henderson's suggestion that the Pagan shorthorn cattle may be derived from humped stock.

General Characteristics of the West African Shorthorn Cattle

The head of the West African Shorthorn is relatively large and slightly dished in the interorbital region. The ears are fairly large, the horns often very small, with a rough surface, and generally directed outwards and downwards, sometimes outwards and slightly forwards. In many instances the horns are devoid of cores and hence loose; some animals have one loose and one normally developed horn. Polled animals are rather rare. In Ghana, Stewart encountered only six polled cattle (all cows) out of approximately 200,000 (Curson and Thornton, 1936); he remarked that the smallest and most degenerate members of a herd were those most commonly devoid of horns.

In the bull the neck is short and thick, and the dewlap moderately developed; in the cow there is hardly any dewlap. The forequarters show better development than the narrow hindquarters. The loins and hips are narrow, and the thighs thin. The rump is prominent, the tail strong at the base, with the high tail head characteristic of the shorthorn type, and an abundant switch reaching nearly to the ground. The legs are short, suggesting a moderate degree of achondroplasia (Montsma, 1959). In general, the body shows pronounced beef conformation, as Montsma (1959) has pointed out; but the dressing percentage does not exceed 40 per cent. The meat is tough and almost devoid of fat. The udder is very poorly developed, and milk production is negligible, the cows suckling their offspring until they dry off in the course of nature.

The coat colour is generally black or dark brown, often with white markings; red animals are rare. The darker shades of colour are associated with forest environment, the lighter with open country. The skin is commonly black, thick and usually, but not always, tight.

The Skeleton of the West African Shorthorn

The skeleton of a dwarf Nigerian Shorthorn bull, about three years of age, in the Museum of the Veterinary Research Laboratories at Onderstepoort, measures 89 cm at the highest point of the dorsal spinous processes. The spinal column shows a slight depression in the lumbar region. The ribs are narrow, and the bone is of a fairly dense material. The tips of the spinous processes of the dorsal vertebrae do not show any sign of bifidity, that characteristic indication of zebu influence. This is significant since zebu cattle and zebu crossbreds extend as far west as Ghana and the region behind, in the bend of the Niger.

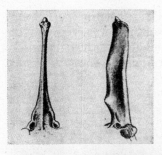

317. Ninth dorsal vertebra of West African Shorthorn bull (after Curson and Epstein)

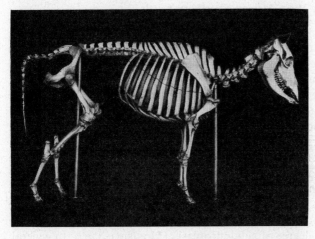

318. Skeleton of a Nigerian Shorthorn bull, Onderstepoort

Cranial Characteristics of the Nigerian Shorthorn

The skull of the West African Shorthorn resembles that of the brachyceros (palustris) breed of the ancient lake dwellers of Switzerland, i.e. Rütimeyer's type specimen (Curson and Epstein, 1934). The forehead is 197 mm long and very uneven, mainly owing to a central bulge which steeply descends in all directions. Between the orbits this bulge passes into a distinct dish. Its descent to the nasals is interrupted by the broad supraorbital fossae. From the fairly high and very wide Torus frontalis the frontal elevation in the centre of the forehead is separated by a shallow depression. The parietal extension into the forehead is well marked. The torus slopes rather steeply to the horn cores. The uneven surface of the forehead has been described by all authors on this subject as a typical shorthorn feature; it has been stressed particularly by Adametz (1895, 1898a) in his descriptions of the primitive Illyrian and Albanian shorthorn breeds.

The horns in the Onderstepoort specimen are short and relatively thick, proceeding from the skull in an outward and upward direction, with the tips pointing slightly forwards. The length of the horn is 78 mm, the basal girth 156 mm and the basal core girth 132 mm (Epstein, 1934). The colour of the horn is greyish black throughout. The horn cores are not attached to necks but directly to the skull. They are only slightly compressed and turned forwards on their axes.

The angle formed by the forehead and the occipital plane is 80°, i.e. more acute than the average figures for primitive shorthorn cattle which vary from 85 to 90°. The exact measuring of this angle is difficult owing to the unevenness of the forehead and the height of the torus; in fact, the size of this angle to a large extent depends on the development of the Torus frontalis, which, again, depends on the moment of rotation of the horns. It seems therefore of less importance in the systematics of shorthorn skulls than has been generally assumed. The nasals are fairly broad and slightly bent at the posterior end. The outer edges are curved, and the exterior processes well developed. The malar point is prominent and the palate flat. The lacrimals run in a straight line from the orbits to the nasals. The premaxillae end at a distance of 20 mm from the latter. There is a triangular aperture at the junction of the frontals, nasals and lacrimals. Laterally the orbits stand out considerably. Their temporal edges are sharp and rugged. The orbits are slightly turned forwards and almost circular in outline. The moderately deep temporal fossae are narrow in front, but markedly broaden aborally; their length is 117 mm, the width at the narrowest point 28 mm, and the depth 26 mm. The hind edge of the posterior mandibular ramus ascends almost

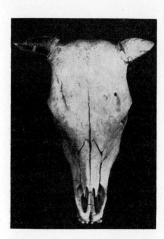

319. Skull of a Nigerian Shorthorn bull, frontal and lateral surfaces

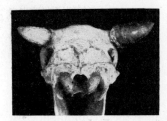
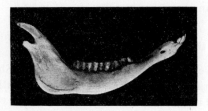

320. (left) Skull of a Nigerian Shorthorn bull (nuchal surface)
321. (right) Mandible of a Nigerian Shorthorn bull

322. Manjaca cow, Portuguese Guinea (after Da Costa)

323. Baoulé shorthorn cow, Ivory Coast (after Mason)

vertically. The anterior ramus begins to ascend gradually from the region of the first molar. The dentition is of moderate length. In the maxilla it forms a flat arch, with the concave side towards the median line; towards the mandibular dentition it forms a convex. The upper molars and premolars are nearly vertical, the lower premolars are bent slightly backwards. The lower dentition forms a concave fitting into the convex upper dentition.

Local Varieties of the West African Shorthorn

Mason (1951a) gives particulars of the West African Shorthorn cattle in the various territories of their range. In Gambia they still occur south of the Gambia river, mainly in forest areas, but are becoming extinct, chiefly owing to the economic superiority of the N'Dama longhorn which, purebred or crossed with zebu, is the predominant type. In Casamanca, the area between Gambia and Portuguese Guinea, the small N'Dama, occurring along with a larger variety (Larrat et al., 1948), may represent a remnant of the shorthorn. In the Brames region of Portuguese Guinea the shorthorn cattle are known as Manjaca (Fig. 322); they are small, well proportioned and of typical shorthorn conformation, distinguished by the black colour of the coat (Da Costa, 1933). In Guinea and Sierra Leone the shorthorn cattle, if they ever existed there, have been absorbed by the N'Dama longhorn (Mason, 1951a). Similarly in Liberia where "the black and white or brown and white dwarf small horned cattle" were described by Johnston (1906) as "the dominant breed in the coast regions", they are superseded by longhorn cattle brought by the Mandigo down from the north to the coast.

The Baoulé tribe of the central Ivory Coast breeds a humpless variety of cattle, which has been described as a sub-breed of the Lagoon cattle adapted to the north and slightly crossed with the N'Dama longhorn (Doutressoulle, 1947). The average height of these cattle is 100 cm, the weight 225 kg. The typical coat colour is black-pied.

In the southernmost part of Mali, i.e. the district of Sikasso, close to the northern border of Ivory Coast, a small humpless dark-coloured shorthorn is bred, "approaching the brachyceros rather than the N'Dama" (Doutressoulle, 1947).

The Ghana Shorthorn

In Ghana the Dwarf Shorthorn is rather rare in a pure form, but it does occur, especially in the lagoon area around Keta and the eastern part of the coastal plain, in the Brong (Abron) country to the west of Yeji and along the Black Volta river, and in Appolonia (Stewart, 1937; Faulkner and Epstein, 1957).

324. Ghana Shorthorn cow (after Faulkner and Epstein)

Stewart (1931) has described the Ghana cattle as rather similar to the Kerry in conformation. They are inclined to be light in the bone, but are well developed in the best beef points, and well ribbed up. In the Northern Territories the average live weight of a bullock is 5 cwt (250 kg), and they go up to 7.5 cwt (380 kg) live weight (Curson, 1934). Montsma (1959) characterises the shorthorn cattle in the lagoon area of southeastern Ghana as low-set, deep-chested and short-legged in relation to their withers height. In 40 adult dwarf cows he recorded a mean height at withers of 87.8 cm, a body length of 107.3 cm and a live weight of 166 kg; the average milk yield was 225 kg per lactation. On the Accra plains the shorthorn cattle are larger and heavier than the dwarf cattle of the lagoon areas, due, as Montsma (1963) suggests, to a moderate amount of zebu blood. "Almost the exact appearance and markings of well known British breeds are often seen. The following are common:—Jersey, Ayrshire, Friesian. Where Fulani herdsmen are employed, selection is largely practised and special 'types' are evolved. Red appears to be a recessive character and red herds are always the result of Fulani selection" (Stewart, 1931).

Faulkner regards the Ghana cattle as predominantly humpless shorthorn, except in the north of the Northern Territories where the type is in contact with humped Fulani cattle (Faulkner and Epstein, 1957). Stewart (1938) suggests that, in addition to zebu admixture, cross-breeding with N'Dama cattle may be responsible for the larger size of the Ghana Shorthorn the further north it is found. Mason (1951a), however, points out that, while the N'Dama and Dwarf Shorthorn "can be called breeds in the accepted sense of the word, there is no direct evidence that they preceded and are the origin of the intermediate types. It may be that the former arose from a heterogeneous population by selection and isolation, the N'Dama in the mountains of French Guinea and the Dwarf Shorthorn in the forests and lagoons of the coast. Elsewhere, in the more open country, there would be less chance of isolation and formation of 'breeds'. On this theory the intermediate types would be contemporaneous in origin with the extreme types."

This may be correct with regard to some of the West African intermediate types, but it cannot be applied generally. As shown in the following, the true shorthorn cattle of West Africa are only an offshoot of the great shorthorn branch extending from Egypt to Rio de Oro, while the longhorn cattle of West Africa, most certainly those of Lake Chad, are a remnant of the original Hamitic longhorn, ubiquitous throughout North Africa prior to the introduction of shorthorn cattle from Asia.

The shorthorn is found north of the Ghana border, in the Upper Volta (Fig. 324). In Togo and Dahomey it is bred in the lagoon region of the coast belt. During the rainy season, when the lagoon area is flooded, the animals are housed on rafts and fed on straw. In the north of Dahomey the shorthorn is encountered in the Atacora Mountains, where it is known by the name Somba. In Nigeria, where the animal is called Muturu, i.e. the Hausa term for humpless, it occurs mainly in the eastern and western regions south of latitude 8° N, except for some isolated enclaves in the hilly areas of the interior as far north as latitude 12° N (Mason, 1951a). Formerly it was common in the north. Prior to the Fulani conquest around 1820, small humpless shorthorn cattle were ubiquitous among the Agaie in eastern and northern Nupe, north-west Nigeria, and among the related Gbari in Zaria province (Nadel, 1942). Similarly, in

325. Muturu bull, Nigeria (withers height 100 cm)

326. Muturu bull, Nigeria (after Faulkner and Epstein)

327. Muturu cows, Nigeria

the Yola and Bornu provinces of north-east Nigeria they were widespread before the Fulani, on their migration from west to east, introduced vast herds of humped stock (Migeod, 1927). Barth (1857-58), who in 1851-55 made the first thorough geographical investigation of the vast tract of land between Lake Chad and the neighbourhood of Timbuktu, recorded that, while cattle had been introduced into the Kingdom of Fumbina by the Fulani, an indigenous dark-grey type, called Muturu, was still in evidence.

Simpson (1912), who encountered Nigerian shorthorn cattle in 1910, wrote:—"The most noteworthy feature in this district (Ifon and Benin) is the presence of large herds of a dwarf variety of cattle, which, according to native evidence, supported by a low rate of mortality, seems to be immune from trypanosomiasis. In the districts of Ondo, Ilesha, Ifon, Ishan in South Nigeria, all forested regions and tsetse habitats, this peculiar dwarfed variety with short legs may be seen in numbers. Their appearance is remarkable. The predominant colours are black and white, and more rarely brownish; there is no dorsal hump and the forequarters are generally lower than the hind...". Ferguson (1966) suggests that the Nigerian dwarf cattle's hig htolerance to trypanosome infections may be due to their predilection for shade which they share with tsetse-flies, thus ensuring regular exposure to the infection.

Bakosi Cattle

In the mountains of Cameroun, where the West African Shorthorn is on the way to extinction, it still exists in small herds in the Mandara, in Namji–Allan–Tikas and in the south (Curson and Thornton, 1936). (The Namji has been described by Weidholz (1939) as long-horned—see p. 207-208). The cattle of the south which are bred by the Bakosi, a Western Bantu people living in the mountain region north of the lower Sanaga river, and also by the Bakwiri, Balundu, Bakundu, Kumba, Ngolo, Bali and the inhabitants of the Rumpi mountains and the Cross river region of Cameroun, north, west and south of the Bakosi territory, are dwarfed, their withers height rarely exceeding 100 cm. They resemble the Tux–Zillertal (brachyceros) cattle of Tyrol, and, similar to these, they show a moderate degree of disproportionate (achondroplastic) dwarfism (Staffe, 1937). The head is short and broad. The skull, of pure 'brachyceros' character, is characterised by marked width, the great length of the neurocranium and the relative shortness of the splanchnocranium, more especially of the nasals, the unevenness of the forehead, the high and wide frontal torus, large circular orbits and long narrow temporal fossae. The profile shows a slight depression below the Torus frontalis, which passes into a moderate elevation orally; the latter slopes down to a marked depression between the anterior edge of the orbits and the root of the nasals, rising again to the very high posterior part of the nasals. The short horns are commonly black, occasionally light-coloured with dark tips. The long cylindrical body is supported by short slender legs. The skin, in contrast with other West African Shorthorn cattle, is markedly wrinkled. About 90 per cent of Bakosi cattle are black, the remainder black-and-white, greyish brown, yellow or variegated. The Bakosi never milk the cows, and keep the cattle only for lobola (bride price) and for the meat and hides used in the cult of the dead. Bakosi cattle are late-maturing, the cows usually calving for the first time when nearly five years old (Staffe, 1938).

329. Head of a Bakosi cow (after Staffe)

328. Bakosi cow (after Staffe)

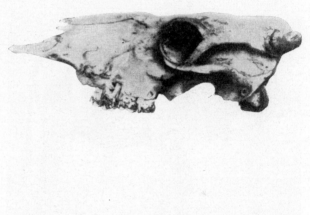

330. Skull of a Bakosi cow, frontal and lateral views (after Staffe)

Shorthorn Cattle in the Western Congo

In the western border regions of the Congo (Mayombe) the West African Shorthorn is encountered in small herds which lead a solitary life in the dense undergrowth of the forest. They closely resemble the Lagoon cattle of the coastal area of Dahomey, from which they are believed to be descended (Flamigni, 1951). They are small and sturdy, always in good condition and free from external parasites. The live weight varies between 150 and 200 kg; adult bulls may occasionally reach or even exceed a weight of 300 kg. These cattle are furnished either with short or with medium-sized horns which are often curved downwards; polled animals are frequent. The colour of the coat is commonly black, less often grey, black-and-white or speckled, rarely red or white.

Leplae (1926) referred to a herd of variegated, humpless, polled or short-horned cattle at Zambi near the mouth of the Congo river as 'Angola cattle'. While Angola cattle have indeed been imported into the Congo, the origin of the Congo shorthorn

CATTLE

from Angola is quite unlikely, as shorthorn (brachyceros) cattle do not occur in Angola. According to Flamigni (1951), they were at one time imported into the Congo from the north (Dahomey), but he does not say whether this importation took place by land by way of Cameroun or by sea. At any rate, the western equatorial Congo region now constitutes the southern limit of the range of the shorthorn in West Africa.

Borgu Cattle

While in Dahomey pure shorthorn (brachyceros) cattle are found in the lagoon area of the coast, north of Kotonu, in the valleys of the Oneme river and its tributaries, and in the Atacora Mountains, and nearly pure shorthorn cattle in a few native villages at the Zu river (Flamigni, 1951), four-fifths of the cattle of Dahomey are of the Borgu breed. "These same animals are encountered in the corresponding regions of the neighbouring .... Togo, Nigeria, Ivory Coast (Mere–Lobi), with small differences due to

331. Short-horned Bayanzi bull from Kasai, Congo (after Johnston)

332. (left) Borgu bull (after Pierre)
333. (right) Borgu cow (after Mason)

environment or to selection" (Doutressoulle, 1947). They are known by various names, such as Borgowa, Kettije, Ketaku, Keteku, Ketari, Kaiama, in addition to Borgu.

The Borgu may formerly have been pure shorthorn, as many recent animals are still humpless and short-horned. Yet, as in the case of the shorthorn cattle of the northern parts of Ghana, the general conformation of the body as well as the occasional occurrence of relatively long horns in the Borgu indicate the influence of White Fulani and possibly N'Dama longhorn blood. In western Nigeria this type of cattle, called Keteku, is derived from a mixture of the humpless dwarf Muturu of the south and the zebu of the north. Most of the Keteku cattle of western Nigeria are in the hands of Fulani who have become settled, or partly settled, in this area (Hill and Upton, 1964). In Gourma, bordering north-western Dahomey, the Borgu type has been crossed with humped cattle to such an extent that the brachyceros (shorthorn) part of its ancestry has been nearly eliminated. Gates (1952), who regards the Borgu as a humped breed, describes the hump as follows:—"Hump varies from small to very small, muscular cervical, although the greater the proportion of White Fulani blood that is evident the larger and further back is the hump." In body size, conformation and colour the Borgu displays considerable variability. The height at withers ranges from 90 to 115 cm. The body is strong and muscular, the head well proportioned, either straight in profile or dished; the vertical or crescent-shaped horns are 15–30 cm long and 15–18 cm in basal circumference, round in cross-section, with thick, relatively short cores. The neck is short, and the dewlap, extending from the chin to the chest, small. The umbilical fold is little developed. In bulls the withers are prominent. The back is long, the rump short but well muscled, and the legs are short. The tail is of medium length and ends in a

334. Biu cow and calf

full brush. The basic colour of the coat is white with black skin and points; in addition to this most frequent pattern, black, black-and-white, fawn, fawn-and-white, or grey-spotted animals are encountered. Borgu cattle fatten readily, but the milk yield of the cows is negligible.

Biu Cattle

A type similar to the Borgu but somewhat smaller in size is bred in the hilly volcanic area of Biu Emirate in Bornu Province of Nigeria. The Biu cattle, numbering not more than one or two thousand head, are commonly white with black points and black markings. Like the Borgu, they are the result of cross-breeding between humpless Dwarf Shorthorn and humped White Fulani cattle (Gates, 1952); this accounts for their relatively long horns (Fig. 334).

ii. The Geographical Origin of the African Shorthorn Cattle

Records of Shorthorn Cattle in Ancient Egypt

Shorthorn (brachyceros) cattle began to enter Africa by way of the Isthmus of Suez towards the middle of the 3rd millennium B.C. In Egypt they are depicted in tombs of the Vth Dynasty, VIth Dynasty, XIIth Dynasty and onwards (Lucas, 1948). At first their number was small; only a few ancient wall paintings of the type are in existence. A shorthorn cow, suckling her calf, is depicted in a rock tomb at Meir III (Fig. 335). On a wall in one of the tombs at Saqqara a dark-brown shorthorn ox is painted in company with several light red longhorn cattle; the shorthorn, considerably smaller than the longhorns, resembles the recent shorthorn cattle of Algeria (C. Keller, 1896). From the XVIIIth Dynasty dates a relief of a shorthorn bull with the high withers, approximating to a hump, which are characteristic of many recent shorthorn cattle in Egypt (Fig. 336). Another bull with a very high crest (Fig. 337), resembling a recent Baladi bull (Fig. 300), is represented on the stele of Nebwawi from Amarna (XVIIIth Dynasty). Zeuner (1963a) regards the animal as a zebu; but its conformation, more especially the unbroken backward slope of the crest, high tail head, straight underline and short head, render this somewhat doubtful.

For a considerable time longhorn cattle prevailed over the shorthorn type. But gradually the latter grew in number, and its influence became more marked. Every successful military expedition of the pharaohs against the rulers of Syria and Palestine brought new shorthorn herds into the Nile country. Remnants of dispersed Asiatic tribes entered the Nile valley as settlers, their shorthorn cattle swelling the ranks of the breed in Egypt. During the Hyksos period (c. 1700–1580 B.C.) shorthorn cattle became the predominant breed. The Egyptians evolved from them also a polled type, as they had formerly evolved a hornless type from the longhorn. In an ancient wall painting a

335. (left) Shorthorn cow with calf, from the rock tombs of Meir III (about 2000 B.C.)

336. (right) Shorthorn bull in sacrificial procession (XVIIIth Dynasty)

337. Bull with a prominent crest on the stele of Nebwawi, Amarna (XVIIIth Dynasty) (after Davies)

group of such small polled animals is depicted, including two oxen with short horns, indicating the original type from which the polled strain was derived.

While in daily life the shorthorn came to occupy the place formerly held by the longhorn, in the cult of the dead the original longhorn type of cattle continued to be depicted in paintings and frescoes on the walls of tombs well into the time of the New Kingdom (Pia, 1941).

Shorthorn Cattle in the Ethiopian Region

Mural bas-reliefs from Hatshepsut's temple at Deir el-Bahari, recording her expedition to Punt between 1486 and 1468 B.C., depict the cattle of Punt as belonging to two different breeds—long-horned and short-horned. Apparently shorthorn cattle had already penetrated to the Ethiopian region by the middle of the 2nd millennium B.C. (Clark, 1954). In Nubia (southern Egypt and northern Sudan) humpless longhorn cattle remained the dominant type during the period of the New Kingdom of Egypt, as indicated by the tribute herds depicted during the reigns of Tuthmosis III (1490 to 1436 B.C.) and Tuthmosis IV (1406–1398 B.C.).

The Introduction of Shorthorn Cattle into the Atlas Countries

In the coast region of North Africa the change from longhorn to shorthorn stock seems to have been accomplished soon after the change in Egypt. Sculptures of cattle recovered from the debris of the Phoenician colony of Carthage show only the shorthorn type.

Ducloux (1930) has expressed the opinion that the recent Brown Atlas cattle of Tunisia are derived from Spanish as well as Asiatic sources, because the type is to be

338. Small polled cattle from Ancient Egypt

339. Shorthorn cattle grazing in the temple garden of Amen (mural bas relief at Deir el-Bahari—about 1500 B.C.)

found in the south-west of France, in the Spanish peninsula, and in North Africa. In view of the fact that North Africa received also some of its sheep and horses from southern Spain, it is possible that the shorthorn cattle of Morocco and Algeria at some time or other interbred with imported Iberian stock. But owing to the great similarity in racial type between the shorthorn breeds of North Africa, western Asia and southern Europe it is very difficult to decide how far interbreeding with Iberian shorthorn cattle has influenced the racial type of the cattle of the Atlas countries.

The Introduction of Shorthorn Cattle into West Africa

"West Africa," Curson and Thornton (1936) write, "represents, as does South Africa, a cul-de-sac and it is the furthest point which could be traversed by tribes migrating from the north and east. Unlike South Africa, however, the unfortunate nomads could not continue following the coast-line southwards (as did the Hottentots in South Africa) for their advance was blocked by the immense equatorial barrier of Glossina. Presumably what happened was that the earliest people with their Hamitic cattle did not penetrate the dense forest region bordering the Gulf of Guinea, but dispersed in the more open country between French Senegal and Northern Nigeria. The succeeding wave of migration, with Brachyceros herds, on finding only the littoral unoccupied were accordingly compelled to inhabit the Glossina infested jungle along the Gulf of Guinea. These territories are known to-day as French Guinea, Liberia, Ivory Coast, Gold Coast, Dahomey, Southern Nigeria and French Cameroons. Through living in such an unfavourable environment Brachyceros cattle have in the course of centuries deteriorated and are now considerably smaller in size than their relatives in Europe, e.g. the Jersey, Guernsey and Kerry. While dwarf-like are of no value at present for dairy purposes, the type is at least resistant to Nagana, a malady which would kill European cattle in a few weeks."

Theory of the Origin of West African Shorthorn Cattle from Europe

This conjecture on the origin and dispersal of the shorthorn cattle of West Africa is acceptable. On the other hand, several authors have expressed the opinion that the recent shorthorn cattle of West Africa originate from Europe. Thus Johnston (1906) suggested:—"The black and white or brown and white dwarf small-horned cattle which are the dominant breed in the coast regions of Liberia seem to be entirely of European origin and to have come from Holland"; and further:—"The parti-coloured short-horned cattle, though they may have been brought direct from Northern Europe, have in the course of three centuries become a well established local breed." Pierre (1906), on the other hand, noted that "this animal presents all the characteristics of the Iberian breed imported into Madeira, the Azores and Angola long ago". Forde (1934) writes that among the Yoruba speaking African peoples who occupy the greater part of south-western Nigeria "a few head of non-humped cattle are to be found in most villages. The cattle, which appear to be immune to the tsetse fly, are something of a mystery, since all other native west African stock are of the humped zebu type; they may, however, be descendants of cattle introduced by the Portuguese in the fifteenth century. But before this time other domestic animals had arrived overland from the north."

None of these authors, however, has produced any evidence in support of the European origin of the West African Shorthorn. No records exist of early importations of Dutch cattle into West Africa. It would appear that Johnston arrived at his view purely from a consideration of the black-and-white or red-and-white coloration common to the cattle of the Netherlands and some West African Shorthorns.

The theory that shorthorn cattle were introduced into West Africa by seafarers from the Iberian peninsula is denied by the antiquity of their existence in West Africa, as indicated by their presence in rock paintings reported from the Bauchi Plateau of Nigeria (see p. 218) and the high degree of their acclimatisation to the moist forest climate of the tropics, and also by their range which is not restricted to West Africa but extends from there, in a nearly uninterrupted line, to the shores of the Mediterranean. Therefore it is more likely that the shorthorn cattle of West Africa are an offshoot of the shorthorn cattle of the Atlas region than of relatively recent European origin. In this connection it is also notable that the present-day cattle of Portugal and Spain, whose seafarers are so often credited with the introduction of livestock into almost every part of Africa, are of longhorn, a very few of medium-horned type, although the former occurrence of shorthorn cattle in the Iberian peninsula is evidenced by a cranial fragment from the early bronze age in the Museo Archeologico of Madrid, and their more recent presence in the mountain regions of northern Spain (Galicia, León, Pyrenees) has been claimed by Stegmann von Pritzwald (1924).

Theories of an Autochthonous Origin of the North and West African Shorthorn Cattle

Cabannes and Serain (1963) assert that the Brown Atlas type is derived from the domesticated Bos taurus brachyceros ibericus, the geographical distribution of which extends from the Atlantic to Tripolitania and Fezzan. Bos taurus brachyceros ibericus, in its turn, is a descendant of the brevilinear, concavilinear and elliptometric Bos taurus brachyceros (longifrons Owen), a variety dating from the beginnings of the palaeolithic era. The latter, along with the rectilinear, mediolinear and hypermetric Bos taurus primigenius mauritanicus, was descended from the fossilized Bos taurus which formerly existed in North Africa.

A similar view has been expressed by Stewart (quoted by Curson, 1934) on the origin of the West African Shorthorn. As the shorthorn cattle of West Africa are owned solely by the settled tribes, while the humped stock are mainly the property of the Mohammedan nomad tribes, of alien origin, Stewart believes that "these shorthorn cattle are the oldest domesticated breed in West Africa", domestication having taken place in the Sahara "when it was a fertile well-watered country". All those tribes, Stewart proceeds, claim as their original home Ghana or Walata, a dynasty which had its centre in the western Sudan to the W.N.W. of Timbuktu on the Upper Niger (2000 years ago). According to tradition the people were pastoral. The people of Ghana were driven south by the later Melestines and Songhais. The descendants of Ghana are African types and are definitely not Hamite, Berber or Semitic as are the owners of the humped cattle of the western Sudan. They now occupy a territory far to the south of the ancient Ghana, some of them even on the coast of the Gulf of Guinea.

Stewart's theory that the shorthorn cattle are the oldest domesticated breed in West Africa cannot be supported. In West as in North and East Africa the shorthorn was preceded by the Hamitic Longhorn, remnants of which are still found among the tribes of the Fouta Djallon plateau and adjacent territories. The earliest rock paintings in West Africa depict the humpless longhorn type (see p. 218). When the true African peoples of West Africa acquired their first cattle from Hamitic tribes from North Africa during the second half of the last pre-Christian millennium, the shorthorn had already superseded the Hamitic Longhorn throughout the Atlas countries. In a cave at Geji, on the Bauchi Plateau of Nigeria, are paintings of small short-horned cattle without humps, in addition to humpless longhorn cattle. The shorthorn cattle may have been owned by the people of the Nok culture, called after a village of the Jaba people in Zaria province. This iron age culture is believed to have flourished from about 900 B.C. to A.D. 200, reaching its full development during the last two or three centuries B.C. (Davidson, 1961). With the disintegration of the Nok culture about A.D. 200, the cattle were dispersed to the south and west (Kennedy, 1958).

Further, Stewart's belief that the shorthorn was domesticated in the Sahara when this was fertile well-watered country is without foundation. During the neolithic wet phase (Subpluvial II or second wet phase), which lasted from about 5000–2500 B.C. and temporarily halted the general desiccation of North Africa since the Allerod (about 9000 B.C.), cattle, sheep and goats were present all across the Sahara (Reed, 1959). But the cattle represented in innumerable rock engravings and paintings in the desert are exclusively of the humpless longhorn type. There is therefore no evidence to show that the shorthorn was domesticated or evolved in North Africa rather than introduced from Asia.

This also applies to the above-mentioned view expressed by Cabannes and Serain (1963), with the addition that not only is there not a shred of evidence of the presence in North Africa of an autochthonous, early palaeolithic Bos taurus brachyceros, but, as will be shown later on, a wild Bos taurus brachyceros, from which the domesticated shorthorn (brachyceros) type is claimed to be descended, did not exist anywhere else either.

Theory of the West African Shorthorn's Introduction from Asia in pre-Hamitic Times

Staffe (1938) doubts that there is a connection between the shorthorn cattle of the North African littoral and those of West Africa, as north of latitude 14° the contiguity of their range is interrupted by the intrusion of humped cattle. Since the pagan owners of the shorthorn cattle in West Africa do not milk their cows, Staffe believes that the shorthorn entered Africa from western Asia at a period of great antiquity, prior to the al-'Ubaid period of early Sumer (c. 3900 B.C.) when the milking of cows became known. Also the shorthorn must have been introduced into Africa by a pre-Hamitic people, because the Hamites had already learnt to milk their longhorn cows. Longhorn cattle arrived in Africa later; they proved so adaptable that the shorthorn type became restricted to a relatively small area (Staffe, 1944).

There are several factors which Staffe has failed to consider: in the first place, the tremendous changes in population that have taken place in West Africa in the course of four thousand years or more as a result of the movements, invasions and incursions of different peoples in possession of different types of cattle. Humped cattle were carried across the savanna belt of the Sudan into northern Nigeria, the western Sudan and upper Senegal to the shores of the Atlantic at a relatively recent date. Secondly, the fact that the majority of the settled cultivators of West Africa do not exploit the milking properties of their cattle is due to the cultural and economic peculiarities of pagan society rather than to their ancestors having received the shorthorn before the practice of milking had been developed in Asia and Africa. This is illustrated by the absence of milking among the Namji and Pape, true negro cultivators who are in possession of a longhorn breed in a small isolated enclave in northern Cameroun, between the Fula settlement of Ngaundéré in the highland of Adamawa and the trading centre of Garoua on the Benue river (Weidholz, 1939). When the negroes of the Sudan came into contact with the neolithic peoples of North Africa, they readily borrowed cattle, goats and sheep from the latter; they did not, however, adopt milking, which even today occurs in West Africa only where introduced in recent times by the Fulani (Murdock, 1959). (A parallel phenomenon occurred in China: sheep, goats and cattle were introduced from the west at an early date, but milk was not used.) Finally, there is no evidence to show that shorthorn cattle preceded the longhorn in Africa, or that shorthorn cattle were evolved in western Asia before the art of milking was developed. On the contrary, the available evidence indicates that longhorn cattle were bred in Africa long before the shorthorn entered the Continent, and that the shorthorn was developed in west Asia as a dairy type.

Theory of the Descent of the Shorthorn from the Zebu

According to another theory, propounded by C. Keller (1896), the progenitors of the shorthorn cattle are to be traced to the zebus of Africa. This theory, based on a number of Somali zebu skulls distinguished by several shorthorn features, has been rejected by the majority of authors on this subject. Were it correct, we should be forced to conclude that those short-horned zebus which were taken from Somaliland to the north lost their humps, the bifidity of the neural spines of the anterior dorsal vertebrae, and all other zebu features, changing into true shorthorn (brachyceros) cattle, whereas those passing southwards and westwards preserved their zebu character. Nor could this theory be supported by reasons of environment and adaptation, since climatic conditions in North and West Africa are not very different from those prevailing in several areas of the continent where zebu cattle are maintained. Again, if the shorthorn cattle of Egypt had come from the south, we ought to find at least some ancient records in support of this theory; yet there are none, all existing evidence indicating that the shorthorn cattle entered Egypt from the north through the Isthmus of Suez whence they spread to the south (Nubia and Ethiopian region) and west (Atlas countries and West Africa). Similarities between shorthorn (brachyceros) skulls and some crania of short-horned zebus are due to the similarity in the mechanical effect of the short horns in both types on cranial, more especially neurocranial, conformation.

The Breeding Area of Shorthorn Cattle North-East of Egypt

In western Asia, in the countries that flank the route of the shorthorn to Egypt, i.e. Palestine, Syria and Asia Minor, the cattle are of true shorthorn type to this day. The Arab cattle of Palestine (Israel and Jordan) form the southward extension of the Oksh or Chaissi of Syria (in Hama this breed is called Klaiti, and in Homs Anatolian), which is the common native breed of that country, constituting about 80 per cent of its total cattle population. In addition, three other shorthorn breeds are bred in Syria and the Lebanon: the Jaulan, in the southern parts of Syria, in the hilly country east of Lakes Tiberias and Huleh (Jaulan), the Shami or Damascus cattle, and the Beirut or Lebanon cattle.

The Oksh Cattle of Palestine

In southern Palestine (Negev) the Arab (Oksh) cattle are of very small size and commonly black in colour; in Galilee and in Syria they are heavier and more compact in build, and fawn, brown, black, more rarely brown-and-white or black-and-white. The head is rather heavy, with small horns. The withers are high, particularly in the bull in which they may outwardly resemble a zebu hump while being due solely to the high neural spines. The chest and rump are narrow, the back is weak, and the tail setting high. The legs are poorly muscled. The average live weight is 220 kg, ranging from 180 to 250 kg. The cows yield from 400 to 800 kg milk per lactation, testing 4–5 per cent butter-fat.

Cranial Characteristics of the Oksh Cattle

The crania of a bull and a cow (Figs. 342 and 343), in the author's possession, show the true shorthorn character of the Oksh cattle. The upper portion of the frontals rises to

340. Oksh bull from Palestine

341. Oksh cow from Syria

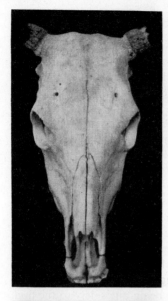
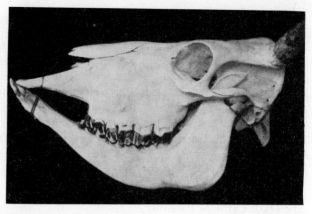

342. Skull of Oksh shorthorn bull from Palestine, frontal (left) and lateral (right) views

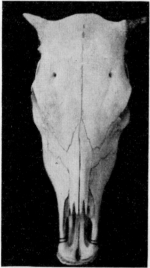
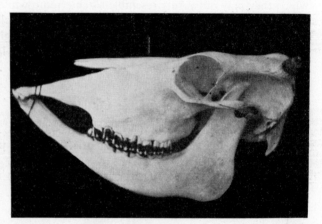

343. Skull of Oksh shorthorn cow from Palestine, frontal (left) and lateral (right) views

a torus, more prominent in the female than in the male skull, and exceeding the bases of the horn cores in height, with a slight concavity in the centre. In the male cranium the intercornual ridge is wider than in the female. The forehead is devoid of a central elevation. The orbital ridges are more prominent in the male than in the female skull. The central depression of the forehead starts below the frontal ridge and ends at the root of the nasals. The supraorbital grooves are shallow, and the orbital ridges of the male skull stand well out laterally. There is no triangular aperture between the lacrimals, frontals and nasals. In the male skull the outer edges of the nasals are curved; in the female less. The maxillae are narrow; the malar point is not marked; the palate is flat. The temporal fossa is shallow, broadening slightly aborally. The occiput shows true shorthorn character. The nuchal portion of the Torus frontalis is only slightly higher than the squama. Duerst (1899), in an analysis of the skulls of a bull and a cow from Syria, drew special attention to the angle of 90° formed by the squama and the forehead, as determined by lines joining the orbits with the Torus frontalis; according to Adametz, this is one of the most important characteristics of all pure shorthorn cattle; but, as pointed out in connection with the Nigerian Shorthorn skull (p. 279), the significance of this feature has been much overrated. The direction of the short horns in the Oksh skull is outward, upward and forward. In animals with very small horns the latter occasionally point backwards. In some animals one or both horns may be loose. In the cranium of the bull (Fig. 342) the horn is 91 mm long, with a basal circumference of 127 mm; the horn of the cow (Fig. 343) measures 52 mm in length, with a basal girth of 81 mm. The ascending ramus of the mandible forms an obtuse angle with the horizontal ramus which markedly rises in front of the first molar. The teeth are set obliquely, the upper cheek-teeth pointing backwards. The length of the first and second molars exceeds their width, while the third molar and the second and third premolars are of greater width than length.

The Jaulan Cattle

The Jaulan breed is superior to the Oksh, being of sturdier build, the cows giving more milk and the oxen being stronger for draught purposes (Curson and Thornton, 1936). The colour of the Jaulan cattle is usually black with a few white patches on the underline and the extremities. The head is commonly white, with black ears, black muzzle and nostrils, and black circles around the eyes, protecting the tender parts of the head from damage by intense solar radiation. The live weight of Jaulan cattle varies between 300 and 400 kg. Under their home conditions cows yield about 700–1200 kg milk per annum, testing 4.3 per cent on an average.

344. Jaulan cow

The Damascus (Shami) Cattle

The Shami or Damascus cattle are by far the highest milk producers of all the indigenous breeds of the Near East. This breed has been evolved in the fertile Ghuta flat both sides of the Barada river, but it is now bred also in the Kuweik valley in the vicinity of Aleppo, and the irrigated lands of Homs and Hama. It is notable that the shorthorn breed most closely related to the Damascus cattle in type, conformation and colour is found in Cyprus (Fig. 349). The Damascus is a single-purpose dairy breed, representing the extreme Typus respiratorius—very leggy, narrow and thin of bone. The colour ranges from light to dark brown to nearly black, with the head, neck and hindquarters a darker, and the inside of the legs a lighter shade. White patches on the underline are rare. Bulls are commonly darker than females. The skin is very thin, and the coat, of fine shiny hair, light in weight. The head is very long and narrow, with a large dark muzzle which is frequently surrounded by a narrow ring of white hair. The horns are only a few centimeters long; mostly flat, thin and deformed. The neck is very thin, with a moderately developed dewlap. Adult bulls develop a prominent muscular crest on the neck, approximating to a cervico-thoracic hump; but the tips of the neural spines of the anterior dorsal vertebrae are not cleft (Hirsch and Schindler, 1957). The withers are high, the chest is narrow and fairly deep, the back and loins are long, narrow and poorly muscled. The rump is very high and narrow, with the high tail head characteristic of the brachyceros type. The tail is thin and of medium length, ending in a dark brown or black tuft. The legs are long and thin, with weak muscles. The udder is large, often badly shaped, and brown or white in colour. The withers height varies between 130 and 150 cm in cows, reaching up to 160 cm in bulls. The latter weigh approximately 700–750 kg. The live weight of cows ranges from 340 to 500 kg, and the milk yield from 2000 to 4500 kg, testing approximately 4 per cent butter-fat.

The Lebanon (Beirut) Cattle

The Beirut or Lebanon cattle are commonly bred by small-holders in the villages in the mountains and valleys of the Lebanon, and on the coastal plain in the vicinity of Saida,

345. Damascus bull

346. Damascus cow

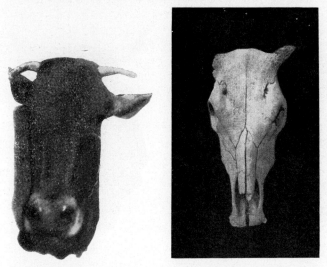

347. (left) Head of a Damascus cow (after Hirsch)

348. (right) Skull of a Syrian shorthorn cow (with one loose horn) (after Duerst)

Beirut and Tripoli (Hirsch, 1932). The Beirut is larger than the Jaulan, but smaller and sturdier than the Damascus. It is not a uniform breed, but rather a crossbred type derived probably from Damascus, Oksh and Red Anatolian cattle introduced from Turkey. The colour is red or fawn, occasionally black. The inside of the legs is frequently white, while the head and neck are darker than the body. The dark muzzle is surrounded by a narrow circle of white hair. The horns, yellowish white in colour with black tips, are larger than those of the Jaulan and the Damascus cattle. The head, while of pure shorthorn character, is shorter and wider than that of the Damascus.

349. Cyprus bull from the plains

350. Beirut cow (after Hirsch)

351. Head of a Lebanon cow (after Hirsch)

352. Statuette of a shorthorn bull from Ugarit, northern Syria (c. 1400 B.C.)

The skin is of medium thickness, the hair rather coarse. The neck is short and moderately broad, with a well developed dewlap. The chest is wide and deep, the back rather weak, the hindquarters are short and well muscled, and the rump is high, with the high tail head characteristic of the brachyceros (shorthorn) type of cattle. The tail is long and rather coarse. The legs are short and strong. The live weight ranges from 350 to 450 kg. Under pasture conditions the cows produce from 1000 to 1500 kg milk per lactation, testing 3–4 per cent butter-fat.

Shorthorn Cattle in Ancient Syria

The statuette of a shorthorn bull from Ugarit (Ras Shamrah), on the coast of northern Syria, dated to approximately 1400 B.C., indicates the ancient lineage of the shorthorn cattle of Syria.

The Anatolian Cattle

In Asia Minor both longhorn (Plevna or Thracian) and shorthorn (Anatolian) cattle are encountered, the former having probably been introduced from the southern Balkans during the period of Turkish expansion in Europe, the latter indigenous since the beginning of the Hittite period, approximately 4000 years ago. At Osmankayasi near Boghazköy, remnants of small cattle, similar to the recent shorthorn cattle bred in this area, have been found in Hittite graves dated to the 17th–14th centuries B.C.; the withers height of a cow, represented by a metacarpus, is estimated at 109 cm (Vogel, 1952; Herre and Roehrs, 1958). Among the recent shorthorned Anatolian cattle two colour varieties are distinguished: black and red; Central Anatolian Black, East Anatolian Red (Çildir and Göle) and South Anatolian Red (Aleppo, Çukurova, Dörtyol, Kilis). Undoubtedly the Anatolian shorthorn constitutes the parent type of the Oksh breed of Syria and Palestine. Duerst's (1899) description

353. Black Anatolian shorthorn cow

of a shorthorn skull from the Dardanelles resembles in all important points our description of the Oksh skulls.

The Anatolian is a small sturdy animal, the cows weighing approximately 220 kg and producing from 500 to 800 kg of milk. The size of the head is proportionate to the body. It is widest between the orbits whence the neurocranium markedly narrows aborally. The profile is straight, but the front view shows a slight depression between the orbits, below the well marked bump of the forehead (the so-called 'Illyrian bulge', caused by the prominence of the brain case at this point of the shorthorn skull). The intercornual line is straight. The horns are slender, crescent-shaped and about 15 to 20 cm long. The ears are of medium length, and carried slightly above the horizontal. The neck is short or of moderate length, rather weakly muscled, with the skin thrown into numerous wrinkles. A moderately developed dewlap, free from folds, extends from the throat to the brisket. The withers are very high and sharp, the back is weak, rising to a fairly high sacrum. The barrel is deep and full. The rump may either be straight or, owing to the high sacrum, droop slightly to the prominent tail setting. The tail is coarse. The hindquarters are weakly muscled. The legs are slender but not long. In the Anatolian Black the hair is blackish brown in colour or a very dark dun; in the East Anatolian Red and the South Anatolian Red—red, brown, occasionally greyish brown.

Asia Minor—Starting Point of Two Shorthorn Migration Routes

The fact that the migration route of the African shorthorn cattle can be traced back to Asia Minor does not imply that this was the original centre of evolution of this bovine type. Asia Minor was the starting point of two important migration routes of shorthorn cattle: one to Egypt, and thence along the southern shores of the Mediterranean to the west coast of Africa; the other to Greece, the coastlands of the Adriatic, Switzerland, France, the Channel Islands and the British Isles. But the area of distribution of the shorthorn cattle is more extensive than the countries mentioned above.

Shorthorn Cattle in Northern Iraq

The earliest records of shorthorn cattle are encountered in Mesopotamia, suggesting that Asia Minor received the shorthorn from the south-east. To this day the (Kurdi) cattle of Kurdistan, including northern Iraq, are of shorthorn type. The coat colour of the Kurdi has been described as "black, often with light markings" (Mason, 1951b), and that of the small, humpless hill cattle of northern Iraq as black-and-dun (Williamson, 1949). Of several thousand head of shorthorn cattle from northern Iraq, which the author handled in 1944, nearly all were cream or golden coloured, of very slender yet proportionate build, certainly the most beautiful type of shorthorn met with in all of western Asia. They probably were of the Dishti breed, for a very few of them had small humps due no doubt to contact with zebu cattle from southern Iraq. But in cranial and body conformation even these showed the pure shorthorn type of the humpless majority of these cattle.

Early Breeders and Diffusers of Shorthorn Cattle

Texts from the archive of cuneiform tablets at Boghazköy testify to the presence in Asia Minor during the last pre-Christian millennia of a great mixture of human races. These belonged to two main stocks: Indo-European and either mongoloid or alpine. Of these the Indo-Europeans were comparative late-comers. If Barton (1934) is correct in regarding the Proto-Hittites as Mongoloids, the latter may have to be credited with the introduction of the shorthorn into Asia Minor. Moreover, should the Swiss pile dwellers, whose palustris cattle have been described as the prototype of the shorthorn, actually have been mongoloids, as Lydekker (1912a) and others believed, the first westward movement of the shorthorn into southern and central Europe would likewise be ascribable to Mongoloid migrations. There is, however, no unanimity among authors as to the racial affinity of the Swiss lake dwellers. Myres (1933) regarded them as of alpine stock, the westward counterpart of and continuous with the anatolian or armenoid type of man. The introduction of shorthorn cattle into Britain and, later, Ireland is ascribed to immigrants of this alpine race. They came across the North Sea at the beginning of the second pre-Christian millennium—a round-headed people with fairish or chestnut hair, who owed their strength to the fact that they had learnt to make implements of bronze. They were followed by the Celts, a great congeries of peoples, partly of alpine stock, partly blended with the big-boned fair-haired Nordics of northern Europe, who spoke an Aryan language and had weapons of iron (Muir, 1936). The Celts also were in possession of shorthorn cattle; and from them this early type of British cattle has derived the name—Celtic Shorthorn. Thus there is overwhelming evidence in support of the view that the westward movement of the shorthorn is to be ascribed to armenoid and alpine migrations.

Nor has the theory of the occurrence of mongoloid folk in pre-Hittite Asia Minor remained unchallenged. As Myres (1933) wrote:—"On the tableland of Asia Minor the earliest portraits of Hittite peoples (about 1285 B.C.) have been thought by some to be Mongoloid; but the evidence is still scanty and inconclusive: on Hittite monuments bearded figures are frequent, and the type is Armenoid." The acceptance of Myres' view would imply that an armenoid people introduced the shorthorn into Asia Minor.

But the importance of this question should not be overrated. The reasons for the great expansion of the shorthorn cattle and their superseding the longhorn cattle over large parts of western Asia, Europe and North Africa should not be sought merely in the movements and migrations of peoples. More important in the diffusion of the short-horned type were doubtless economic and physiological factors to which generally but little attention is paid by scholars interested in the history of domesticated animals.

Evolution of the Shorthorn Type in Ancient Elam

The evolution of the shorthorn cattle falls into the period of the urban revolution in Sumer and Elam, where the fertility of the soil and the cooperative effort of a planned society enabled farmers to produce a surplus of foodstuffs. Villages expanded into cities, such as Ur, Erech, Lagash, Eridu and Susa, surrounded by gardens, fields and pastures. The wealth was concentrated by the priesthoods of the temples, and distributed to artisans, labourers, transport workers and traders, withdrawn from direct food production. The temple provided tools, farm equipment, boats and breeding stock. At Lagash the temple kept stud bulls which were imported from the Elamite hills, since the local stock was liable to deteriorate on the sultry plains, if not crossed periodically with mountain breeds (Childe, 1942). The demand of urban society for a regular supply of milk and dairy products called for the development of a type of cattle physiologically adapted for milk production. To this day the outstanding dairy breeds of the world are either entirely or predominantly of shorthorn type (Amschler, 1956).

A connection between the increased reliance on milk production and the small size of brachyceros cattle has been suggested by Howard (1962). In neolithic Europe the domestic cattle of the primigenius cranial type appear to have been allowed considerable freedom in grazing, which would account for their relatively large size. In iron age sites there is a great preponderance of the remains of cows of the brachyceros cranial type, suggesting a concentration on milk production, which would entail the keeping of the animals in byres during the winter months. Shortage of feed and poor living conditions would account at least in part for the small size of these cattle. As Troels-Smith (1953) has shown for Denmark, the early peasants fed their stabled cattle mainly on leaves, especially of the elm tree.

Shorthorn Cattle in Ancient Mesopotamia

In Mesopotamia shorthorn cattle began to replace the original longhorn breed approximately 3000 years before the Christian era. They are represented on cylinders and reliefs; at Tello and in other sites also by figurines. The oldest relief of shorthorn bulls is found on an early Sumerian steatite vessel (Fig. 354), dated to the second half of the 4th millennium B.C. A slightly later bronze figure of a shorthorn bull bears the name of King Dungi of Ur, whose reign is dated to about 3000 B.C. A similar figure is known from Erech, while another shorthorn bull is represented on a reign ring from Ur

354. Early Sumerian steatite vessel showing shorthorn bulls in relief (second half of 4th millennium B.C.)

355. (left) Statuette of a shorthorn bull from Ancient Babylonia (after Meissner)

356. (right) Shorthorn bull on a reign ring from Ur

357. Shorthorn cows and calves in a milking scene on a temple frieze at al-'Ubaid (c. 2900 B.C.)

358. Shell figures of bulls on the temple frieze at al-'Ubaid

(Fig. 356). Shorthorn cows with their calves in front are depicted in a milking scene on an early dynastic mosaic frieze from A-Anni-Pad-Da's temple of Nin-Kharsaf at Tell al-'Ubaid, Ur (c. 2900 B.C.). At that period cows were still milked from the back, a practice which suggests that the milking of goats, and possibly also of sheep, preceded that of cattle; in sheep and goats it has been retained to this day. The shell figures of bulls from the frieze at Tell al-'Ubaid show horns of medium length; but in view of the

20 Epstein I

short horns of the cows in the same scene, Friederichs' (1933) and Amschler's (1956) designation of these cattle as short-horned (brachyceros) is acceptable.

It is notable that whereas the longhorn cattle of Sumer, as those of Egypt, were spanned into yokes fastened to their foreheads, the people who introduced the shorthorn cattle from Elam, east of the Tigris river, into Mesopotamia, and those who subsequently took the breed to Egypt, used the collar which in the course of time completely replaced the yoke.

Shorthorn Cattle at Ancient Anau, Transcaspia

At Anau in Transcaspia the original longhorn breed of domesticated cattle was superseded by shorthorn cattle during the copper age. Duerst (1908a) suggested that the short-horned type was either evolved at Anau during the arid period at the end of the aeneolithic culture, or introduced simultaneously with the camel, hornless sheep, and dog of matris optimae type.

Although favoured by Duerst, the former theory may be discarded. The arrival at Anau of a number of new breeds as well as of the hitherto unknown camel and goat, together with the first appearance of the shorthorn cattle, strongly suggests the introduction of the latter from elsewhere. But whence these animals came we can only guess. Duerst (1908a) wrote:—"Considering the localities of fossil remains thus far found, and the present geographical distribution of these animals, it is possible that the camel came from the south or east, and the goat from the south or west, since the wild form now lives in Persia and the Caucasus. The dog, however, may have come either from the sphere of Indian culture or from Russia." In the case of each of these three domestic animals of the copper period of Anau two alternatives are offered in regard to the direction whence they might have come: the camel from the east or south, the goat from the west or south, the dog from the north or south. Since it is unlikely that these animals arrived at Anau simultaneously from two or three different directions, they must have come from the south, one of the alternatives in each case.

359. Seal impressions of shorthorn bulls from Mohenjo-daro, Indus Valley, (left) and Ur, Sumer (right)

At the same period a hornless sheep made its first appearance at Anau. It seems to have come from the Iranian plateau, and to have entered Turkestan likewise from the south.

Since, then, it is probable that the domestic animals of the copper period of Anau reached Turkestan from the south, we may assume that the shorthorn cattle too came from the south. South of Turkestan and east of Asia Minor and Mesopotamia lie the northern parts of Iran, Afghanistan and the valleys of the headwaters of the Indus. As shorthorn cattle appear in Mesopotamia and the Indus valley at a similarly early date (first half of the 3rd millennium B.C.), the original centre of their evolution may have to be sought in the mountainous regions of Iran, the western part of which once formed Elam.

iii. The Phylogeny of the Shorthorn Cattle

Comparison between Shorthorn (Brachyceros) and Zebu Skulls

Considerable differences of opinion exist on the ancestry of the shorthorn cattle. Keller's (1896) theory of their zebu descent has already been refuted for geographical and historical reasons (p. 294); it also lacks anatomical support. The comparison of zebu and shorthorn skulls shows that there is little general resemblance between the two types, although skulls similar in horn weight and direction, i.e. the horns' moment of rotation, may show similar effects in particular features (conformation of forehead, Torus frontalis, horn necks, etc.). The general narrowness of the zebu skull, especially marked in the interorbital distance, contrasts with the conspicuous width of the interorbital region in the shorthorn skull. In zebu cattle the facial part of the skull is much longer than the neurocranium, whereas in the shorthorn the neurocranium is long in comparison with the shortened splanchnocranium, as Owen's term, Bos longifrons, indicates. In the zebu, generally, the horn cores are attached to the skull posteriorly to their attachment in the shorthorn. Owing to the frequent backward direction of zebu horns, the zebu's forehead is usually convex, whereas the shorthorn's forehead is dished. In some shorthorns, as among the Tux and Jersey breeds, the stop is so well defined that the head obtains a bulldog-like expression (niatism). In zebu cattle the skull is never bulldog-like, not even in the chondrodystrophic dwarfs of India. Similar differences prevail in the occiput; whereas in the shorthorn the squama forms a right angle with the ideal surface of the forehead, in zebu cattle this angle is decidedly acute.

Since the theory of the zebu descent of the shorthorn cattle is without foundation, it is superfluous to discuss the relationship between the shorthorn and the banteng (Bibos javanicus) to which Keller traced, through the link of the zebu, the ultimate descent of the African shorthorn.

Theory of Genetic Connection between the Jersey and Damascus Shorthorn Breeds and the Zebu

Recently, several authors have claimed a genetic relation between certain shorthorn (brachyceros) breeds and the zebu, basing their claim on the distribution of haemoglobin and blood group types. Two electrophoretically distinct types of adult bovine haemoglobin, namely the slow-moving A and the fast-moving B, are controlled by a pair of autosomal allelomorphic genes which are both fully expressed in the heterozygote. The haemoglobin type B is common in the zebu, but rarer in humpless cattle. It is absent in the N'Dama longhorn and Muturu shorthorn of West Africa (Bangham and Blumberg, 1958), and also in the Friesian, Black Pied Danish, Red Danish and 10 different British breeds of cattle (Shorthorn, Ayrshire, Hereford, Welsh Black, Aberdeen-Angus, North Devon, Red Poll, Sussex, Dexter, Galloway). However, it is well represented in the Jersey, Guernsey and South Devon breeds (Bangham, 1957), in the Charolais, Brown Swiss, and the Brown Mountain, Spotted Mountain and Yellow Hill cattle of Germany (Schmid, 1962), and it has also been established in unselected Algerian hill cattle (Cabannes and Serain, 1955). From the common presence of the haemoglobin B gene in both the Jersey breed and the zebu group, Stapleton (1953) inferred that the Jersey was related to the zebu.

A number of authors have attributed the pronounced crest of the Damascus bull and the long narrow head characteristic of this shorthorn breed to an admixture of zebu blood, although the development of a very prominent crest is characteristic also of several Italian (Marche, Piedmont, Chiana, Romagna) and other European breeds either only in the male or in both sexes.

Recently, Volcani (1960) has supported this theory on the basis of an analysis of 48 blood samples of purebred Damascus cattle by Professor Moustgaard of the Danish Royal Veterinary College, who identified several of the alleles of the B blood group system. In the presence of the Z' factor and the frequency of the Z', U_1, V_2 and A factors, the Damascus resembles the Jersey and south-east European (brachyceros?) breeds, and is intermediate between other European cattle and Indian zebus. Hence Volcani claims that there is some relationship between zebu and Damascus cattle, and that the absence of certain anatomical features, such as the bifidity of the neural spines of the anterior dorsal vertebrae does not disprove such relationship, as any hereditary characteristic may vanish in time, especially if several crosses have been made to that purpose.

It would appear that the only valid conclusion that may be drawn from the blood typing of the Damascus is that it is nearly related to the Jersey and certain south-east European breeds of cattle. For this agrees with the anatomical similarity of the brachyceros breeds of south-west Asia, south-eastern Europe (Albanian, Buša, Greek Shorthorn) and Jersey Island. A similar conclusion may be drawn from the presence of the haemoglobin type B in the Jersey and Algerian hill cattle. This again agrees with their anatomical similarity. Bangham (1963) regards the high frequency of the haemoglobin B gene in the Island Jersey as an extreme example of genetic drift, due to the importation of a small original shipment followed by strict inbreeding.

The occurrence of the haemoglobin B gene in the Jersey may be due to one of several factors: the presence of zebu blood, the descent of the Jersey breed and zebu group from a common ancestral stock, or convergence.

The supposition of the presence of zebu blood in the Jersey and south-east European brachyceros breeds may be discarded. If the classification of domesticated cattle on anatomical grounds is of any sense and significance at all, the humpless brachyceros and humped zebu must be referred to two different basic types. Accordingly, the Jersey is clearly distinct from the zebu, even though it does possess the haemoglobin B gene common in zebu cattle. It is distinct from the zebu not only anatomically but also in the alleles of the gene-controlled β-globulin fraction and type, which involves the iron-binding glycoprotein, transferrin. The genetic control of transferrin synthesis is by a system of multiple autosomal alleles with full expression of each allele in the heterozygote combinations. Several alleles – Tf A, D, E, B, F, G – have been analysed. Of these, the Jersey, in common with other European breeds of cattle, has no transferrin B and F, which are confined to the zebu group (Ogden, 1963). Similarly, the Jersey and Guernsey lack the transferrin E gene, which is highly frequent in zebu (0.3 in the Brahman) and partly zebu (0.6 in the Africander) breeds (Ashton, 1959). The Jersey cattle are also devoid of the transferrin G gene, which is carried by cattle of zebu type (Rendel, 1967). Rendel (1967) confirms that the relationship, suggested by the distribution of the blood factors Z', U_1 and V_2 and the haemoglobin B, between zebus and central and southern European cattle, including the Channel Island breeds, and a particularly close connection of the Jersey with the zebu, as deduced by Bangham and Blumberg (1958) from the remarkably high frequency of the haemoglobin allele B among Jerseys, are not supported by the transferrins, lactalbumins and albumins.

The second conjecture, namely that the occurrence of the adult bovine B haemoglobin gene or of certain factors of the B blood group series in both the zebu group and several brachyceros breeds, including the Jersey, may be attributed to derivation from a common ancestor, is invalid for similar reasons. It is invalid although all basic types of domesticated cattle originate either solely or in part, directly or indirectly, from the Asiatic urus. In particular, the origin of the shorthorn cattle of North Africa, Spain, south-eastern Europe and Switzerland from an Asiatic source is evidenced by the historical and chronological sequence of the appearance of the shorthorn type along the routes leading from Syria and Palestine through North Africa to the Iberian peninsula, and from Asia Minor through the southern and south-western Balkans to Switzerland and Brittany. The Jersey's racial alliance to the shorthorn breeds in the countries flanking these routes is indicated by their similar conformation as well as the presence of certain blood group factors in Jersey, south-east European and Damascus cattle, and of the B haemoglobin type in Jersey and Algerian hill cattle.

But this argument cannot be extended to include the zebu group, because of the fundamental anatomical differences between the humpless shorthorn and humped zebu, and also because the presence or absence of the above-mentioned blood characteristics is an unreliable criterion of racial connection. This is illustrated by two closely allied African shorthorn breeds, namely the West African Muturu and the Algerian hill cattle, which differ in adult bovine haemoglobin types, B being present in the Algerian but absent in the Muturu. The two breeds, while distinct in haemoglobin types, are descended from the same population of domesticated cattle and the same wild ancestor. Theories on racial relationships cannot, therefore, be based solely on frequencies of a few genes controlling haemoglobin, globulin or blood group factors. At the best, similarities in such factors may provide a working hypothesis, but this is all that it can ever amount to (Hodges, 1956). It would be necessary to obtain figures on all the protein polymorphisms and blood groups before these can be used as more than support for theories of descent derived from anatomical characteristics and historical facts. This may be illustrated by the following two examples:—

Osterhoff and Van Heerden (1965) recorded the following gene frequencies of the transferrin E type in several South African dairy, beef and dual-purpose breeds of cattle of zebu and sanga, European, and crossbred origin:—

Gene Frequency of Transferrin E

Zebu and Sanga Breeds		Sanga × European Breeds		European Breeds	
Nguni	0.33	Drakensberger	0.20	Red Pol	0.27
Africander	0.27	Bonsmara	0.19	Ayrshire	0.14
Boran	0.22			Hereford	0.06
				South Devon	0.05
				Friesian	0.03
				Aberdeen-Angus	0.01
				Sussex	0.00
				Beef Shorthorn	0
				Dairy Shorthorn	0
				Brown Swiss	0
				Guernsey	0
				Jersey	0

The Red Poll, with a Tf^E gene frequency of 0.273, differs radically from the other breeds imported from Europe and approaches the highest figure in the indigenous zebu-sanga group. On this, Osterhoff and Van Heerden (1965) who, following Ashton (1959), regard the relative high frequency of Tf^E in zebu, sanga and zebu-European crossbred cattle as an indication of their climatic and ecological tolerance, soberly remark:—"... the high frequency of Tf^E in the Red Poll would be difficult to explain along the same lines." And, we may add, any attempt to attribute this phenomenon to descent of the Red Poll from the zebu or their racial alliance would obviously be utterly misleading.

The second example refers to the phenotypic frequencies of the Z', U_1 and V_2 factors of the A, SU and FV systems respectively and the frequency of the B allele of the haemoglobin system, recorded by Hesselholt et al. (1965) in 147 Cyprus and 116 Damascus shorthorn cattle and 98 Tharparkar and 85 Sahiwal zebus.

Taxonomically the Cyprus and Damascus cattle are closely related, and this is supported by the frequencies of the four factors examined. Again, the Tharparkar and Sahiwal zebus are taxonomically related, but their blood factors are far from showing this. While Cyprus and Damascus cattle obviously differ from both zebu breeds in the

Breed	Frequency (%)			
	U_1	Z'	V_2	Hb^B
Cyprus	40	17	16	12
Damascus	64	22	22	11
Tharparkar	9	56	2	11
Sahiwal	18	13	24	40

phenotypic frequency of the U_1 factor, they approximate to the Sahiwal in the frequencies of the Z' and V_2 factors and to the Tharparkar in the frequency of the Hb^B allele; they differ from the Tharparkar in the frequencies of the Z' and V_2 factors and from the Sahiwal in that of the B allele of the haemoglobin system. Were we to base our conclusions regarding racial relationships or descent on the Hb^B allele alone, the Cyprus and Damascus cattle would have to be considered as closely related to the Tharparkar zebu of Sind and unrelated to the Sahiwal of the central and southern Punjab. Were we to base our conclusions on the U_1, Z' and V_2 factors, the Cyprus and Damascus cattle would have to be regarded as racially radically different from the Tharparkar zebu; but solely on the basis of the frequencies of the Z' and V_2 factors they would have to be considered as racially allied to the Sahiwal.

For these reasons we conclude that the parallel occurrence of certain haemoglobin, globulin or blood group factors may in some cases, as in those of the Damascus and Cyprus cattle or of the Jersey and other shorthorn breeds, be due to racial relationship, and in others, e.g. shorthorn and zebu cattle, to convergence. This convergence may either be fortuitous or due to a selective advantage connected with certain haemoglobin, globulin and blood group factors. Such advantage may be due to pleiotropy affecting the above factors as well as physiological characteristics, or to linkage.

Genetic associations between blood groups and blood and milk proteins on the one hand and production traits, disease resistance and climatic tolerance on the other hand undoubtedly exist. Studies in red cell antigens in cattle have established the existence of significant correlations between some blood group genes and milk yield or butterfat content (Rendel, 1967). Ashton (1965) found relatively strong associations between transferrin type in cattle and milk production. Evans (1963—unpublished data) suggests that in sheep the genes for certain haemoglobin types have an adaptive significance. Most breeds of sheep, if maintained in one environment, appear to exhibit a balanced polymorphism for the Hb and $[K_e^+]$ types. If flocks or breeds are moved from one environment to another, this balance is altered towards the gene frequencies found in flocks indigenous to the new environment. In man certain haemoglobin variants influence resistance to malaria (Ingram, 1963), and there is evidence that the ABO blood groups may affect resistance to some diseases such as smallpox, and that the frequencies of these blood groups have been greatly influenced by the occurrence or absence of particular diseases in particular areas (Vogel, 1961). In poultry heterozygosity at some blood group loci has a favourable influence on production, viability and fertility. Further, in a two-way selection experiment for high and low initial egg weight in White Leghorns the gene frequency of blood groups A_1, A_2, B_2, B_4 and of the egg white protein components EW III A and EW III B changed significantly, while the frequency of the blood groups B_3, B_6 and B_7 remained unaffected. In the blood groups A_1 and A_2 a significant drifting apart of the two breeding lines occurred in the course of the experiment (Hilfiker–Hengartner, 1967).

In the light of these findings it would appear that directional changes in gene frequency in blood groups and blood, milk and egg white protein components may occur in animals in changed circumstances, and that similarities between breeds for certain genes may therefore indicate common or similar environments or common or similar breeding aims rather than common origin. This calls for great care in the use of marker genes to trace breed origin and relationship between breeds and should serve as a warning of hasty conclusions. However, coution in their application does not gainsay the usefulness of marker genes as support of theories of descent and relationship of breeds, wich are based on well founded historical and anatomical evidence (see pp. 377, 424–425, 461, 551 and 553).

Theory of the Descent of the Shorthorn from a Dwarfed Urus

There are two other theories on the descent of the shorthorn cattle: One regards them as derived from a wild dwarf species or subspecies of ox; the other considers them to be descended from the Asiatic urus (Bos primigenius namadicus).

Cranial size in Bos primigenius is not uniform. There are skulls of gigantic dimensions, as one in the British Museum, the profile of which, from the centre of the Torus frontalis to the frontal ridge of the maxillae, measures 912 mm, while others from adult animals measure less than 600 mm (Nehring, 1889). But the latter are of a similar conformation to the large skulls, and distinct from the typical shorthorn skull.

V. d. Malsburg (1911) regarded the small Bos primigenius as a separate species—Bos taurus minutus. Others have considered these small skulls as those of domesticated longhorn or of degenerate uri. Hilzheimer (1926) ascribed their small size to malnutrition, while Duerst (1926a) attributed them to crossbreds of Bos primigenius and domesticated longhorn stock (see p. 229). However, Von Leithner (1927) has produced convincing evidence of the pronounced sexual dimorphism in size characteristic of Bos primigenius, concluding that the small urus skulls, referred to above, are those of cows.

A few skulls belong to a different category. Owen (1846) described the remnants of a wild ox, which he believed to originate from pliocene or pleistocene strata, referred to as fresh water deposits or fresh water newer pliocene deposits, where they were found together with bones of Elephas, Rhinoceros and Bison priscus. The horns were shorter and weaker than those of Bos primigenius primigenius, the forehead was not even and square in outline but dished, the Torus frontalis was markedly convex, the distance between the horn bases was smaller than the interorbital width, the splanchnocranium was shorter, and the posterior mandibular ramus more nearly vertical. Dawkins (1867), who re-examined Owen's material, arrived at the conclusion that the dating was erroneous and that the bones were actually derived from domesticated stock.

Owen's erroneous designation of the brachyceros cranial fragments from Britain, and a similar error by Fraas (1869) with regard to a shorthorn skeleton from Schussenried, misled Naumann (1875) to postulate the existence of wild brachyceros cattle during the period of the pile dwellings of Switzerland.

Bos Mastodontis and Bos Brachyceros Europaeus

A dwarfed ox, represented by the upper portion of a skull, has been described by Pohlig (1912) with the name Bos mastodontis. Gypsum copies of this fragment, manufactured by a commercial firm in Bonn, are designated as Bos brachyceroides Pohlig. The cranium of Bos mastodontis is indistinguishable from a typical recent domestic shorthorn skull. It has been claimed to belong to a wild species because, according to the record of an amateur collector, it was found by him in pliocene deposits of Asti, Italy, together with molars of Elephas meridionalis and Mastodon avernensis. But La Baume (1947) has made it clear that the fragment belongs to a domestic animal, and that the terms Bos mastodontis or Bos brachyceroides are synonyms of Bos taurus brachyceros Rüt.

A similar cranial fragment from early recent (alluvial) deposits of Krzeszowice, Galicia, was described by Adametz (1898) with the name Bos brachyceros europaeus. Duerst (1899), rejecting the idea that this fragment belonged to a wild beast, pointed out that all those features on which Adametz based its wild derivation exist to the same extent, and occasionally more strongly pronounced, in domesticated shorthorn cattle. This is supported by La Baume (1947), while Von Leithner (1927) regarded the skull of Bos brachyceros europaeus Adametz as that of a female urus, attributing the peculiar conformation of the forehead, occiput and temporal fossae to the static pressure of the female's relatively weak horns on the skull.

The Pamiatkowo Skulls

Doubts have been expressed also on the wild origin which Adametz (1925) claimed for two crania from a peat bog at Pamiatkowo, Poznan, found at a depth of nearly 5 metres in the vicinity of remnants of Cervus megaceros. Their measurements (total length 481 mm; basilar length 420 mm) are less than those of wild uri, including female and immature animals. The asymmetry in the shape and direction of the horns of one of them renders it unlikely that the skulls belonged to wild animals. Again, it has proved impossible to ascertain the exact geological age of this find. La Baume

360. Cranial fragment of Bos mastodontis Pohlig

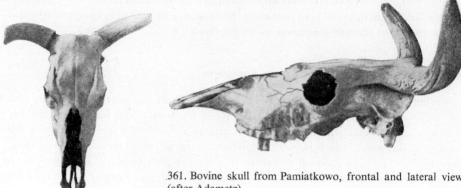

361. Bovine skull from Pamiatkowo, frontal and lateral views (after Adametz)

(1947) has pointed out that numerous remnants of Cervus megaceros occur also in early recent strata; in addition, he regards it as improbable that the Pamiatkowo bog itself is of pleistocene age. The peculiar character and appearance of the bone, which Adametz believed to indicate the wild state of the animals, may be due to the natural environment in which these cattle were kept, or to the feral state to which they had reverted. However, this assumption is not essential, for crania with similar peculiarities occur also in domestic cattle, especially in male longhorn stock. La Baume (1947) refers the Pamiatkowo skulls to domestic bulls of brachyceros type influenced by primigenius (longhorn) blood, similar to crania sporadically found in late neolithic and early bronze age stations of Switzerland.

From a historical point of view at any rate it is impossible to consider a European origin of the shorthorn cattle. All historical and archaeological records indicate that the shorthorn cattle of West and North Africa, as well as those of the lake dwellings of Switzerland and the Balkan peninsula, originated in Asia where they had been evolved at the beginning of the third pre-Christian millennium.

Bos Brachyceros Arnei from Shah Tepé, Northern Iran

Excavations at Shah Tepé, in the Turkmen steppe of northern Iran, yielded bones of a single bovine type—the shorthorn (brachyceros). 35 pieces were recovered from the lowest strata, provisionally dated to about 3000–2500 B.C., 18 from intermediate strata, and only one fragment from the highest (recent) horizon. With one exception, the shorthorn bones from all three levels resemble in measurements and conformation the respective bones of typical shorthorn cattle from Europe and Asia, both ancient and recent. They are derived from immature animals, as indicated by the incompletely ossified symphyses of the long bones. However, one group of bones of an adult shorthorn bull from the lowest horizon—comprising a cranial fragment with the horn cores, fragments of the left mandible, metacarpal and radius-ulna, and the second and third upper molars—is distinguished from the other shorthorn bones of Shah Tepé by a darker coloration, denser and harder material, keener edges and higher specific weight.

Also, the length and width of the horn cores of this bull considerably exceed the measurements characteristic of domesticated shorthorn cattle, while the intercornual line, frontal widths, frontal length and occipital width fall within the range of variation of the typical shorthorn. In view of the peculiar material, specific weight and coloration of the bone, in addition to the thickness of the horn cores, Amschler (1939) considered these fragments to be derived from a wild bull of shorthorn type—Bos brachyceros arnei, which he regarded as the wild ancestral species of all shorthorn cattle of Asia, Africa and Europe, as well as of the orthoceros cattle of the Kalmyks and Mongols. Bos brachyceros arnei was believed to have ranged, along with Bos primigenius primigenius or Bos primigenius namadicus, throughout Europe and Asia, particularly the southern mountain chain including the Alps, Balkans, Caucasus, Taurus and the Iranian tableland with its folded mountain ranges. Amschler further suggested that Bos brachyceros arnei had its forerunner in a pleistocene (diluvial) urus furnished with horns which, although more powerful, resembled brachyceros horns in shape and direction (Fig. 363).

An analysis of Amschler's arguments reveals several contradictions. To begin with, the reputed pleistocene (diluvial) ancestors of Bos brachyceros Arnei, represented by a cranium of Bos primigenius Boj., illustrates two points previously discussed: the great variability of horn shape in the urus, and the influence of the direction of the horns on

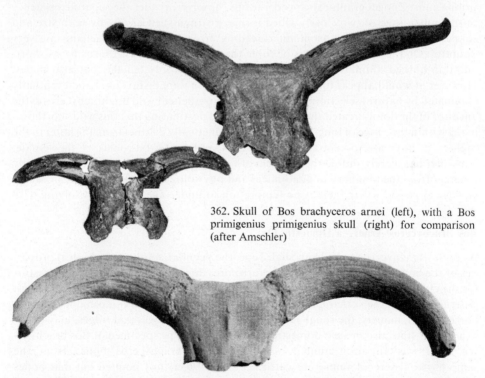

362. Skull of Bos brachyceros arnei (left), with a Bos primigenius primigenius skull (right) for comparison (after Amschler)

363. Skull of Bos primigenius primigenius from pleistocene levels, showing similar horn direction to that of Bos brachyceros arnei (after Amschler)

the conformation of the neurocranium—horns of a similar shape result in a similar neurocranial conformation. But the similarity in horn direction offers no indication of the descent of the shorthorn (brachyceros) type from a particular urus race.

Horn Size in the Urus

Horn size in the urus is a variable character, with pronounced sexual dimorphism (see also p.227). While crania with relatively short horns have been recorded, the occurrence of thin horns in the urus, at any rate in the adult male, has so far not been established with certainty. In support of the conjecture of their occurrence, Amschler (1939) mentions the genus Syncerus in which short-horned subspecies are found in addition to long-horned, and the genus Bison which existed in both long-horned and short-horned forms. But the different bisons occupied geographically separate ranges, and the same holds of the recent wild buffaloes of Africa. For the wild Bos brachyceros, on the other hand, Amschler claims a common range with Bos primigenius namadicus or Bos primigenius primigenius. The short-horned crania from early neolithic levels of Preston Docks, England, reputed to be derived from wild uri, were found in the neighbourhood of a long-horned Bos primigenius primigenius skull. Similarly, the Pamiatkowo skulls come from the general area of distribution of Bos primigenius primigenius. Zoogeographical considerations, however, render the separate existence in the same range of two closely allied forms, distinguished merely by horn size and shape and not prevented from interbreeding by any isolating mechanisms, very doubtful. On the other hand, it is possible that within long-horned urus herds short- and-thin-horned animals did occur occasionally, not only females but also males. However, it would appear that in the wild state these were disfavoured and eventually eliminated by natural selection. In regarding a hypothetical wild Bos brachyceros as the ancestor of the domesticated shorthorn cattle, and postulating that this wild shorthorn, living within the range of long-horned uri, was genetically distinct from the latter in the higher—at any rate, potentially higher—milk and butter-fat yields of the female, Amschler has merely shifted the problem of how the shorthorn came to be genetically different from the longhorn in anatomical and physiological properties into an earlier geological period where there is no temptation to undertake the task of solving it.

Bos Brachyceros Arnei—a Domestic Ox

As to Bos brachyceros arnei from Shah Tepé, the peculiar colour and physical properties of the bone and the relatively large horn cores do not seem to be sufficient reasons for the claim that it was a wild beast. Hilzheimer (1910) has confirmed that the main criteria by which the skeleton of the wild ox is believed to be distinguishable from the domesticated, namely, the rough insertions and clear-cut edges and ridges, may occur in a similar state also in aged domesticated cattle. The type specimen of Bos brachyceros arnei was an adult animal, while the other brachyceros bones from the same horizon were of immature cattle. Zeuner (1963a) has pointed out that bones of domesticated brachyceros cattle were found both above and below the level of the supposedly wild specimen. The bull, as Amschler believed, had been used as a sacrifice and been given ritual burial. This was done by people not in a pure hunting stage of culture, but breeders of domesticated shorthorn cattle as well as of sheep, goats and swine. It was done during a period when cattle had already been domesticated for several thousand years. Why should the cattle breeders of Shah Tepé have taken the trouble of catching an adult wild bull for their ritual, when they had their own domesticated beasts at hand? But the weightiest argument against the wild state of the Shah Tepé bull is the complete absence of any skeletal material of a wild Bos brachyceros at Anau, situated only a short distance east of Shah Tepé. The only wild ox encountered at Anau is Bos primigenius namadicus, while domesticated cattle are represented by both long-horned and short-horned forms.

La Baume (1947) has pointed out that the cranial measurements of Bos brachyceros arnei fall within the range of variation of urus cows, save for the horn core length which falls below the shortest length found in Bos primigenius cows. In spite of this peculiarity of the horn core, he refers Amschler's type specimen to a female urus. Nobis (1954), on the other hand, from a craniological comparison of Bos brachyceros arnei with wild urus bulls and cows as well as with prehistoric and early historical domestic bulls, cows and oxen arrived at the conclusion that Amschler's type specimen belonged not to a wild bull or cow but to a domestic brachyceros ox (castrate).

Theory of Descent of Domesticated Cattle from Several Species of Urus

In his critique of the mode of research into the origin of domesticated cattle, Szalay (1930) called the brachyceros question the most difficult part of the theory of their descent and at the same time the key to it. He suggested that five or six true species of wild oxen, descended from Bos trochoceros Meyer, formerly occurred in Asia, Africa and Europe; the northern urus was darker than the southern, and the Egyptian more slender than the European. These uri were domesticated during the stone age.

The theory of the descent of domesticated cattle from several races of wild oxen is acceptable, although Szalay may err in describing these as separate true species. Indeed, judging from the variability of Bubalus and Syncerus as well as from the vast range of the urus, we may assume that several geographical races of the urus formerly existed in Asia, Europe and northern Africa. The great variability in domesticated cattle may partly be due to the different genetic complexes built up in them through the interbreeding of distinct races. This variability has provided man with the material for selection in accordance with his various needs and conditions. It is not Szalay's theory of the polyphyletic origin of cattle, therefore, that is under dispute; disputed is his attempt at connecting this theory with the brachyceros question, his assumption that the brachyceros (shorthorn) breeds are derived from a brachyceros species of urus, as the primigenius (longhorn) breeds are derived from a primigene urus. It is notable that Szalay offers no evidence whatsoever in support of his view.

Theory of Descent of the Shorthorn Type from Bos Primigenius Namadicus

The last theory on the descent of the shorthorn cattle regards these as a stunted or dwarfed domestication form of the urus. The scientific foundation of this view has been laid mainly by Klatt (1913) in his study on the influence of total body size on the skull, and by Duerst in his works on the effects of the size and direction of the horns on cranial conformation.

Bohlken (1962) has pointed out that crania of European cattle are very much smaller than those of Bos primigenius primigenius. Other than in the domestic horse, dog and rabbit, the higher values of the size (basilar length) range of cattle skulls do not exceed those of the wild form. In particular this applies to male skulls which fall considerably below the size range of those of urus bulls. The marked sexual dimorphism in size, characteristic of the urus, has become reduced under domestication.

According to Klatt (1913), the preponderance in the shorthorn skull of the neurocranium over the splanchnocranium, and in the longhorn skull of the splanchnocranium over the neurocranium, is due to the different basilar lengths of the crania of these two basic types of domestic cattle. In a European longhorn and an Algerian shorthorn skull the basilar lengths are in the ratio of 640:418, or approximately as 3:2. All other differences between shorthorn and longhorn skulls, such as the relative height of the occiput, the broad palate, large orbits, shallow temporal fossae, and steep ascent of the posterior mandibular ramus in the former, are ascribed to the fact that the shorthorn skull as a whole is smaller than the longhorn skull. From similar cranial conditions obtaining in large and small breeds of domestic dog, Klatt deduced that the shorthorn was a dwarfed form of cattle.

Objection to the Theory of the Shorthorn's Descent from the Urus

Laurer (1913) did not accept the urus as the ancestor of the shorthorn cattle for the following reasons. The superior length of female Bos primigenius primigenius skulls averages 66 cm, of female domestic shorthorn skulls 41 cm. The circumference of the horn amounts to 29 cm in the urus, and 10 cm in shorthorn cattle. The difference in horn length is still greater; urus horns are approximately 56 cm long, those of domestic shorthorn cows 11 cm. In the female Bos primigenius primigenius the length of the horns amounts to 191 per cent of the basal girth, in shorthorn cows to 103 per cent. Laurer claimed that these differences were so vast as to render it difficult to attribute them to physical degeneration. Again, in the shorthorn (Laurer's measurements and descriptions refer to the shorthorn cattle of the ancient Swiss lake dwellers) the horns grow sideways, upwards and inwards, but not, as in Bos primigenius primigenius, forwards as well. The angle formed by the forehead and occiput is approximately 90° in shorthorn cattle, but only 58° in the urus. In the conformation of the occiput the differences are also pronounced. In Bos primigenius primigenius the occiput ascends obliquely forwards, whereas in shorthorn cattle it is nearly vertical. These conditions are accompanied by differences in the relative measurements of the superior and inferior cranial lengths. In Bos primigenius primigenius the inferior skull length amounts to 85 per cent, in domestic shorthorn cattle to 91 per cent of the superior cranial length. Again, the parietals and interparietals are much less developed in Bos primigenius primigenius than in shorthorn cattle. In the urus the mandible, in relation to the superior cranial length, is considerably shorter than in the shorthorn. From these differences Laurer concluded that the shorthorn cattle could not be descended from the urus. The absolute as well as the relative measurements of their respective crania were too much at variance. On the latter point Laurer laid special stress, for he believed that the relative cranial measurements in animals of the same species were identical, irrespective of variations in the absolute size of body and cranium.

Attribution of the Shorthorn's Cranial Characteristics to Small Body Size

Klatt (1913), on the other hand, held the contrary:—"If the relative measurements of two skulls of different sizes are identical, there must be some peculiarity in one of them; but if these measurements are not in concord, then the assumption of racial identity is more justifiable, although even such a case does not necessarily call for this assumption." Laurer's argument that the differences in the relative measurements between Bos primigenius primigenius and shorthorn skulls refute the theory of the shorthorn's descent from the urus indicates rather the contrary. For, apart from the effects of the different horn development on the cranium, these differences are to a considerable extent due to the relatively large brain of the small domestic shorthorn as compared with the brain of the large wild Bos primigenius.

Nor do the differences in the absolute cranial and skeletal measurements between the urus and the shorthorn cattle refute the theory of their close relationship. At an early stage of domestication all bovines were smaller than their wild ancestors. This was due to natural selection in relation to the unfavourable influence of the primitive conditions, commonly obtaining during the early stages of domestication, on their growth and development. To this day no domesticated breed of buffalo reaches the average size of the wild Indian buffalo, while the differences in size between the wild and domesticated yak, the wild banteng and the domestic Bali cattle, the wild gaur and the domestic gayal are still greater. The smaller absolute size of the shorthorn cattle as compared with the urus is therefore not exceptional but the rule.

Attribution of the Shorthorn's Cranial Characteristics to the Small Size of Horns

The difference in size and shape of the horns between wild and domestic Bovini is also typical. In wild Bovini, more especially in the bulls, the horns are of much greater importance for defence and attack in the struggle of life than in domesticated animals. The horns of domestic bovines rarely reach the size of those carried by their wild ancestors. This is particularly evident in the wild banteng and the domestic Balinese cattle. Where selection for gigantic horns is not practised by breeders, the horns of domesticated bovines are usually shorter, simpler in shape and lighter in weight. In this respect the domesticated descendants of the wild yak, whose horns are similar in shape to those of Bos p. primigenius, display practically the same changes that distinguish the shorthorn cattle from the urus.

The differences in the shape and position of the occiput between shorthorn cattle and the urus are partly due to the differences in horn development, affecting the extent of the parietal and occipital cavities. Nor are the frontal cavities unaffected by variations in horn development. Hilzheimer (1926) has illustrated these conditions in a drawing of shorthorn and polled crania on a longhorn skull (Fig. 364), showing the

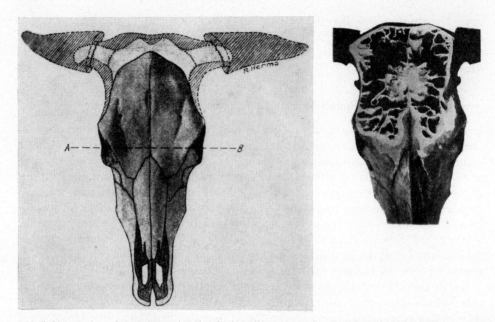

364. (left) Drawing of short-horned and polled skulls on a longhorn skull (after Hilzheimer)

365. (right) Longhorn skull after removal of external frontal lamella, showing extent of the Sinus frontales (after Hilzheimer)

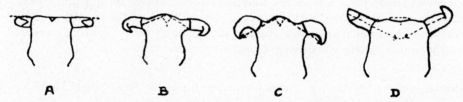

366. Variations in the shape of the Torus frontalis and intercornual ridge of shorthorn skulls (after Duerst)

extent to which the neurocranial sinuses remain undeveloped in short-horned and polled animals. Again, the so-called 'Illyrian bulge', an elevation in the forehead of the shorthorn, especially of small shorthorn specimens and breeds, such as the Illyrian, is due to the prominence of the brain case which, owing to the disappearance of the frontal sinuses, shows up in the form of a bump.

Duerst (1926a) has explained the effect on the shape of the Torus frontalis, the intercornual ridge and the Processus cornu ossis frontalis (horn necks) of the small, thin and light horns of the shorthorn cattle, characterised by a small moment of rotation (Fig. 366):

    A. If the horns are directed laterally, with the tips in the plane of the horn axes, the intercornual line is straight and horn necks are absent.
    B. If the horns point sideways and forwards, their moment of rotation acting

downwards at an angle of approximately 25° (negative angulation), the Torus frontalis is broad and prominent and the horn necks are moderately developed.
C. If the horns are bent markedly downwards (negative angulation), the intercornual line is strongly convex (similar to polled skulls) and the horn necks are scarcely developed.
D. If the horns project laterally upwards at an angle of approximately 25°, with the tips curving inwards (positive angulation), the intercornual line is slightly concave in the centre, the central depression being flanked by two moderate lateral elevations, and there are no horn necks.

While in general Duerst's analysis of the correlation between the direction of the horns and the shape of the intercornual ridge holds true, exceptions due to individual variability have been recorded (Bohlken, 1962).

Secondary Causes of the Shorthorn's Cranial Peculiarities

The difference in the anatomical conditions of the occiput is expressed by the ratio of the lower to the upper length of the skull, which is 658:541 in the urus, and 407:363 in shorthorn cattle (Gans, 1915). This feature is not an exception but the rule since in all domesticated bovines the difference between these two measurements is smaller than in their wild ancestors.

Owing to the relatively large eyes in shorthorn cattle, their orbits are considerably larger in relation to the superior length of the skull than in the urus. Again, this feature is not exceptional, but due to physiological factors (Dubois, 1897). Smaller animals have relatively bigger eyes than large animals of the same species, as a comparison of large and small antelopes or dogs shows (Gans, 1915).

The shortness of the mandible in the urus, in relation to the superior length of the skull, is due to the fact that in the urus the occiput extends more forwards than in the shorthorn. The more vertical position of the posterior mandibular ramus is a feature characteristic of all small animals of a species. Owing to the preponderance in a small animal of the neurocranium over the splanchnocranium, the latter is restricted to a smaller space than it would occupy, had all cranial parts proportionately decreased in size. In small animals the upper molars are, therefore, closer to the maxillary joint than in large beasts. In order to preserve contact with the upper dentition, the mandibular molars have to follow suit, occupying a place very close to the maxillary joint. Since the ascent of the posterior ramus begins immediately behind the last molar, it has to be more vertical in order to facilitate the normal contact of the teeth (Klatt, 1913). The photograph of the mandibles of a large goat and an African dwarfed specimen (Fig. 367), reduced to the same size, illustrates these conditions.

The shortening of the splanchnocranium in relation to the neurocranium is peculiar to all domesticated Bovini with the exception of the zebu. The skull of the Balinese cattle has a shorter nasal part than has the banteng skull, and similar anatomical conditions distinguish the domesticated gayal from the wild gaur, and the domestic yak from its wild ancestors (Antonius, 1922). In dogs the extreme corresponding types are the bulldog in which the nasal part of the skull is considerably shortened in relation

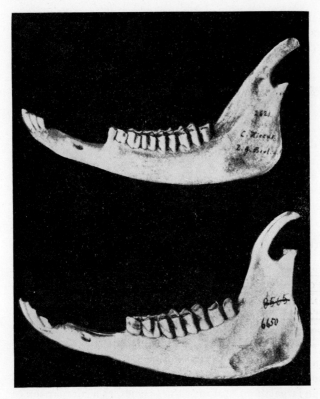

367. Mandibles of a normal (above) and of a dwarf goat (below) (after Klatt)

to the neurocranium, and the greyhound in which the total length of the skull has remained practically the same as in its wild ancestors, while the upper portion has become narrower. The shorthorn, Bali cattle and domesticated yak approach the bulldog, the zebu the greyhound, type.

Although the decrease in the size of the splanchnocranium is common to all shorthorn cattle, it is especially marked in the Tux breed of Switzerland and in the Jersey. However, it does not reach its highest degree in these two breeds, but in the Niata cattle of Uruguay, descended from Iberian longhorn stock; in the Niata the splanchnocranium is so short and the frontal sinuses are so undeveloped that the head has a decidedly pug-like appearance.

The Combined Effects of Small Body and Horn Size on the Shorthorn Skull

The majority of features and measurements of the skull, whether absolute or relative, by which the shorthorn type of cattle differs from the urus, may thus be explained by the small body and horn size in the former, the natural outcome of the domestication process. These differences do not refute the theory of the close relationship between the shorthorn cattle and the urus; they support it. In our quest for the ancestry of the shorthorn cattle we have, therefore, no reason to look for a special short-horned race of wild ox, or to assume that the shorthorn is descended from a dwarf species or subspecies of urus. For the crania of small uri have a different conformation from the shorthorn skull; they are small, but in other respects similar to larger skulls.

Large Cattle of Shorthorn Type, and Small Cattle of Longhorn Type

The last-mentioned fact, however, is a weighty argument against Klatt's theory that the anatomical peculiarities of the shorthorn skull are mainly due to the influence of unfavourable environmental conditions on the body size of this bovine type. Were this theory correct, the larger shorthorn breeds, evolved under favourable conditions, should revert to the ancestral cranial type; actually the larger shorthorn breeds retain their shorthorn character under the most favourable conditions. Again, if small body size were the sole reason for the cranial shorthorn type, no dwarf breeds of the cranial longhorn type should exist. Yet, dwarf domesticated cattle of longhorn type occur in north-western Anatolia and some parts of the Balkans (Fig. 370) (Antonius, 1922); and even small urus skulls, as we have seen, are known.

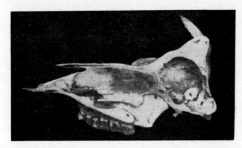

368. (left) Median section of Jersey skull (after Adametz)
369. (right) Niata skull

370. Dwarfed longhorn cattle from the Balkans (after Antonius)

The Effect of Horn Width on the Cranium

Duerst (1903; 1926a; 1931) ascribed the shorthorn cranial type almost entirely to the small size of the horns. Klatt, agreeing in a later work (1927) that small horn size was a contributory factor in the evolution of the brachyceros cranial type, stressed particularly the effect of the basal circumference of the horns on cranial conformation. The small Turkish (Plevna) cattle, he held, were of the primigenius cranial type owing to the considerable length and relatively large basal girth of their horns, so that an important element, i.e. the short thin horns, affecting the cranial conformation of the typical shorthorn, was absent.

The Effect of Other Genetic Factors on Cranial Conformation

Although the influence of the small horns on the cranial conformation of the shorthorn type of cattle should not be underrated, since the degree of pneumatization of the frontals, and hence the width of the aboral part of the cranium, are markedly influenced by horn development (see Fig. 364), other factors, too, have important effects. Bohlken (1962) has pointed out that in different breeds of cattle the various cranial features, such as infraorbital width, occipital width, jugal width, etc., are very differently affected by horn development. In urus bulls he could not find any influence of the horns on the skull. European cattle, he writes, have inherited from the urus a cranium that is constructed to carry larger horns than are actually present; and horn size influences cranial conformation in cattle much less than do body size and growth type (the term 'growth type' in the urus and domesticated cattle, as in other mammals, refers to broad or slender cranial conformation, irrespective of body or horn size). In hornless Aberdeen–Angus × Jersey and Holstein–Friesian cross breds, Cole and Johansson (1948) found the principal cranial characteristics of the horned parent stocks preserved. Again, in several antelopes cranial morphology and horn shape or hornlessness are not in harmony (Lenz, 1952). Antonius (1922) referred to a number of short-horned and polled breeds of cattle which showed in their cranial characteristics that their immediate ancestors were not of short-horned but of long-horned type. Duerst (1931) asserts that only such breeds may be compared that are genetically polled for a similar length of time. This implies that the variations in the size and shape of horns or the absence of horns are not the only factors determining cranial character—in other words, that the latter is influenced not only mechanically, but that additional hereditary factors contribute to the shaping of the skull. As far as the splanchnocranium is concerned, Duerst (1931) agrees that it is partly independent of the horns' weight, size and direction; accordingly, he proposes to apply Kollmann's (1882) principle for the classification of man into leptoprosopal (broad-faced) and euryprosopal (narrow-faced) types also to cattle.

While conceding the great influence of the size, weight and direction of the horns on cranial conformation, we arrive at the conclusion that not all important cranial features can be explained by it. Nor is the great influence of body size on the shape of the skull—combined with the small size of the horns—sufficient fully to explain the evolution of the shorthorn from the longhorn type of cattle. Nor is this completely accounted for by a consideration of the adjustments of various cranial bones to the direct effects of both body size and the moment of rotation of the horns on other bones of the skull.

Theory of Termination of Growth of Different Cranial Bones at Different Age Stages

Referring to the differences between shorthorn and longhorn skulls, not explicable by the influence of body and horn size, Hilzheimer (1926) asserted that certain parts of the cranium cease their growth at an early period of calfhood, while others continue theirs to a later juvenile stage, or reach full maturity. While the brain with the brain case is bound to keep up a certain size in order to provide for the activities of the nerves, it does not affect the viability of the domestic animal if the maxillae are a little shorter than normal, or if the frontal sinuses do not reach their full development. At birth the skull of the longhorn (primigenius) calf is spherical. Later the bones grow in length, the initially insignificant horn cores gradually increase in size; simultaneously the superior part of the forehead, which at first tapers aborally, widens (von Lengerken, 1953). In the fourth or fifth month of its life a longhorn calf resembles cranially an adult shorthorn. The skull of the latter terminates its development when the air sinuses have just reached the horn bases. This applies not only to their cavities; it applies also to the shape and position of the mandible of the adult shorthorn, which resembles a longhorn mandible at a juvenile stage. The early termination of growth is held to explain certain features typical of the majority of shorthorn skulls, such as the dished forehead and the angulation of the profile at the posterior end of the nasals.

While Hilzheimer's preference for expressing variations in the size of cranial bones in terms of termination of growth at different age stages seems of questionable value, it is evident that certain cranial features characteristic of the shorthorn type, which are not accountable by the small body size of these cattle, their short thin horns, and the mechanical adjustments of the parts of the cranium not directly affected, owe their origin to changes in the hereditary constitution, more especially in the relative growth rate during different phases of foetal and postnatal development. Since these inheritable variations include certain features which in the majority of Bovini seem to occur only in domestication, it is probable that those responsible for the shorthorn type appeared only after the domestication of the wild Bos p. namadicus, or, should some of these changes, such as short weak horns, have sporadically occurred already in the wild beast, that they were eliminated by natural selection.

371. Median sections of longhorn (left) and shorthorn (right) skulls (after Adametz)

Conclusion

In other words, it is unlikely that the shorthorn cattle are descended from a small shorthorned subspecies of the urus. Rather it would seem that their origin is to be traced, through the link of domesticated longhorn cattle, to the wild Bos p. namadicus, the various shorthorn features being due to the combined effects of hereditary changes in body size, horn weight and shape, and allometric cranial growth promoted by intentional selection.

In the light of these circumstances it would be unreasonable to assume that changes in body size, length and shape of horns, and allometric cranial growth towards the brachyceros type occurred only once in the history of domesticated cattle. While the origin of the shorthorn cattle of Africa and of several shorthorn breeds of Europe from south-west Asia is attested by archaeological and culture-historical evidence, some of the numerous breeds of cattle in the huge expanse of land between the Atlantic and Pacific Oceans may owe their cornual and cranial approximation to the brachyceros prototype, not to blood relationship and transmission, but to an independent development from domestic longhorn cattle.

IV. The Humped Cattle of Africa

i. On the Classification of the Humped Cattle of Africa

Africa harbours a great variety of humped breeds of cattle, some resembling the types represented on ancient Mesopotamian, Persian and Indian monuments, others closely allied to the recent humped breeds of Asia, others again peculiar to Africa.

Earlier Classifications

Epstein (1933) and Curson and Epstein (1934) proposed the classification of African humped cattle into true zebu and crossbred (sanga) types. This classification has been generally accepted.

Curson (1936) and Curson and Thornton (1936) class the humped cattle of Africa into True Zebus, of Asiatic origin, and Pseudo-Zebus, of African origin, sub-dividing the true zebus into the lateral-horned zebu (Africander) and the short-horned zebu, essentially of Asia, but well represented in East Africa.

The term 'lateral-horned zebu' is not a fortunate one, as the direction of the horns is only a minor anatomical characteristic easily influenced by directive selection. But since other types of cattle, e.g. the orthoceros, are classed in accordance with horn shape, there is some justification for this terminology. The author originally proposed the term 'long-horned' instead of 'lateral-horned zebu' for the Africander, in distinction from the zebu breeds of West and East Africa, which commonly carry smaller horns. This term is no longer upheld because the Africander cannot be regarded as a zebu proper (see: 'The Status of the Africander and Hottentot Cattle'—pp. 548 to 551), and also because its horns are not conspicuously large as compared to the horns of some other African cattle types, but are rather of medium length, while some of the African zebu breeds included in the short-horned zebu group have horns not shorter than those of the Africander. Moreover, Curson and Bisschop (1935) have described a more important feature distinguishing the different types of humped cattle, namely, the situation and anatomical structure of their humps. Bisschop (1937) says:—"In the study of African types of cattle, the hump is of special significance, not only because its anatomical nature and situation are the main differentiating features between two important basal types, but also because in derived types of these parent stocks the hump forms one of the principal clues by which the parentage of such crosses can be traced."

Theories on Hump Functions

Both in size and shape the hump differs in different animals and in the same animal at different times (Milne, 1955). According to Curson and Bisschop (1935), humps may be classified:—a) According to situation—cervico-thoracic or thoracic; b) according to structure—muscular or musculo-fatty; c) according to function—traction (locomotion) or storage of reserve fat.

It has been suggested that humps in cattle, like fat tails and fat rumps in various types of sheep, the humps of the camel and dromedary, and the fat buttocks of Hottentot and Bushwomen, represent stores of reserve energy, peculiar to steppe areas where nutritional conditions vary from seasonal abundance to scarcity. Duerst (1931), sharing this view with regard to the humps of zebu cattle, suggested artificial selection with a view to fat production, and traumatic muscular hypertrophy brought about by the draught function of the cattle, i.e. by pulling up against a yoke, as additional factors. Bisschop (1937) in rejecting this theory remarks "that functional adaptation as Duerst suggests would produce a muscular cervico-thoracic in addition to, and not as a modification of, the fatty thoracic hump", and adds that both explanations fall under the theory of the 'inheritance of acquired characteristics'—a principle not accepted in modern genetics. If the hump had any significance with regard to the work performed by the animal, Slijper (1951) says, its development in oxen would exceed that in bulls, since oxen are employed for draught; in reality, the hump is better developed in bulls than in oxen and cows. For this reason, Slijper regards it as a secondary sexual character. This may be accepted for breeds, in which the male has a pronounced crest approaching a cervico-thoracic hump, while the female is humpless. But it cannot be applied to zebu cattle in which both the male and the female are humped. Here it may at most be said that hump *size* is a secondary sexual character.

In contrast with the humps of the camel and dromedary which, consisting purely of fat, represent a kind of lipoma, the zebu hump is composed of muscle or muscle and fat tissues. Being an indication of the general condition of the animal and of its ability to survive during periods of drought, hump size may have been a materialistic aim of selection. In addition, an aesthetic purpose, similar to breeding for coat colour or horn size, may have been involved in its development. Hump evolution, as Duerst (1931) suggested, is connected with the shortening of the spinous processes of the dorsal vertebrae, brought about by selection under domestication.

Musculus Trapezius and Musculus Rhomboideus in Humpless Cattle

The muscles forming the zebu hump are M. trapezius and M. rhomboideus. Curson and Bisschop (1935) gave the following description of these muscles in an unhumped beast:—

M. trapezius

"This is superficial and fan-shaped. Its fibres extend from the dorsal aspect of the cervico-thoracic region (atlas to about the tenth thoracic vertebra) to the scapula. As they descend, caudally from the cervical region and cranially from the thoracic, they converge and terminate in an aponeurosis, which is inserted into the spina scapulae.

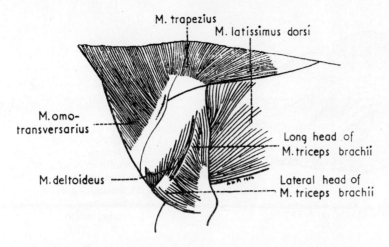
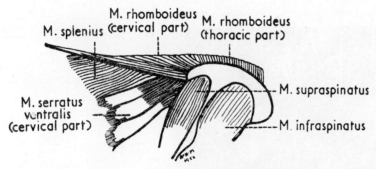

372. Unhumped bovine: superficial dissection of shoulder region (above); deep dissection of dorsal part of shoulder region (below) (after Milne)

The cervical and thoracic portions are, however, not clearly separated. The attachment of the fibres along the mid-dorsal line becomes more intimate as one follows the origin of the muscle backwards. The cranial border is firmly adherent along its anterior half to the M. cleidooccipitalis, and posteriorly to the M. omotransversarius. The caudal border is attached by fasciae to the M. latissimus dorsi which in this region is covered by the sheet-like M. cutaneus scapulae et humeri."

M. rhomboideus

"This muscle may be divided into a pars cervicalis and a pars thoracalis, the former being most conspicuous and having fibres pursuing generally a longitudinal direction, whereas the latter is relatively insignificant and has fibres running obliquely downwards and backwards. The origin is the pars occipitalis of the ligamentum nuchae and its caudal prolongation, the ligamentum supraspinale, while the insertion is the medial surface of the cartilago scapulae. The cervical part is pointed cranially, but as one proceeds caudally so does the muscular tissue become expanded."

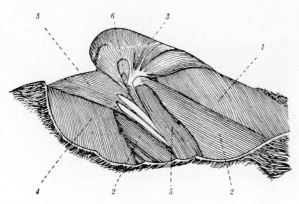

373. Hump of a 3-day-old zebu (after Duerst)

1 = M. splenius
2 = M. serratus ventralis (pars cervicalis et thoracalis)
3 = M. trapezius
4 = M. latissimus dorsi
5 = M. deltoideus
6 = M. rhomboideus

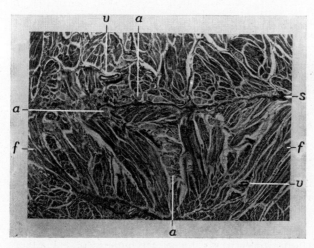

374. Horizontal section through the hump of a zebu embryo from Madagascar (10 times enlarged) (after Pettit)
s = Median fibrous septum
f = Muscle fibrils
a = Fat tissue
v = Blood vessels

Musculus Trapezius and Musculus Rhomboideus in Humped Cattle

In humped cattle the hypertrophy of M. trapezius and especially of M. rhomboideus is already clearly visible in the newborn calf. In a foetus, 32 cm long, M. rhomboideus contains a median fibrous septum which disappears at a later stage of foetal evolution. Duerst (1931) considers this septum as the last rudiment of the spinous processes which originally formed the median division between the lateral parts of M. rhomboideus.

Below the skin the hump is covered by M. trapezius which in Ceylon cattle is clearly separated into the cervical and thoracic portions. In Madagascar cattle the two parts of M. trapezius remain connected and are not significantly enlarged. M. rhomboideus, on the other hand, is modified; already during calfhood it forms a fold in which a small amount of fat may be deposited closely above the nuchal ligament. The muscle itself may be lean or fatty, the amount of fat varying in adult stock according to the general body condition. In lean beasts the hump contains very little fat; but in animals in good condition its hollows may be filled with 15–20 kg of fat, corresponding to approximately the twentieth part of the live weight of the animal (Duerst, 1931).

Differences in Hump Structure

Kronacher (1928), quoting Ramm (1901), has described the hump as "a remarkable fat deposit the basis of which is formed by a correspondingly modified dorsal muscle in which larger or smaller quantities of connective tissue and fat are deposited". This description does not distinguish between fundamental anatomical differences in hump structure, but refers merely to differences in size. Duerst, on the other hand, stressed certain differences in the structure of the humps between Ceylon and Madagascar cattle. Similar differences have been analysed by Curson and Bisschop (1935) in African humped cattle, particularly in the Africander of South Africa and the (short-horned) zebu of East Africa, and by Milne (1955) in Ankole sanga and East African short-horned zebu cattle.

The Anatomy of the Africander Hump

The Africander cattle have a large prominent hump set closely to the withers (see p. 476). "Viewed from the side the hump is situated in front of the withers. It occupies the posterior two-thirds of the upper border of the neck. The structure is somewhat pyramidal in shape and cranially it rises at an angle of about 40° to a rounded apex situated well in front of the point of the withers. From this rounded apex the hump falls at an angle of just over 30° on to the withers and merges with them to form a uniform and strong attachment. Viewed from the front the hump appears to sit snugly over the upper border of the neck and is firmly attached to its sides, from which it rises steeply at an angle of about 60°–65° to a rounded summit. For this reason the hump of the Africander is sometimes spoken of as being well attached. In the bull the hump is decidedly prominent, rising sometimes to 8 inches above the top-line. In oxen the structure is a little less prominent and in females it is relatively smaller still. The hump is well developed even in a foetus" (Curson and Bisschop, 1935).

The Anatomy of the Tanganyika Short-horned Zebu Hump

"The hump of the Shorthorned Zebu is less constantly pyramidal ... Very frequently the structure is not only more prominent but also actually dome-shaped. Another

375. Thoracic hump (Zanzibar zebu bull)

376. Cervico-thoracic hump (Africander Bull)

common feature is the greater mobility of the hump, especially the posterior part due to its fatty nature. Most striking of all features is the more caudal situation of the hump, a vertical line through the summit passing at least through two vertebrae behind a line similarly drawn in the Africander" (Curson and Bisschop, 1935). Thus Slijper (1951) describes the position of the thoracic hump in an Ongole zebu bull as "exactly dorsal of the foreleg". In an about 50-day-old East African zebu foetus, examined by Milne (1955), the hump was already quite distinct and definitely thoracic in position.

"In extreme cases of hump development," Curson and Bisschop (1935) write, "the term 'loose' may be applied, but it must be emphasised that anatomically it is nevertheless securely attached, particularly by the M. trapezius...". Slijper (1951) has pointed out that whereas the cranial part of the muscle forming the hump (M. rhomboideus cervicis) is firmly attached to the underlying connective tissue between the left and the right M. splenius, there is only a small layer of very loose connective tissue

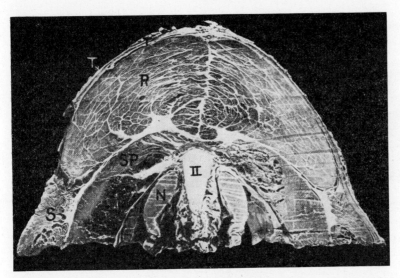

377. Transverse section (at summit) of hump of Africander bull, over 2nd thoracic vertebra (II). Cranial view (after Curson and Bisschop)

T = M. trapezius
R = M. rhomboideus
S = M. serratus ventralis
SP = M. splenius
N = L. nuchae

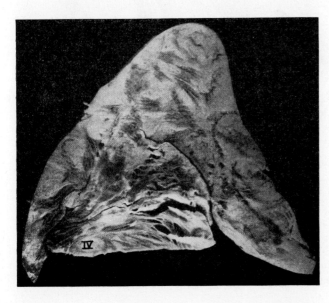

378. Transverse section (at summit) of hump of Short-horned Tanganyika Zebu bull calf, over 4th thoracic vertebra (IV). Cranial view (distorted appearance due to tight packing in drum) (after Curson and Bisschop)

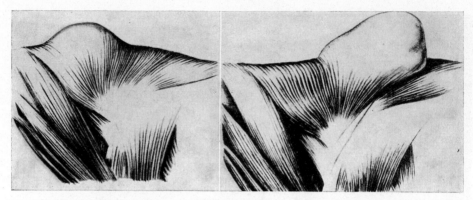

379. (left) Africander bull, showing cervico-thoracic hump with skin removed but M. trapezius intact

380. (right) Tanganyika Short-horned Zebu bull calf, showing thoracic hump with skin removed but M. trapezius intact (after Curson and Bisschop)

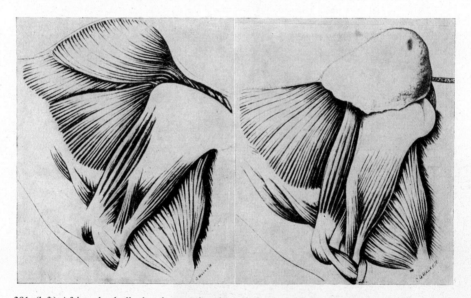

381. (left) Africander bull, showing cervico-thoracic hump with M. trapezius removed

382. (right) Tanganyika Short-horned Zebu bull calf, showing thoracic hump with M. trapezius removed. (It will be observed that the M. rhomboideus is quite different from that of the Africander) (after Curson and Bisschop)

between the caudal part of M. rhomboideus cervicis and the underlying ligamentum nuchae and tips of the neural spines. Thus the hump is attached to the body of the animal mainly by the aponeuroses of M. trapezius and M. latissimus dorsi.

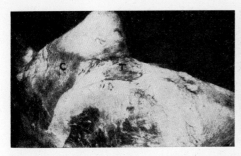 

383. (left) Short-horned zebu. Dissection showing M. trapezius
C = M. trapezius cervicalis
T = M. trapezius thoracalis
(after Milne)

384. (right) Short-horned zebu. Dissection showing hump with M. trapezius removed (after Milne)

Comparison between Africander and Tanganyika Short-horned Zebu Humps

The striking differences in the anatomical structure of the humps of Africander and short-horned zebus are explained by Curson and Bisschop (1935) in the following comparison of the two humps:—

Comparative table concerning humps of African zebu types

Feature	Africander Bull	Tanganyika Short-horned Zebu
M. trapezius	The muscle is well developed and may be arbitrarily divided into cervical and thoracic parts, the former overlying the hump and the latter caudal of the hump.	The muscle is distinctly separable into a cervical and a thoracic portion, the latter being darker in colour and more strongly developed. The cervical portion instead of arising entirely from the funicular part (pars occipitalis) of the ligamentum nuchae is closely associated with hump development. In fact it may be considered as a superficial cranio-lateral anchor or attachment of the hump. The thoracic portion is well developed and except for a small cranial slip does not act as a stay.

Feature	Africander Bull	Tanganyika Short-horned Zebu
M. rhomboideus	The M. rhomboideus has distinct cervical and thoracic portions. The cervical part is extraordinarily well developed, and forms the hump. The thoracic part is made up of short muscular fibres having generally a ventro-caudal direction.	This is characterised by the entire absence of the longitudinally arranged cervical part.[1] Instead, the anterior portion of the thoracic part is very much developed, in fact the fatty hump having not only a foundation, but also a framework of muscular tissue. The fibres run generally in a ventro-caudal direction, especially at the base of the hump. This structure is best described as a musculo-adipose development of the M. rhomboideus, pars thoracalis (anterior portion).
Situation of hump	The hump lies cranial to the angulus cranialis of the scapula over the 6th/7th cervical to 4th/5th thoracic vertebrae (Crista spinosa)—a cervico-thoracic hump.	Dorsal to the margo vertebralis of the cartilago scapulae. In terms of vertebrae it lies over the 1st to 9th thoracic vertebrae (Crista spinosa)—a thoracic hump.

Variability in the Structure of Thoracic Humps

Milne (1955) has described the hump of the East African short-horned zebu as a "predominantly fatty hump, ... a musculo-fatty development of part of the rhomboideus muscle... It appears that this type of hump represents an accumulation of reserve food material. Examination of sections of foetal humps ... tends to indicate that fat begins to be laid down in the hump at about the third month of foetal life." This is confirmed by Hornby's comments on the hump of the Tanganyika short-horned zebu: "At birth the musculature is well marked, but as the animal develops this becomes more and more obscured by fat deposition. Even when the adult loses condition and

---

[1] Since the cervical and thoracic parts of the M. rhomboideus are not separated by any well defined structure, e.g. of fibrous nature, it is of course possible to consider the hump of the Short-horned Zebu as a development of the pars posterior of the cervical part of M. rhomboideus. Duerst (1931) considers that a fusion of the two parts has taken place (Curson and Bisschop, 1935). Slijper (1951) is definite on this point:—"The hump of the zebu consists of Musculus rhomboideus cervicis". In dissecting this muscle it can be separated into two different parts; the cranial part comprises about one-fourth, and the extremely well developed caudal part about three-fourths of the muscle.

the hump shrinks very greatly, the hump never regains the lean muscular state of calfhood—a certain amount of fat and fibrous tissue persist."

Epstein (1955a; 1956) examined several hundred longitudinally halved frozen humps of Eritrean and Ethiopian zebu cattle. In the Ethiopian cattle the hump was all fat, on a foundation and in a framework of muscular and connective tissue which conveyed the impression of complete adipose degeneration. The forequarters of the Eritrean cattle were exceedingly poor; in consequence, the muscular base and fibrous framework of the small humps were more apparent than in the well covered Ethiopian forequarters; still, the humps retained a perceptible amount of fat, more especially in the uppermost posterior part, and not even in the leanest specimens did they warrant the term muscular instead of musculo-fatty. In the thoracic humps of Eritrean and Ethiopian zebu cattle the infiltration of fat commences at the highest hindmost part of the hump, then extends downwards, and finally forwards. When an animal loses condition, the fat recedes first from the anterior part of the hump, then from the lower posterior region. It persists in the uppermost caudal part of the hump even in the leanest specimens encountered. Thoroughly cooked the thoracic hump of adult animals retains the consistence of a tough meatless compound of fasciae, ligaments and tendons with a variable quantity of fat, somewhat similar to, but very much tougher than, the fatty part of the brisket.

Contrary to these findings, Bettini (1940) has described three different structural types in the thoracic humps of southern Somali cattle:—1) Muscular, of a pale veal colour—the most common type; 2) muscular, of a bright red colour; 3) musculo-fatty, with a large quantity of fat and a light muscular framework—the rarest form (one only). Hill's (1961) observations indicate that in the Azaouak and Shuwa zebu cattle of West Africa the hump is muscular, but in the Sokoto and Adamawa zebus it is invariably musculo-fatty. Recently the present author handled the forequarter of an Eritrean zebu in which the hump section was entirely muscular (Fig. 385).

In Indonesia, De Moulin (1922; 1924) observed a considerable deposit of fat at the base of a zebu hump, while Musculus rhomboideus itself was lean. Thorpe (1953) found a purely muscular hump in an Indian zebu cow of the Gujerat type. Similarly, in an adult Ongole zebu bull in good condition, Slijper (1951) has described M. rhomboideus thoracis as "almost entirely fleshy", and M. rhomboideus cervicis as "an almost entirely fleshy muscle containing no more fat and no more connective tissue than most of the other muscles of the animal". But in African thoracic-humped cattle

385. Longitudinal section of a muscular hump of an Eritrean zebu

he found an altogether different condition:—"I was especially struck by two humps of East African zebus that were sent to me by the Antwerp Zoo. They contained a large amount of fat" (personal communication—20.11.1957). It has been observed that the introduction of Sindhi and Sahiwal blood into East African short-horned zebu stock "significantly reduces the amount of fat in the hump" (Milne, 1955).

Individual or seasonal differences in structure are demonstrated by the different descriptions of Toposa–Murle humps: "muscular" by S.D.I.T. (1955) and with "a large amount of fat" by Joshi et al. (1957).

Hence we may conclude that the structure of the thoracic hump in zebu cattle is not uniform; it may be either musculo-fatty or muscular, and in this respect there is variation both within and between breeds.

In the cervico-thoracic hump of the Africander cattle "fat may also occur, but it is usually distributed in layers, first subcutaneous, then between the M. trapezius and M. rhomboideus, and finally beneath the latter" (Curson and Bisschop, 1935).

Classification of African Humped Cattle according to Hump Situation

Epstein (1955b) formerly based the classification of humps on structure, i.e. muscular and musculo-fatty, distinguishing between flesh-humped and fat-humped cattle. This has now been abandoned because lately several authors have stressed the muscular framework of the thoracic hump and the general muscular condition of some thoracic humps, and also because, as Mason has pointed out, structure is not discernible in the live animal, still less in an illustration. Classification of humps according to situation is therefore preferable. Accordingly, humps are divided into two main groups, viz. thoracic and cervico-thoracic. Cattle with thoracic humps are occasionally referred to as chest-humped, and those with cervico-thoracic humps as neck-humped.

The chest-humped cattle breeds of Africa are for the greater part zebu proper, as defined on p. 200. Neck-humped zebu breeds are known only from ancient representations from Egypt, Sumer, and the Quetta–Pishin and Indus valleys.

Zebu, Sanga and Fulani

In many parts of East Africa where, during recent centuries, the thoracic-humped zebu has spread out at a phenomenal speed, native stock, formerly cervico-thoracic-humped, have become thoracic-humped, while in the region between the Republic of the Sudan and the Zambesi river there is considerable variability in hump situation among and within the different cattle breeds.

The cervico-thoracic-humped cattle of Africa are derived from a mixture of zebu and humpless cattle. Cervico-thoracic-humped breeds are found in East and South Africa. South of the Zambesi river, whither the range of thoracic-humped zebus does not extend, they are the only type of native cattle.

In order to distinguish the zebu breeds proper from the breeds showing a mixture of zebu and humpless longhorn features, the breeds of such mixed type are referred to as 'sanga', an Abyssinian word meaning 'bull'. This term, which seems to have been first employed by the traveller Salt with reference to giant-horned Galla oxen (Vasey, 1851), has been generally accepted for breeds south of the Zambesi, and for some north of the Zambesi.

In West Africa the impact of thoracic-humped zebus on humpless longhorn cattle or on cervico-thoracic-humped breeds derived from zebu and humpless longhorn crossbreds has produced a well defined racial type which Curson and Thornton (1936) have called 'Lyre-horned Zebu'. This terminology was followed by Mason (1951a) and Joshi et al. (1957). Since the shape of the horns in the humped cattle of Africa is considered as anatomically and phylogenetically less significant than the hump, the author has avoided reference to the lyre shape of the horns in the terminology employed for this type. Breeds of this type are called Long-horned Fulani after their owners and the most conspicuous feature distinguishing them from the short-horned zebus of West Africa.

Curson and Thornton (1936) grouped the sanga and Fulani together under the heading of Pseudo-Zebu. The term sanga might conceivably be used to include the sanga breeds of East and South Africa and the Fulani of West Africa. Mason (1951a) has rejected the term sanga for West African cattle because it was misapplied to zebu-humpless shorthorn crosses, while Curson and Thornton (1936) used it for the Kuri, Bandara and Jakoré. In its stead Mason proposes the general term 'zeboid' which would include breeds of cattle whose characteristics are intermediate between zebu and humpless cattle, such as the sanga and Fulani. It would also include cattle of a similar racial mixture, e.g. the cervico-thoracic-humped Siri cattle of Bhutan, Sikkim and Darjeeling and the similarly humped Kiangsu, Shantung, Chin-chuan (Shansi), Huang-Po (Hupeh) and several other 'Yellow Cattle' breeds of central China, as well as recently developed breeds such as the Santa Gertrudis and Bonsmara.

The difficulty confronting the taxonomist in the classification of African humped cattle is that the majority do not live in the geographical and genetic isolation characteristic of different races of wild Bovini or of domesticated breeds in economically advanced countries. Therefore it is often impossible to draw a clear line of demarcation between breeds and individuals representing borderline specimens between zebu and sanga; and unless one is inclined, in view of the large number of intergrading links, to give up classification of African humped cattle on the basis of racial characteristics as a hopeless task, restricting it to geographical divisions, the break between the zebu and sanga has to be made arbitrarily at some point of the chain.

In the contact areas between thoracic-humped zebu and humpless shorthorn (brachyceros) cattle, as in the Nuba mountains and some parts of West Africa, cattle of mixed type, representing transitional stages from the thoracic-humped zebu to the humpless shorthorn, are encountered. But as these crossbreds have not consolidated into a definite type, they have not been placed into a class of their own, but are dealt with in connection with the humpless shorthorn or the thoracic-humped zebu group, depending on the degree of their racial approximation to either parent type.

ii. The Zebu Cattle of Africa

Zebu cattle are found in East and West Africa. They display considerable variability in colour and conformation, as their ancestors entered the African continent over a period of many centuries, from Arabia and the countries beyond, bordering the Arabian Sea. In Africa artificial as well as natural selection and adaptation to different environmental complexes and intermixture with other bovine types have added to their variability. Every shade of colour known in domesticated cattle is encountered: red, brown, white, black, grey—whole or in different patterns of variegation. In the majority the horns are short—hence the term 'short-horned zebu' in common use; but some have horns of medium length. Horn direction is usually lateral and upward, occasionally lateral and downward. Some have loose horns devoid of cores, or one loose and one firm horn; finally, polled animals are encountered. Hump development varies considerably. Some of the African thoracic-humped zebus, as the Lugware (Mangbattu) cattle of the eastern Congo Republic and the Adamawa Banyo, have exceedingly prominent humps, in others, such as the Somali cattle, the hump is of negligible size.

1. The Zebu Cattle of East Africa

Distribution of Zebu Cattle in East Africa

In East Africa zebu cattle are found in a large area extending practically over the whole eastern and central part of the continent, from latitude 20°N in the Northern Province of the Republic of the Sudan in the north to the Zambesi river, latitude 15°S, in the south. In this region the zebu forms by far the largest proportion of the cattle population.

The Influence of Other Types of Cattle on the East African Zebu

In the course of the phenomenal extension of its range in East Africa the zebu has widely absorbed or superseded the other cattle types formerly bred there: humpless shorthorn cattle in the northern Sudan, sanga cattle to the south, and possibly also humpless longhorn cattle in Ethiopia where the earliest rock paintings depict cattle with long horns and no humps. This intermixture and absorption has left its stamp on some of the zebu breeds in East Africa, particularly, but not only, in the present contact areas with other types. Thus, while the great majority of East African zebu cattle are short-horned, some of them have horns of medium size, exceeding 20 or even 30 cm in length. This may not always be due to the influence of long-horned sanga cattle, and the geographical proximity or remoteness of sanga territory should be decisive. More important is the variability in hump situation. The typical zebu cattle of East Africa are thoracic-humped. Wherever cervico-thoracic-humped animals occur —as they do in some East African zebu breeds in considerable numbers—this must be attributed to the influence of humpless shorthorn cattle (Sudan) or of cervico-thoracic-humped sanga cattle (Eritrea, central Sudan, south-western Somalia, Uganda, Malawi). Some of these breeds, such as the Nganda of Uganda and the Baggara cattle of the central Sudan, show so even shares of zebu and sanga in their conformation that it is a matter of choice whether to deal with them in connection with the zebu or the sanga type.

Large and Small Types of East African Zebu

Faulkner and Brown (1953) roughly divide the zebu cattle of East Africa into a large and a small type. The large is believed to originate from an arid hot environment, the small from the higher rainfall areas. Between the extreme types there are numerous gradations in size which largely depends on the natural environment of the present breeding area and the wealth or poverty of the breeders. In well managed herds African zebu cattle may attain live weights of over 450 kg, whereas in the poorer cattle districts they do not scale more than 150 kg. Some, like the improved Boran cattle, are large-framed beasts, while others, like the Wachagga, on the slopes of Kilimanjaro, in close vicinity to the tsetse-fly belt, are dwarfed, light-boned and weedy, being kept under miserable conditions inside the dwelling houses of their owners and fed largely on banana leaves, couch grass and manioc, while the calves are consistently starved. In some areas in Africa zebus have acquired the browsing habit of goats and have grown so unaccustomed to grazing that when moved to grass country such animals invariably fall away in condition during the first few months (McCall, 1928).

The Zebu Cattle of Somalia

From Bab el Mandeb the area of distribution of zebu cattle in East Africa extends south-eastwards along the coast of the Gulf of Aden and thence throughout Somalia; north-westwards zebu cattle range along the Red Sea coast through Eritrea.

In the former Italian Somaliland three different zebu breeds were recognised, in addition to the Jiddu (see pp. 379–383), an intermediate zebu-sanga type:—(1) The Gasara, known also as Aria and 'the little breed of the sand dunes', is bred in several parts of the country, foremost in the coastal areas near Mogadishu, in the poorest areas of upper Juba, on the right bank of the upper Webbe Shibeli, in Mudugh and Nogal; (2) the Gherra or Dauara, found chiefly in Ghel del Dafet, at the upper and middle course of the Webbe Shibeli, decreasing in number towards the lower Shibeli and Juba rivers; (3) the Boran (see pp. 374–378), bred by the Harti and Mohamed Zubier tribes in the more southerly districts west of the Juba river; (in Somalia they are occasionally called Avai from the town of this name near the confluence of the Juba and Webbe Shibeli rivers) (Bozzi and Triulzi, 1953).

The Gasara Zebu of Somalia

In Somalia the small type of East African zebu is represented by the Gasara, the large type by the Gherra and the Boran. The small type also includes a few cattle in the

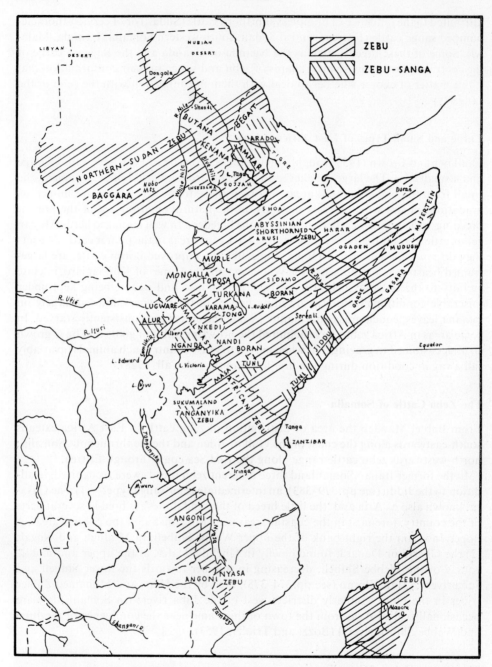

386. DISTRIBUTION OF ZEBU AND ZEBU-SANGA INTERMEDIATE TYPES IN EAST AFRICA (after Mason and Maule)

extreme north and on the east coast of Somalia, which are of the Arab type ranging into Eritrea. Gasara (Abgal or Aria) cows average 110 cm in withers height and 218 kg in weight. The head is of moderate size, the facial profile convex, the horns are short and thin, sometimes loose and pendulous or absent, and the ears are large and semi-pendulous. The hump is thoracic in situation and pronounced, the dewlap small with few folds, and the prepuce is little developed. The rump exceeds the withers in height; it slopes backwards to the long tail. The udder of the cow is relatively large, the Gasara being used mainly for dairy purposes. The typical colour is leaden grey due to a mixture of red, black and white hairs, often dark grey or dark red. Pied and fawn animals are also encountered. The face is commonly white or shows a white blaze on the forehead. The skin is thin and the mucosa invariably black. A variety of the Gasara, called Magal or Correi, is bred near the Juba river (Gadola, 1947); but Bozzi and Triulzi (1953) have pointed out that 'magal' means black, and that this is the colour of a cross between the Jiddu and the lead-coloured Gasara.

The Singhi, at one time numerous in the Webbe Shibeli area, is now being absorbed by other breeds, notably the Gasara (Bozzi and Triulzi, 1953).

The Gherra Zebu of Somalia

The Gherra (Garre or Dauara), called after the Garre or Gherre tribe, is larger and of a better conformation than the Gasara. With careful management the cows give more milk than those of any other native breed of Somalia; in addition, the Gherra also dresses out well at slaughter. It stands 120–132 cm at the withers, the anterior part of the rump being approximately 10 cm higher. Bulls average 320 kg, cows 285 kg in weight. The Gherra has a long narrow face with a convex profile; the ears are large and somewhat pendulous, the horns small, sometimes loose or absent. The neck is long and thin, the hump relatively small; the dewlap and umbilical fold are well developed. The rump slopes only a little to the high tail attachment. The skin is fine,

387. Gherra cow

388. Somali zebu cattle (after Drake-Brockman)

389. Begait (Barca) zebu bull from Eritrea (photograph: Dr. Y. S. Goor)

390. Begait (Barca) zebu cow from Eritrea (note the more cervical situation of the hump) (photograph: Dr. Y. S. Goor)

and the udder large. The colour of the coat is a deep red, often with indistinct spots; red-pied or black-pied animals are also seen (Gadola, 1947). A variety of the Gherra, named Bimal, which Bozzi and Triulzi (1953) class with the Gasara, is found in the zone of the coastal sand dunes between Mogadishu and Merca. Owing to the poor nutritive conditions of its habitat, it is smaller than the Gherra proper and weighs about 20 kg less.

The North Somalia Zebu

The cattle of northern Somalia are generally small and unsightly, and of a similar type to those of southern Arabia, having in all probability been imported at a relatively recent date (Drake–Brockman, 1912). The thoracic hump is very small, the skin is thin and elastic, and the hair fine. The coat is light brown or chestnut in colour, sometimes black or white. In the Borama–Hargeisa area in the west, roan and spotted cattle are common. The horns are black in colour, thin and short, rarely exceeding 20 cm in length. Many animals are polled or have loose horns. In loose-horned cattle the horn sheaths are normally developed, but there are no cores to support them. In their stead very short rough elevations project from the skull. Such bony elevations are occasionally found also in polled animals. The skull is long and narrow, the profile in some animals slightly, in others markedly convex. In polled and in loose-horned individuals the forehead rises to a high poll. The effects of different horn direction, loose horns or the absence of horns on the skull are responsible for variations in the conformation of the forehead, apparent from Keller's (1896) and Duerst's (1899) contradictory descriptions of Somali Zebu skulls.

The Arab (Bahari) Zebu of Eritrea

The same breed, under the name of Arab, occurs in Eritrea, especially along the coast and on Massawa island. The Arab is known also by the name of Bahari or Sea Cattle, a term referring to their origin from across the sea. In the coast belt of Massawa the Arab is found in its purest racial form; sporadically it occurs also in other parts of Eritrea, but seldom purebred. The withers height of the Arab does not exceed 130 cm, and is often considerably less, and the live weight averages little more than 200 kg, so that the oxen are too weak for draught. The colour is usually chestnut, rarely pied.

The Begait Cattle of Eritrea

In the western lowlands of Eritrea a larger type, i.e. the Begait, is bred by the Barca and Beni Amer tribes. These cattle are also found across the Sitit river in the lowlands along the Sudanese border of the Amhara Province of Ethiopia. While in its unadulterated form closely related to the Arab, the Begait is larger in size, averaging 130–135 cm in withers height (Bonelli, 1938). Bulls weigh approximately 390 kg, and cows 295 kg. Size largely depends on the district and the extent of crossing with the Aradò, a zebu-sanga crossbred type (see pp. 347–350). The head of the Begait is long and narrow, especially the facial part, with short horns; many animals are polled. The ears are long and frequently somewhat pendulous, the dewlap is moderately developed, and the hump is fairly large. As in the North Sudan zebu, which it resembles, the Begait frequently has the hump more cervically situated than is common in East African zebus, possibly owing to cross-breeding with sanga or humpless shorthorn cattle in an earlier generation. The legs of the Begait are long and the rump is drooping. The animals are often lower in front than behind. The coat colour is white, black-and-white or red-and-white, the coloured areas being fairly symmetrically distributed over the body. Like the Arab, Begait cattle make poor draught beasts and do not thrive in the cold climate of the high plateau owing to their thin hide and short coat (Gadola, 1947).

391a. Head of a begait Bull (after Marchi)

391b. Head of a Begait cow (after Marchi)

392. Polled Begait cow and calf

Zebu Crossbreds in Eritrea

The intermingling of the three major Eritrean cattle types, viz. the Arab and Begait zebus and the Aradò zebu-sanga, has given rise to a considerable number of intermediate types. Indeed, such animals are numerically superior to all purebred Eritrean cattle collectively. Among their great variety Marchi (1929) has singled out the Baria for its superior conformation and racial uniformity. It is bred by the Shukria in the western lowlands, and by the Bileri of Keren. The Baria is a strong large-framed beast, with rather heavy horns of medium size and a slightly convex facial profile. It is believed to be derived from a mixture of Begait zebus and Aradò zebu-sanga crossbreds; but in conformation it is a pure zebu.

The Aradò Group of Northern Ethiopia

Aradò is a collective term employed for a number of different local breeds, mostly of zebu-sanga crossbred type, but some pure zebu and others pure sanga. Mason and Maule (1960) restrict the term Aradò to the cattle of the highlands of northern Ethiopia (including Eritrea), which show characters intermediate between short-horned zebus and long-horned sangas. The Aradò is found on the entire high plateau of Eritrea, including the districts Maria, Mensa and Sachel. In Ethiopia proper the Aradò is bred throughout the territory contiguous with the Eritrean habitat of the type; but in the north-eastern part of this area the sanga component predominates to such an extent that the Danakil and Galla Azebo cattle, included by several Italian authors in the Aradò group, are practically pure sangas.

The various types of cattle, classed by Italian writers with the Aradò group, differ in body size and proportions and are either long-horned, short-horned or devoid of horns; occasionally the horns swing freely at the sides of the face owing to the absence of cores (Masoero, 1938). Generally the body shows good depth and length, the dewlap is well developed, and the rump is less drooping than in purebred zebu cattle. The hooves are large and heavy. The hump is thoracic in situation, large in bulls, less developed in oxen, and hardly noticeable in cows (Marchi, 1929). It is composed of muscular tissue with variable deposits of fat between the fibres (Nieri and Robotti, 1939). The thick hide, characteristic of the majority of Aradò cattle, offers protection against the cold climate of the high plateau. In its resistance to cold and its great capacity for work the Aradò is superior to the Begait and Arab zebus of Eritrea.

In Eritrea, the Aradò is represented by a mixture of different types, some small, others large. They are predominantly zebu, with only a slight admixture of sanga and possibly also a humpless shorthorn strain. Their horns differ in size and shape. Some have long, others medium-sized horns, still others are short-horned or polled. Animals with one or both horns loose are also encountered. The average withers height is 118 cm, the height at the summit of the hump 121 cm, and rump height 126 cm. Of the Assaorta varieties of the Eritrean Aradò, the largest are found in Hamasen and Libam, those of medium size in the possession of the Mensa and Waira tribes, and the smallest in Habab and Akele Guzai (Marchi, 1929). For the Akele Guzai, Bonelli (1938) gives

393. Aradò bull from Eritrea

394. Aradò ox

a withers height of 123–130 cm and an average live weight of 230 kg. The larger types are suitable for work and beef production (Gadola, 1947).

Beyond the Marab river the Aradò cattle in the Quarà district of north-west Amhara are known as Galla, according to the region from where their ancestors came. The Galla area of Ethiopia lies well to the south and east beyond the Abbai and Bashilo rivers. However, in the past the Galla used to make frequent excursions to-

395. Aradò cows

396. Aradò Galla ox

wards Lake Tana. The Galla Aradò are medium- to long-horned; they are quite different from the Galla cattle of the Ethiopian Kenya border (see p. 378).

The Zebu Cattle of Ethiopia

In Ethiopia zebu cattle constitute the majority of the cattle population. In Amhara the principal breed, in addition to the long-horned Aradò zebu-sanga, is the Begait zebu (Nieri and Robotti, 1939). In the central highlands of Shoa, Harar and Galla Sidamo a small short-horned zebu type is encountered. In Shoa four different zebu varieties are distinguished: Walega, Shoa, Arusi and Harar (Girardon, 1939). In the south-east, towards the Kenya border, the Boran predominates (Roetti, 1939). Sandford (1964) distinguishes among the zebu cattle of Ethiopia, excluding Eritrea, three main types, namely, Fogara (Wagara), Black Highland and Boran, in addition to the Adal (Keriyo or Dankali) sanga.

The Wagara (Fogara) and Walega Zebu Cattle of Ethiopia

The Wagara type is bred in Amhara and Gondar, mainly east and north of Lake Tana. It is tall but narrow-framed and slab-sided. The horns are usually short and blunt, often loose, and occasionally absent. The hump is thoracic in situation; the dewlap, navel flap and prepuce are long and pendulous; the rump is short and steeply sloping, and the tail long and fine. Udder development is poor. The coat colour varies; a black-and-grey or black-and-white coloursided pattern is the most frequent. The skin, hooves and teats are pigmented black (Mr. R. H. D. Sandford, FAO—personal communication).

Sandford (1964) suggests that the Wagara (Fogara) cattle originated from India during the period of Portuguese influence in Ethiopia. While the ancestral stock of the Wagara, in common with that of nearly all other zebu cattle of Africa, was indeed introduced from Baluchistan and India, it is very doubtful that the Wagara as such was imported during the 16th or 17th century, that is the period of Portuguese influence in East Africa, rather than developed locally from a population of East African zebu cattle.

Along with the Wagara, the Walega variety of western Ethiopia was tentatively included by Mason and Maule (1960) in the Aradò group, but it is actually of nearly pure zebu stock. Walega cattle are of large size and furnish good work and meat animals. The live weight averages 300–350 kg, but may reach as much as 600 kg. The horns are of moderate length, the hump is small to medium in size, the dewlap small, the skin fine (in this respect the Walega differs from the Aradò group), and coat colour is very variable. "To this type of zebu is related that of Gojjam, Agaumdir and Lake Tana, especially Dambiya and Fogara" (Girardon, 1939).

The Zebu Cattle of the Central Highlands of Ethiopia

The zebu cattle of the central highlands of Ethiopia are small; bulls weigh approximately 290 kg and cows 230 kg (Salerno and Congiu, 1939). The head is small and short, with a straight or convex profile and semi-pendulous ears. The horns are very variable in size and shape, usually short and stout or of medium length, sometimes very short, and occasionally loose. The hump is musculo-fatty in structure and thoracic in situation with a tendency to lean over backwards. The dewlap is large, the body thick-set and the rump has a steep slope. The tail is long and thin. The coat colour is extremely variable: self dark and light fawn, black, brown, grey, roan, red-pied or black-pied, tricolour and spotted.

397. Wagara (Fogara) bull

398. Wagara (Fogara) cow

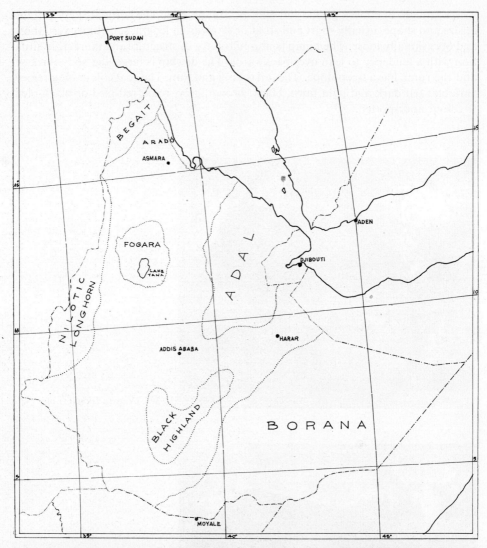

399. BREEDING CENTRES OF PRINCIPAL ETHIOPIAN CATTLE TYPES (after a sketch by Mr. R.H.D. Sandford, FAO Addis Ababa)

The Black Highland Cattle of Ethiopia

The Black Highland cattle, which constitute a more or less distinct type among the very mixed highland cattle, are found mainly in the northern parts of Sidamo, in Wollamo and the highland of Bale, being concentrated in the area between Goba in Bale province and the highlands round Ageresalaam (Yirga-Alam) in Sidamo. Their existence as a separate type is indicated by the fact that whereas the majority of the

400. Zebu cattle from the central highlands of Ethiopia (after Stang)

401. Highland cattle from Sidamo-Arusi, Ethiopia

402. Black Highland cattle, Ethiopia

cattle in northern Sidamo, Wollamo and Bale are black or black with a white face or white markings on other parts of the body, the cattle in the surroundings of these highlands are rarely black. In their purest form Black Highland cattle are encountered at altitudes above 2500 metres, being adapted to the wet and cold conditions of their habitat. They are commonly small and compact, with slender limbs. The horns are of short to medium length, fine, and sharp-pointed. The hump is of moderate size; in cows it is often so small that they may appear almost humpless (Mr. R. H. D. Sandford—personal communication).

Main Types of Sudan Cattle

Several types of cattle exist in the Sudan, as may be expected in a territory of so vast an area. There is little doubt that in many districts these types are descended from cattle which in the not far distant past virtually constituted different breeds in the European sense of the term. Although in the contact areas their identity is becoming progressively lost as general freedom of movement increases, three main types, differing widely from each other, can be distinguished. The northern or Arab cattle are of zebu type. The approximate boundary of demarcation between these and the Nilotic sanga of the south is formed by the rough east to west line which divides the northern Mohammedan from the southern pagan areas of human habitation. The negligible degree to which these two group types have mixed in historic times may be ascribed to the same causes that have kept the human inhabitants of the two areas equally distinct (Bennett, 1938). The differences between these two principal groups of Sudan cattle are more than merely morphological, since neither type thrives in the other's environment; a large proportion of zebu cattle will die if maintained in the sanga area during the rains. The economic potentialities of the two types also differ; the zebu does not generally reach the size and weight of the sanga, but the zebu cows are superior milkers (Bennett et al., 1948).

The third type includes the Toposa–Murle and Mongalla zebus in the extreme south.

The Zebu Cattle of the Northern Sudan

The zebu cattle of the Mohammedan areas of the northern Sudan have horns of medium size and well developed humps. Boyns (1947) stresses their close similarity to Indian thoracic-humped zebus and the relatively low degree of conformational variability. But other authors emphasise their variability. Generally they are thoracic-humped, though in many of them the cervico-thoracic situation of the hump suggests an earlier admixture of a humpless shorthorn or a neck-humped sanga strain. Indeed, Bisschop (personal communication to I. L. Mason) regards one breed (Butana) as sanga proper. Bennett (1938) believes them to be derived from a cross of African humpless shorthorn cattle and cervico-thoracic-humped zebus; this is contradicted by the large size and generally thoracic situation of their humps.

The zebu cattle of the northern Sudan have been divided into the Kenana (White Nile) cattle, the Butana, Dongola and Shendi (Red Desert) cattle, and the Baggara.

403. Thoracic-humped zebu bull from the Sudan (photograph: Dr. Y. S. Goor)

404. Thoracic-humped zebu cattle from the Sudan (photograph: Dr. Y. S. Goor)

Kenana Cattle

The Kenana or Rufa'ai cattle, whose name is taken from the semi-nomadic pastoral tribe which owns the major herds of this type (Hattersley, 1951), are found along the Blue Nile river as far north as latitude 15° N.

Boyns (1947) distinguishes between two separate types: the Kenana of the Fung area of the north-central Sudan, between the White and Blue Niles, and the White Nile type in the Kosti district of the White Nile valley (between latitudes 12° and 14° N, and to a lesser extent to the east between the same latitudes), which is influenced by Baggara cattle. While similar in general conformation, size and late maturity, the Kenana of the Fung, as a result of the more secluded nature of their habitat, approach nearer to a separate breed than the White Nile cattle. Among the latter coat colour varies considerably, red, fawn, black, white and variegated animals being common. The Kenana-Fung type is more uniform in colour which is commonly steel grey, shading towards the extremities and terminating in black points. The bulls are darker in colour than the cows. As the hairs are banded black at the root, the shade of the coat colour depends on the colour of the hair tips which are red or white, and on the degree of extension of the black basal band. The skin is black or brown, with the exception of the inside of the ears, the udder and teats and, to some extent, the belly, which are of a yellowish colour. The calves are born brown, changing to steel grey after three to six months (McLaughlin, 1955).

The Kenana is a medium to large animal of dairy rather than beef type; the average height of bulls is 142 cm and of cows 132 cm; the average weight of bulls 500 kg and of cows 400 kg. The general conformation is characterised by lightness of bone and leanness of musculature. The head is long and somewhat narrow below the eyes, but wide across the forehead; in profile it is convex, straight or slightly concave in the male,

405. Cow of Kenana-White Nile type, Northern Sudan

406. Cow of Kenana-Fung type, Northern Sudan (after Boyns)

407. Kenana bull, Sudan

 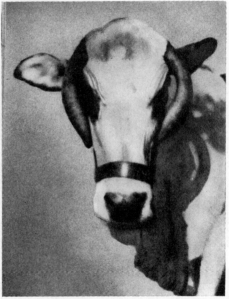

408. (left) Head of a Kenana-Fung cow with normal horns (after McLaughlin)
409. (right) Head of a Kenana-Fung cow with coreless horns of normal length (after McLaughlin)

convex or, more rarely, straight in the female. The ears are 14–22 cm long and generally carried horizontally. The horns are 15–20 cm long, rarely exceeding 35 cm, and broad oval in cross-section; they project upwards and outwards approximately in line with the profile, and turn inwards and forwards at the tips. Animals with loose horns, scurs or short stubs, or without horns, are also fairly frequent. The well developed dewlap commences under the chin often as a double fold, and runs well back along the floor of the chest, frequently joining the large umbilical fold. The hump varies considerably in position and development. In the cows it is generally small, muscular, and in many animals distinctly cervico-thoracic in situation. In bulls it is larger and may be long and rounded antero-posteriorly or pyramidal in shape; its tendency to overhang to the rear makes it appear to be more nearly thoracic in position (McLaughlin, 1955). The body is long and of moderate depth, but sometimes narrow and flat of rib behind the shoulders. The topline rises from the base of the hump to the sacrum. The rump is fairly long and markedly sloping to the low-set tail. The pelvis is short, and narrow posteriorly. While the upper thighs are often well developed, the lower thighs lack width and fullness. The bone, though light and fine, is of hard material. The udder is often well developed, of good shape and well attached, occasionally pendulous and poorly attached behind (Faulkner and Epstein, 1957).

Red Butana Cattle

The Butana cattle are bred along the banks of the Nile and the Atbara, north of latitude 14° N, and in the semi-arid scrub and desert areas (North Kassala and the Gash

410. Red Butana bull

411. Red Butana cow (after Faulkner and Epstein)

412. Cow of Dongola type, Northern Sudan (after Boyns)

413. Cow of Shendi type, Northern Sudan (after Boyns)

Delta) to the east and west of the rivers. The Butana, representing a large part of the desert cattle, is owned by the Batahin and Shukria, the two main tribes in the more northern parts of the Butana east of Khartoum (Faulkner and Epstein, 1957).

Butana cattle in size and type resemble the Kenana. They are distinctive among Sudan cattle in their general conformity to a good dairy type (Hewison, 1945); under good management cows produce approximately 1900 kg milk per lactation, with a butter-fat test of 5.5 per cent. Butana cattle are commonly dark red, occasionally light red with some white on the head, udder and other parts of the body. The red is sometimes associated with black points or black areas over the neck, shoulders, hump, forelegs and hindquarters. The head is well proportioned but narrow especially below the eyes, with a convex profile. The horns are usually short and moderately thin, projecting upwards and slightly forwards, or downwards. Polled or loose-horned individuals are also common. In most animals the hump is thoracic in situation, but it may be cervico-thoracic. It is small in the cows, larger in bulls in which it is rounded from front to back with hollows at the back near its attachment to the withers. The dewlap is only moderately developed, with few folds. In some cases the umbilical fold may run back to the udder which is well developed although sometimes inclined to be pendulous. The body is long and not well fleshed, often lacking in depth, especially behind the shoulders. The back is of good width and fairly well covered, the topline rising from withers to sacrum. The rump is of good length but poorly filled, sloping steeply from hooks to pins so that the tail head is very high. The upper thighs are thin and the lower thighs long, lacking width and fullness (Faulkner and Epstein, 1957).

Dongola and Shendi Cattle

Among the cattle of the riverain areas of the Northern Province of the Sudan, Boyns (1947) distinguishes between two types, viz. Dongola and Shendi, which he believes to have originated from the Red Butana. Both mature at an earlier age than the nomad cattle—3 years as against between 4 and 5. The Dongola cattle, bred in the irrigated areas of the extreme north, have been isolated from foreign blood for many generations; their predominant colour is red, with a few red-and-whites. Among the cattle of the Shendi district, which originate from the nomadic Butana cattle, red of different shades predominates. The cultivators of the irrigation lands select their breeding stock with a view to milk production, with the result that the Shendi cows are among the best dairy cattle of the Sudan.

Bambawa Cattle

Another strain of the Red Butana or Red Desert cattle, called Bambawa, is found along the Sudanese frontier of Eritrea. It differs from the Begait (see pp. 345–346) in its greater weight and milk yield, and in the prevalence of a red or red pied coat (Bonelli, 1938; Mason and Maule, 1960).

Ingessana Cattle

The Ingessana Hills in the south of the Blue Nile Province of the Sudan, near the Ethiopian border are inhabited by an isolated negroid tribe. Their cattle were formerly of a small zebu type; but as they have been considerably crossed with the surrounding Kenana cattle, the original Ingessana type is now rare (Mason and Maule, 1960).

The Baggara Cattle of the Central Sudan

West of the White Nile, in the Kordofan and Darfur provinces of the central Sudan, the cattle present a great mixture of types (Figs. 414–416). Evidently, thoracic-

414. Long-horned Baggara bull, Sudan (after Faulkner and Epstein)

415. Short-horned Baggara bull, Sudan

humped zebus have interbred with cervico-thoracic-humped sanga, thoracic-humped sanga and, possibly, with humpless shorthorn cattle (see also p. 264). They are divided into Darfur and Kordofan cattle, and variously called Baggara, Darfur, Kordofan, Arab or Western cattle, indicating the area of their origin rather than a specific type. Lydekker (1912a) recorded:—"The Bagara tribes of the Berbera district of Nubia ... own cattle ... very similar to those of ancient Egypt, but ... characterised by the huge size of their horns. They are stated to be slightly-built animals; the hump varies in size, being in some instances large, and in others almost obsolete ...; but the Bagaras ... also possess a shorthorned breed." During the Mahdi wars Darfur and Kordofan were practically cleared out of cattle by Fulani (Fellata) marauders from the west and Dinka raiders from the south. The present cattle population has been built up from remnants of original herds driven into the mountains and from cattle obtained from adjacent territories, which accounts for the mixed type of the Baggara (Faulkner and Epstein, 1957). In the west the influence of Fulani (Fellata), in the east of Dinka, and in the south of Nilotic, cattle is apparent.

416. Baggara cattle

The Baggara cattle are basically of zebu type. Duerst's (1899) description of a skull from Bahr Sennar approximates to that of a Somali Zebu, the similarity being particularly pronounced in the shape of the frontals which are markedly vaulted so that the skull is convex in profile. Baggara cattle are of medium size; the average weight of bulls is approximately 350 kg, although animals of up to 600 kg have been recorded. The head is well proportioned but tends to be narrow. The profile is straight or convex at the orbits. The ears are relatively small. The light horns are short or of medium length; in the cows they average 38 cm in length. There is a tendency for horns to be longer in the west, probably owing to the influence of Fulani (Red Bororo) cattle. The horns usually project outwards and upwards with the tips tending to curve inwards. The dewlap is very well developed, particularly over the sternum, and runs well back to between the front legs. The prepuce and umbilical fold are also generally well developed. The hump is pronounced; in the bulls it is often pyramidal in shape and tends to be cervico-thoracic in situation in contrast with the hump of the pure East African zebu type. In females it is smaller than in bulls, but also often cervico-thoracic. The body is fairly compact, of good depth and capacity, but commonly narrow in front. The topline rises from the base of the hump to the moderately high sacrum. The rump is fairly long with a marked slope; like the thighs it is poorly muscled. The limbs, of moderate length, show good quality of bone. The hooves are of very hard material. There is considerable variation in coat colour; whole colours include white, grey, dun, brown and black. White and grey with black or other coloured points are frequent. Similar patterns with speckling or splashes on the neck and sides of the body are also

417. Nuba Mountain bull

common. Deep red is characteristic of the area surrounding the Nuba mountains, and white of the far west. The cows are poorer milkers than the Kenana and Butana cattle of the Sudan, but superior to the Nilotic Dinka cattle (Faulkner and Epstein, 1957).

Nuba Mountain Cattle of Short-horned Zebu Type

While the cattle of the Nuba mountains of southern Kordofan, Sudan, seem to have been formerly of pure humpless shorthorn (brachyceros) type (see p. 264 and Fig. 303), they have been crossed with, and graded up to, the surrounding Baggara zebu cattle to such a degree that most of the present-day Nuba cattle are "proper little zebus" (Mills, 1953). They measure only 90–125 cm to the top of the hump or 97 cm to the middle of the back, and are of a harmonious conformation. The head is short, with a broad muzzle and short horns which may be upright, lateral or lyre-shaped, but not loose or absent. The dewlap is well developed. The hump is commonly thoracic in position and very variable in size. There is a great diversity of colour. The milk yield is very small, and the animals are not slaughtered for meat (Mason and Maule, 1960).

Southern Sudan Zebu

The zebu cattle of the southern Sudan are divided into the cattle of the Toposa, Murle and Mongalla tribes.

The Toposa-Murle Cattle of the Sudan

In the Murle area of Ethiopia, near the northern shore of Lake Rudolf, and in the Pibor Post district in Upper Nile Province of the Sudan, the Murle (Beir), of Nilo-Hamitic origin, breed a small type of zebu cattle believed to have been introduced from the Maji area of south-western Ethiopia. A similar type of zebu, albeit differing in size and coat colours, is owned by the Toposa and Boya (Longarim) tribes in the Eastern district of Equatoria Province of the Sudan, the area to the west and north of Kapoeta. The Toposa tribe is considered to be an offshoot of the Jie of north-west Uganda, that

418. Toposa bull at Kapoeta, Equatoria

419. Murle cow

moved with their cattle from Jieland northwards into the present area approximately 200 years ago. At the periphery of the breeding areas the Toposa–Murle cattle are in continuous or seasonal contact with (Nilotic) sanga and different types of zebu cattle, which, together with the different origin of the Toposa and the Murle, accounts for the considerable degree of variation encountered among different herds (S.D.I.T., 1955; Joshi et al., 1957; Mason and Maule, 1960). Toposa and Murle cattle cannot be considered "as distinct breeds now, in spite of past very expert opinion to the contrary. The cattle owned by the Taposa tribe no doubt originated from the south-east and belong to the same family as cattle of the Jie area of Uganda. Also the Murle, Mongalla, Boya, Latuka, Lafit and Lafon tribes have, in general, a smaller type of animal, which may have originated from the small East African Zebu type. Nevertheless, it is at present almost impossible to distinguish one type from the other, except as regards location. There is no distinct characteristic left apart from size, which I think is due to locality" (H. T. B. Hall, 1960—personal communication).

Toposa–Murle cattle have fairly long bodies of moderate depth. Murle cows stand approximately 115 cm at the withers and average 240 kg in weight. In the Toposa area the cattle are considerably larger, and good Toposa bulls may exceed 400 kg in weight. In general, the head of the Toposa–Murle is of medium length, with a broad muzzle and a straight or concave, occasionally slightly convex profile. The forehead is dished between the eyes, and the orbital arches are prominent. The ears are up to 30 cm long. The horns are of medium length, rather longer than is common in East African zebus, though rarely exceeding 46 cm in length in bulls or 30 cm in cows. They rise from distinct pedestals in an upward and outward direction with a tendency to grow inwards and forwards at the tips. Occasionally the horns are artificially trained inwards until the tips meet. The hump is pyramidal in shape and thoracic rather than cervico-thoracic in situation; it is described as 'muscular' by S.D.I.T. (1955) and with 'a large amount of fat' by Joshi et al. (1957). The dewlap is of moderate size, the umbilical fold little developed, and the prepuce in the males seldom pendulous. The topline rises slightly from the withers to the rump, but the sacrum is not greatly accentuated. In general, there is great variability in coloration: red, black, white, grey, dun, and patterns of red and black, together or separately, on a white ground are common. Among Murle cattle the lighter coat colours predominate (S.D.I.T., 1955).

The Mongalla or Southern Sudan Hills Zebu

West of the breeding areas of the Toposa–Murle, mainly on each side of the Bahr el Jebel, in the eastern part of Equatoria Province of the Sudan as far north as latitude 5° N, lies the breeding area of the Mongalla or Southern Sudan Hills zebu. The Mongalla cattle are owned by a number of tribes of mainly Nilo–Hamitic origin, which include the Didinga, the Latuka–Lango group and the Bari on the east of the Nile. They are believed to have entered their present habitat from territories further to the east (Joshi et al., 1957).

Mongalla cattle are small, stocky, well fleshed animals, the cows standing 100 to 105 cm at the withers and weighing approximately 150 kg. They are characterised by a moderately short head, with a straight or slightly concave profile in the cow, and

420. Mongalla bull

421. Mongalla heifer, Government dairy herd, Torit, Equatoria

convex profile in the bull. Generally the horns are short; except in the cattle owned by the Bari, they seldom exceed 20–30 cm in length. Their direction is commonly outward and upward, forward of the line of the profile. The hump is cervico-thoracic or thoracic in position, indicating the lack of racial uniformity in the Mongalla type. The dewlap is moderately developed, sometimes slightly folded, and the umbilical fold is small. The body is fairly deep and broad, of good girth, with the topline sloping only slightly upwards to the rump. The latter is of moderate length and slope. The hindquarters and thighs are well fleshed. There is a wide range of coat colours; grey and dun are common, but other colours and patterns on a white ground are also frequent. Black-and-white is a common pattern among the cattle of the Kajokaji area of Yei district. The skin is dark (S.D.I.T., 1955; Joshi et al., 1957).

The variable position of the hump, and the longer horns encountered in the cattle of the Bari in the Juba area to the east of the Nile indicate the influence of Nilotic sanga blood on the Mongalla. The cattle of the Didinga hills and the Dongatona mountains, on the other hand, probably carry Toposa blood (Mason and Maule, 1960). "Undoubtedly the Didinga cattle are a good deal smaller than any of the other breeds down there," writes Mr. H.T.B. Hall (personal communication); "they spend their lives in the hills—very steep hills—and again this is more likely to be an environmental effect. Among the other cattle of the (Murle, Mongalla, Boya, Latuka, Lafit and Lafon) tribes ..., the smaller size than the Taposa is so slight as to be almost non-existent, and there are no other characteristics which would serve to distinguish them apart. Perhaps they came from different original stock, but I would question the ability of anybody to pick out a random selection of Mongalla–Murle type from a random selection of Taposa type if they were mixed together."

The Lugware Cattle of the Congo

Cattle of the same type, called Lugware (also Lugbari, Mangbattu or Bahu) are bred by the Mangbattu, who dwell at the headwaters of the Uele, in the Eastern Province of the Congo, contiguous with the breeding area of the Mongalla in the Kajokaji area of the Yei district of the Sudan, and by other tribes in the region around Aru in the Kibali–Ituri district of the Congo to the north of Lake Albert. The Lugware cattle are closely related to the Mongalla of the Sudan, as their ancestors are believed to have been brought south from the Sudan about two centuries ago (Mason and Maule,

422. Mongalla cattle in the southernmost part of the Sudan (after Faulkner and Epstein)

CATTLE

423. Lugware zebu bull of the Mangbattu, Congo (after a drawing by G. Schweinfurth)

424. Well-bred Lugware bull
Live weight: 630 kg
Heart girth: 200 cm

425. Lugware cows

1960). Ninety years ago Schweinfurth found the Mangbattu in possession of pure zebu cattle which had been given to King Munsa by a friendly ruler beyond the south-eastern border of his kingdom. They probably came from the vicinity of Lake Albert and the zebu country north-east of Victoria Nyanza where cattle of a similar type (Nandi) occur to this day. A closely related type of zebu is found also in the western part of the West Nile district of the Northern Province of Uganda.

The Lugware, like the Mongalla of the Sudan, belong to the small type of East African zebu. Adult bulls stand approximately 107 cm at the withers, cows 104 cm. Bulls average 300–350 kg live weight and cows 230–250 kg (Mason and Maule, 1960). But selected well-bred and well-fed animals may reach considerably higher weights. Lugware cattle are rather thick-set and of good conformation. The general appearance is strongly zebu, and in bulls the hump is often very well developed. Coat colour varies, the most usual pattern consisting of small black, red or yellow patches on a white ground, frequently crowded on the flanks (Curson and Thornton, 1936).

The Zebu Cattle of Uganda

In Uganda several zebu breeds are encountered. The Bukedi (Nkedi, Teso, Lango, Tesse Island) occupies the Eastern and Northern Provinces, mainly low-lying savanna country of an average altitude of slightly over 1000 m (Curson and Thornton, 1936). The Bukedi is also prevalent in the Buganda Province and Western Province (Joshi et al., 1957).

In the Teso district of central Uganda several minor types occur, among these the Kyoga and Usuku. An allied type, the Serere, is found along the lake shore in the Serere Peninsula of Lake Kyoga, in the south of Teso. The Karamojo-Turkana (Elgume) area of north-east Uganda and north-west Kenya is the breeding centre of the Karamajong zebu; the same type is encountered among the Suk of north-west Kenya, south of the Karamojo-Turkana area.

426. Bukedi bull, Uganda

427. Head of a Bukedi ox from Lango Country, Uganda (after Johnston)

428. Bukedi zebu cow, Uganda

The Bukedi

The Bukedi and Nandi belong to the small type of East African zebu. They are so similar that "it would be difficult to pick out distinctive characteristics which distinguish one type from the other. The variation between the two types (the Bukedi and the Nandi) is in fact no greater than that between the individuals of each type" (Faulkner and Epstein, 1957). The weight of adult Bukedi bulls averages approximately 360 kg and of cows 285 kg; the withers height of bulls 104 cm, of cows 101 cm. The horns are short and rather thick at the base, round in cross-section, and lyre-shaped or with a slight outward and inward curve. The prominent musculo-fatty hump is thoracic in position and elongated with a varying posterior overhang. The dewlap and umbilical fold are only moderately developed. Coat colours include grey, greyish white, light red, black, and black-and-white (Joshi et al., 1957; Ross, 1958).

The Kyoga

The Kyoga type is bred by the Kumam tribe in the Kyoga Peninsula in Lango and also in the southern tip of Kaberamaido county, which is an extension of Kyoga Peninsula in Lango. The Kyoga is larger than the Bukedi, with a deeper chest and shorter legs. It has a broad forehead and characteristic wrinkles above the eyes. The ears are often pendulous. The horns are usually short and thick, curving outwards, slightly backwards, and upwards; polled animals occur but are rare. The hump is small and rounded, occasionally dome-shaped. The dewlap, prepuce and umbilical fold are well developed. Black and red self colours are common (Faulkner and Epstein, 1957; Ross, 1958).

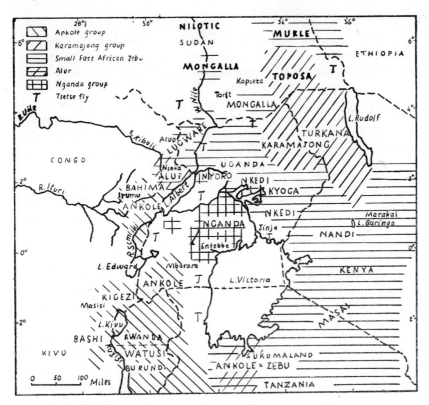

429. DISTRIBUTION OF CATTLE TYPES IN UGANDA AND ADJACENT REGIONS (after Mason and Maule)

The Kyoga is believed to be derived from sanga (Ankole)-zebu crossbred ancestors (Nganda), introduced from Bunyoro district by Kaberaga in 1895. In their new home the Nganda cattle interbred with Karamajong zebus brought by the Iteso to Kyoga and Kaberamaido Peninsula from the north-east in the course of their migrations between A.D. 1700 and 1800 (Ross, 1958).

The Usuku

The Usuku type is found in north and north-east Teso in areas adjacent to the Karamoja border. Usuku cattle vary greatly in conformation, but are usually larger and heavier-boned than the Bukedi. They are believed to be derived from a fairly recent cross of the Bukedi and Karamajong. The hump is larger and higher than the Bukedi hump; it does not overhang posteriorly but is commonly upright and dome-shaped, similar to the Karamajong hump. The dewlap, prepuce and umbilical fold are not quite as large as in the Kyoga type. Whole colours predominate; red, grey and roan are frequent (Faulkner and Epstein, 1957; Ross, 1958).

The Serere

The Serere type is found in the Serere Peninsula, in south Teso near the lake shore. Serere cattle are very similar to the Kyoga type; they also resemble the Bukedi zebu in general conformation, but are slightly larger and the hump is small and dome-shaped, in cows sometimes inconspicuous. The dewlap, prepuce and umbilical fold are only moderately developed. Typical horns are similar to Bukedi horns, but animals with long thin horns of the Ankole type and polled cattle are quite common. According to Ross (1958), the occurrence of long horns and the very small hump of the Serere suggest that the latter originates from Ankole-zebu crossbreds (Nganda), introduced through Bunyoro from Western Province, Uganda. In the Serere Peninsula the Nganda were crossed with small zebu cattle of the Bukedi type.

The Karamajong

The Karamajong cattle are slightly larger and heavier-boned than the Bukedi zebu, but are less deep in the chest and longer in the leg. They belong to the large type of East African zebu, and closely resemble the Toposa in the south-east corner of the Sudan, and are also similar to the Boran cattle of Lake Rudolf. Bulls weigh approximately 400 kg, cows 335 kg. They have broader and longer heads than the Bukedi, and less erect, occasionally pendulous, ears. The horns are short to medium in length, sometimes sloping backwards, or lyre-shaped. Large horns are encountered occasionally. The skin has a loose appearance; the dewlap is very well developed, often in folds, and the prepuce and umbilical fold are large. The hump is rounded or dome-shaped. The predominant colours are red roan, grey and white; black or brown markings are also common.

The Zebu Cattle of Kenya

In Kenya all indigenous cattle are of zebu stock. Holm has pointed out that they are similar in appearance to the Tharparkar and Sindhi breeds of India. Small differentiations appear in different districts. The various sub-types are named after the tribes breeding them:—(a) Nandi, on the western side of Kenya almost on the equator, a few miles north-east of Lake Victoria. (b) Boran, in the Northern Frontier Province (Tanaland) and west of Mt. Kenya; (the improved Boran is bred mainly in the Central and Rift Valley Provinces of Kenya). The range of the Boran further includes the Borana Province in the southern part of Ethiopia, the western part of Somalia (Jubaland and inland along the Ethiopian border), a small area in the former British Somaliland Protectorate, and a few recently established herds in Tanzania. (c) Watende, in the tribal area on the borders of Kenya and Tanzania, adjacent to the Masai. (d) Kamba, in the Ukamba African Area, north-east of Nairobi. (e) Kavirondo, in the south-west of Kenya, on the shores of Lake Victoria. (f) Masai (see pp.387–379), in southern Kenya, extending to central Tanzania (Masai Steppe).

The Nandi

The Nandi people, according to tradition, came from the north-west, eventually settling in their present area at the beginning of the 17th century. They have a long history

430. Nandi bull and cow

431. Nandi cows

of raiding and fighting, especially with the Masai. Hence their cattle are considerably influenced by Masai and other stocks.

The Nandi resembles the dairy type rather than the beef type in conformation, cows kept under good conditions producing an average of 1300 kg milk, testing 5.8 per cent butter-fat. Bulls weigh 340–430 kg live weight, cows 250–320 kg. The average withers height of bulls is 118 cm, and of cows 111 cm. The head of the Nandi is long, especially from the eyes orally, being widest between the eyes. The profile is straight or slightly convex. The horns are generally short, fine, round in cross-section, pointed at the tips, with only a very slight curve. Occasionally horn direction is lateral. Nandi cows with loose horns or without horns are not uncommon. The dewlap is well marked, thin, somewhat pendulous, and passing well down between the legs. The prepuce and umbilical fold are little developed. The hump varies greatly in size and prominence. In bulls it is large, rounded from front to back and with a backward fall. In cows it is only slightly developed, sometimes hardly noticeable. In the majority of animals the hump is thoracic in situation, but occasionally it appears more cervico-thoracic. The body is long but of good depth throughout. The shoulders are little prominent. The back slopes slightly up from hump to sacrum, but is strong and fairly wide; ribs are long and well sprung. The rump is of moderate length, with a marked slope from hook bones to pin bones. The legs are of medium length and distinguished by the extreme fineness of bone. The udder is moderately well developed, strongly attached, fairly well carried forwards and level. The teats are rather small and placed close together. Colour markings and patterns vary greatly, and include black and red, also fawn, white and grey, and a mixture of these (Faulkner and Epstein, 1957).

The Boran

The Boran is the largest of the zebu cattle of Kenya, being typical of the large type of East African thoracic-humped zebu. It differs in size and type in different areas. The Somalia Boran are probably the smallest representatives of the type. The Tanaland Boran, which originally came from Ethiopia and are now found south and west of the Tana river in Kenya, are also lighter than the improved Kenya Boran which have been

432. Boran cattle, southern Ethiopia

433. Boran bull, Kenya (after Faulkner and Epstein)

434. Boran cow, Kenya (after Faulkner and Epstein)

bred and selected mainly for beef (Mason and Maule, 1960). The withers height of well-bred bulls averages approximately 125 cm, of cows 120 cm; the live weight of bulls in good condition ranges from 550 to 675 kg and of cows from 400 to 475 kg. In Somalia average weights are 318 kg for adult bulls, and 257 kg for cows. In Tanzania, where Boran cattle have recently been introduced, lower weights than in Kenya have been recorded—about 400 kg for adult bulls. The Boran is exceedingly well adapted to hot dry areas. It is essentially a beef type, but is used also for milk and draught.

The head is fairly long but well proportioned, wide between the orbits, with a straight or convex profile and slightly drooping ears of medium length. The horns display great

435. Boran cattle, northern Uganda

436. Boran cattle (improved type), Kenya

variability. Generally they are short, fine, and round in cross-section, projecting upwards and slightly outwards with a moderate curve at the upper two-thirds and blunt tips. They may, however, be short and thick throughout, except near the tips. Boran cattle with larger horns of considerable basal thickness (up to 84 cm in length and 46 cm in basal girth) are encountered, doubtless as a result of interbreeding with sanga cattle. Sanga influence is suggested also by the gene frequencies of α-lactalbumin A and β-lactoglobulin A in the Boran, as compared with humped White Fulani and Indian zebu and humpless European cattle:—

Gene frequencies of α-lactalbumin A and β-lactoglobulin A in different breeds of cattle

Breeds	Gene Frequencies		References
	α-lactalbumin A	β-lactoglobulin A	
Indian Zebus	0.24	0.09	Bhattacharya et al., 1963
White Fulani	0.15	0.18[1]	Blumberg and Tombs, 1958
Boran	0.13	0.19	Aschaffenburg, 1963
British breeds	0	0.28	Aschaffenburg and Drury, 1955
Iceland	0	0.34	Blumberg and Tombs, 1958
Holstein-Friesian	0	0.52	Plowman et al., 1959

The dewlap of the Boran is well developed, often pendulous, but rarely with skin folds, passing from the chin down to well between the legs. The prepuce and umbilical fold are highly developed. In the improved Boran the body is compact and of excellent depth, width and capacity. The shoulders are well placed. The hump is thoracic in situation and composed of a light muscular framework filled with fatty tissue. It varies greatly in size, shape and prominence. In many bulls it has a marked backward fall, while in others it is pyramidal. The topline slopes slightly up to the rump. The back is wide and well covered, as are the loins which pass smoothly on to the rump. The latter shows good length, the slope from hook bones to pin bones varying considerably. The upper thighs are well developed, plump and thick with a deep twist, while the lower thighs, although less developed, are larger and fuller than in other East African cattle. The tail setting is low. The legs are of moderate length, the hooves well formed, small and hard. The great majority of Boran cattle have pigmented skins, either black or light red. The most common hair colour is white; in numerous animals, more especially in males, the white merges into fawn, dark grey or even black on the neck and shoulders and down the thighs. A pale wheaten to medium red, or black self colour is also common. Sometimes red is accompanied by black. An occasional broken pattern consists usually of irregular mottled patches of white on red cattle. Boran cattle of red or black coloration seem to originate from southern Ethiopia, whereas white with dark points is the most common colour in the Galla herds of Boran stock in Somalia and the adjacent eastern part of the Northern Frontier Province of Kenya (Faulkner and Epstein, 1957).

[1] Blumberg and Tombs quote 0.21, but Mason (personal communication) has corrected this figure to exclude some Friesian-White Fulani crossbred animals.

In the Samburu area (Northern Frontier Province of Kenya) the cattle are predominantly of Boran stock, with an admixture of zebu cattle acquired from the Suk and Turkana tribes, and of long-horned Ankole sanga (McKay, 1957).

The Tanaland Boran

The Tanaland Boran is inclined to be leggy and narrow-chested, with a rather hollow back and sloping rump. The head and neck are finer and lighter than in the improved Boran. The dewlap is large, and the hindquarters are well developed. The colour is typically white with a black or a white skin, occasionally fawn in different shades, but not always with the black points characteristic of the Somalia Boran. The horns vary considerably; in general they are larger than in other Boran strains, often very thick and slightly oval at the base, and upright (McKay, 1957; Mason and Maule, 1960).

The Watende Cattle of Kenya

The Watende cattle of Kenya are characterised by a long though reasonably deep body, well sprung ribs, rather long limbs, clean and fine of bone, a small hump, small dewlap free of folds, small umbilical fold, and short, moderately thick horns with blunt tips. Whole black is by far the commonest colour; brown is less frequent. With both these colours some white may appear.

The Kamba and Kavirondo Cattle of Kenya

The Kamba (the name Kamba with regard to cattle is practically synonymous with Kikuyu) and the Kavirondo cattle of Kenya are slightly larger than the Nandi (see

437. Watende zebu cattle, Kenya

pp. 373–374); bulls weigh approximately 375 kg, and cows 305 kg. The most frequent coat colours are black, white or black-and-white.

The Jiddu

The Jiddu (Surco, Tuni or Serenle) in its purest form is bred in the semi-desert areas of Somalia, especially in the region of Ballei on the Juba river, and in the Juba districts of

438. Kamba cow, Kenya (after Faulkner and Epstein)

439. Kamba bull, Kenya

440. Nyanza zebu from the Kavirondo District of Kenya (after Faulkner and Epstein)

441. Kikuyu bull

Gosha and Lugh, whence it extends into the south-eastern part of Ethiopia. It occurs also along the Benadir coast between the lower Webbe Shibeli and the sea, and in the Rahanuin territory of southern Somalia (Gadola, 1947). This is the area occupied by the Jiddu, Tuni and neighbouring tribes. During recent years many Jiddu cattle have been imported from the south-western part of Somalia into Kenya where they are thriving under a range of conditions varying from the hot drier areas at around

442. Kikuyu cow

443. Jiddu bull (after Faulkner and Epstein)

444. Group of Jiddu cows

1000 m to the high rainfall subtropical areas at an altitude of over 2000 m. A few are found also in Tanganyika (Faulkner and Epstein, 1957).

The Jiddu cattle have been classed as sanga by Italian authors and related to the Aradò of Eritrea, which is of mixed zebu-sanga derivation (see pp. 347–350). "In British East Africa, however, they have been contrasted with the very long-horned Ankole and classed as short-horned zebus" (Mason and Maule, 1960). According to Somali tradition, the Jiddu came from the west, i.e. from Ethiopia, which would explain the sanga influence found nowhere else in Somalia.

In body size the Jiddu is smaller than the improved Kenya Boran, but not smaller than the Somalia Boran, and larger than the Somali Zebu. In Somalia the weight of adult bulls averages 336 kg and of cows 277 kg. The latter average 118 cm withers height (Rossetti and Congiu, 1955). In Kenya Jiddu bulls weigh 495–585 kg and cows 360–400 kg. Though thick-set and well proportioned, with good depth and fairly short legs, the Jiddu tends more to the dairy than the beef type; yet the average milk yield of the cows is only 650 kg per lactation, testing slightly over 5 per cent. The head of the Jiddu is heavy but well proportioned, straight in profile, with small eyes, fairly long semi-pendent ears, and horns highly variable in size, indicating the Jiddu's mixed zebu-sanga ancestry. In many animals, closely akin to the East African Zebu, the horns are short and not very thick, outward and forward in direction, often with only a slight curve in them. In others, suggestive of the sanga type, the horns project from a thick pedestal-like base outwards and upwards in a lyre or sickle-shaped curve, with the tips pointing forwards or backwards. Their length along the curve may be 90 cm and the basal girth 42 cm (Gadola, 1947). The cross-section is generally round. The dewlap is moderately developed and free from folds and fleshiness. The hump varies markedly in size and position—evidence for the origin of the Jiddu from thoracic-humped zebu

445. Jiddu cow, front view

and cervico-thoracic-humped sanga cattle. In the female it is sometimes very small. It generally appears long and rounded antero-posteriorly. Its position is thoracic or cervico-thoracic, its structure mainly muscular. The anterior part of the body is long, the lumbar region of the back short and strong, the rump short and less drooping from hook bones to pin bones than in typical zebu cattle. The Jiddu shows good depth and width, well sprung ribs and a capacious barrel. The limbs are of moderate length and fine of bone. In colour the Jiddu is one of the most distinctive breeds in Africa. All shades of white and light fawn to dark mahogany are found. The darker colours are usually concentrated on the head, over the forequarters, along the top of the back and under the abdomen, the lighter colours on the sides of the body. Whole colours are found mainly in the darker shades, such as deep red, rarely in the lighter ones. In nearly every case Jiddu cattle have white muzzles, white eyelashes, white hairs around the eyes and white hair inside the ears. The strong, folded skin is usually red, and the horns and hooves are also dark red (Faulkner and Epstein, 1957).

The Zebu Cattle of Tanzania

The main areas in Tanzania in which the zebu is found are the drier parts of the great central plateau and to a lesser extent the open mountain grassland stretches of the northern and southern highlands of the Lake Province. The best Tanzania zebus are encountered in great numbers among the Masai, especially in the higher country around Ngorongoro and Balbal, where abundant salt deposits, perennial streams, sufficient rainfall, the presence of limestone in the soil and clover in the grazing make for more or less ideal ranching conditions. To this may be added the very important factor of an ample milk supply assured to the calves during their youth, as the Masai are a purely pastoral people and own large herds with the result that only a very little milk is taken from each cow (McCall, 1928). In several areas, as in Masailand and in Musoma District, east of Lake Victoria, the zebus attain live weights of over 400 kg, while in the poorer cattle districts, e.g. in the hilly breeding grounds of the Mbulu and Wachagga, average specimens do not scale more than 160–180 kg. Hutchison (1955) has pointed out that at the present time "no specific breeds of indigenous cattle are recognised in Tanganyika". The most common type, representing perhaps 90 per cent of the cattle population, is a small light-framed zebu of an average withers height of 113 cm, and an average weight of about 295 kg. It is characterised by a fairly well developed thoracic hump which may reach a height of 30 cm and has a tendency to sag (Curson and Thornton, 1936). The rump is sloping, the hair short and smooth, the skin dark and loose, and the dewlap, prepuce and umbilical fold are moderately developed. The horns are usually very short, 7–15 cm, and lateral in direction, occasionally longer with an upward turn. The ears, about 15 cm long, are carried horizontally. As in all varieties of small East African zebu, there is a wide range of colour. The cattle of Wanyaturu, Sekenke, Mkalama and many parts of Singida are in the great majority white animals with dark skins, relatively small horns, black tips to their tails and ears, often showing black on the extremities of the limbs. Farther south in Dodoma among the Wagogo herds black-and-white and brown-and-white animals predominate, whilst in Iringa and the Southern Provinces of Tanzania red is the common-

446. Wachagga zebu cattle, Mt. Kilimanjaro, Tanzania

447. Polled zebu oxen from Mbulu, Tanzania (after McCall)

est colour encountered. A few decades ago, McCall (1928) selected a number of characteristic animals from the major cattle areas of Tanzania, namely Singida, Iringa, Mkalama and Ugogo.

The Singida Type of Tanzania Zebu

Singida, one of the highest districts north of the Central Railway, has long been famous for its cattle, the most prevalent colour of which was silver-grey to white, with the brush of the tail, the skin, hooves and muzzle black. The cows were excellent milkers, and the type was so distinctive and reproduced itself so consistently that it was thought worthy of being regarded as a breed. While resembling the Boran cattle in many respects, the Singida showed more symmetry but was not so large (McCall, 1928).

448. White Singida ox (after McCall)

449. White Singida cow (after McCall)

450. Red Iringa cow (after McCall)

The Iringa Type of Tanganyika Zebu

"In the Iringa highlands, in the enzootic East Coast Fever country, nearly all the cattle are more or less deep red in colour. Large-framed animals of rather striking appearance, of late years their numbers have been sadly depleted by wars and disease. According to many of the old men time was when all the cattle in Iringa, Uhehe and Njombe were of a uniform blood-red colour, and it has only been within comparatively recent years that importations from Ugogo have led to deterioration in the original strain. These cattle are rather indifferent milkers and more attention seems to have been paid to the size of their humps than to the shape of their rumps, but nevertheless, as a hardy breed suited to the cold bleak uplands of the southern area their ability to stand up to the long hard months of the dry season is undoubted (McCall, 1928)."

The Mkalama Type of Tanganyika Zebu

In the Mkalama district cattle of a deep golden dun colour were frequently encountered. The golden dun colouring tended to turn black at the muzzle, the switch of the tail, and the extremities of the legs. The skin was dark. Good specimens showed well sprung ribs, were short of leg, and low to the ground. The milk from these animals was invariably rich (McCall, 1928).

451. Golden dun Mkalama bull
(after McCall)

452. Golden dun Mkalama cow
(after McCall)

The Ugogo Type of Tanganyika Zebu

In the drier regions of the Masai steppe and throughout Ugogo a peculiar type of zebu was frequently met with. Its body colour varied from iron to steel grey, but irregularly splashed upon this were areas of red or black. These animals breed exceptionally true to colour, and a peculiar feature was the frequency with which grey eyes were encountered. Of a large deep-framed conformation, they were credited by the dry area tribes as being particularly suited to inhospitable drought stricken country (McCall, 1928).

The Masai Cattle

The present Masai area extends from south of Lake Naivasha in Kenya to central Tanganyika. The Masai cattle show considerable variation in size, colour, conformation and horn development; for as a result of numerous predatory raids on their neighbours and on other tribes through whose territory they advanced, nearly every variety of cattle in East Africa has found its way into the Masai kraals (Forde, 1934). There is still evidence of sanga influence among some Masai herds; this derives from the period before 1889–92, during which time the great rinderpest epidemic devastated the Masai herds, the sanga cattle suffering most heavily (Mason and Maule, 1960). In former times the Masai preferred the long-horned sanga to the short-horned zebu type, although their zebu cows yield nearly twice as much milk as the sanga. This preference was part of their cattle complex which developed farther north among the Hamitic Galla of Ethiopia and was carried south by the advances of the Nilo-Hamitic pastoralists, of whom the Masai are the largest and most powerful group (Epstein, 1955b).

Generally, the Masai zebu cattle have a bigger frame than the Nandi, for example. They are distinguished by a small slender head, with a straight or convex profile and short thick horns varying from 15 to 46 cm in length, and usually directed outwards and upwards or forwards, occasionally downwards; polled animals are also encountered. The ears, of medium length, are directed laterally and slightly backwards. The neck is moderately long and furnished with a well developed dewlap. The chine is broad, and the hump approximately 30 cm high. In bulls the withers height reaches 140 cm, in cows 125 cm; the latter average 360 kg live weight. The shoulder is oblique and well muscled, the chest relatively deep but sometimes narrow, the barrel deep and wide, and the prepuce prominent. The back and loins are straight and of medium length, the rump is drooping, and the tail long and thin, ending in a full brush. The thighs are fairly muscular, the legs rather short, slender and well placed, with small hard dark-coloured hooves. The udder is small and fleshy, and the teats are short.

453. Grey Ugogo cow with calf (after McCall)

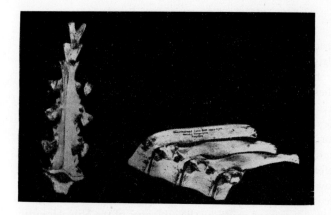

454. 7th–9th dorsal vertebrae of a Masai zebu bull, front and lateral views (after Curson)

455. Skulls of a Masai zebu bull (left) and cow (right), frontal surface

The skin is thick and strong, covered with short smooth lustrous hair, red, yellow or brown in colour, occasionally black, grey or white, rarely spotted or variegated (Schroeter, 1914). In the Kenya Masai the typical colour is black, but black with white patches on the flanks and belly, and brindle are also occasionally found, and are popular. Other colours occur, e.g. red, red-white, red with black points, grey and fawn.

As in other zebu cattle, the superior spines of the dorsal vertebrae from the sixth vertebra caudally are bifid, the tip being divided medially and also compressed anteroposteriorly (Fig. 454).

The cranium, more especially the splanchnocranium, of the Masai zebu is distinguished by the general length and narrowness. The latter is particularly marked in the interorbital width which amounts to about 45 per cent of the basilar length of the skull. The orbits are little prominent and outwardly directed (Figs. 455 and 456).

The Zebu Cattle of Zanzibar and Pemba

From Tanganyika zebu cattle were brought to Zanzibar and Pemba where they furnish meat, milk and draught power, the ox cart providing the most common form of

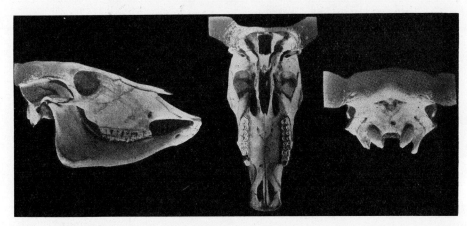

456. Skull of a Masai zebu cow, lateral (left), palatine (middle) and vuchal (right) views (after Curson)

457. Zebu cow from Pemba Island (after Faulkner and Epstein)

458. Zambia Angoni bull (after Faulkner and Epstein)

459. Zambia Angoni cow and calf (after Faulkner and Epstein)

transport. More recently considerable numbers of Boran cattle have been introduced into Zanzibar for breeding purposes.

The cattle of Zanzibar and Pemba (Figs. 375 and 457) are similar to other East African zebu types, such as the Nandi, Bukedi and Tanganyika cattle. They are small neat animals whose height at the hook bones averages 111 cm. The Pemba type is lighter than the Zanzibar type, finely boned, short-legged and short-bodied (Mason and Maule, 1960). The thoracic hump of the Zanzibar and Pemba Zebus is pronounced. The horns are usually small and flat. The commonest colours are light red, dun, black and grey; roans and brindles are fairly frequent. Adult bulls weigh approximately 320 kg, and cows 250 kg. A good cow will produce close to 700 kg milk in a lactation of one year. There has unquestionably been some admixture of Indian blood in the past and even today an Indian appearance is often quite noticeable in individual cattle (Tidbury, 1954).

The Angoni Cattle of Zambia

Zebus, known as Angoni cattle, are found on the Tanganyika plateau and in the Fort Jameson district of north-eastern Zambia, north of Abercorn and east to Petauke. From the north-eastern corner of Zambia the breeding area of the Angoni cattle extends south into the adjacent border region of Mozambique and east into the western parts of the Northern Province of Malawi.

The Ngoni, a mixed people of Venda and Jena stocks from south of the Zambesi, and of conquered tribes from the Songea district of Tanganyika and the Ngoni highlands, united under a ruling aristocracy of Zulu-Swazi origin, received their present breed of cattle at a relatively recent date. Originally their stock was of cervico-thoracic-humped sanga type, composed of the small Makalanga and the large Mangwato cattle (Wilson, 1941). But they lost all their beasts in the tsetse-fly belt after crossing the Zambesi near Tete in November 1835 on their trek from the south. Before finally settling in the Ngoni highlands, in the Dedza and Ncheu districts of Malawi and the Vila Cortino division of Mozambique, they crossed the southern end of Lake Nyasa in the hope of renewing their herds. However, they found no cattle until they reached Songea in the south-west corner of Tanganyika, east of Lake Nyasa, where they remained for a quarter of a century, fighting, raiding and building up the foundation stock of their present herds. Their raiding parties penetrated as far north as Lake Victoria. The cattle captured from the Awankonde, Batumbuka and other tribes in the northern parts of Malawi and in Songea seem to have been of East African Zebu stock, while those acquired in the region of Lakes Tanganyika and Victoria were of Ankole and similar sanga stock, "the larger and longer horned plateau animals" (Cullen-Young, 1932). In 1865 the Ngoni returned with their herds to the Ngoni highlands where the long-horned sanga cattle were concentrated around the villages of chiefs. These cattle suffered heavily during the rinderpest epidemic in July 1892, but sufficient remained to impose a lasting mark on the Angoni cattle, that accounts for the difference between the Angoni breed and the Nyasa Zebu proper (Faulkner and Epstein, 1957). After the Ngoni war of 1896, the Ngoni lost about half the cattle that had survived the previous rinderpest epidemic to British troops (Read, 1938).

The Angoni cattle of Zambia belong to the medium-sized type of East African thoracic-humped zebu. The live weight of adult bulls (at Mazabuka) is given as 568 kg, that of cows as 477 kg; the withers height of bulls is 127 cm and of cows 120 cm. The horns are usually short and stout, occasionally fairly long and upright; the ears are of medium size. The hump and dewlap are well developed in both sexes. The coat is short; its colour is very variable and may be red, brown, dun, black, red-and-white, black-and-white or brindle. The skin is darkly pigmented. The Angoni is used mainly for meat, but also for draught and milk (Faulkner and Epstein, 1957; Joshi et al., 1957).

The South Malawi Zebu (Nyasa Zebu)

The South Malawi Zebu is found in the Central and Southern Provinces and the Lake Shore area of the Northern Province of Malawi. Extending from the littoral, it is gradually displacing the original sanga stock of the country (Curson, 1936; Curson and Thornton, 1936). In the Central and Northern Provinces the cattle are almost entirely of zebu type (Faulkner and Brown, 1953). But quite pure zebu herds are found only in Karonga district on the north-western shore of Lake Nyasa, where, owing to the surrounding Misuku and Nyika highlands, native migrations have been limited (Wilson, 1941). Malawi and Mozambique north of the Zambesi river represent the most southerly habitat of the East African Zebu (Faulkner and Epstein, 1957).

The South Malawi Zebu is typical of the small short-horned East African Zebu and essentially of the same type as the Angoni of Zambia and Mozambique. The live weight of adult bulls averages 375 kg and of adult cows 275 kg; the withers height of bulls 124 cm and of cows 114 cm.

460. South Malawi Zebu bull

461. South Malawi Zebu cow

462. North Malawi bull

The North Malawi Zebu (Nyasa Angoni)

The Angoni (Ngoni) cattle of Mzimba and the western parts of the Northern Province of Malawi, adjacent to the Ngoni territory of Zambia, are bigger, more slender in conformation and longer in body and legs than the cattle in the major part of Malawi. They show the influence of sanga blood in their poor milking qualities, with low butter-fat content, and poor resistance to tsetse-fly infection, and also in their skeletal development and their horns and humps. In contrast with the short horns of the typical South Malawi Zebu, their horns are medium to long, projecting outwards and upwards, usually slightly forwards, with a characteristic sudden curve inwards about one-third from the tips. The hump displays considerable variation in shape and position. It may be quite small or large and pyramidal in shape, or large and long with a backward fall. Its situation is either thoracic or cervico-thoracic. The dewlap of the North Malawi Zebu is generally less developed than in the South Malawi Zebu. The dominant coat colour is red; other colours are in the minority. In spite of the defects of their cattle, the Ngoni have continued to encourage the breed and Mzimba has long been known as the home of big cattle (Wilson, 1941).

463. North Malawi Zebu heifer (after Faulkner and Epstein)

464. North Malawi Zebu cow

The Angoni Cattle of Mozambique

The Angone zebu cattle of Mozambique occur in Angonia, north of the Zambesi river, contiguous with the breeding area of the South Malawi Zebu in the Dedza and Ncheu region of Malawi. They are very small in stature, too small to serve as draught animals. The horns are short and stout, often rudimentary or absent. The hump is well developed, thoracic in position, with a nearly vertical posterior border. The most common coat colours are black, red, brown, broken or coloursided.

The Zebu Cattle of Madagascar

The cattle of Madagascar are for the greater part (80 per cent) of zebu type, the remainder being crossbreds of local zebus and improved breeds recently imported from France and South Africa. Prior to the middle of the 1st millennium A.D., all immigrants into Madagascar were of Malaysian race and culture; these immigrants seem to have introduced thoracic-humped zebu cattle from India or southern Arabia into the island. Around the beginning of the 2nd millennium A.D. the early Malaysian immigrants were reinforced by Indonesians with an admixture of Hamitic and Bantu elements from the Azanian coast and by Bantu from the east coast of Africa. With these immigrants an early humpless type of cattle (see p. 266) as well as the sanga cattle mentioned by C. Keller (1898) and Lydekker (1912a) (see p. 471) reached Mada-

465. Angoni cattle, Vila Coutynho Zootechnical Station, Mozambique

466. Madagascar Zebu cattle (photograph: Service Général de l'Information de Madagascar)

467. Ninth dorsal vertebra of a Madagascar cow

gascar. The Bantu also introduced the words of Bantu origin, e.g. Omby (cf. Ngombe) = cattle, which the Madagascans use for their cattle (Mason and Maule, 1960).

The Madagascar zebu is a thick-set animal, kept primarily for meat production, in addition to work. Adult bulls average 125 cm in height behind the hump and weigh 360–420 kg; cows are 117 cm high and weigh 320–340 kg. The head is short with a straight profile and crescent-shaped, more rarely lyre-shaped horns of medium size and circular cross-section. Cattle without horns or with loose horns also occur. The neck is light and the dewlap moderately well developed. The hump is prominent, especially in the bull, with a large amount of fatty tissue. Its position, according to Mason and Maule (1960), is thoracic. Joshi et al. (1957) describe the hump as cervicothoracic; but this seems to be erroneous as the hump, more especially of the bull, is situated above and well behind the shoulder, leaving the neck free save for the anterior attachment of the hump at the posterior cervical border. The body is of moderate length, the hindquarters tend to be narrow, and the rump is rather short and sloping. The legs are short and coarse, the udder and teats small. In the skeleton of a thoracic-humped Madagascar cow in the Onderstepoort Collection the neural spines of the anterior dorsal vertebrae are distinctly bifid (Fig. 467). Also, they are long and slender, markedly compressed antero-posteriorly, and distinctly curving backwards, as are the respective spines in purebred zebu cattle (Epstein, 1955c). The short soft coat of the Madagascar zebu displays a wide range of coloration: black, red, fawn, yellow and white in whole colours or variegated; common patterns are red-and-white, black-and-white or grey with black points (Keller, 1898; Joshi et al., 1957; Mason and Maule, 1960).

2. The Zebu Cattle of West Africa

Distribution of Zebu Cattle in West Africa

From the western part of the Republic of the Sudan and the northern territory of Congo short-horned zebu cattle extend in a narrow belt, flanked by the Sahara desert in the north and, west of Lake Chad, by the long-horned Fulani cattle in the south, into West Africa. They resemble the zebu cattle of East Africa in general conformation.

The Shuwa Arab Zebu

In Chad, contiguous with the short-horned zebu territory of the Sudan, short-horned zebus are found all over the country except in the south-west. They are called Arab cattle either with reference to their home country or the traders who introduced them into the Chad region. A variety of these Arab cattle, known as Fellata (= Fulani but distinct from the long-horned Senegal, Sudanese and White Fulani types—see pp. 501–505), is bred by the Fulani of Chad. From Chad the Arab cattle extend into the extreme north of Cameroun and into Bornu Province (around Maiduguri and Dikwa) and Adamawa Province (around Yola), north-eastern Nigeria, where their name is Shuwa (Choa).

The Shuwa Arab (Choa, Arab, Waddara or Tour) cattle of Chad and the northern parts of Nigeria and Cameroun are used mainly for baggage work and milk, selected cows producing approximately 1200 kg milk per lactation. The Shuwa is rather heterogeneous in conformation, but generally well formed and of medium size, the withers height varying between 125 cm (cows) and 140 cm (bulls). Adult bulls weigh about 400 kg and cows 325 kg. The head is long, straight in profile, with short, moderately thick, round or flat horns with blunt tips, which commonly curve slightly outwards and upwards or downwards; occasionally the horns project laterally or lie close to the head or are loose. The ears are long but not pendulous, and are carried horizontally with the inner surface to the front. The neck is short, tending to carry the head low. The dewlap is only moderately developed; it shows some folds but is not markedly loose or pendulous. The hump is muscular in structure (Dr. D. H. Hill—personal communication); small in cows, and of medium size in bulls. The body is fairly deep and compact with well sprung ribs and good capacity. The top line is fairly

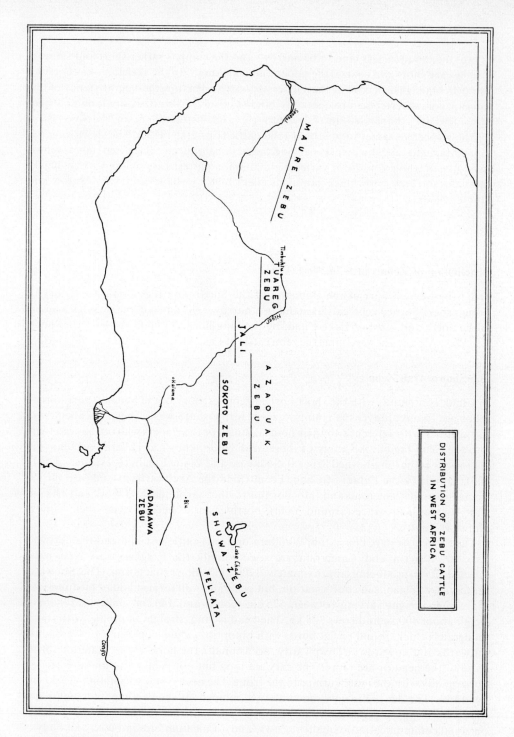

468. DISTRIBUTION OF ZEBU CATTLE IN WEST AFRICA (after Doutressoulle)

strong, rising slightly to the moderately high sacrum. The rump is long and sloping, the tail thin and long. The legs are short and fine-boned, with small, hard, well formed hooves. The coat is generally chestnut or dark red-and-black, less frequently reddish brown or black, with or without small white patches on the underline. In the eastern part of the habitat coat colours are usually paler. The skin and hooves are black or dark red (Curson and Thornton, 1936; Faulkner and Epstein, 1957).

The Fellata Zebu

Cattle of the same (Shuwa) type bred by the Fulani (Fellata) of Chad are slightly larger than the average Shuwa, with bigger humps and larger dewlap, also lighter in colour (Mason, 1951a).

469. Shuwa zebu bull

470. Shuwa zebu cow (after Mason)

471. Sokoto zebu bull (after Faulkner and Epstein)

472. Sokotograph zebu cow (Photograph: Mr. T. C. Okoro)

The Sokoto Zebu

In the north-eastern part of Sokoto Province of Nigeria, but also in various areas over the north-western part of Nigeria, the Fulani and Hausa keep a closely allied short-horned zebu breed, known as Sokoto or Gudali.

The Sokoto is highly valued as a dairy animal, selected cows producing over 1500 kg milk per lactation, testing approximately 6.5 per cent butter-fat. It is used also for work and meat production. In conformation, type and size, it closely resembles the Boran cattle of East Africa, the average height at withers being 137 cm in selected bulls and 127 cm in cows, the live weight 540 kg in bulls and 335 kg in cows. The head is long but wide between the eyes and across the forehead, narrow below the eyes. In profile it is straight or slightly convex. The ears are large and sometimes a little pendulous, the horns commonly short, often very short, longer in the cow than in the bull, round, flat or oval in cross-section, with blunt tips; they project from the head in a lateral and upward direction. The dewlap is highly developed, with pendulous neck folds, frequently passing into the large umbilical fold or prepuce. The thoracic hump is well developed, especially in the male, firmly placed over the withers and rounded antero-posteriorly with a slight fall at the back. It is invariably musculo-fatty in structure (Dr. D. H. Hill—personal communication). The body is well balanced, deep, wide and compact. The rise of the top line from hump to sacrum is only slight, but the rather short rump shows the typical zebu slope. The upper thighs are of fair width, tending to be narrow lower down. The legs are short, with clean strong bone. The tail is long. The udder shows good development, but the teats are rather rarely well placed. In purebred Sokoto Gudali cattle the coat colour is white, light dun, cream or fawn. In the bulls dark grey areas usually extend from the neck over the shoulders and hump. Dark patches are occasionally seen around the eyes, and small dark spots on the legs just above the hooves. The skin, muzzle, tail switch and hooves are black (Faulkner and Epstein, 1957).

The Adamawa Zebu

In Cameroun, in addition to the Shuwa (Arab) cattle, there is another zebu breed, i.e. the Adamawa. It is concentrated mainly in the Ngaundéré, Banyo and Bamenda regions and in the Mabila Highlands in the southern tip of Adamawa district. In Nigeria this breed has been subdivided (by Gates) into the pure Banyo and Ngaundéré zebu types and the mixed Yola type, the latter being most common in Muri Emirate and the central division of Adamawa Province.

473. Adamawa (Banyo) bull

The Banyo Type of Adamawa Zebu

The Banyo type of the Adamawa zebu of Cameroun is used for milk and meat production and to a limited extent for work. It is believed to have been influenced by Red Bororo (Fulani) blood (see pp. 489–501). The Banyo Adamawa averages 124 cm in height; adult bulls weigh a little over 400 kg, and cows 360 kg. Selected cows produce an average of slightly more than 1000 kg milk per lactation. The head of the

474. Adamawa (Banyo) cows (after Faulkner and Epstein)

Adamawa Banyo is well proportioned with good width between the eyes and at the muzzle. The horns are either short and thin or of medium length and moderately thick, projecting outwards, upwards and slightly forwards, with the tips occasionally backwards. They are generally round in cross-section, sometimes flat or oval. The dewlap is moderately developed and not pendulous. Similarly the umbilical fold and prepuce are not particularly large. The hump is well developed in both sexes, firm, pyramidal in shape and thoracic in situation. The body is long but well balanced, deep and wide. The top line slopes only slightly from hump to sacrum. The rump is of fair length but sloping, with a prominent tail head and high setting. The upper thighs are broad, fairly full and thick, but narrow and lean lower down. The legs are moderately long with fine bone. The coat colours are white and a deep but bright red. The usual pattern is a white face with coloured muzzle and eye rings, white dewlap, underline and legs, the rest of the body red. In some cases red may extend over the whole face, or white may be present in splashes over the back or on the sides of the body.

The Ngaundéré Type of Adamawa Zebu

The Ngaundéré type of Adamawa zebu is somewhat heavier than the Banyo type, bulls weighing 450 kg and cows 400 kg. The head shows good proportions, and a straight or slightly convex profile. The horns are small to medium in length and not thick; they usually project outwards, upwards and slightly forwards, but may also be devoid of cores and hang downwards. The dewlap, umbilical fold and prepuce are only

475. Adamawa (Ngaundéré) bull (after Faulkner and Epstein)

moderately developed. The thoracic musculo-fatty hump is very large and pendulous, hanging over one side and having the appearance of being broken; this is a characteristic feature of the Ngaundéré type of Adamawa and one distinguishing it from the Banyo type. The body of the Ngaundéré is slightly shorter, deeper and more compact than that of the Adamawa Banyo, the height behind the shoulders averaging about 117 cm (Mason, 1951a). The back is level and broad, especially over the loins, and very well fleshed. The rump is less sloping than in the Banyo, and the upper thighs are wide and full and, for African zebu cattle, well let down to the second thighs. The legs are somewhat shorter than the Banyo's, and the bone, while clean and strong, is less fine. The coat colours include white-and-red and reddish brown-and-white, the colours being more broken and showing far more white than in the Banyo type. Brindle and roan animals also occur (Faulkner and Epstein, 1957).

The Yola Type of Adamawa Zebu

The Yola type of Adamawa, named also Adamawa Gudali or Tattabareji (the Fulani term for speckled), is believed to have resulted from the cross-breeding of the Banyo and Ngaundéré types of the Adamawa zebu with White Fulani and humpless dwarf shorthorn cattle (Gates, 1952). It is used for milk, meat and work. Considerable variability in conformation is encountered in Yola herds, pointing to a mixed origin. In general the Yola resembles more closely the Banyo type of Adamawa zebu than the other types held to be involved in its origin. The Yola is smaller than the Banyo and Ngaundéré; the average height at the withers is 122 cm in bulls and 120 cm in cows, the average live weight 350 kg in bulls and 335 kg in cows. The horns resemble the Banyo's, being short or medium in length, and outward, upward and slightly forward in direction. The dewlap, prepuce and umbilical fold vary from being small to moderately well developed. The hump is moderately large, upright and variable in position. The coat colour consists of a mixture of white and red, black, dun, brown or blue roan, with the white either in patches or speckled areas (Faulkner and Epstein, 1957).

The Azaouak Zebu

Short-horned zebu cattle of a similar type to the Shuwa are bred by Arabs and Tuareg in the Azaouak basin in eastern Mali and central Niger. The same type is also found along the northern border of Nigeria, especially the western part, extending into northern Sokoto, Nigeria. From the Upper Volta it has recently been imported also into the Northern Territories of Ghana. In eastern Mali and central Niger this breed is known as Azaouak, and in Nigeria under various names: Adar, Azawal, Azawaje, Wadara, Tur.

The Azaouak zebu is used largely for transport purposes, but the cows are also good milkers; in fact, the Azaouak is said to be the best milker in West Africa (Mason, 1951a). It is a small to medium-sized type, standing about 123 cm at the withers and

476. Adamawa (Ngaundéré) cow

weighing 400–450 kg in the case of the adult bull. The head, long, narrow below the eyes and straight in profile, is carried somewhat low. The horns are short, moderately thick at the base, but becoming thin as they project outwards and upwards with only a very slight curve. The dewlap varies greatly in development, in some animals showing heavy folds, in others few or no folds. The hump also varies considerably in size, shape and position. While small in cows, it is often large in bulls, rounded from front to back with a long slope up the front from quite far foward on the neck and ending in a steep fall at the back. Generally the hump is narrow, being about 14 cm thick in the males and 12 cm in the females (Joshi et al., 1957). It is muscular in structure (Dr. D. H. Hill —personal communication). The body is somewhat long, lacking muscular development, often shallow and deficient in heart girth. The topline slopes up to the high sacrum, and the rump is sloping down from hook bones to pin bones. The hind-

477. Azaouak zebu bull

478. Azaouak zebu cow

479. Jali bull from Upper Volta

480. Jali cow (after Doutressoulle)

quarters are fairly well developed, but the lower thighs are lean and narrow. Legginess observed in many Azaouak cattle is due to the fairly long limbs as well as to the shallowness of body. The coat colour is variable; most common is a dark wheat colour; in addition fawn, black-and-white, red-and-white, and other colours and colour combinations are encountered. The pigmentation of the slightly loose and moderately thick skin is usually dark brown, more rarely black (Doutressoulle, 1947).

The Jali

The Jali (Diali) cattle, encountered in Ilorin, south Sokoto (Nigeria) and in the portion of the Niger valley situated in Niger, constitute a rather ill-defined type, composed of a variable mixture of short-horned zebu and long-horned Fulani stocks. They may also have been influenced by humpless shorthorn cattle, but of this there is no evidence.

Doutressoulle (1947) classes them with the lyre-horned Fulani cattle, Mason (1951a) with the medium-horned zebus; Faulkner and Epstein (1957) group them with crossbreds of humpless shorthorn and humped short-horned zebu and longhorned Fulani parentage. The Jali is very variable in conformation, but it is basically of short-horned zebu stock. It is a good beef but poor dairy animal, and is only little used for work. In the south of Niger, along the border of Nigeria, the Jali cattle, bred by semi-sedentary Fulani are generally of a superior type. In this area two local Jali varieties are distinguished: Yakanaya and Bokoloji, both of them crossbreds of short-horned zebu and long-horned Fulani parentage. Typically Jali cattle stand 115–130cm at the shoulder; bulls weigh 300–350 kg, cows about 50 kg less. They have a slender head, with long facial part, broad muzzle and straight profile, occasionally with a receding forehead. The horns vary greatly in size and shape. In the bull they are usually small, in the cow 25–30 cm long. Horns of larger size indicate a former or recent cross with long-horned Fulani cattle, whose habitat borders the Jali's. Horn direction is commonly lateral, with an upward and forward sweep. Animals with loose horns or polled Jali cattle are rare except in herds of Bokoloji type. The neck is short, and the chest fairly wide and deep. In the bull the dewlap and hump are well developed, the hump being irregularly shaped, indented and furrowed, bending always sideways, to the right or the left. Behind the hump, the back rises to the high sacrum from which the short rump steeply descends to the long well attached tail. The thighs are rather flat but fairly well muscled, the legs slender. The udder is poorly developed, with small teats. The skin is fine and supple. The coat is commonly wihite in colour with pigmented mucosa, occasionally white with black spots (Yakanaya), black-and-red or roan (Doutressoulle, 1947).

The Tuareg Zebu

The short-horned zebus of the Tuareg in the Niger bend of Mali are known as Tuareg cattle. They are of compact build, suitable for draught and meat production, but poor in milk yield (Doutressoulle and Traore, 1949). The withers height averages 135 cm

481. Tuareg zebu bull

482. Maure zebu bull and Sudanese Fulani cows

in the bull and 115 cm in the cow. The live weight of bulls varies between 300 and 350 kg and of cows between 250 and 300 kg, while oxen weigh up to 400 kg. The head is short and wide in the bull, narrower in the cow, with fairly prominent orbits, a straight profile and broad muzzle. The horns are small and fine, very variable in shape, but generally projecting from the skull in a lateral direction. Loose-horned animals are occasionally encountered, but polled Tuareg cattle are very rare. The neck is short, heavy in the bull, more slender in the cow, with a thin dewlap extending from the throat to the brisket. In the bull the hump is fairly prominent, but in cows and oxen it is little pronounced. The chest is deep but narrow. The top line rises from the hump to the sacrum, which exceeds the withers in height. The rump is short and sloping, and the thighs are flat and short. The long fine tail extends well below the hocks. The coat is very variable in colour; whole black, fawn or dark grey, and variegated patterns predominate (Doutressoulle, 1947; Mason 1951a).

The Maure Zebu

In Mauritania the Maure or Mauritania zebu cattle are bred by nomad tribes. They are large animals, used for transport and milk production. Loosely built and more leggy than the Tuareg, Maure bulls and cows stand 125–130 cm at the withers, and oxen 140–150 cm. The live weight is 350–400 kg in the bull and 250–300 kg in the cow. The head is long and narrow, with a straight profile and prominent orbits. The horns are short and slender in the bull, longer but slender in the cow, round in cross-section, commonly lateral and upward or lateral and forward in direction, and grey or brown in colour. The neck is thin and flat, of moderate length, with only a slight dewlap. In the bull the hump reaches a height of 10–20 cm; in the cow it is less marked. The chest is long and shallow. The flat ribs acutely slope from the sharp back; the loins are thin and flat, the rump is drooping, and the tail is thin and set on high. The thighs are poorly muscled. The legs are strong of bone, with large, flat hooves. The skin is fine and supple but not wrinkled, the hair short. The udder is relatively well developed, with large teats. The colours most frequently encountered in west Mauritanian herds are black or black-and-white, and in the east dark red. The occurrence of white spots in the coat and of a very prominent hump in Maure zebus in the eastern part of their range is regarded as a sign of cross-breeding with Fulani cattle (Doutressoulle, 1947).

iii. The Sanga-Fulani Group

1. The Sanga Cattle of Eastern and Southern Africa

Definition of the Term 'Sanga'

The term 'sanga' denotes a group of East and South African cattle which in one or more of the following characters, namely situation or size of hump, horn size, or cranial or body conformation, suggest that they are neither of pure or nearly pure zebu nor of pure or nearly pure humpless longhorn type, but represent a variable mixture of these two parent stocks.

Distribution of the Sanga Cattle

The area of distribution of the sanga cattle extends from Ethiopia through the lake district of East Africa to South West Africa. Before short-horned zebu cattle made their appearance in East Africa, the sanga was ubiquitous in this area. This is illustrated by its early introduction into the island of Madagascar (Keller, 1896) (see also p. 471), Burton's record of the presence of giant-horned sanga cattle on the eastern shore of Lake Tanganyika in 1858 (see p. 434), and also by Orde Brown's (1925) reference to their former occurrence in Kenya:—"There was a breed of cattle which has now died out, which possessed far larger horns than the existing breed (zebu), and are said to have been finer animals in every way. They are said to have been very numerous, and to have been killed off in the epidemic of rinderpest, which occurred apparently about 1890; ... There are no traces of these animals now left, except the horns, which are occasionally to be met with, made into drinking vessels; these nearly always came from Emberre."

It would appear that these large-horned cattle are not completely extinct in Kenya, for as late as 1964 Mason saw an ox with gigantic horns projecting from enormous frontal necks (Fig. 484) at a market place near Marakwet on the Elgeyo Escarpment in western Kenya.

General Characteristics of the Sanga Cattle

Sanga cattle vary in size from very small breeds like the Makalanga or Mashona to large animals such as the Barotse. The head is usually long, of moderate width, with a convex profile as in the majority of the Nguni cattle, or a straight profile as in the Barotse (Bisschop, 1937). The horns vary greatly in length, basal girth and direction; polled sanga cattle are comparatively rare. Among the horned breeds two main types can be distinguished: one with long, relatively slender horns, and the other with gigantic horns of great basal circumference.

The vast majority of sanga cattle are humped. The tendency to be humped is more general than is horn gigantism. In bulls the hump is more prominent than in cows; it is better developed in some breeds (e.g. the Bahima) than in others (e.g. the Barotse).

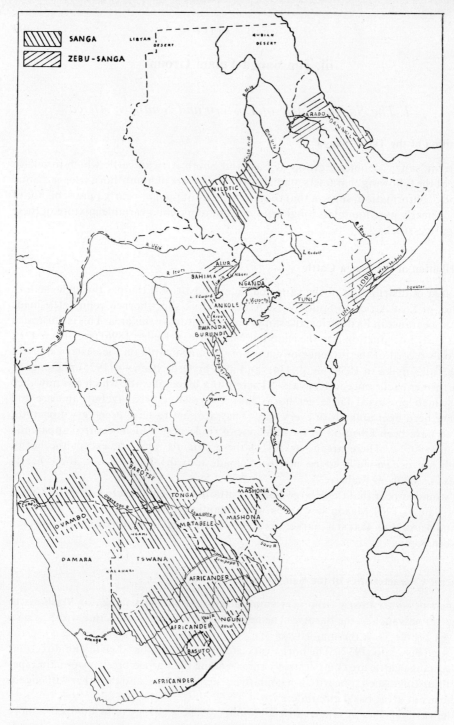

483. DISTRIBUTION OF SANGA AND ZEBU-SANGA INTERMEDIATE TYPES IN EAST AND SOUTH AFRICA (after Mason and Maule)

484. Giant-horned sanga ox, western Kenya

Hump position varies. The typical sanga hump, as characteristic of all sanga breeds south of the Zambesi whither zebu cattle with thoracic humps have not penetrated, is cervico-thoracic. North of the Zambesi line, few sanga breeds have remained unaffected by thoracic-humped zebu stock which, during recent centuries, has spread throughout Africa south of the Sahara at a phenomenal speed. And, as Gates (quoted by Mason, 1951a) has pointed out, the greater the proportion of (thoracic-humped) zebu blood that is evident, "the larger and further back is the hump". A postage stamp from the former Belgian Congo schematizes the distinction between the cervico-thoracic and the thoracic type of sanga hump admirably (Fig. 485).

The Anatomy of the Sanga Hump

The typical sanga hump is situated in front of the withers and is well defined. The anterior axial border passes up from the upper border of the neck to the apex of the hump at an angle of approximately 40°. From the apex to the withers the posterior axial border falls at an angle of about 30° to the horizontal. The hump is well attached to the neck (Bisschop and Curson, 1935).

Curson and Bisschop (1935) give the following description of the hump of a sanga (Ambo) bull, 2 years 4 months old:—

M. trapezius

"It is arbitrarily divisible into cervical and thoracic portions. There is no development of adipose tissue resembling the hump of the Shorthorned Zebu."

M. rhomboideus

"The muscle is clearly represented as a cervical portion and as a thoracic portion. The former is markedly enlarged, in fact constitutes the hump. The fleshy fibres are disposed as a rule in a longitudinal direction. The thoracic part is relatively poorly developed and its fibres run in a ventro-caudal direction."

Situation of hump

"From the sixth to seventh cervical vertebra caudally to the fourth-fifth thoracic bone. The hump is definitely cranial to the angulus cranialis of the scapula—a cervico-thoracic hump."

485. Postage stamp from the former Belgian Congo, showing cervico-thoracic-humped and thoracic-humped sangas

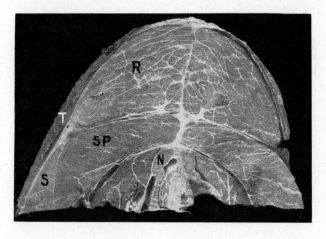

486. Transverse section (at summit of hump of South African sanga (Ambo) bull, over approximately 2nd thoracic vertebra. Cranial view (after Curson and Bisschop)
(T = M. trapezius; R = M. rhomboideus; S = M. serratus ventralis; (SP= M. splenius; N = L. nuchae)

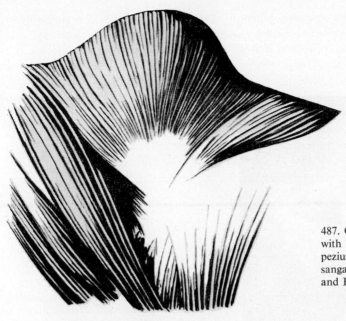

487. Cervico-thoracic hump with skin removed but M. trapezius intact, of South African sanga (Ambo) bull (after Curson and Bisschop)

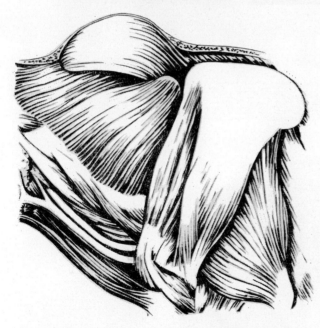

488. Cervico-thoracic hump with M. trapezius removed, of South African sanga (Ambo) bull (after Curson and Bisschop)

Milne's (1955) description of the Ankole (sanga) hump in the main tallies with Curson and Bisschop's account of the Ambo (sanga) hump. The Ankole hump is cervico-thoracic, lying, in relation to the vertebral column, from the 4th or 5th cervical to the 4th thoracic vertebra. It is "purely muscular in structure, not musculo-fatty...

In view of its firm and double attachment to the scapula, its function is probably to assist in the movement of the fore-limb." This function is termed "locomotory".

From this description it is evident that the typical cervico-thoracic sanga hump and the hump of the Africander (see p. 476) are anatomically alike.

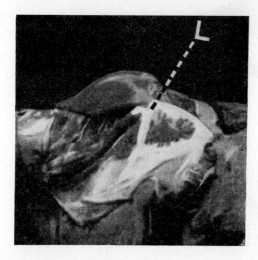

489. Ankole hump with M. trapezius removed
L = Ligamentous insertion on to scapula
(after Milne)

490. Giant-horned Galla Azebò ox
(after Müller)

The Sanga Cattle of Ethiopia in Earlier Times

Since Ethiopia appears to be the centre of origin and dispersal of the sanga cattle, it is convenient to refer to the Ethiopian sanga first. In former times the sanga appears to have been the prevalent type in Ethiopia. As Lydekker (1912a) wrote:—"... in Abyssinia and Gallaland are numerous breeds of humped cattle remarkable for the enormous girth of their huge horns; but many of them have of late years been more or less decimated by rinderpest. The breeds with the largest horns inhabit the lowlands, those from the mountains having these appendages of smaller dimensions. Some of the largest horns of all are met with in the cattle of the Arusi–Gallas and the Shilla tribe, especially those members of these tribes who inhabit the neighbourhood of Lake Zuay or Zwei. These Galla or Sanga cattle are generally white and have small or no humps, their muzzles being black, the legs relatively long, and the bones small. In stature these oxen are very large, and their horns, which rise vertically, and are often more or less nearly lyre-shaped, frequently measure close on four feet in length, and will hold four or five gallons of water." Of a pair of Abyssinian sanga horns, which the traveller Salt presented to the Museum of the Royal College of Surgeons, each horn measures 119 cm in length and 38 cm in basal circumference, while the tip-to-tip distance is 101 cm. This is not a record; some Ethiopian sanga horns measure 120 cm in length and 53 cm in basal girth (Adametz, 1895).

The Danakil and Galla Azebò Cattle of Ethiopia

At the present time the sanga is bred in a pure form only by the Danakil and Galla Azebò peoples. Beyond the borders of this breeding area and extending into northern Eritrea, it is found greatly mixed with short-horned zebu blood. This crossbred type, known as Aradò, has been discussed together with the East African zebu (see pp. 347

491. Adal cow

492. Adal ox

493. Danakil cattle – en face

to 350). Similarly, sanga blood is conspicuous in some of the Jiddu cattle of southern Somalia, which have likewise been mentioned in connection with the East African zebu group (pp. 379–383).

The breeding area of the Danakil is in the north-east (Afar or Dancalia), that of the Galla Azebò or Raya in the extreme north of Wollo, just north-west of Lake Asangi. The Danakil (Dankale) is also known as Adal or Keriyo (Sandford, 1961—personal communication)—names of Danakil tribes.

Between this rather restricted breeding area of sanga cattle in Dancalia, along the Red Sea coast between Zeila and Tajura and thence inland in a narrow triangle nearly to Ankober in the centre of Ethiopia, and the still more restricted sanga territory in the south-west of Ethiopia, that is in the major part of the country, zebu or zebu-sanga crossbred cattle are encountered. Indeed, in view of the proximity of Ethiopia to the gates of entrance of thoracic-humped zebus from Arabia, it is surprising, and illustrative of the conservative attitude of the breeders to their cattle, that the Danakil and Galla Azebò have preserved the sanga type.

The Danakil and Gallo Azebò cattle have small humps, moderately developed dewlaps, and immense horns, lyre-shaped or forming a complete circle or oval (Marchi, 1929; Bonelli, 1938; Salerno, 1939). The height of the Danakil and Galla Azebò varies between 125 and 145 cm, and their weight between 250 and 375 kg. The colour is usually light chestnut, blond or ash-grey (Mason and Maule, 1960).

Cranial Characteristics of the Ethiopian Sanga Cattle

In cranial conformation the Ethiopian sanga cattle display a remarkable resemblance to the humpless longhorn cattle of ancient Egypt; their skulls show hardly any zebu characteristics. In a long-horned Ethiopian sanga skull, described by Duerst (1899), the forehead is broad and flat, and the facial part comparatively short. The frontal ridge slopes upwards to the Torus frontalis, the summit of which is slightly depressed. In the centre of the forehead there is a depression flanked by two lateral elevations. The orbits are flat; the deep supraorbital fossae do not reach the horn cores; at the other extremity they end at a distance of 1.5 cm from the nasals. The frontals cut slightly into the lacrimals. The nasals are bent inwards in the centre. The nasal branches of the premaxillae almost touch the nasals. The malar point is moderately developed, the palate narrow and decidedly arched. The edge of the jugals is very sharp. The temporal fossae are deep and narrow behind the orbits, then broaden considerably above the jugals, and are wide open aborally. The conformation of the occiput is very similar to that in the ancient Egyptian humpless longhorn. A triangular projection of the parietals extends into the Torus frontalis. The frontals project powerful necks towards the horn cores. The bases of the latter are surrounded by wreaths of horn pearls which bulge over the edges of the horn necks. Deep furrows run from the occipital crest towards the cores. The horns of this skull are sickle-shaped and dark in colour. In another Ethiopian sanga skull, described by Duerst (1899), the horns approach each other at the tips, forming a complete circle.

The Nilotic Sanga Cattle

In the Sudan the sanga cattle are named Nilotic or Sudanese Longhorn, also Dinka, Nuer or Shilluk cattle. They are found in the southern part of the Sudan, chiefly the Bahr el Ghazal and Upper Nile Provinces and the Mandari area of Equatoria Province, from where they extend into the Akobo–Gambeila area in the south-west of Ethiopia. Here they are bred mainly by the Anuak tribe, but are not numerous, as the Anuak cannot keep cattle in the east of their territory owing to tsetse-fly.

Of the two principal cattle types of the Sudan, i.e. zebu and sanga, the sanga is undoubtedly the older. Being bred mainly by the pagan Dinka, Nuer and Shilluk tribes, it was pushed back behind its present northern boundary by the steady pressure of Moslem invasion that persisted throughout the Middle Ages.

Among the Nilotic cattle several local varieties, such as Aliab Dinka, Nuer, Aweil Dinka, Eastern Nuer, and the Ethiopian Anuak, could formerly be recognised; but

494. (above) Ethiopian sanga skull, frontal surface (after Duerst)
495. (right) Ethiopian sanga skull, lateral surface (after Duerst)

496. Nilotic Sanga bull (after Faulkner and Epstein)

with the cessation of inter-tribal strife and with improved communications these subtypes are no longer clearly distinguishable (Mason and Maule, 1960).

Geographically Nilotic cattle vary in size from west to east, small light-bodied refined animals being found in the west, and large heavily built cattle weighing up to 550 kg live weight among the Eastern Nuer along the Ethiopian border in the northeast. While larger than the Aweil, the cattle of the Eastern Nuer are less robust than the Aliab. In the south-western part of Ethiopia, their humps are often thoracic, resembling the hump of the typical East African zebu, rather than cervico-thoracic in situation. Near the eastern border of their range, where they are influenced by the Abyssinian zebu, Anuak cattle are smaller and shorter-horned than in the west. The average withers height of Nilotic cows is 115 cm, the live weight 254 kg (S.D.I.T.,

497. Nilotic Sanga cows

498. Sanga ox of the Nuer

499. Sanga ox from the Blue Nile (after Lydekker)

500. Nuer ox

1955). Differences in size are due primarily to environment. In general, body conformation in the Nilotic sanga cattle varies greatly. Some have an excellent conformation with a strong topline, straight rump, good heart girth, and good width throughout, being smoothly covered and well-fleshed, while others are leggy, shallow of body, slab-sided and lacking in width and spring of rib, with a high sacrum and steeply sloping rump. Commonly the head is well proportioned with good width of forehead and muzzle. In profile the face is straight. In the cattle of the Bor district and the Aliab valley in the south-eastern part of the breeding area the horns tend to be immense, often exceeding 150 cm in length with a basal diameter of 20 cm. They project from a very thick pedestal-like base outwards and upwards, occasionally curving inwards so as to form a perfect oval. Smaller horns, 30–40 cm long and 8–10 cm in diameter at the base, are encountered in the herds of the Aweil Dinka and Nuer in the northern part of the breeding area. Polled cattle are occasionally seen throughout the Nilotic area; but animals with loose unattached horns are restricted to the northern part of the area. The occurrence of this anomaly is attributed to the influence of zebu cattle on the Shilluk and northern Dinka herds by the Nile north of Malakal (Joshi et al., 1957). The horns are either light cream or dark in colour. The hump varies from very small to moderately large in size. In the cow it is placed cervico-thoracically and is of average sanga type, size and shape. In the bull it may lean over at the back and resemble a zebu hump rather than the hump of a typical sanga. It is in this respect that Dinka cattle are not typically sanga. The hump is also in width more zebu than sanga (Bisschop, MS). The dewlap which commences as a single fold varies greatly in size and length; in the cows it is 25–30 cm long and up to 23 cm deep. The rump slopes less steeply, the tail setting is higher and the tail shorter than in zebu cattle. In the Aliab Dinka a light coat colour is typical, often white with a pattern in red or black. In other regions the most common coat colours are grey, cream, light red, light brown, dun, and black. Whole colours are as common as variegated patterns. The hide is pigmented. The Dinka select their breeding stock on the basis of horn shape and coat colour. Nilotic cattle appear suitable for meat production. The milk yield of the cows averages 750 kg per lactation, testing 6.5 per cent butter-fat (Faulkner and Epstein, 1957).

501. Head and neck of a Shilluk Sanga (after a drawing by H.R. Millais)

502. (Head of a Dinka bull (after a drawing by Schweinfurth)

503. Dinka cattle (after a drawing by Schweinfurth)

Artificial Deformation of Nilotic Sanga Horns

Certain Nilotic tribes, such as the Dinka and Nuer, but not the Shilluk, practise artificial deformation of the horns of their cattle, a custom which Seligman (1932) attributed to ancient Egyptian (Vth Dynasty—c. 2500–2350 B.C.) origin. However, it may be of even greater antiquity (before 3000 B.C.), as Bovidian rock drawings at Bardaï, Enneri Debassar and Areun, in the desert between Tassili-Oua-N-Ahaggar, Fezzan and Tibesti, show cattle with one horn trained upwards and the other downwards (Huard, 1957). Among the recent Dinka the usual deformity is with one horn forwards and the other back; among the Nuer one horn is trained to grow upwards and slightly backwards, while the other is brought in a curve forwards across the beast's forehead. The left horn is trained down to resemble the left arm holding a shield, while the right horn is trained upwards to resemble the right arm raised to hold a spear. Cattle with artificially or naturally deformed horns have a social and emotional value, and the interest in such animals extends far beyond the Nile region into Kenya where it is known among the Suk (Seligman, 1932).

The Indigenous Cattle of the Lake Area of East Africa

The indigenous cattle of the Congo are found in the extreme east of the territory in the highlands bordering the Great Lakes. Their area of distribution is divided into a northern and a southern zone, which are separated by the mountains of the Ruwenzori range and the forests of the Semliki river basin. In the north of this region cattle are found north-west of Lake Albert, i.e. between 1° and 3°30′ N. latitude in that part of the Kibali-Ituri province east of longitude 30° E. The southern area is in Kivu province between 1° and 4° S. latitude. This is an extension westwards to longitude 28°30′ E. of the breeding area of Ruanda–Urundi.

In the northern area there are two breeds: one, the Lugware (see pp. 367–369), is a typical short-horned zebu, while west of Lake Albert, in the Bunia, Irumu and Gety areas, the Bahima breed sanga cattle. Between these two areas there is an intermediate type, termed Nyoka (see below), which is closely similar to the Nganda cattle of Buganda. The cattle of the Ankole district in the south-east of Uganda are a branch of the Bahima of the Congo.

The southern group of breeds are all of sanga type. The principal breed is the Watusi which is found in Ruanda–Urundi and the adjacent area of Tanganyika. North-west of Lake Kivu in the Ruchuru Masisi region is the Barundi type of the Ruanda breed (see pp. 430–432), introduced by Watusi tribesmen. The same breed was brought by the Barundi into the Ruzizi valley (between Lake Kivu and Lake Tanganyika) and the mountains west thereof (Uvira–Fizi–Mwenga area). Between these two areas the Bashi, who live in the Kalehe–Kabare areas west of Lake Kivu, have a much smaller sanga type (see pp. 432–433), which since 1900 has been much influenced by Ruanda blood (B.A.C.B., 1952).

The Nyoka Cattle of the Congo

From Ethiopia and the Sudan sanga cattle passed through Uganda and followed the Great Lakes southward to the Zambesi. However, north-west and north-east of Lake Albert, zebu (Bukedi, Mongalla, Lugware) cattle form an intrusion, in fact break the continuity of the sanga line from the southern Sudan to the northern shore of Lake Tanganyika. They have influenced also the sanga cattle east and west of Lake Albert. West of this lake, the sanga cattle, mixed with short-horned zebu (Lugware, Bukedi) blood to a greater or lesser degree, are known as Nyoka, Alur or Blukwa cattle. In its purest form the Nyoka is localised in the broken plateau area of the upper Ituri, Nyoka, Mahagi and Djugu regions in the Eastern Provinces of the Congo (Joshi et al., 1957). Cattle of Nyoka type extend also farther west into Buta, where they are bred by the Ababua (Duerst, 1899), and also south into the Lake Kivu region.

Nyoka cattle are basically of Ankole sanga stock (see pp. 426–429) with a variable degree of Lugware zebu admixture. Owing to their descent from two distinct parental types, the Nyoka cattle vary considerably in general conformation; the hump, for instance, varies from the rudimentary to a prominence similar to that in the zebu. They are, on an average, compact, medium-sized animals with a fairly well developed dewlap, a soft pliable skin, strong hooves and short hairy coat. Adult bulls weigh 540 kg and cows 340 kg on an average. The withers height of cows averages 116 cm. The usual coat colours are red, red-and-white, and black, with a dark skin.

The formation of the horns varies greatly (Joshi et al., 1957). On their variability Leplae (1926) has based his classification of the Nyoka cattle into three sub-types:—
1) Polled Nyoka, of small body size, fairly good dairy qualities, and easily fattened. This type is rare. 2) Nyoka cattle with horns of medium length and well developed

504. Nyoka (Alur) bull

humps. These represent the most common type. The body is superior in conformation to that of the polled type, and larger in size; but the milk potentialities of this type are generally poor, save for the cows of Chief Blukwa's herd. The coat is red in colour, black or variegated. 3) Long-horned Nyoka cattle, of large size, standing 130–140 cm at the withers, with a long narrow body, deep chest, well developed dewlap, and small hump. The coat is white or of a reddish colour, or red-and-white. The milking qualities of this type are negligible.

Leplae's classification has been criticised by Curson and Thornton (1936):— "Classification of cattle types in the Congo has been based largely on the absence or presence of horns and in the latter case, whether they are long or short. While the horn length is an important feature, it seems preferable also to consider the general conformation, especially the shape of the skull and the position and structure of the hump. As an example we may mention the cattle of Kivu, which vary not only to the extent of possessing polled as well as horned representatives, but also among the latter in having members with short, medium and long horns. Yet if we group these animals according to conformation they are seen to belong to the Sanga type, there being a uniformity in all respects except the horns."

The Nganda Cattle of Uganda

East of Lake Albert, in Buganda and the Central Province of Uganda, that is the region between Lakes Albert, Kyoga and Victoria, interbreeding of (Ankole) sanga with East African zebu cattle, mainly of the smaller Bukedi type, has resulted in the Nganda breed which is similar to the Nyoka. The Nganda is bred also in an enclave, surrounded by tsetse-fly, to the south of Lake Albert and east of the Semliki river.

In the gene frequency of the B type of adult bovine haemoglobin, the Nganda, with a frequency of 0.23, approximates more closely to the Ankole sanga (frequency 0.20) than to the Uganda zebu (frequency 0.32) or the North Sudan zebu (frequency 0.47) (Lehmann and Rollinson, 1958). There is considerable variation both between herds and between individuals in a herd.

The head is well proportioned, of good width at the muzzle and between the eyes, and straight or slightly concave in profile, not convex as commonly in the zebu. The horns, which are round in cross-section and white in colour with dark tips, vary greatly in length, thickness and direction. They are often thick at the base and fairly long, but not so long nor so thick as in some giant-horned sanga cattle; a maximum length of 79 cm has been recorded. They generally grow upwards and slightly outwards. Polled Nganda cattle are quite common. The neck is moderately long, and the head is carried somewhat low. The hump is small, especially in cows, and like the typical sanga hump, cervico-thoracic in position. The dewlap varies, but is generally only moderately developed; it may be fleshy but has few folds. The umbilical fold and prepuce are not markedly developed. A preponderance of Ankole sanga blood in Nganda cattle results in higher withers and longer legs, whereas a larger share of East African zebu blood produces a lower-set animal. Good specimens are long in body, but deep with good width. The shoulders are somewhat prominent; the back is strong with a level top line. Some Nganda cattle are rather flat of rib, but usually they have well sprung ribs, wide and well fleshed loins and excellent capacity. The sacrum is a little high, but the rump slopes less than that of the pure zebu type. The upper thighs are well developed, and the lower thighs are wide and moderately well carried down. The most frequent coat colour is red, followed by black and brown. Whole colours are commoner than broken patterns, of which red-and-white is the most frequent. The Nganda is used mainly for milk production and work, but also for meat. Adult bulls weigh approximately 400 kg and cows 335 kg. The average milk yield is slightly over 1000 kg per lactation, testing between 6 and 6.5 per cent butter-fat (Faulkner and Epstein, 1957).

505. Nganda cow, Uganda

The Ankole Cattle of Uganda

Pure sanga cattle, known as Ankole, Ankole Longhorn, Bahima, Watusi or Nsagala, are found in Uganda mainly in the south-western part, where they are bred chiefly by the Bahima of Ankole. From Uganda, according to Carlier (1912), these cattle have been introduced into Ruanda–Urundi and the Lake Kivu region of the Congo, where they are kept by the Watusi people on the foothills of the Virunga volcanoes on vast stretches of fertile pasture at an elevation of 1500–1800 m above sea level. South of Lake Kivu they are bred mainly by the Warundi and Wawira, only occasionally by the Wanyabungu. They were introduced by the Bahima, a pastoral people from the north, in or around the 14th century A.D. (Davidson, 1961); but according to Murdock (1959) the Bahima Nilotes infiltrated the territory of the interlacustrine Bantu tribes soon after the close of the 1st millennium A.D., and Payne (1964) was informed by Posnansky that there was evidence that the Bahima settled in western Uganda as early as A.D. 1200. When the Bahima arrived in the neighbourhood of Lake Victoria, one branch proceeded farther south, west of the lake, and then spread eastwards around its southern end and down the eastern border of Tanganyika, where there were pastures and freedom from trypanosomiasis. The Ankole cattle were being pushed further south and west by the advances of Glossina morsitans both from the West African fly belt coming down from the north and by the East African fly belt coming in around the south and west of Lake Victoria. During the last 50–70 years the gap between the two belts closed from over 300 miles to less than 50, with the result that the Ankole cattle survived only in a few isolated herds surrounded by tsetse-fly, and in the names of long-dead kings (Ford, 1960). The tsetse-fly is now being pushed back by extensive spraying and much of the Mbarara area has been recently restocked with cattle (I.L. Mason, 1970—personal communication).

In Tanzania, McCall (1928) reported that Ankole cattle constituted less than one-fourtieth of the herds of the country; they now only number 100 000 out of 7 million head (Hutchison, cited by Mason and Maule, 1960). They are found in a greater or lesser degree of purity in the north-western part, that is in the Bukoba and Biharamulo districts, contiguous with their habitats in Uganda and Rwanda-Burundi. A few herds

506. Ankole Sanga bull and cow

are found on the eastern side of the Tabora belt, but here they are in contact with and tend to be swamped by the more virile zebu which can adapt itself to almost every kind of vegetation, whereas the Ankole is essentially an animal of mountain grassland (Hornby, quoted by Curson and Thornton, 1936).

The Ankole are among the most typical sanga cattle, and generally free from the thoracic-humped zebu influence conspicuous in some of the sanga cattle of Ethiopia and the Sudan and still more in the Nyoka and similar breeds. Ankole cattle are medium to large in size, and tall at the withers, but not particularly heavy. Adult bulls average about 145 cm in withers height and 500 kg in weight, and cows 118–130 cm in height and 300–400 kg in weight. In an aged Ankole bull, McCall (1928) recorded a shoulder height of only 130 cm and a live weight of 385 kg.

The head of the Ankole is moderately long, with fair width throughout its length and with a wide muzzle. In profile the face is straight or slightly dished below the eyes. The horns are of very large size; they project from a thick pedestal-like base outwards and upwards and may then make a regular curve which would form a perfect oval if the tips met. They may also be wide-spread, with the tips curving forwards or backwards. Their colour is white or cream with a reddish hue. Hornlessness is seen among herds of this breed, and manifests itself in the occasional appearance (about 2 per cent) of polled animals; very seldom is a compromise met with; the animals are either beasts with abnormally long thick horns, or are completely devoid of even a vestige (McCall, 1928). "Polled animals form a very small minority of the Ankole cattle and are not liked... The horns are in every way an important feature of this breed, and individual animals are known by the shape of the horns. It is possible to divide the Ankole breed into local races which vary in size of horn: one, the Ishesha, has horns resembling and no larger than the Africander's but other races have enormous horns which, typically, grow outwards, upwards and backwards" (Curson and Thornton, 1936).

The Ankole has a small hump; in the female it is often hardly noticeable. It is cervico-thoracic in situation, resembling the Africander hump. The dewlap is moderately developed, free from fleshiness and with few folds. The prepuce and umbilical

507. Head and horns of Ankole cow (after Lydekker)

508. Giant-horned Ankole bull and polled cow (after McCall)

folds are also not well developed. The udder is very small. The body shows fairly good depth and a strong topline in the middle of which there may be a slight depression. However, many animals are flat-sided and narrow. The rump is of fair length, with a moderate slope. The thighs are poorly developed and tend to be narrow at the gaskins. The skeleton is thin and fine of bone. The milk production of the cows is, as in many other pure sanga breeds, negligible, Ankole cows yielding only about 300 kg milk per lactation, though higher yields (up to 1000 kg) have been recorded in Uganda as well as in the Congo (Faulkner and Epstein, 1957).

The Ankole is usually dark red, but animals of every ordinary colour common among cattle may be found, including red, black, white, grey, brown, yellow, and dun. Nevertheless a whole dark red is preferred or dark red with small spots or with large white splashes. Whole white animals are fairly often encountered. Lydekker (1912a) has quoted the Ankole as an example of sex-dimorphism in colour; but this is erroneous, as not all Ankole bulls are white nor all females red.

McCall (1928) has pointed out that the resemblance between the Ankole and the Africander cattle is striking in certain respects. "In shape the body shows many affinities whilst the unusual height and length of leg accentuate the resemblance. The wide spread of the horns is seen in both breeds, although in the Ankole these are exaggerated and have a greater tendency to curve. The general cylindrical conformation, restricted hump, tendency to whole colours in which dark red predominates still further accentuate the similarity."

Cranial Characteristics of the Ankole Cattle

The skull of the Ankole is characterised by the tremendous horn necks to which the powerfully developed horn cores are attached. The forehead is generally flat, and distinguished by the triangular projection of the parietals into the Torus frontalis. However, the shape of the forehead and the extent of the parietal lap depend entirely on the direction of the gigantic horns (Epstein, 1958). Duerst (1899) distinguished between two different types of Ankole (Bahima) cattle according to the shape of the frontals and Torus frontalis. In the one type the intercornual line is concave, while the frontals bulge in the centre; in the second type the frontal ridge is straight, and the frontals are flat. From Duerst's own subsequent studies on the horns of the Cavicornia it is evident that the above differences in the shape of the Torus frontalis and frontal bones cannot be explained on racial grounds; they are due to individual differences in the shape and direction of the horns, i.e. their moment of rotation. The supraorbital fossae in the Ankole skull are very deep above the orbits, but shallow near the lacrimals where they converge considerably. The upper edge of the broad lacrimals is nearly straight. There is an aperture at the junction of the frontals, lacrimals and nasals. The nasal branches of the premaxillae are very short and do not reach the lateral edges of the nasals. The latter are broad at the posterior end, narrowing orally and ending in two short tips. The lateral edges are slightly compressed, a feature common to most sanga cattle. The prominent malar point sends a ridge to the third premolar, and a rough line to the edge of the jugal. The palate is vaulted, and the lower edge of the jugals sharp. The orbital processes are narrow. The orbits are oblique and rise only slightly above the temporal ridge. The temporal fossae are deep and narrow, and strongly compressed aborally by the large stalk-like horn necks. The Torus frontalis rises high above the squama. The latter forms an angle of nearly 90° with the forehead. The formation of the occiput resembles that in humpless longhorn cattle (Adametz, 1894). The posterior mandibular ramus is broad and nearly vertical, and the anterior markedly oblique. The horn cores are of large dimensions. Duerst (1899) measured a diameter of 12 cm and a basal girth of 36.5 cm. They are almost completely round, relatively smooth and slender, and in spite of their great size very light, being practically hollow. They are deeply furrowed and pearled at the bases. The horns, light-coloured with dark tips, are 100–125 cm long, with a diameter of 13–15 cm and a basal girth of 40–50 cm. The horn substance is rough and fibrous.

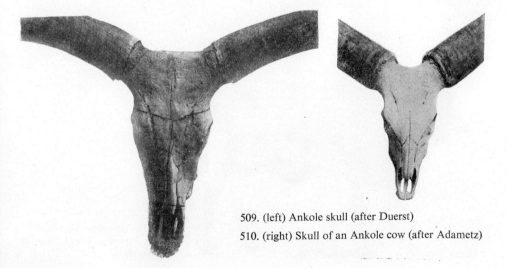

509. (left) Ankole skull (after Duerst)
510. (right) Skull of an Ankole cow (after Adametz)

The Kigezi Cattle of Uganda

In the Kigezi district, in the extreme south-western corner of Uganda, a slightly smaller, more refined, and paler-coloured type of cattle than the typical Ankole is encountered. Bulls weigh approximately 360 kg; cows weigh 320 kg and stand close to 120 cm at the withers. The horns of the Kigezi cattle are finer, of smaller basal girth, also usually shorter and more upright than those of the typical Ankole cattle of the Bahima. Ford (1964—personal communication) believes that the Kigezi is derived from a mixture of Ankole sanga and thoracic-humped zebu. It nearly resembles the medium-horned Bashi cattle of Kivu (see pp. 432–433) (Mason and Maule, 1960).

The Sukuma Cattle of Tanzania

In Sukuma, the region of Tanzania due south of Lake Victoria, the zebu cattle show the influence of Ankole sanga blood in their relatively small humps and fairly large upturned horns which reach about one-third the length of Ankole horns. The predominant colours are red, light dun-and-red, and grey roan (Mason and Maule, 1960).

The Ankole Cattle of Rwanda-Burundi

The sanga cattle of Rwanda and Burundi and the northern shore of Lake Kivu represent a local sub-type of the Ankole, bred by the Watusi (Batutsi). The sacred cattle (Inyambo) of the king of Rwanda are of this type. The horns of the Watusi cattle are of gigantic dimensions, ranging from 70 to 110 cm in length; occasionally the horns are so huge, as in the Inyambo strain where they reach a span of 230 cm between the tips, that the animals are seriously incommoded. Horn shape differs; most common are

511. Ankole-zebu crossbred cow, Tabora, Central Province, Tanzania

512. Long-horned Kivu sanga ox (after Carlier)

513. Cranium of giant-horned Kivu ox (after Leplae)

U-shaped and lyre-shaped horns, less common horns which project outwards, backwards or forwards. Occasionally the horns grow downwards, or are devoid of cores. Owing to the tremendous development of the horn necks, the head of the giant-horned cattle sharply narrows orally. The bone of the cranium and the entire skeleton is weak, and muscular development poor. The animals are not as large as the Bahima, and almost always in a very poor condition except on the rich pastures of the volcanic district. The neck is short, the chest flat, the back weak and the rump drooping. The hump is of negligible size, and generally cervico-thoracic. The legs are weak. The usual coat colours are brown, fawn, red and black, often variegated with white. The Warundi and Wawira tribes value these cattle highly because of the large size of the horns, neglecting all other purposes of their stock; but the Wanyabungu do not hold them in favour because they are hard to fatten and the cows are generally poor milkers.

In Rwanda and Burundi some wealthy tribal chiefs keep giant-horned sanga herds of a uniform coloration—red, white or variegated (Kroll, 1929). This custom, which is known also among several Bantu tribes of southern Africa, is traceable to Ethiopia where it may have originated.

The Bashi Cattle of the Kivu Region

In Kivu district two basic sanga types are recognised, namely the giant-horned Rwanda in the north and south, and the Bashi in the centre of the livestock area (Hendrickx, 1953).

The Bashi is named after an agricultural Bantu tribe that arrived in its present habitat west and south-west of Lake Kivu in the 17th century. Bashi cattle are smaller than

514. Watusi bull from Rwanda-Burundi

515. Watusi cow from Rwanda-Burundi

the Ankole and have a finer skeleton. The head varies in shape, depending on the presence or absence of horns, their size and direction. The latter is commonly upward with a slight twist, occasionally lateral; some animals are polled or have pendulous horns devoid of cores. The ears are 18–20 cm long. The hump varies in size but is generally small; it may be cervico-thoracic or thoracic in position. The dewlap and umbilical fold are well developed. The body is short and compact, the chest fairly wide and deep, the belly capacious, the sacrum slightly higher than the withers, and the rump is wide with a moderate slope. The legs and joints are strong, the udder and teats fairly well formed, and the skin is fine and pliable. The coat is fawn, red or black, frequently variegated with white (Carlier, 1912). Bashi cattle are rather slow-maturing and light in weight. Adult bulls weigh 290 kg and cows 240 kg on an average; the withers height averages 118.5 cm in bulls and 116 cm in cows. The cows are poor milkers, many having hardly enough to feed their calves adequately (Kevers, 1952).

516. (left) Polled Kivu cow (after Carlier)
517. (right) Horned Kivu cow (after Carlier)

518. Long-horned Ruzizi Sanga bull (after Tondeur)
519. Long-horned Ruzizi Sanga cow (after Tondeur)

The Ruzizi Cattle

Pure sangas, known as Ruzizi cattle, are found in the Ruzizi valley as well as in the mountainous region west of the Ruzizi river, between Lake Kivu and Lake Tanganyika. They are regarded as a local variety of the Ruanda sanga. Ruzizi cattle are distinguished by long sickle- or lyre-shaped horns and a very small cervico-thoracic hump. In weight, conformation and milk production the Ruzizi excels the sanga and intermediate cattle in other parts of the Kivu region. Under good management bulls exceed 500 kg in live weight (Tondeur, 1955). The coat is brown, often lightened to a bay shade, while red, red-and-white and black-and-white animals also occur (Joshi et al., 1957).

Former Occurrence of Giant-horned Sanga Cattle at Ujiji, Tanzania

In 1858, Burton (1961) encountered cattle with unusually large horns at Ujiji (Ugoi), on the eastern shore of Lake Tanganyika, where giant-horned sanga cattle are now no longer found; their stature combined with the smallness of the hump rendered them rather like English than Indian or African cattle. They were rarely sold by their owners, except for an enormous price. These cattle were said to have been derived originally from Karagwe, between Ruanda on the west and Lake Victoria on the east; in Karagwe Ankole cattle with small humps and large horns are bred to this day.

The Path of Sanga Diffusion to the South

The Bantu of the middle Zambesi region are separated from the closest cattle-raising people to the north by a gap, approximately 300 miles in width, which is inhabited by agricultural peoples of Bantu speech, who lack cattle owing to the wide prevalence of tsetse-fly in the area between the Rufiji and Zambesi rivers (Curven and Hatt, 1953). But in the middle of this gap, namely in the Fort Jameson region, large herds of cattle are kept by the Mpezeni branch of the Ngoni (Murdock, 1959); indeed, these (Angoni) cattle extend from Lake Nyasa to Lake Tanganyika.

The region lying between the southern end of Lake Tanganyika and the northern end of Lake Nyasa is believed to be the bridge through which sanga cattle were diffused from the north. For if the assumption that the native cattle of southern Africa have come from the north is sound, a path of diffusion must be found. It is likely that the gap was crossed at its narrowest point in Zambia between the present habitats of the Iwa and Tumbuka tribes in the north-east and the Ila–Tonga cluster of peoples on the middle Zambesi river (Herskovits, 1926; Murdock, 1959).

The Setswana (Bechuana) Group of Sanga Cattle

South-east of Lake Tanganyika the sanga line is interrupted by the intrusion of the Angoni and Nyasa zebus. However, save for the Angoni zebu's restricted range in the north-eastern border area of Zambia, this territory constitutes an important section of

520. Approximate distribution of Tsetse-fly in Africa (after Leeson)

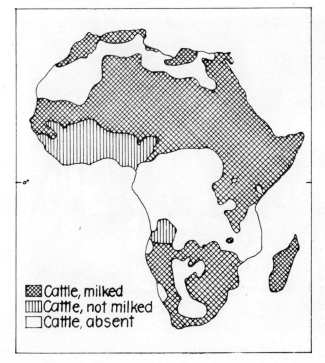

521. Distribution of cattle and milking in Africa (after Murdock)

one of the principal breeding grounds of the sanga, formerly known in this area as Setswana, Bechuana or Sechuana cattle. In the same way that the term East African zebu denotes a large group of basically similar cattle breeds in East Africa, the term Setswana has been used to a group of similar sanga breeds which occupy the immense arid grassland region in the middle of the northern part of South Africa and the southern part of Central Africa. Botswana, including the greater part of the Kalahari desert, is the key region in the location and distribution of Setswana cattle. The Tswana group of the south-eastern Bantu advanced westwards and occupied the region now known as Botswana after A.D. 1720.

The Setswana group of cattle includes several sub-types or varieties which have received different names in the different territories lying within this area:—

Barotse	– Zambia
Ngwato (Mangwato), Amabowe, Tuli	– Rhodesia (Matabeleland)
Bamangwato	– Botswana
Batawana, Ngami (Ngami Lake area)	– Botswana
Makalanga (Francistown–Plumtree area)	– Botswana
Damara (Rakops–Latkone area)	– Botswana and South-West Africa

522. Barotse oxen, Zambia (after Faulkner and Epstein)

The Barotse Cattle of Zambia and Angola

The Barotse cattle of the Barotse (Lozi) tribe, inhabiting the western part of Zambia, bordering the Zambesi river and its tributaries and extending into Angola as far west as 20° E. latitude, are large in size and frame, and rather coarse-boned. Under good management adult bulls reach a weight of over 600 kg and cows 450 kg. The withers height of 3-year-old bulls averages 137 cm and of cows 129 cm (Walker, quoted by Faulkner and Epstein, 1957). The head is long and moderately broad, with little variation in width down its length; in profile it is straight. In former times Barotse cattle of the West or Plains type had much larger horns than are now common, those of the East or Forest type smaller horns, 60–70 cm long. In the north-west, towards the Angola border, Barotse cattle still tend to have longer horns (Walker, 1957). Livingstone (1857) wrote of the Barotse cattle north of the Chobe river as follows:— "They stand high on their legs, often nearly six feet at the withers, and they have large horns. Those of one of a similar breed that we brought from the Lake measured from tip to tip eight and a half feet." The usual horn direction is lateral and slightly backward, then forward and upward, the last 10–15 cm of the horn frequently curving back to form a lyre shape with wide-spread tips. "Horns that hang down and swing or that are otherwise distorted excite high admiration" among the pastoralists of Northern Rhodesia (Zambia) (Smith and Dale, 1920).

While the withers of the Barotse are sometimes prominent, the muscular cervico-thoracic hump is either very small or not apparent. Curson (1934a), refuting a statement in the Annual Report of the Department of Animal Health, Northern Rhodesia (1930), in which the old (prior to 1915) Barotse type is described as "moderately tall, big-horned and non-humped", says:—"It is certain that the description non-humped is an error, for all pure indigenous cattle in the Sub-continent are humped, the hump in

523. Barotse cattle, Zambia

many cases being exceedingly small." The dewlap of the Barotse is only moderately well developed, not pendulous, and free from folds. The prepuce and umbilical fold are also little developed. The body tends to be long though of fair width, with the top line sloping down between withers and rump. A high sacrum with a marked slope between hook bones and pin bones is also very characteristic of the breed. Although fairly wide at the hook bones, the Barotse is narrow at the pin bones. The thighs are narrow and lack depth and fullness. The legs are long or of medium length, enhancing the rangy appearance of the breed. In former times the most common coat colour was red; now black and brown are more common, while dun and fawn are also occasionally seen. Chief Lewanika of the Barotse people maintained a herd of pure white cattle with black points. These were all slaughtered when Lewanika died in the year 1916 (Curson and Thornton, 1936). The skin is loose and of medium thickness with dark pigmentation (Joshi et al., 1957). Barotse cattle are docile and make good working animals; under their home conditions they have proved of great value for beef production; the cows are milked.

The Setswana Cattle of Botswana

In Botswana all sanga cattle are basically of Setswana type. Several regional sub-types have been distinguished, such as the Batawana, Bamangwato, Damara, Sengologa, Seshaga. Owing to the movements of pastoralists within this large region during the past century, several names by which different sub-types of cattle were formerly known have been superseded by new names, in some instances resulting in a somewhat confusing terminology. Cross-breeding with Africander bulls has been widespread even prior to the Anglo–Boer War, and with many European breeds in addition to the Africander since then.

In northern Botswana it is still possible to distinguish between a Western and an Eastern type, roughly corresponding to the Western or Plains type and the Eastern or Forest type of Barotse cattle in Zambia (see p. 437). The Western type of Botswana cattle is found in Ngamiland and the Western Bamangwato Reserve and approximately corresponds with the Batawana; it is large and long-horned, and has a pyramidal hump. The Eastern type is smaller and has a rounded hump (Mason and Maule, 1960).

524. Skull and horns of a Barotse ox

525. Setswana cattle, Botswana, (left) western-type cow, (right) eastern-type cattle

526. Ngami ox (after Lydekker)

The Batawana or Ngami Cattle of Botswana

The Batawana or Ngami cattle are found north of Lake Ngami in northern Botswana. Formerly these cattle were very large and heavy-boned. But already half a century ago Lydekker (1912a) wrote of the old Ngami breed that it was on the verge of extinction, adding that the development of tremendous horns, especially in old oxen, was characteristic of the region. In a Ngami ox skull in the British Museum the horns measure 4 feet 8 inches (142 cm) along the curve, with a span of 8 feet 5 inches (256.5 cm). Darwin (1868) mentioned the horns of a Ngami ox with a tip-to-tip distance of 8 feet $8^1/_4$ inches (265 cm) and a total span along the curve of 13 feet 5 inches (409 cm). In a Batawana beast from Lake Ngami the length on the outside curve of the horns measured $81^1/_2$ and $78^1/_2$ inches (207 and 199.5 cm) respectively, the basal circumference $18^1/_4$ inches (46.3 cm), and the tip-to-tip distance $103^1/_2$ inches (263 cm). The greatest basal circumference of the horn of a Ngami ox recorded appears to be $19^1/_2$ inches (49.5 cm) (Curson, 1934a). The horns of these Ngami cattle were round in cross-section and tended to be horizontal, projecting from the skull in a moderate backward sweep. The moment of rotation resulting from this peculiar direction of the heavy horns gave rise to a prominent median longitudinal ridge along the frontal suture, from which the frontals sloped away caudally, giving the skull a unique roof-like appearance.

527. Batawana oxen and bull (after Curson and Thornton)

528. Batawana oxen, Tsotsoroga Pan

J. H. R. Bisschop (personal communication to I. L. Mason) holds that Batawana cattle with lateral horns are Africander grades. But in view of the representation of giant-horned cattle with the same type of laterally projecting horns in rock drawings in the Horn of Africa (Fig. 241), the general applicability of this assertion appears to be doubtful.

In more recent times by far the more common type of cattle in Ngamiland, in addition to a few Damara cattle introduced at the beginning of this century (see pp. 449 to 451), is the large-horned Tswana which originally came from the Bamangwato Reserve at the beginning of the 19th century, when Chief Tawana seceded from the main tribe. Through grazing for generations on the luxuriant pastures of Lake Ngami and the Okavango Delta, the type of stock has improved considerably, so much so that for slaughter purposes the Batawana or Ngami ox has no equal among stock of native origin. The cattle are well built, perhaps long in the limb, but nevertheless well supplied with muscle. The horns, circular in cross-section at the base, are of a large size, but the remarkable skulls still encountered at the end of the 19th century are now rarely seen (Curson, 1934a).

The Bamangwato Cattle of Botswana

The Bamangwato tribe of the Suto–Tswana group of Southern Bantu migrated with its herds from the southern part of Botswana (Mochudi) to the north-eastern part. Their cattle were formerly called Makalanga after the Shona tribe of that name. But they were quite distinct from the recent Makalanga (Mashona) cattle of Mashonaland, for they were of pure Setswana stock, very large in size, heavy-boned, and furnished with large horns, although not quite as heavy nor as thick at the base as Barotse horns. The breeding area of the old Mangwato type included the south-west corner of (Southern) Rhodesia (see p. 454), and some of these cattle were taken by the Matabele on their way north to what is now Matabeleland, Rhodesia (see pp. 460–461) (Faulkner and Epstein, 1957).

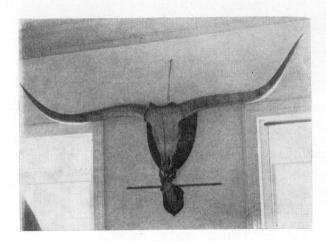

529. Skull and horns of a Tswana ox, Mafeking Club. Tip-to-tip distance $102^{3}/_{4}$ inches (261 cm); basal right 17 inches (43 cm)

530. Skull and horns of a Tswana ox (after Antonius)

531. Batawana bulls

532. Batawana cow (after Curson)

533. Sengologa oxen

The Sanga Cattle of the Kalahari Desert

Two distinct, but now very rare, Setswana sub-types, namely the Sengologa and the Seshaga, are restricted to the centre of the Kalahari desert. The Sengologa cattle, named after the Bangologa people of the vicinity of Lehututu, are tall well-fleshed animals with gigantic horns, well developed dewlaps and muscular cervico-thoracic humps. The Seshaga cattle of the Bashaga people from the Kang area (east of Lehututu) are leggier and lighter-built than the Sengologa, with thinner and longer, commonly laterally projecting horns (Mason and Maule, 1960). A century ago, Chapman (1868) described the Bakalahari cattle as "immensely tall and lanky ... rather higher than an Africander ox, with immense horns."

Short-horned Tswana Cattle

As among other sanga breeds ordinarily long-horned or giant-horned, individual animals that are either short-horned or polled are occasionally encountered among the Batawana and other Setswana sub-types of Botswana. Nobbs (1927) mentioned a large black polled type of Ngami cattle that occurred among the herds of Lobengula. Curson (1934a) observed a few black polled cattle in Ngamiland in 1930–31, but they were not more noticeable than elsewhere in the Setswana cattle region. While polled animals may appear spontaneously in herds of long-horned cattle, the occasional absence of horns or the presence of short horns in Botswana cattle, as the absence of a hump, "may also be attributed to the persistent introduction of pure-bred European bulls by Khama and other Chiefs" (Curson, 1934a).

The Sekgatla Cattle of Botswana

In the southern part of Botswana, the original Tswana has been superseded by a smaller, more compact and shorter-limbed type of cattle, evolved upon the introduction of considerable numbers of Africander cattle since 1888. The horns are comparatively slender and usually lateral in direction, often curving forwards or upwards. The Sekgatla cattle of the Bakgatla Reserve, also called 'Dikgomo tsa Borwa' (= cattle of the south) represent this type (Fig. 536) (Mason and Maule, 1960).

534. Seshaga ox

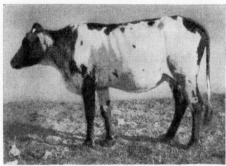

535. Short-horned Tswana cow (after Antonius)

536. Sekgatla cow ('Dikgomo tsa Borwa')

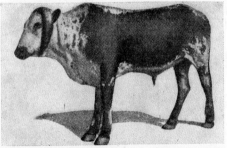

537. Cervico-thoracic-humped sanga ox with loose horns from Benguella, Angola (after Da Costa)

The Distribution of Sanga Cattle in Angola and South West Africa

On the west coast sanga cattle are found in Angola and South West Africa.

In Angola two different types are encountered: the Barotse (Setswana) in the east, more especially the south-east (east of Bié), and the Ovambo in the south-west. In the north of Angola there are few cattle (of Setswana stock) on account of tsetse. They have been reported as far north as Loango. Everywhere in Angola the influence of imported breeds on the original native types is much in evidence.

In South West Africa there were two types of sanga cattle prior to the European occupation: the Ambo (Ovambo) in the north, and the Damara or Herero (the Damara and Herero are two clans of the same Bantu tribe) in the south (Schinz, 1891). A few Ovambo cattle have survived in the north of South West Africa. South of their habitat, in the central districts of South West Africa (Damaraland), lies the breeding area of the Damara cattle, originally brought hither by the Ovaherero. Damara cattle are found also scattered throughout the arid area around Lake Ngami in Botswana, where some of the Herero from South West Africa sought refuge during the German–Herero campaign of 1904–06. The Damara of Damaraland was the southernmost sanga breed on the west coast of the Continent.

The Setswana Cattle of Angola and Western Congo

The Setswana cattle of Angola are distinguished by large horns, long coarse legs with large hooves, a long neck and conspicuous vertebral spines. But they are rarely as giant-horned as the Barotse of Zambia or the Batawana cattle of Botswana. Their horns are

538. Setswana cow, southern Angola

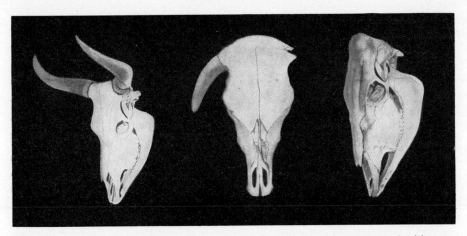

539. Sanga skulls from Benguella, Angola (after Da Costa); (left) horned; (centre) with one firm and one loose horn; (right) polled

more variable in size and shape, and may be loose or missing altogether. Loose pendulous horns occasionally grow very long, so that they hinder the animals in grazing. Great variability is also encountered in the size of the hump which may be large in bulls, but hardly noticeable in some of the cows. It is always cervico-thoracic in situation. The withers height of these cattle reaches approximately 120 cm. They are strong of bone, but poorly muscled. The most common colour pattern is tan or brown and white with the coloured areas extending over the sides of the body and legs. Whole herds are found in which this pattern is general (Da Costa, 1933; Sousa Dias, 1954).

The cattle imported from Angola into the Lower Congo and Kasai, in the most western part of the Congo, are described as large and coarse of bone, with long horns; they mature late and the cows are poor milkers, but they are hardy and disease-resistant (Leplae, 1926).

Cranial Characteristics of Angola Cattle of Setswana Type

Crania of sanga cattle of Setswana type, brought north into Loango, are distinguished by the flat frontal region typical of long-horned or giant-horned Setswana skulls. There is only a very slight interorbital depression. The orbits are completely lateral in situation and do not rise above the plane of the forehead. The supraorbital fossae are shallow. The Torus frontalis is only slightly raised and has a wide but shallow depression in the centre. The parietals extend deeply into the forehead in the shape of a triangular lap. The frontal edge of the lacrimals forms a straight line. There is a triangular aperture at the junction of the frontals, nasals and lacrimals. The nasals are broad and flat, each ending in two long tips. The nasal branches of the premaxillae do not reach the lateral edges of the nasals. The malar point is not prominent, and the palate is only slightly vaulted. The jugals are broad, with a narrow masseteric surface. The temporal fossae are deep and narrow in front, broad and shallow behind. The occipital crest, rising high above the squamosals, is markedly furrowed and very broad. The frontals extend deeply furrowed stalk-like processes to the horn cores the bases of which are surrounded by pearled horn wreaths rising above the horn necks. The cores are smooth. The horns are of a very scaly material, and slightly curved in the form of a sickle; in one of the skulls examined by Duerst (1899) they are yellow with black tips, in another skull black throughout.

The Ambo Cattle of Angola and South West Africa

The Ambo cattle, bred by the Ovambo of south-west Angola and the northern half of South West Africa (Ovamboland, Kaokoveld), are generally small in size but well proportioned; in the Kaokoveld they tend to be larger, in Ovamboland smaller and lighter. They are shorter in the leg than Setswana cattle and sturdier in appearance. The height at the withers is approximately 123 cm in the adult bull and 115 cm in the cow. The head is of moderate size and length, with a wide, nearly flat forehead and rather narrow muzzle. The supraorbital arches are fairly heavy but do not hang over the eyes nearly as much as in the Africander cattle, for example. The ears are of moderate size, thin in texture, somewhat pointed but with an open auricle, and carried horizontally. The long massive horns are lyre-shaped and practically circular in cross-section with a basal diameter of about 10 cm. The neck is of moderate length and fair depth; its upper border is straight from poll to hump. The latter is situated in front of the withers; it measures approximately 30 cm in length, 20 cm in width and 10 cm in depth, and weighs about 3 kg; in the cow it is considerably smaller than in the bull. The dewlap is heavy and folded, thick of skin, somewhat filled and of saccular appear-

540. Sanga skull from Loango, Congo (after Duerst)

541. Cattle in the Kaokoveld, northern South West Africa

542. Ambo bull

ance. The naval has a loose skin fold; the prepuce of the bull is prominent. The withers are of fair width and rather loose, and the shoulders are slightly heavy. The chest is small, of medium width but of fairly good depth. The body is of good length and moderate depth and capacity. The top line is narrow and roofy, lacking in musculature; in the bull it is fairly straight, but in the cow it appears weaker owing to the downward curvature of the chine and the high sacrum. The rump is narrow, sloping from hook to pin bones as well as towards the tail head which is set on rather low. The thighs are small and narrow. The gaskins are long, narrow and poorly muscled. The legs are of moderate length, showing good hard bone and dry tendons. The hoofs, however, have been described as soft owing to the stoneless country where the animals graze (Schlettwein, 1914). The tail is long and slender. The udder consists of a small tight meaty bag with small teats. The common coat colour is a uniform light or dark dun (game

CATTLE

543. Ambo bull, front and caudal views (after Bisschop and Curson)

544. Ambo cow in the Kaokoveld, northern South West Africa

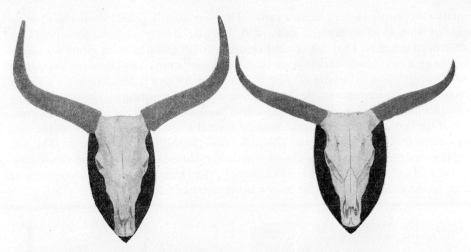

545. Skulls and horns of an Ambo bull (left) and cow (right) (after Groenewald and Curson)

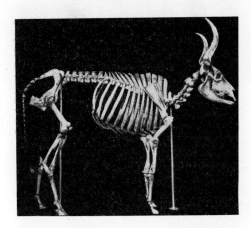

546. Skeleton of an Ambo cow (withers height 120 cm) (after Groenewald and Curson)

colour), taking on a slightly darker tint on the head and over the shoulders and withers. White areas are occasionally seen on the underline. The skin is dark in colour, and blue-black spots show up through the hairy coat which is short and laid smoothly close to the body (Groenewald and Curson, 1933; Bisschop and Curson, 1935).

The Damara Cattle of South West Africa

The Damara or Herero cattle of South West Africa were originally a Setswana subtype. A century ago, Andersson (1856) described them as "big-boned, but not particularly weighty; their legs are slender, and they have small, hard and durable feet. The hair on the back is short, smooth and glossy, and the extremity of the tail is adorned with a tuft of long, bushy hair, nearly touching the ground... The horns are the most remarkable feature in the Damara cattle. They are usually placed on the head at an angle of from 45 to 90 degrees, and are at times beautifully arched and twisted, but rarely bend inwards. They are of an incredible length and one often meets with oxen the tips of whose horns are from seven to eight feet apart. The Damaras prize their oxen in proportion to the size of their horns. Some African tribes take much pains in forming them to a certain shape. This is effected either by sawing off the tips, splitting them, bending them forcibly when yet tender, and so forth." That artificial deformation of the horns of oxen was practised by several South African Bantu peoples has been confirmed by Le Vaillant (1790), Barrow (1801–04), Lichtenstein (1811–12), Livingstone (1857), Fritsch (1872) and others. In addition to horn size and shape, coat colour used to be an important aim in selection. The Damara possessed whole herds of cattle of the same colour, bright brown being popular (Curson, 1934a).

547. (left) Damara bull (after Adametz)
548. (right) Damara ox (tip-to-tip distance of horns 160 cm) (after Schultze)

549. Damara cow and calves

Remnants of these cattle still survive, as the Herero pay careful attention to the selection of purebred stock (C. Schlettwein, Otjitambi—personal communication dated 28.1.1931). But in general the cattle of the Damara, owing to the introduction of bulls of European breeds, now show signs of better breeding, judged by modern standards. They are less coarse, and their horns and dewlaps are smaller than formerly. They are larger in size than the Ambo cattle, and more zebu-like in appearance. The head is long and narrow, with a large broad muzzle; the length of the forehead exceeds its width; it exceeds also the length of the facial part of the head. The orbits are flat, the eyes small, and the nasals narrow and convex in profile. The horns are long, moderately thick at the base with thin pointed tips, sickle- or lyre-shaped, often widespread. Although Damara oxen with horns up to 90 cm long are still encountered, the average horn length is now considerably less. The neck of the Damara is short and strong, the dewlap little developed. The cervico-thoracic hump is of moderate size; in cows it is often rudimentary. The chest is deep and fairly broad, with a good spring of ribs. The back is strong and straight, the sacrum prominent, and the rump markedly sloping. The thighs are well muscled, resembling those of a draught horse. The legs are long, slender and strong of bone, the hooves small and very hard. The tail setting is low, the tail itself long and thin.

The Nama Cattle of Great Namaqualand

South of Damaraland, in the northern part of Great Namaqualand, South West Africa, a crossbred type, known as Nama or Bastard cattle, was common until the beginning of the present century. It came into being as a result of the northerly migration of the Rehoboth Bastards from the Cape Colony (Hermann, 1902). But the devastations of rinderpest, 1899–1900, and later of the native war, 1904–1906, says Dr. G. Schmid, Government Veterinary Officer, Omaruru, in a report to the Director of Veterinary Services, sadly depleted the herds of native cattle in South West Africa. Rohrbach states: "After the war, 1906, the whole native stock was exterminated with the exception of about 3000 cattle, which were distributed to the farmers. Also nearly all the cattle in possession of the old farmers were lost. Only the Rehoboth Bastards saved about 600 female cattle. Consequently the herds of the farmers and newcomers had to be built up eventually almost entirely from imported cattle from the Cape and Europe" (Groenewald and Curson, 1933).

550. Nama cow (after Schlettwein)

The Nama or Bastard cattle were not a purebred sanga breed, although they retained a small hump cervico-thoracic in situation. They had been evolved by crossing Ambo, Damara and Hottentot cattle with European breeds, chiefly Friesian cattle imported from the Cape. Nama cattle were larger than the average Damara and Ambo beasts, of a better beef conformation, and the cows were superior milkers. A Nama ox could also pull a heavier weight than a Damara ox; but owing to its rather soft hooves it could be employed only on good roads (Schlettwein, 1914).

The Sanga Cattle of the Eastern Half of Southern Africa

While the Setswana group of cattle includes a number of sub-types which were all formerly of the giant-horned type, the cattle in the eastern part of South Africa, namely those of the Shona and Zulu-Xosa peoples, are merely long-horned, apparently because their owners did not pay special attention to and select for very large horns.

The Tonga Cattle of Zambia (Northern Rhodesia)

In Zambia a long-horned sanga breed, i.e. the Tonga, is encountered in addition to the giant-horned Barotse sanga of the Setswana group. The Tonga cattle belong to the Tonga and related tribes in the southern and central provinces of Zambia, extending from the Zambesi valley north as far as Broken Hill, east to the Sunemfwa river, and west as far as the Manyeke river.

In the 19th century the Baila and Tonga cattle were virtually identical, and prior to 1950 Baila, Ila, Tonga, Mashuk and Bashukolumbwe were synonymous terms used for the cattle of all the Ila–Tonga peoples of the Southern Province of Northern Rhodesia. Since then the cattle of the Ila have been so extensively crossed with the Barotse that they have markedly increased in size and become so similar to the Barotse, and different from the Tonga, that they are considered a Barotse sub-type with traces of Tonga blood. The name Baila is now restricted to the cattle of the Ila, Sala and Lenje tribes, living in the Kafue flood plain between Namwala and Mumbwe and extending to the Lukanga swamps (Mason and Maule, 1960).

Since 1905 Tonga as well as Barotse cattle have been imported from Northern Rhodesia into Katanga–Lomami in the southernmost part of the Congo (Leplae, 1926).

Tonga cattle are somewhat smaller in size than the Barotse. The weight of adult bulls averages 500 kg and of cows 330 kg. The average withers height of 3-year-old bulls is 127 cm and of cows at the same age 122 cm. Tonga cattle are used mainly for work, occasionally for milk. Only old and decrepit animals are slaughtered. The head is of moderate length and well proportioned, with a straight or slightly convex profile and a moderately concave forehead due to the somewhat prominent orbital arches. The horns, in contrast with those of the Barotse, are only 30–35 cm long, fine and dense of structure, and round in cross-section; sometimes they are attached to large necks. Commonly their direction is outward, upward and slightly forward, more rarely the tips curve backwards so that the horns form a lyre shape; polled or loose-horned Tonga cattle are also encountered. The neck is short, the hump small, muscular, and

551. Tonga cows, Zambia (after Faulkner and Epstein)

cervico-thoracic in position; in cows it is sometimes not apparent. The very small size and distinctly cervico-thoracic situation of the Tonga hump refute the theory, mentioned but not supported by Curson and Thornton (1936), that the Baila (Tonga) is a crossbred type derived from cervico-thoracic-humped Bechuana (Barotse) sanga and thoracic-humped Angoni zebu cattle. The dewlap, prepuce and umbilical fold of the Tonga are little developed. The body is long but of good depth and capacity, although often narrow behind the prominent shoulders and at the pin bones. The forequarters tend to be light. The sacrum is high, and there is a slight depression in the top line between the withers and hook bones; but the slope from hook to pin bones is not marked. The thighs lack fullness and are somewhat narrow at the gaskins. The limbs are relatively long and fine of bone, with hard durable hoofs. The most common coat colours are whole black and whole red or patterns of black or red with white. Light brown or dun in whole colours are also seen. The skin is loose and darkly pigmented (Faulkner and Epstein, 1957).

Dispersal of the Sanga Cattle South of the Zambesi

"As the Zambesi river, constituting a formidable barrier to migration especially during the summer months, flows just south of the southern extremity of Nyasaland, it was probably in this vicinity that the Bantu hordes halted and subsequently dispersed in all directions, but chiefly south-east to west in a fanlike fashion. Obviously reliable information will never be obtained regarding the details of those prehistoric folk-wanderings, but judging from the present distribution of cattle, the tribes accompanied by the cattle known as Ovambo to-day must have migrated westwards... The territory they occupied comprises now Angola and South-West Africa Protectorate. Then following the Ovambo herds were the nomads possessing the Bechuana cattle which are to be found to-day in no less than six countries, viz.: Angola, Northern and Southern Rhodesia, South-West Africa Protectorate, Bechuanaland Protectorate and Transvaal. Proceeding due south were the owners of what are now the Makalanga and Zulu sub-types, the former keeping to the plateau of Southern Rhodesia while the latter preferred the warmer climate of the coast. It is, of course, also likely that the original owners migrated in the reverse order to that just suggested. So many factors, e.g. distribution of Glossina, stock diseases, wars, further migrations, and climatic conditions, are concerned in this problem that it is doubtful whether the veil of obscurity will ever be raised from the events of the past millennium" (Curson and Thornton, 1936).

The Sanga Types of (Southern) Rhodesia

In (Southern) Rhodesia three different sanga types were formerly recognised (Nobbs, 1927):—1) Ngwato (Mangwato) or Amabowe, of the Setswana group, found particularly in the south-western part of Matabeleland, (Southern) Rhodesia, in the south and west of Gwanda district, in the southern end of Bulalima–Mangwi, and across the border to the south of Tuli, that is in the most easterly corner of Botswana and the adjacent border area of the Northern Transvaal. 2) Kalanga or Mashona, bred in (Southern) Rhodesia in an area bordered on the north by the Zambesi valley, on the east by the (Southern) Rhodesia–Mozambique border, on the south by the Limpopo valley and Devuli river, and on the west by the Shangani river. 3) Matabele, owned by the Matabele tribe of (Southern) Rhodesia. At the present time the three recognised types are the Tuli, Mashona and Manguni.

The Ngwato or Amabowe (Tuli) Cattle of (Southern) Rhodesia

The name Amabowe was formerly applied to cattle of Ngwato Setswana type in the south-west corner of (Southern) Rhodesia, from Tuli to Plumtree. The breeding area of these cattle is contiguous with the main Bamangwato territory across the Botswana border, north-west of the Limpopo river to Mochudi (see p. 441). Owing to the disastrous consequences of the 1896 rinderpest, which destroyed approximately 95 per cent of African-owned herds in Southern Rhodesia, and the indiscriminate crossing of the remnants with Africander cattle and European breeds, the original Amabowe type has become very rare.

Typically Ngwato cattle are of the large heavy-boned sanga type, furnished with long wide-spreading horns. They stand on long strong legs, and make heavy powerful trek-oxen of great endurance. Maturity is only reached at eight or nine years. Red, red-and-white and golden brown are the most frequent colours encountered (Nobbs, 1927).

552. Ngwato sanga ox, (Southern) Rhodesia (after Faulkner and Epstein)

553. Ngwato sanga ox, (Southern) Rhodesia (after Faulkner and Epstein)

A few years ago a herd of Amabowe cattle was collected in the Tuli area, and these cattle are now called Tuli and have become a flourishing breed. They are selected for golden-brown colour and absence of horns. The live weight of adult bulls averages approximately 800 kg, of cows 500 kg and above (Mason and Maule, 1960).

554. Tuli cow

555. Horned Mashona bull (after Faulkner and Epstein)

556. Mashona cow (unimproved)

557. Mashona heifer

The Mashona Cattle

The Mashona (Makalanga, Kalanga or Makaranga) are the cattle of the Shona people of Mashonaland, the eastern half of (Southern) Rhodesia. Their breeding area extends eastwards over the Mozambique border into a small tsetse-free area south-west of Tete, and includes also several parts of Matabeleland, particularly the Matopo Hills.

In the early decades of this century the term Makalanga was used for cattle scattered in small numbers through the north-west Sibasa area of the Northern Transvaal. These are now called Sibasa. They are of mixed Mashona–Nguni type, in some areas pure Nguni (Mason and Maule, 1960).

While the Makalanga (Ngwato) cattle owned by the Bamangwato tribe of Botswana were large and heavy-boned, the Makalanga (Mashona) cattle encountered in Rhodesia at the present day are of quite a different type. They are very similar to the Tonga cattle of Zambia. At the turn of the 19th century two pandemics, namely, rinderpest in 1896–98 and East Coast fever in 1900–06, decimated the cattle of Southern Rhodesia, and this has considerably influenced the racial composition of the Mashona breed. In order to replenish the African-owned herds in Mashonaland, large numbers of cows, mainly of the Angoni breed, were imported from across the Zambesi. The Angoni is larger than the Mashona and thoracic-humped. It is however not a pure zebu breed but includes a sanga strain which is still discernible in the occasional occurrence of fairly long horns (see p. 391). The relatively small effect of the Angoni cows on the size and conformation of the Mashona is attributed to the use of the original Mashona bulls on the imported females (Oliver, 1965).

Mashona cattle, though small and fine of bone, are generally well proportioned, strong and sturdy. The live weight ranges from 275 to 350 kg. The breed is used mainly for meat and draught.

The head is shapely, wide in the upper part, but relatively narrow and long from the eyes downwards, widening again at the muzzle; seen from the front, it has a coffin-

558. Polled Mashona bull and heifers

559. Mashona cattle at Bawe, Mozambique

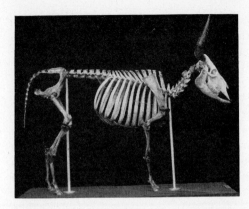
560. Skeleton of a Mashona cow

shaped outline. The eye is large and prominent, and very different from the small, sunken, protected eyes common in zebu cattle. The orbital arches are also prominent. The forehead is slightly dished, but the facial part from just below the eyes to the muzzle is convex in profile. In proportion to the size of the animal, the horns are long and heavy, their length varying between 45 and 60 cm. Until the early years of this century, the true Mashona had still longer and much heavier horns than the recent type. In cows horn direction is generally outward, upward and backward with the pointed tips turning forwards. In the bull the horns are thicker at the base, shorter, coarser, stronger and less curved than in the female. The cross-section of the horn is circular. Polled animals are quite common. They used to be called 'Izuma' and were highly valued by the Mashona who thought that polled cattle were better able to maintain their condition in the dry winter and fatten more readily than horned animals (Nobbs, 1927). The neck of the Mashona is of moderate length and good depth. Its upper border is horizontal from the poll to where it passes on to the hump. The latter, cervico-thoracic in situation and placed well in front of the withers, is of moderate size in the male and only slightly developed in the female, muscular and well attached. The dewlap is thin and loose but devoid of folds. The prepuce and umbilical fold are little developed. The top line shows neat withers, a broad and well muscled back with a characteristic depression a little way back from the withers, and exceptionally good loins, wide, flat and well muscled and passing over evenly into the rump. The sacrum is nearly always slightly higher than the top of the withers. The rump is of fair length with the sacrum running back horizontally in line with the loins and ending in a rather heavy and high tail root. On either side of the sacrum the rump falls away badly. The hook bones are of good width, fairly prominent, and level with the loins and sacrum. The thurls are rather narrow and low-placed, the pin bones very narrow and set on low. The thighs show fair length dorsally, but become very narrow over the long gaskins. The hooves are small and very hard. The tail starts from the coarse, high set on tail root, is rather heavy, and ends below the hocks in a large brush which almost reaches the ground. The udder of the cow is placed far forward on the belly and shows very little udder tissue, small teats and practically no milk veins. The coat is short, with straight, close and glossy hair. Mashona cattle may be of any colour except roan. Most common is black, followed by red; brown, dun, yellow, cream and brindle also occur, and red-and-white and black-and-white in various combinations. Black-and-tan merging into each other is a very frequent and characteristic pattern. In general the division of two colours on a coat is not clear-cut but mottled and uneven (Bisschop and Curson, 1933; Faulkner and Epstein, 1957; Oliver, 1965).

The Cattle of Matabeleland, Rhodesia

The cattle of the people known as Matabele originate from a blend of many stocks and are a mixture of many breeds the enumeration of which reflects the history of the Matabele nation. Having broken away from the Zulu in Natal in 1817 during Chaka's regime, the Matabele settled first in the Transvaal, but later, owing to the arrival of the Boers, moved farther north into that portion of Southern Rhodesia now known as Matabeleland, overpowering the indigenous Mashona tribes (Seligman, 1930). On their way to Southern Rhodesia they acquired cattle in the Eastern Transvaal, southern Botswana and the northern parts of Botswana. Nobbs (1927) has pointed out that the Matabele cattle "are of a very mixed derivation, so recently collected and amalgamated that the original types are known and distinguished, and they possess no common characteristics distinguishing them as a breed". The following have contributed to the make-up of the Matabele cattle:—1) Zansi cattle which came with the Matabele from the south and may have been of Zulu stock. (From descendants of these the Rhodesian Department of Native Affairs has established selected herds of colour-sided type, formerly called Inkone and now Mangoni—Fig. 561); 2) Boer cattle captured by the Matabele on their way north; 3) Ngwato (Amabowe) and Ngami (Batawana) cattle captured from the Bamangwato and Batawana tribes in Botswana; 4) Mashona cattle; 5) Barotse cattle; 6) Govuvu (Kavuvu or Kwavovu) cattle (Figs. 562 and 563), derived from Northern Rhodesian Tonga cattle which had been raided from the Mashukolumbwe tribe across the Zambesi river; 7) Africander cattle and various European breeds recently introduced into the Matabele herds.

Govuvu cattle are found scattered in small numbers over the whole of (Southern)

561. Manguni (Matabele) cow, Inkone type, Rhodesia

562. Matabele bull, Govuvu type, Rhodesia   563. Binga (Govuvu) cow

Rhodesia. In north-west Rhodesia, adjacent to the Zambesi river, they are locally known as Binga cattle; these have been described as dwarfed (Payne, 1964). While of small size, the Govuvu cattle are distinguished by a compact conformation and hardy constitution. The bulls have a moderately developed cervico-thoracic hump, but in cows the hump is very small and sometimes barely noticeable. The coat is usually either black or red, occasionally variegated with some white on the brisket, belly and thighs. The cows are tame and comparatively good milkers. A herd of Govuvu cattle was maintained by the Native Affairs Department of Rhodesia at Tjolotjo (Mason and Maule, 1960).

The Nguni Cattle

On the east coast of South Africa the sanga type is represented by the Swazi and Zulu cattle, collectively called Nguni. Nguni cattle are found wherever the descendants of original groups of the Nguni tribe have settled, that is in the eastern part of Swaziland, the eastern part of Zululand in the Republic of South Africa, and in Mozambique south of the Save river, where they are called Landim. The main area of their distribution is bounded in the north by the Komati river in northern Swaziland, and extends southwards to the Tugela river in Natal, including Zululand proper. The eastern boundary is formed by the coast and the western boundary by the Drakensberg in Swaziland, the line continuing south, east of Vryheid, through Babanango to Mapumulo. Typical Nguni cattle are also encountered in Pondoland (Brown, 1959). Blood group tests have shown that the Nguni cattle of Zululand and those of Swaziland are genetically identical (Annual Report, 1960-61).

Bryant (1949) considers it to be probable that the Zulu settled on the east coast of Natal, which ranks with the most trying and unhealthy cattle country of South Africa, owing to the arrival of the Europeans. As far back as 1689, shipwrecked sailors described Natal as being 'full of cattle'. Robinson (1872) recorded that "along the coast a hardy breed of small size, known as the Zulu, is found to suit the country best".

In those parts of South Africa where temperature and humidity are high (muggy climate), Nguni cattle and their crosses have proved to be suitable for beef production, whereas in hot and dry regions (scorching climate) Africander cattle and their crosses are preferable. This refers to ranching areas with a high mean annual atmospheric temperature (above 65° F or 18°C), in which British beef breeds would deteriorate even if adequate feed were available (Bonsma, 1951; Bonsma et al., 1953).

The Nguni cattle are of medium size; the withers height of bulls averages 133 cm and of cows 122 cm. Under fair management bulls reach an average live weight of approximately 500 kg and cows 340 kg. But in sourveld and overstocked areas adult cows do not exceed a height at withers of 105 cm and a live weight of 225 kg. The Nguni, although tending in body build to the milk rather than the beef type, is used mainly for meat and work. The average milk yield of Nguni (Landim) cows tested in Mozambique was 572 kg in a lactation period of 139 days; one cow gave 1100 kg and another one nearly 1300 kg in 300 days (De Pinho Morgado, 1954).

The head of the Nguni is well proportioned, broad over the eyes and at the muzzle, and straight or slightly convex in profile, with a moderately dished forehead. The latter is constricted just below the horns. These are of medium length and round in cross-section, projecting horizontally outwards, upwards and slightly forwards, then upwards and backwards, frequently forming a lyre shape in adult cows. In oxen the horns are often twisted around their axes. Polled animals are common. The ears are small and

564. Nguni bull (after Faulkner and Epstein)

565. Nguni (Swazi) cow

566. White Zulu bull with black points

567. White Zulu cow with black points

pointed. The neck is moderately long. In bulls the hump is fairly well developed, cervico-thoracic in position and muscular in structure. In females it is very small or absent. The dewlap, prepuce and umbilical fold are little developed, Nguni cattle being in general rather tight-skinned. The chest is fairly deep, but relatively narrow, especially over the anterior ribs. The shoulder is often short and straight. The body is long but of good depth and capacity, although sometimes lacking in width at the shoulders and pin bones. The topline, partly owing to the height of the sacrum, shows a slight depression behind the withers. The rump has a roofy appearance, with a moderate slope; generally it is a little short in relation to the back. The thighs tend to be narrow and flat. The limbs are of medium length, fine and dense of bone, with hard, durable, pigmented hooves. The tail is long, frequently somewhat coarse, and set on high. The udder and teats are small to moderate, and the milk veins are not prominent. The hide is very tough, fine in texture and pigmented; the coat is short and glossy (Faulkner and Epstein, 1957; Brown, 1959).

Nguni cattle show a great range of colours. White, black, brown, red, dun, and yellow are common; these may be found whole, mixed (black-and-tan or brindle), or in several specific patterns and markings. Whole black cattle play an important part in the ceremonial life of the Swazi and Zulu. White hair in combination with a pigmented skin and coloured hair around the muzzle and eyes and on the inner surface of the ears, occasionally with a few spots near the tail setting, is a very frequent pattern. Cattle of this coloration were called 'Cetewayo's Cattle' or 'Nyoni-ai-pumuli', that is 'The bird does not rest', a name associated with the constant trekking of the Zulu on their migration from the north. At one time animals of this pattern were regarded as royal cattle and treated with great respect.

Cranial Characteristics of the Nguni

The skull of the Nguni is distinguished by an uneven forehead. The frontal ridge slopes upwards to a considerable elevation into which the parietals extend. The Torus frontalis lies about 5 cm above the squamosals and is slightly depressed in the centre whence a clearly defined ridge extends into the forehead. This ridge is flanked on both sides by moderate depressions which become shallower towards the horn cores and make the profile of the female cranium slightly concave. The orbits are heavy and strongly curved, the supraorbital fossae shallow. The stalk-like horn necks are slightly furrowed and of considerable length. The horn cores are porous, but relatively smooth. The nasals are broadest aborally, slightly compressed in the middle, and end in two conspicuous tips. The nasal branch of the premaxilla just reaches the edge of the nasal. The maxillae are broad and long. The malar point is not very prominent. The palate is vaulted. The jugals have a broad orbital surface, but their lower edge is thin and sharp. The temporal fossae are deep and broad, and wide open aborally.

The Bapedi Cattle of Sekhukhuneland

The cattle of the Bapedi tribe of the Nguni, in the Sekhukhuneland area of the Eastern Transvaal, are a sub-type of the Nguni cattle. The predominant colour is black with

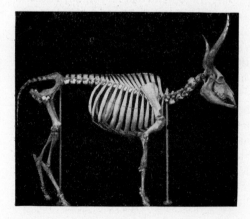

568. Skeleton of a Zulu cow

569. Basuto bull

570. Basuto cow

white on the belly and lower neck border; some animals are blue roan, or white with black points. Adult Bapedi bulls weigh up to 680 kg, cows average 440 kg. The facial profile is slightly concave in bulls and straight in cows. The horns point upwards and slightly forwards in an incomplete lyre or crescent, being shorter and thicker in the bull than in the cow. Polled cattle or animals with scurs or slugs are common. The dewlap is thin but well developed. The cervico-thoracic hump varies considerably in size and shape. The majority of Bapedi cattle are of good length, spring of ribs and capacity. The legs are fairly long, often faulty, the triangular rump is sloping, and the sacrum is sharp and higher than the withers. The udder and teats are small (Mason and Maule, 1960).

30 Epstein I

The Basuto Cattle

South of the breeding area of the Nguni, there are several enclaves of sanga cattle in a sea of nondescript grades. In Lesotho (Basutoland) a high proportion of Basuto cattle are of sanga blood, although the introduction of European breeds over many years has considerably influenced the original type of the Basuto.

Basuto cattle are smaller than the Nguni, and resemble the Mashona of (Southern) Rhodesia. They are used mainly for work, but also for meat and milk although their muscular development is meagre and their milk potentialities are poor. The head is fairly well proportioned, but tends to be long in cows and oxen. The forehead is wide and slightly dished, and the profile straight or convex. The poll is as wide as the forehead and often heavily fringed with hair. The ears are of medium size, carried horizontally, and well fringed. The horns are 45–60 cm long, rather thick, round, oval or flat in cross-section, and first lateral, slightly upward and backward in direction, then curving forwards, upwards and outwards; they retain their basal thickness for the greater part of their length and only become finer and sharper near the tips. A small percentage of Basuto cattle are polled. In bulls the hump is relatively well developed for a sanga breed, but in cows and oxen it is small. Its position is cervico-thoracic, its structure muscular. The dewlap is little developed and free from folds. Similarly the prepuce and umbilical folds show poor development. The body is of fair length and moderate depth and capacity; but it often lacks width behind the shoulders and at the pin bones. The top line is level save for the rather high sacrum. The rump shows a

571. Basuto cattle (after Faulkner and Epstein)

572. Rock paintings of a sanga from western Natal (above) and of Basuto cattle at Wepener, Orange Free State (c. A. D. 1820–1830) (after a drawing by J. Walton)

marked slope from hook bones to pin bones, and the tail head is set on high. The legs are of moderate length, light of bone, but hard and of good quality. Legginess is rare in Basuto cattle. The predominant coat colour is black; other colours such as dun, red, roan, red-and-black in a variety of broken colours and colour patterns, tan-and-white with the pigmented areas located on the sides of the animals, are also found (Bisschop, 1952).

The original sanga cattle of the Basuto (Fig. 572), from about 1830 on, became mixed with stock stolen from Bantu and European farmers in the Orange Free State. In 1896 most of the Basutoland cattle were wiped out by rinderpest, and the country was restocked with black sanga cattle from the Drakensberg, which are now regarded as the Basuto type (Figs. 569–571). The similarity of the Basuto and Drakensberger cattle rests on black colour and horn shape. The Drakensberg cattle, together with the remnants of the original Basuto breed and subsequently introduced Africander and Friesian cattle, constitute the present cattle population of Lesotho (personal communication by Mr. James Walton to Mr. I. L. Mason).

Drakensberger Cattle

Drakensberg cattle are bred near the eastern borders of the Orange Free State and Transvaal and in the north-western part of Natal, that is on the escarpment in the districts of Harrismith, Memel, Vrede, Volksrust and Ermelo and also at the foot and on the slopes of the escarpment in the districts of Estcourt, Ladysmith, Newcastle, Utrecht, Vryheid and Paulpietersburg. These high and lower-lying districts are closely connected by farming activities; in autumn many of the highveld farmers still trek with their herds to the warmer, lower-lying districts, where they remain until the spring.

The Drakensberger breed traces its origin to three foundation stocks, namely Uys cattle, Kemp cattle and Tintern Blacks (Van Rensburg et al., 1947).

The Uys cattle were named after one of the Voortrekkers, Swart Dirk Uys, who crossed a black Friesian (Vaërlandsche Friesch) bull with a white spot on the forehead, white hind-legs and white switch with two red heifers of Africander type. In the Umkomaas Valley, Uys probably also introduced Zulu and Basuto cows into his herd, the original Basuto stock being responsible for the marked similarity between the Basuto and the Drakensberger, but he never used bulls from outside sources. Subsequently the breeding of Uys cattle was continued by the sons and grandsons of the founder (Reinecke, 1964). These cattle were distinguished by hardiness, large size, a black coat, loose pliable dark skin, sturdy legs, strong hooves and a broad muzzle. Later breeders of Uys cattle repeatedly used Kemp bulls in their herds.

The Kemp cattle were called after the founder who, in 1911, started farming with a herd of good Africander-type cattle, in which first a Friesian bull and subsequently an Africander bull were employed. Later on two black bulls, one of these from the Tintern herd, were introduced into the Kemp herd, but no further foreign blood was

573. Drakensberger cattle

used. In general appearance this large herd (1300 head) was of black Africander type; it differed from the Uys cattle in its superior beef type and greater resemblance to the Africander.

The Tintern Blacks originated from cattle looted from Africans, which were selected for black colour and bred to an Africander-type bull. After many years of exclusive use of black bulls from the same herd, another Africander bull was introduced; thereafter no further blood was brought in from outside.

Drakensberger cattle are large animals with pronounced sexual dimorphism in size. While supplying beef, milk and draught power, meat is the primary product. The head is well proportioned, with a broad muzzle and short horns which project outwards and forwards; the horns are usually white in colour with black tips. The bull has a small cervico-thoracic hump, but the cows are humpless. The body is long, the back straight, the sacrum often prominent, and the rump slightly sloping. The legs are sturdy, with strong hooves. The udder is small but well shaped. The skin is loose, pliable and dark in colour, the coat is black, only occasionally with some white on the underline (Mason and Maule, 1960). In very cold highveld conditions the Drakensberger, similar to the Basuto, develops a woolly coat. In the warmer middleveld region of Natal the woolly hair does not develop to the same extent, and in the lowveld it does not develop at all. The adaptability of the breed to extremely hot, humid conditions is attributed to its sleek, shining black hair covering and relatively large skin area (Reinecke, 1964).

The Bolowana Cattle of Bomvanaland

In Bomvanaland, in the district of Elliotdale, Cape Province, two herds of sacred cattle were maintained until recently, remnants of many similar sacred herds that existed at one time throughout the Transkeian Territories. One herd was kept at the kraal of Tyelinzima, chief of the Ama-Tshezi, the right-hand house of the Bomvana, the other at the seat of the reigning house, in safe keeping of Ngubezulu, the son of Gwebindlala, for his tribe of Ama-Bomvana. The herd at Tyelinzima's was named Ondongolo, and that belonging to Gwebindlala's family was named Bolowana or Zankayi. The origin of these cattle is traceable as far back as the year 1650, when part of the Bomvana tribe left Natal and settled in Pondoland owing to differences over the sacred Izankayi cattle. The Bomvana remained in Pondoland until the early years

574. Bolowana cattle, Transkeian Territories (after Thompson)

575. Xosa cow (after Soga)

of the 19th century, when they were compelled to quit the country, and obtained permission from the Ama-Xosa to occupy the territory between the Bashee and Umtata rivers and along the coast south-east of Tembuland, the country now known as Bomvanaland.

The Bolowana cattle of the Bomvana were inclined to be small, and most of them were polled or had short horns pointing downwards. Hump position and hump development were similar to the Basuto's. The coat colour varied from red to dark brown and dun. Bolowana cattle were employed for ploughing and draught, also for racing purposes. The cows were poor milkers. The principal use of the breed was for sacrifice in times of calamity (Thompson, 1932).

The Ama-Xosa Cattle of Cape Province

Among the Ama-Xosa, located in the region of the Fish, Kei, Keismana and Bashee rivers, in the eastern part of Cape Province, the last remnants of sanga cattle are still found, very much mixed with stock of European derivation. The sanga cattle once kept by the rest of the Bantu tribes of the Union of South Africa have intermingled with cattle of European origin to such a degree that, in the words of Curson (1936), "the time has long passed for describing the indigenous cattle of the Cape, Orange Free State and Southern Transvaal".

The cattle of the Ama-Xosa were distinguished by a flat barrel, long legs, and long horns variously curved; occasionally the horns were devoid of cores or entirely missing. Pendulous horns were often a deciding factor in the selection of breeding stock. Bulls grew a prominent cervico-thoracic hump; in oxen the hump was smaller, and in cows nearly non-existent. Xosa cattle were of a hardy type, suitable for transport work, but poor in meat and milk potentialities. The variability in colour was remarkable; Soga (1931) lists twenty-five Xosa terms for different colours and colour combinations in cattle. Formerly the Ama-Xosa kept a number of sacred herds in connection with religious worship.

Sanga Influence on the Sakalava Cattle of Madagascar

It is likely that the cattle of the Sakalava of Madagascar, who are the most negroid of all Malagasy tribes in physique, have a strain of sanga blood, for C. Keller (1898), who visited the island in 1886, wrote that it was possible to distinguish between two different breeds, namely the cattle of the central and eastern parts, with horns of medium size, and the giant-horned Sakalava cattle of the west. The latter, which had lyre-shaped horns projecting upwards and slightly backwards, were anatomically most closely related to the Ethiopian sanga. Lydekker (1912a) also stressed the sanga connection of the long-horned cattle of the west coast district of Madagascar, pointing out that these presented a marked similarity to the East African breed. Sanga cattle may have been introduced into Madagascar by immigrants of mixed Indonesian, Hamitic and Bantu stock from the Azanian coast or by Bantu elements from the east coast of Africa, around the beginning of the 2nd millennium A.D. Hence the Sakalava cattle may derive from either a Bantu or Hamitic origin, but, as Murdock (1959) has pointed out, the latter seems more probable in view of the Malagasy custom of keeping livestock in pens to conserve their manure, and of the fact that the north-east coastal Bantu themselves obtained cattle from the Hamites.

None of the recent authors on the cattle of Madagascar refers to the presence of a sanga strain in the Sakalava breed; the latter is now generally classed with the zebu.

2. The Africander Cattle of South Africa

a. Distribution and Characteristics of the Africander Cattle

Distribution of the Africander Cattle

The area of distribution of the Africander cattle comprises a large part of the Republic of South Africa. The Africander breed is also found in large numbers in South West Africa, Botswana, Swaziland, Rhodesia, Zambia and Malawi; and in smaller numbers in Tanzania (Tanganyika) and Kenya. In the southern part of the Congo it is now the dominant breed.

Within the Republic of South Africa it is concentrated largely in three distinct parts of the country; the northern ranching districts, the Cape Midlands and eastern districts, and the western parts of the Transvaal and Orange Free State cropping area (Joubert, 1953). The most important breeding centres are in the districts Hoopstad, Kroonstad and Winburg, in the northern and central parts of Orange Free State. Valuable herds are also found in the southern part of Orange Free State, in the east of Cape Province (Midlands), and in the districts of Pretoria, Potchefstroom and Klerksdorp, Transvaal.

Head, Horns and Cranium of the Africander

Owing to the care bestowed on them by many generations of breeders, Africander cattle at the present time show a high degree of uniformity in colour and conformation, rarely encountered in other African breeds of livestock. The head of the Africander is a characteristic feature of the breed, distinguished by great length, more especially of the nasal part. The forehead, broad and full between the eyes and not depressed at the posterior end of the nasals, reaches its highest point immediately behind the orbits. The nasal part is slightly Roman; the muzzle, mouth and nostrils are large, and the eyebrows prominent and moveable. The eyes are big; bold in the bull and alert but placid in the cow. They are protected in oval-shaped eye-sockets which slant slightly downwards. The lower corner of the eye ends in a clear tear duct, free from hair (Opperman, 1952). The drooping eyelids give the eyes the appearance of being partly closed. The receding forehead narrows considerably towards the poll, which is neatly rounded, slightly raised above the horn bases, and frequently covered with a tuft of hair. The peculiar shape of the frontal crest, caused by the downward and backward direction of the horns, and the decidedly convex profile, due to the Roman nose and the receding forehead, are distinctive features of the Africander cattle, shared by many zebu breeds. The jaws of the Africander are relatively deep and massive. The ears are small, pointed and covered with fine hair.

The long spreading horns, hard, smooth, clean and ivory-like, with a clear seam or ridge at the back, leave the head in a downward and backward direction, then, at maturity, bend gracefully forwards, upwards and backwards. Although modern breeders do not emphasise extreme length, great importance is still attached to the placing of the horns, their uniformity in thickness, and their oval shape. In profile

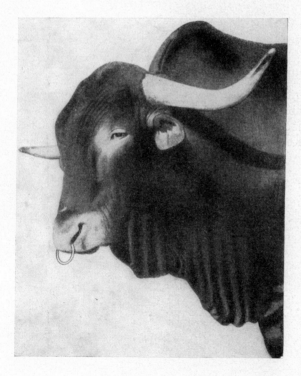

576. Head of an Africander bull

HUMPED CATTLE

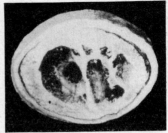

577. (left) Skull of an Africander bull, frontal surface (after Bisschop)

578. (right) Cross-section of Africander (ox) horn; horizontal diameter 93 mm, vertical diameter 65 mm

579. Polled Africander bull, Omatjenne, South West Africa

the horns are usually set well behind the face. The horns of the cow are thinner and placed somewhat higher than those of the bull. In oxen the shape of the horns is less   directed laterally, bending
se.
l necks, are slightly com-
he base, otherwise smooth
ompact and firm.
579) has been developed,
attle are heavier than those
gether with horned cows in
the horned ones.
tals rise above the squama
d shape of the Torus fron-

CORRIGENDUM

On p. 473 of vol. I. Fig. 579 instead of "Polled Africander bull" read "Polled Africander crossbred bull".

talis depend on the direction of the horns. The shallow frontal fossae end, slightly converging, at some distance from the lacrimals. The latter are low and broad; their superior edge is quite straight, and the lower moderately bent. Very occasionally there is a small triangular aperture where the nasals, frontal and lacrimals meet. The comparatively long nasals join the frontals evenly, and show a slight bend orally, so that the whole cranium appears convex in profile. In some specimens the upper branch of the premaxillae reaches the nasals, while in others it ends immediately before these. The palate is nearly flat. The temporal fossae are wide and open, with rounded edges. The jugals are moderately developed. The posterior ramus of the strong mandible ascends steeply, while the horizontal ramus leads straight to the centre whence it begins to rise slightly. The teeth project from the maxillae in an oblique direction. Their grooves are simple, and the enamel is strong.

Epstein (1956) has recorded the following average measurements of the horns of six mature bulls, six cows and six oxen:—

Average Length and Basal Girth of Horns of Africander Bulls, Cows and Oxen (in cm)

Measurement	Bulls	Cows	Oxen
Length of Horn	54.9	53.6	61.7
Basal Girth	25.7	20.5	25.5

The Neck and Dewlap of the Africander

The neck of the Africander is short and deep but not round, with extraordinarily well developed muscles, essential for the pulling of heavy loads in the neck yoke. The skin of the neck has few folds or none. The neck slopes down steeply from the high frontal crest and occiput, gradually rising again towards the hump. It is full and evenly joined to the shoulders. In a normal pose the top line of the neck is lower than the back line, so that the neck appears to be attached to the middle of the front quarter. This constitutes an important difference from European cattle in which the top line of the neck is generally on the same level as the backline (Opperman, 1952).

The dewlap is well developed. It starts underneath the chin, is slightly tied in at the throat, and continues along the underline to far back on the chest. The skin is very loose, the tights-kinned Africander being disfavoured by breeders. The loose skin of the dewlap, resembling a curtain with characteristic folds and waves, gives the neck the very deep appearance. In bulls the dewlap is even more developed than in cows. Bisschop gives the following description:—"The dewlap commences from the chin as two separate folds which converge a few inches further back. In the region of the throat the dewlap shows an indentation, but from this point backwards it hangs evenly and conspicuously to well between the front legs. In the region of the brisket it may be rather pendulous and so create the impression that the thorax is deeper than is actually the case. The dewlap is never 'filled' but consists of two directly apposed layers of skin. Vertical folds of the dewlap, such as sometimes seen in the short-horned zebu, are considered undesirable" (Curson, 1935).

580. Skull and horns of an Africander ox, frontal surface (after Curson)

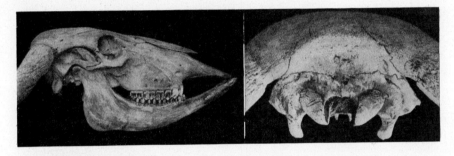

581–582. Skull of an Africander ox, lateral surface (left) and nuchal surface (right) (after Curson and Epstein)

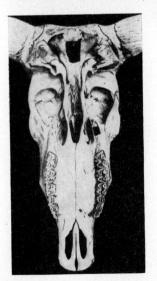

583. Skull of an Africander ox, palatine view (after Curson)

The Hump of the Africander

The hump rests on the border between the neck and withers (see also pp. 331–336). Its formation is devoid of a bone foundation. The official description by the Africander Cattle Breeders' Society is as follows:—"Large, prominent and set closely to withers." The hump begins from approximately the middle of the neck and rises fairly steeply in a parabola-shaped bridge till it reaches the highest point immediately above the shoulder points. Then it falls fairly precipitately for a short distance till it joins the rising chine. The hump has a firm, solid, muscular appearance; it is broad and flat in front, and rises in a slightly slanting manner from the sides (Opperman, 1952). It may reach 10 kg in weight (Faulkner and Epstein, 1957).

The Body and Legs of the Africander

The withers of the Africander are moderately broad, very well covered, and slightly exceed the poll of the head, the back and the hips in height. The shoulders are deep, strong, muscular, and smoothly attached to the withers and ribs. The chest is deep, wide, and well filled behind the elbows; the brisket is deep, broad and full, but not excessively prominent.

There is a distinct contrast in the conformation of the body between the old draught type and the recent beef type of Africander (Joshi et al., 1957). The former is rather flat of rib, while the latter has well sprung ribs so that the chest is considerably rounder in cross-section than in the older type. Again, in the draught type the neural spines of the thoracic and dorsal vertebrae are long and conspicuous, producing a rather narrow roofy back which ascends forwards to the hump and caudally to the prominent sacrum between which the top line is hollow. In the modern beef type the top line is wide, full and nearly straight, and the back broad, strong and well covered. There is

584. Africander cow

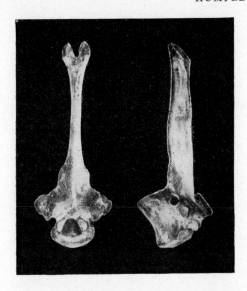

585. Front and lateral views of ninth dorsal vertebra of Africander cow (after Curson)

only a slight depression approximately in the middle, caused by the height of the hump, the descending line of the withers and the prominent muscular development over the loins. The back begins comparatively broadly behind the hump, widening gradually until it ends in the broad, strong, full and bulging loins.

The superior spines of the dorsal vertebrae from the sixth vertebra caudally are bifid, the upper portion being divided medially, and also compressed antero-posteriorly.

The crops are filled, but the flanks and second thighs are inclined to be light; however, this defect is rapidly disappearing in well-bred herds. The hips are wide and well covered, but not prominent. On the whole, the body does not show extreme depth, although the loose skin at the navel creates the impression that it is deeper than is actually the case.

The slope of the rump from hook to pin bones, and the high smooth setting of the tail head are outstanding features which the Africander shares with the zebu breeds. As in the case of heavy draught horses, the drooping rump is of advantage to animals pulling heavy loads. In some Africander cattle the hook bones tend to be too wide and high-set, and this, combined with the tendency of the pin bones to be excessively close together, results in a triangular rump and cow-hocked hindlegs (Joshi et al., 1957). The thighs are full but rounded at the buttocks rather than well let down to the hocks. The twist is deep, full and plump, and of good length from tail head to scrotum or udder. The buttocks are broad, round and well carried down to the second thighs. There is a characteristic notch in front of the tail root. The tail should be smoothly attached, neither coarse nor prominent. Generally it is moderately thick at the root and tapers fairly long and thin down to the hocks, ending in a long, bushy, clearly defined switch. The tail joints are clearly visible.

The legs are comparatively long in the old trek type, but shorter in the modern beef pe; they are lean, slender and straight. The hocks are straight with flat but light bonety and prominent ham strings. The knees are broad and flat, and their joints moderately long. The bone is fine, but the substance dense and hard. The lower parts of the legs never show coarseness or sponginess of bone. The hooves are moderately large, round and deep. The toes are even, well closed and pointing straight forwards.

The Male and Female Organs of the Africander

The male organs are well developed with moderately big testicles of even size which hang level in a thick, loose scrotum which contracts with ease. The skin of the scrotum is 4 mm thick. At high atmospheric temperatures the scrotum of the Africander bull is drawn in; the skin becomes puckered and is a poor conductor of heat. The testicles are retracted against the perivisceral cavity so that their temperature does not rise above the body temperature (Bonsma et al., 1940). The prepuce is very prominent. The udder of the cows is usually small, with soft and moderately big teats. The milk signs are negligible. Bonsma (1951) measured the average milk yields of 10 summer-calving and 10 winter-calving cows; the former was 875 kg in 23 weeks and the latter 1180 kg in 26 weeks.

Body Measurements and Weights

Epstein (1956) has recorded the following average measurements of bulls and cows in show condition, and oxen in working condition:—

Measurements of Africander Bulls, Cows and Oxen

	Height at withers (cm)	Body length[1] (cm)	Percent of withers height (cm)	Girth of chest (cm)
Bulls	149.4	176.5	118.1	235.0
Cows	140.3	154.3	110.0	214.2
Oxen	152.7	173.3	113.5	–

[1] Measured from pinbone to point of shoulder

The live weight of adult bulls averages 745 kg and of adult cows 525 kg. Bulls in show condition may exceed 1100 kg in weight, and cows in show condition 800 kg. The Standard of Excellence of 1957 gives 817 kg as the desirable weight of mature bulls and 590 kg as that of cows.

Six-year-old oxen in good veld condition weigh between 450 and 650 kg. Four 4-tooth steers in show condition averaged 522 kg live weight and 331 kg dressed weight, i.e. a dressing percentage of 64. Eight 22–23-month-old steers from a commercial farm averaged 368 kg live weight and 217 kg carcass weight, with a dressing percentage of 58 (Opperman, 1950).

Type

The general impression conveyed by the old draught type of Africander is of an animal of the typical 'desert type'. Africander cattle bred on modern beef lines show a marked departure from the extreme respiratory type, though their forequarters are still too heavy in relation to the hindquarters.

Coat Colour in Africander Cattle

The colour of the Africander cattle ranges from the darkest to the lightest red. Until a few decades ago, yellow varying in shade from a dark honey-coloured yellow to light cream was the prominent colour in a number of purebred herds. The skin in all cases has a brownish amber pigmentation.

The origin of the yellow colour has been variously interpreted by breeders. The theory, says Reinecke (1933), that it indicates the admixture of foreign blood in the yellow strain of Africander cattle through former cross-breeding with Swiss cattle, and that on that account the dark muzzle and lead-coloured 'points' are more common among yellow Africander cattle, is quite untenable. In point of fact, yellow Africander cattle have usually flesh-coloured muzzles, and not dark but flesh, amber or yellowish brown points. "The progeny of any admixture of Africander and Swiss blood are generally fawn-coloured and not yellow. Such 'throwbacks' usually have dark lead-coloured muzzles, with dark hoofs and dark-tipped round-shaped horns, whereas the true yellow-coloured Africander has a flesh-coloured muzzle with light yellow-coloured hair round the muzzle and eyes, with amber-coloured hoofs and light-coloured horns. This is substantiated by the fact that the red progeny of a mating between a red and yellow Africander, or one with yellow-coloured individuals further back in the pedigree, frequently show a light-coloured band of hair round the muzzle and eyes. In any case the yellow Africander was in existence long before the introduction of Brown Swiss cattle into South Africa. The latter breed was introduced only after the Boer War in 1907" (Bonsma, 1938).

It appears that the creamy or yellow colour does not owe its origin to crossing with a Swiss breed, but rather to the introduction of a modifying diluting factor by one parent, a theory upheld by experienced breeders of Africander cattle. These maintain that if yellow is bred to yellow, there is a tendency for the colour to become lighter in shade, until almost white progeny are produced (Reinecke, 1933). Professor F.N. Bonsma, who has made a genetical study of the light-coloured Africander, considers that there are six shades in the Africander, namely, dark red, blood red, light red, golden yellow, light yellow, and cream (personal communication—December 22nd, 1936). "The yellow colour in the Africander seems to behave in a recessive manner. The differences in colour intensity found with both red and yellow coloured Africanders, however, suggest that probably various modifying factors of a multiple factor nature are involved in the colour inheritance of the Africander" (Bonsma, 1938).

Some white patches or speckles on the underline of otherwise red animals are allowed for herd book registration purposes; silver colour in the switch and scattered white hairs on the body are undesirable but permissible. However, excessive white above

586. White Africander Heifer

587. Africander calves, red-spotted and red with white underline (after Bonsma)

the underline, on the front part of the dewlap, the groins, in the switch or on the hind feet, is objected to; and although such animals sporadically occur even in the best registered herds, they are not liked because breeders fear that such 'points of origin of depigmentation' may crop up in the progeny on other parts of the body. Whole red colour is usually dominant to red broken with white. Before 1912, when red was fixed as the standard colour of the breed, entire herds were found in which a large proportion of the cattle had their red colour broken by snow-white markings. These white markings were usually confined to the underline, particularly the belly, dewlap and flanks, sometimes spreading to the switch and hindlegs. In some cases the white extended to uniform markings along the top as well as the underline, although this pattern was not so frequent. The roan, that is to say, a more or less regular intermixture of red and white hairs, was practically unknown in purebred Africander herds.

Black in any form—black hair, black muzzle, hooves or teats, black pigmentation of the skin, or black-tipped horns—is strongly deprecated and disqualifies an animal from registration in the stud book proper or in the appendix. Such markings are regarded as a sign of impurity and carelessness in breeding. It should, however, be mentioned that some old breeders remembered one or two herds in which the predominant colour was black.

The Colour of the Horns and Hooves of the Africander

The colour of the horns in the Africander is a sound criterion of breed purity. In young bulls the horn sheaths are of an almost transparent clearness, and the background has a reddish gleam. The horns of adult animals are creamy white in colour, with amber, brown or wax-coloured tips. The hooves are slightly darker than the horns, varying from dark amber to brown.

Characteristics of Skin and Hair in the Africander

"The Africander possesses a loose, pliable, tough skin, appreciably thicker and not as elastic as that of European breeds. In texture it is also different; it is more dense and the deeper layer of the skin is more oily. The hair is short, comparatively thick, smooth and glossy, and lies close against the skin" (Bonsma, 1938). The skin of the thick-skinned areas of the Africander is 6 mm thick, and of the thin-skinned areas 3.6 mm. The tough skin and smooth glossy coat are tick-repellent properties, and have a bearing also on the amount of heat penetrating the skin (Bonsma, 1944). The sensitivity of the pilo-motor nervous system, and the paniculus muscles with which the animal moves its skin to free it of ticks and flies, are strongly developed (Joubert, 1953); in the carcase the panicular muscles are seen as flesh-coloured stripes on the outside of the ribs. The hair has an average diameter of 50–53 μ. The winter coat of an adult Africander weighs 129 g, the summer coat only 29 g, i.e. 22.5 per cent of the winter coat (Bonsma and Pretorius, 1943).

The average weight of the wet hide in adult Africander cows (live weight 500 kg) is 38 kg, the mean surface area 5.3 square metres; in young oxen (live weight 470 kg) 42 kg and 5.4 square metres, and in old oxen (live weight 650 kg) 50 kg and 6.2 square metres respectively (Bisschop, quoted by Joshi et al., 1957).

The Bonsmara

Africander cows served as foundation stock in the development of the Bonsmara breed of cattle, begun by J.C. Bonsma at Mara Research Station, Republic of South Africa, in 1936. The Bonsmara is based on 5/8 Africander and 3/16 each Shorthorn and Hereford blood. The breed is distinguished by a good beef conformation and strong constitution. In its adaptation to high temperature, tick-borne diseases, drought and poor nutrition it resembles the Africander; in size, rate of growth, milk yield, fertility and viability it is superior to the parent stocks. The Bonsmara has a red coat and horns of medium size; the male has a small cervico-thoracic hump. Adult bulls weigh between 800 and 1000 kg and cows about 550 kg (Mason and Maule, 1960).

b. The Ancestors of the Africander Cattle

Descent of the Africander Breed from Hottentot Cattle

The Africander breed can be traced to the cattle which the early European discoverers and colonists of South Africa found in the possession of the Cape Hottentots. Kolb (1719), who visited the Cape in 1705 and spent there a period of eight years, recorded:—"As the Hottentots know, the cattle of the Europeans originate and are descended from the Hottentots' own stock." The cattle of the Bantu peoples of South Africa apparently did not contribute to the origin of the Africander, at any rate not to a significant extent.

General Conformation of the Hottentot Cattle

Hottentot cattle were gaunt bony creatures (Hoernlé, 1923). In the log-book of the first voyage of Vasco da Gama, the cattle of the Hottentots encountered at Aguada de São Brás (Mossel Bay) in 1497 are described as "very big like the Alemtejo cattle, very fat, reddish of colour and very wild" (Martinho, 1955). Jan van Herwaerdens in his 'Reizen in Zuid-Afrika in die Hollandsche Tyd' (1652–1686) has given the following account of the cattle of the Hottentot chief Oedasoa:—"There were numerous cattle which were much bigger than those in the homeland where we had never seen cattle of such size; their loins and rumps measured $2^1/_2$ feet across and, in addition, they stood so tall that one could scarcely see over them or rest one's elbow on their backs for the height of the back was as high as a man could reach with his extended elbow." Kolb (1719) described the Hottentot cattle as bigger and stronger than contemporary European breeds, a Hottentot ox weighing about 225–275 kg. They were characterised by a long head, narrow in the nasal part, but fairly wide in the forehead. The body was moderately deep and broad, and the legs were strong, sinewy and well placed. The tail was long and thin, ending in a full switch.

In 1778 the famous naturalist Col. R. Gordon painted a bull, which he called 'Namaqua' bull (Fig. 588), a 'Cape' cow and calf (Fig. 589) and a 'Cape' ox, belonging to a farmer at Camdeboo; these three water colours are now in the Amsterdam Rijksmuseum. Gordon also recorded the body measurements of each animal: height (following the body curves) in the bull 162 cm, cow 132 cm, ox 170 cm; length of horn along the curve in the bull 60 cm, cow 59 cm, ox 90 cm (Jones, 1955).

The Horns in Hottentot Cattle

In the 'Namaqua' bull, painted by Gordon, the horns are long, upright and lyre-shaped, quite similar to those of the 'Cape' cow and ox; but in other illustrations of Hottentot cattle, as in Samuel Daniell's sketch (Fig. 590), the horns are much finer and mainly lateral in direction with the slight twist that is characteristic also of Africander horns. This horn shape may have led Stow (1905) to the conclusion that the Namaqua used to train the horns of their oxen artificially, confining their shape to a spiral similar to that in the koodoo. But Stow's observation may be based on an entirely different

588. 'Namaqua' bull, owned by a Cape farmer in 1778 (reproduced from a painting by Col. R. Gordon in the Rijksmuseum of Amsterdam)

589. Cape cow and calf, owned by a Cape farmer in 1778 (reproduced from a painting by Col. R. Gordon in the Rijksmuseum of Amsterdam)

phenomenon: Sparrman (1789) recorded that in the Zuurevelden the cattle sometimes fell into the habit of gnawing one another's horns when shut up in their kraals at night, which accounted for the carved appearance of the horns, "a circumstance which ought therefore by no means to be ascribed, as it has been, solely to the ingenuity and manual operations of the herdsmen."

The Hump in Hottentot Cattle

Not all Hottentot cattle seem to have been furnished with well marked humps. "The majority of authors," Kolb (1719) noted, "who have described the cows and oxen of the Hottentots, have stated that these are distinguished by large humps. But I can assure my readers that, although I have seen the herds of the Dutch settlers and of many Hottentots, I have never come across a humped beast. The above statement is therefore either untrue, or it may be that the yoke pressing upon the neck of the ox is the cause of the withers appearing raised to a hump. Yet it is certain that by nature their cattle are not humped though they are larger and stronger than European breeds."

590. Korah Hottentots preparing to move (after Samuel Daniell). (The ox may not be typical, as Daniell's sketches were considerably altered in the process of engraving)

It is difficult to accept this statement. Kolb himself recorded that most other writers described the Hottentot cattle as humped. The 'Namaqua' bull, in Gordon's painting, is humped, the Korah Hottentot ox, depicted by Daniell, humpless. The lack of humps Kolb observed in Hottentot oxen may have been due partly to the fact that the western Hottentot tribes lived in one of the driest and poorest parts of the Sub-continent. In all humped cattle the size of the hump is influenced by nutrition. In zoological gardens it is frequently observed that zebu calves born there develop large humps, whereas in their imported parents the hump is hardly noticeable. In poorly fed Africander cattle the hump is barely marked. Yet the hump of the Africander is one of the principal characteristics of the breed; and it is unlikely that this feature would have been developed by the Cape settlers in so short a time or at all, had the genetic factor not been present in the original Hottentot stock. Kolb seems to have mistakenly believed that the cervico-thoracic hump of Hottentot cattle was produced by the pressure of the neck and shoulder yoke, and not by an inherent genetic factor. The photograph of a cow of Hottentot type, reproduced by Schultze (1907), shows an animal which an untrained observer may describe as humpless (Fig. 591). But to the trained eye it is clear that this cow would show a proper hump when well fed and, bred with a bull of a similar conformation, would doubtless be capable of producing a calf that would develop a prominent cervico-thoracic hump if properly reared.

The Hump in Early and Recent Africander Cattle

In several paintings of Dutch settlements, dating from the end of the 18th and the beginning of the 19th centuries, a large proportion, though not all, of the oxen are depicted with humps. O. F. Mentzel, in his description of the cattle of the Cape around 1787, pointed out that the cattle were comparatively small in size; and although some writers declared that the Africander ox had a hump on its back similar to that of the camel, this statement was incorrect. The animals, Mentzel noted, merely had a hump on the neck and shoulders, not extending beyond the first thoracic vertebra (Bonsma and Joubert, 1952).

Antonius (1943–44) was certainly mistaken in his belief that the Africander breed showed slight hump development in the bulls, while Lydekker (1912a), who described the recent Africander as "a breed of long-horned cattle without humps", was still wider of the mark. The Africander hump is large and prominent, not only in bulls but also in well fed oxen and cows, though at the early period of the evolution of the Africander breed hump development may have been as variable as it apparently was in the cattle of the Hottentots.

Coat Colour in Hottentot Cattle

The principal colour of the Hottentot cattle was red through every shade from darkest to light. In the log-book of Vasco da Gama's first voyage, the Hottentot cattle at the Bay of São Brás are described as reddish in colour. The ox represented in Daniell's painting is dark red, and the cow reproduced by Schultze seems to have been of the same shade. Black Hottentot cattle too were fairly common, and there were many red animals which had a white top and underline, or their red colour broken by white markings on almost any part of the body. Yellow stock occurred frequently. In Col. Gordon's paintings the Namaqua bull is black-and-white and the Cape cow light brown with a white, slightly spotted belly and udder; her male calf is of a similar pattern to that of the dam, while the Cape ox is a yellow-dun. In rare instances Hotten-

591. Hottentot cow (after Schultze)

	cm
Distance between horn tips	71.3
Distance from highest point of poll to upper edge of muzzle	53.5
Withers height	129.5
Distance from poll to root of tail	195.0

592. Hottentot riding oxen in South West Africa in 1891/92 (photograph: E. von Üchtritz)

tot cattle were of a creamy white, occasionally of an almost snow-white colour. To facilitate distinction between the herds of neighbours, it was a common practice among the Hottentots to select their breeding stock according to colour. Every stock owner would, as a rule, exchange animals differing in colour from his accepted standard for those of the desired pattern (Schultze, 1907). At the time the European settlers obtained their cattle from the Hottentots, previous selection had thus ensured that most of the animals bred true to colour.

The Fighting Oxen of the Hottentots

The great interest the Hottentot pastoralists took in their cattle and the latter's importance to their masters are recorded in Kolb's (1719) interesting work in which the fighting oxen of the Hottentots are described as follows:—"The Hottentots have a sort of oxen which they call Backeleyers or fighting oxen. These are the biggest, strongest and boldest animals of the whole herd. In each kraal there are about five or six, and in some even more." More than one hundred years later, Hamilton Smith (1827), referring to some Hottentot cattle of extraordinary size, wrote:—"It is from these that their Backeley or war oxen are chosen: they ride them on all occasions, being quick, persevering, extremely docile, and governed by the voice of a whistle of the owners with surprising intelligence."

Historical Evidence of the Africander's Descent from Hottentot Cattle

The history of the European settlement at the Cape shows that the Hottentot cattle may be regarded as the Africander's sole or principal ancestors.

During the first decades after the occupation of the Cape of Good Hope by the Dutch East India Company the European settlers there never encountered Bantu tribes. It was only in 1778 that the Cape Dutch for the first time came into contact with Bantu who were advancing south-westwards (Fitzgerald, 1943). Until then they had met only Hottentots and Bushmen. During the centuries preceding the discovery of the Cape by Europeans, the Hottentots, a nomadic pastoral people, had wandered with their herds of cattle and flocks of sheep to the Cape. The first Europeans found the southern and south-western parts of the Cape inhabited by Hottentots in possession of large herds of cattle and flocks of fat-tailed sheep. The early Portuguese explorers (Bartholomeu Dias) named several of the coastal places where they saw Hottentot pastoralists with their cattle (in 1487) accordingly: 'Cows River' (Gouritz River), 'Cows Cape' (C. Vacca), 'Cape of the Cattle Herders', 'Bay of the Cattle Herders' (Fish Bay), 'Coast of the Shepherds' (Mossel Bay coast). Van Riebeeck, the founder of the Dutch settlement at the Cape, has recorded that the native herds in the surroundings of Table Bay were as numerous as blades of grass in a field. Akembie, chief of the Namaqua Hottentots, possessed, in 1661, approximately 4000 head of cattle and 3000 sheep (Stow, 1905).

Acquisition of Hottentot Cattle by the Dutch East India Company

In 1652 the Dutch East India Company acquired their first herd of cattle from the Hottentots. Subsequently Van Riebeeck organised frequent expeditions into the interior to procure cattle needed for provisioning the ships with fresh meat. However, as the settlers took gross advantage of the Hottentots, the latter eventually refused to barter with the Europeans. For weeks and months they used to graze their herds in full view of the European settlement, but would not part with their animals. A serious situation arose in consequence. The Company had to furnish provisions for the vessels clearing from the Cape on the long voyages to India and Europe. The crews of these ships totalled five to six thousand sailors a year, and the provisioning of these resulted in a severe scarcity of fresh meat. Thus the Company was forced to devote itself to cattle breeding.

This undertaking was started in 1673 when war broke out between the Dutch and the Hottentots of the Cochoqua tribe who had sold the Cape Peninsula to the Company the previous year. Upon the defeat of the Cochoqua their cattle were taken away from them, and all attempts at regaining them during the following years failed (Schapera, 1930).

But the Company did not persevere with its pastoral enterprise for long; a quarter of a century later all agricultural undertakings were abandoned, and the task of provisioning the stores of the Company with foodstuffs was left to the settlers. These obtained their original herds of cattle by various means; by purchase from the Company which, in turn, acquired stock from the Hottentots; or by direct barter with the latter, a practice that could not wholly be stamped out even when private barter was forbidden; or, finally, by cattle raiding (De Kock, 1924). Only fifty years after Van Riebeeck had obtained the first herd of cattle from the Hottentots, Hottentot cattle in the possession of European settlers numbered 13,000.

Extinction of the Hottentot Breed of Cattle

In those days all expeditions of the Dutch into the interior were undertaken solely in order to obtain cattle. Consequently, in spite of their reluctance to part with their animals, the Cape Namaqua, by 1797, were left with sheep only, their large herds of cattle having passed entirely into the hands of the settlers (Stow, 1905).

Today purebred Hottentot cattle are no longer to be found. But during the early decades of the nineteenth century many herds were still in existence. These have since been absorbed into the herds of the Europeans. Pure Hottentot cattle existed longest in South West Africa where they could be found until the time of the rising of the Namaqua against the German administration. After their subjection the Namaqua were not allowed to keep cattle, but only a limited number of sheep and goats (Schapera, 1930).

The evolution of the Africander breed commenced during the first decades of the Dutch settlement. As trek oxen these cattle were unexcelled; and since the transportation of goods in Cape Colony was effected mainly by ox waggon, this quality became the determining factor in the breeding and further development of the Africander.

Theory of the Africander's Descent from Portuguese Alentejo and Indian Zebu Cattle

In former days it was widely held that the Africander cattle originated from stock which early Portuguese seafarers were believed to have brought for bartering purposes or as presents from Portugal to South Africa during the 16th century. Several Portuguese authors continue to hold this belief. Da Costa (1933), for example, is convinced that the Africander traces its origin to three different stocks: the humpless sanga classed together with the Hamitic longhorn under the name Bos taurus asiaticus, the Indian zebu, and the Alentejo or Transtagana cattle, a longhorn breed from Portugal. Da Costa regards the humpless sanga as the original type of African cattle. Portuguese seafarers, he suggests, presented the native chiefs of South Africa with Indian zebus. The crossing of the latter with the humpless sanga produced the humped sanga, which, in turn, was crossed with Alentejo cattle, introduced into Africa likewise by the Portuguese. The Africander breed is claimed to be the outcome of this triple cross. As evidence Da Costa compares a modern beefy Africander with horns of an upward trend, with an Alentejo bull carrying similarly shaped horns. Both animals are red in colour, their profiles are convex, and they resemble each other in general conformation and the shape of their horns, the differences being limited to the hump, shorter neck and the occasional occurrence of white markings on the belly of the Africander. No additional anatomical or historical evidence is adduced in support of this theory. Had its author compared an Africander of half a century ago with an Alentejo beast, he would not have found any similarity even in respect of general type; and had he chosen an Africander with more typical horns, he would have found no similarity in horn shape. It is obvious that a theory based on nothing but the similarity in colour and facial profile between two breeds has not a leg to stand on.

A quite similar view has been expressed by Martinho (1955) who writes that the crossing of Indian zebu and Portuguese Alentejo cattle, brought by the early Portuguese navigators in their vessels to South Africa as presents for the Hottentots, took place in the vicinity of Mossel Bay. The cattle Van Riebeeck encountered 150 years later (in 1652) at the Cape of Good Hope were a mixture of the descendants of these Indian zebu and Portuguese Alentejo cattle and of the original Hottentot cattle of sanga stock. Martinho does not concede to the Hottentot cattle even that meagre share in the evolution of the Africander breed which Da Costa allows them, for although an admixture of their blood to the alleged Indian zebu-Alentejo cross could hardly have been avoided, as he says, it could have been only slight; otherwise the dilution of the zebu blood would have been much more pronounced in the Africander than it is. Martinho's evidence consists again of the similarity in certain features between the Africander and the Alentejo, the obviousness that several conformational and physiological properties of the Africander not shared by the Alentejo could only be derived from zebu stock, and of the historical fact that the Portuguese shipped cattle from their homeland to Brazil where no cattle existed prior to the European occupation.

However, it is improbable that the early Portuguese navigators, who brought Iberian cattle to the New World, imported Alentejo and Indian zebu cattle also into South Africa; for the pastoral native tribes of South Africa may have been in need of many other products Europe then had to offer, but not of cattle. Even if a few Portuguese cattle were occasionally unloaded in South Africa, it is unlikely that they had any share in the subsequent evolution of the Africander breed.

Theory of the Africander's Descent from Bantu Cows and imported Zebu Bulls

It is remarkable how foreign authors are frequently in doubt as to the origin of the Africander cattle. Snapp (1939), for example, described the Africander as a cross of "Kafir cows ... with Zebu or Brahma bulls for the purpose of obtaining an increase in size".

There is not a shred of evidence for this theory on the origin of the Africander. Also, a breed evolved from such a recent intermixture would hardly have proved capable of the Africander's extraordinary preservation of its uniform character in every part of the vast area of South Africa after the profound changes caused by the Great Trek, the rinderpest epidemic of 1896–97, and the destructive consequences of the Anglo-Boer war.

Theory of the Africander's Descent from Native Cattle imported from East Africa

Molhuysen (1911) regarded, not the cattle of the Hottentots, but animals imported from East Africa as the ancestors of the Africander breed. The Hottentots used their oxen solely for riding and carrying, and the cows were poor milkers. "It is probable that the colonists were dissatisfied with these cattle and soon looked around for other breeds. Various breeds were imported, and of these one with exceptionally strong horns found preference. Towards the end of the eighteenth century, a large number of this particular breed appear to have been imported into South Africa. During the Great Trek their descendants drew the waggons of the Boers into the Highveld where they formed the nucleus of the present Africander breed. C. Keller believes that these cattle originated in the vicinity of Lake Tanganyika."

Molhuysen's theory is entirely unfounded. No historical records point to such early importations of cattle by Europeans from the neighbourhood of Lake Tanganyika. In fact, such importations would have been quite inconceivable at that time. Besides, Molhuysen's assertion that the cattle of the Hottentots gave little milk and were used only as beasts of burden proves what it tries to deny, namely, the close relationship between the Africander and the Hottentot cattle.

Theories of Friesian and Red Devon Influences on the Africander Breed

The influence of Friesian cattle on the Africander breed has been denied by Holm (1912) who has pointed out that the black markings, which characterise the crossbred progeny of Africander and Friesian cattle, are difficult to eliminate, and that the rarity of their occurrence in purebred Africander cattle is an indication of the absence of Friesian blood in this breed.

Although Holm's opinion that Friesian cattle had no share in the evolution and further development of the Africander is undoubtedly correct, his argument with regard to colour is not acceptable. Until the middle of the 18th century the cattle of the Netherlands were for the greater part red or red-and-white, similar to the Africander cattle. They belonged mainly to the red type of cattle then common throughout central Europe. Between 1730 and 1760 the red cattle of the Netherlands were so reduced in numbers by rinderpest that new stock had to be imported from southern Denmark and western Germany. The black-and-white pattern came to the fore owing to the influence of the importations from Jutland; subsequently this pattern was favoured by early American importers of Dutch breeding stock (Bakker, 1909). But red cattle are encountered in the Netherlands and East Friesland to this day. We may therefore assume that the majority of the Dutch cattle imported into South Africa during the early period of the evolution of the Africander breed were red or red-and-white. Their crossing with red Africander cattle would not have produced black markings in the progeny.

Several students of the origin of the Africander cattle have attached an importance to the red colour of the breed that is unwarranted. The overestimation of this factor was apparently also responsible for the erroneous belief that red Devon cattle, introduced into South Africa by Lord Charles H. Somerset, Governor of the Cape during the years 1814–1826, were the ancestors, at any rate had a share in the origin, of the Africander breed. As a matter of fact, until the beginning of the 20th century several colours were encountered in Africander cattle: red, red with white underline or with white topline, and yellow (Bosman, 1924); a few farmers even bred black stock. The fact that the Africander profoundly differs also in conformation and production from Friesian and Devon cattle is of much greater significance. Had these exercised any influence on the Africander cattle, there would be some indications of this; but these European breeds have no common characteristics with the Africander that may justify the assumption of their influence on the latter.

This does not imply that cross-breeding of Africander and Friesian cattle has not occurred. Since the early importations of Dutch cattle into South Africa such cross-breeding has played an important role in the animal husbandry of the country. But the crossbred descendants were rarely mated with Africander bulls, but either with similar crossbreds or with purebred Friesians. Crossbreds of indigenous cows and Friesian bulls are the ancestors of the 'Cape Cattle', a peculiar heterogeneous type bred on many commercial dairy farms, and also of the Drakensberger, officially recognised as a breed in 1948.

The early European settlers at the Cape refused to cross the Africander cattle with European breeds. Van Rijneveld (1804), who endeavoured to induce the colonists to improve their stock through the use of purebred bulls from the Netherlands, first imported into South Africa during or, as Abbott (1952) holds, before the time of Van Plettenberg, Governor of the Cape from 1771 to 1785, had to admit his failure, as the Boers regarded such crossbreds as slow and lazy. This indicates that already at the end of the 18th century the Africander represented a well defined breed, valued by the settlers for draught purposes.

The Establishment of the Red Colour in the Africander Breed

The stabilisation of the red colour in the Africander breed seems to have been due to the prevalence of red in Hottentot cattle, as well as to the preference given to red beasts by the first settlers of the Dutch East India Company, who evolved the Africander from the Hottentot cattle. Except for a small number of Huguenots, these colonists were mainly of Dutch and German origin (Walker, 1928). As above mentioned, the cattle of the Netherlands were then predominantly red or red-and-white. In south-west and central Germany, whence the majority of the first German settlers had come, the peasants did not know any other cattle than the indigenous red breeds. The red colour in cattle was therefore most familiar to the Cape settlers.

Epstein (1933; 1956) suggested, possibly erroneously, that a tax of one stuiver per head of black cattle, levied by the Dutch East India Company, might have contributed to the culling of black animals. Originally the tax on black cattle was in force in the districts of Stellenbosch, Swellendam and Graaff Reinet only; but upon the establishment of new districts it was extended to these as well (De Kock, 1924). The error may lie in the contraposition of black cattle to cattle of other colours; in reality *black cattle* may at that time just have meant *cattle* as distinct from other farm animals such as sheep, goats and horses. Sinclair (1907) has pointed out that "the phrase 'black cattle', employed by some early writers, was used to distinguish that class of stock from horses, and did not apply exclusively to the colour of the hair of the animals". The old use of the phrase is illustrated by the wording in the advertisement of William Dick's first course of veterinary lectures in Edinburgh in 1816:—"... lectures on the diseases of horses, of black cattle, sheep and other domestic animals" (Watson and Hobbs, 1951). In China, cattle, in distinction from buffaloes, are called Hwang Niu, i.e. yellow cattle, irrespective of their actual coat colour (Epstein, 1969).

In warm regions, cloudless for the major part of the year, a black coat in working oxen is undesirable, as black hair reflects neither heat rays of shorter wave lengths nor light rays; these are effectively reflected by white, yellow or reddish brown hair (Rhoad, 1940; Riemerschmid and Elder, 1945). Ultra-violet rays, on the other hand, which are toxic to skin and nerve, are resisted by yellow, reddish brown and black hide colours, but not by an unpigmented skin. The red hide and hair of the Africander offer therefore protection against the injurious effects of intense solar radiation of all wave lengths. Experience of South African farmers and transport riders of the weakness of white and black, and the soundness of red in their cattle may have contributed to the elimination of the unfavourable colours in breeding stock.

Formerly animals of other colours were encountered in the Africander breed along with the red. These, however, were gradually eliminated. Holm (1912) attributed this chiefly to the transport riders who, before the construction of railways, conveyed goods by ox waggon for hundreds of miles. These vied with one another for the possession of the most beautiful and uniform spans. Red colour, long finely shaped horns, hard flinty feet, sinewy legs and quick movement were highly valued. It was the pride of the transport riders and many farmers of those days to possess spans of oxen of uniform colouring and appearance. Thus it came about that animals of a definite type and colour were selected for breeding purposes.

Different Types in Early Africander Cattle

However, the role of the transport riders should not be overrated. All of them selected their oxen for speed and endurance, but their tastes with regard to colour and appearance differed. Until the establishment of the Africander Cattle Breeders' Society and the acceptance of a Standard of Excellence in 1912 (amended in 1926, 1932 and 1957), the Africander breed displayed considerable variation in type. C. J. Cloete, a former prominent breeder and judge, distinguished between three basic types of Africander cattle prior to 1912. Two of these he named after the shape of their horns: Cupshape-Horn and Long-Twisted-Horn; the third: Notch Neck, from the high hump producing the impression of a notch in the anterior part of the neck. In the Cupshape-Horn the horns were long and thin and twisted irregularly, sometimes meeting below the throat. In conformation this type is described as big, heavy, clumsy and loose-jointed, with barrel-shaped ribs, broad back and hips, sloping rump and loose skin—a type rejected by modern breeders. The Long-Twisted-Horn type had the longest horns, nearly straight with two twists. This was the typical desert type; tall, with a fairly long neck, long straight back and rump, flat ribs and long legs. It provided the speediest oxen of great staying power. With modern breeders this type is not in favour, owing to its poor beef qualities. The Notch Neck comes closest to the modern breeding standard: horns finely shaped with a downward and backward trend, head fairly broad, neck short and thick, chine and hump well developed, dewlap extraordinarily big, back short, rump sloping, hips and ribs round, legs short (Opperman, 1952).

However sceptical we may be of the soundness of this classification of the early unimproved Africander cattle, particularly in view of the attempt at correlating different horn shapes with different body conformations, Cloete's description serves as a valuable illustration of the great variability of the breed and of the different trends in breeding in former days. The modern breeding standard in a way represents a synthesis of the long-legged nimble animal and the short-legged, slow and loose-skinned type.

c. The Origin of the Cattle of the Hottentots

Theory of Origin of the Hottentots' Domestic Animals from Bantu Stock

The amount of research that has been devoted to the racial origin of the Hottentot's domestic animals is negligible. For this reason practically all theories advanced on this point have proved erroneous. Johnston (1908) inferred from the resemblance of Hottentot and Bantu roots of the terms for cattle, sheep and goats that the goat first, then the ox and the sheep were brought to the Hottentots from the north by Bantu or Nilotic negroes. Yet it is well known that, at any rate in historical times, the Hottentots possessed cattle and sheep and an old pastoral tradition long before they encountered Bantu peoples in South Africa.

Schapera (1930) writes that it is a more debatable question if the Hottentots obtained their domestic animals from a Bantu people in East Africa. Should this be the case, they must have received them from one of the few East African Bantu peoples who had given up goat breeding. For, as Kroll (1929) has pointed out, the goat was one of the characteristic domestic animals of the agricultural Bantu when they still constituted a single group with a single language and culture in their homeland. The domestic fauna of the early Hamites of East Africa, on the other hand, consisted mainly of cattle, sporadically also of sheep. The goat was transmitted to the Hamitic peoples of East Africa by agricultural Bantu. Its rejection by several Bantu peoples was secondary, and in most cases due to the influence of Hamitic pastoralists.

Theory of Derivation of the Pastoralism of the Hottentots from the Bantu Herero

Murdock (1959) believes that the Hottentots acquired their mode of life from their northern neighbours, the Bantu Herero. There are indeed several important similarities in the economic and cultural patterns of these two peoples. In the domestic fauna of the early Hottentots the goat was absent; the same has been reported of the early Herero (Viehe, 1903). The Hottentots are pure pastoralists who do not practise agriculture, while the Herero have nearly abandoned agriculture and pursue a life of independent pastoral nomadism—the only Bantu tribe in Africa to do so. Only in Angola, Herero women also raise millet crops (Forde, 1934). Among the Hottentots, in contrast to the majority of South African Bantu tribes, the milking is done by the women, not the men (Seligman, 1930). Yet the segregation of women from cattle apparently did not spread to the south-western Bantu, for the Herero assign milking primarily to females, while among the Ambo, though men do most of the milking, their wives often assist, and among the Nyaneka of southern Angola a man's chief wife tends his sacred cows and performs the cult activities associated with them. Even among the Sotho, in the eastern part of South Africa, the Lovedu and Venda, in contrast with all other Sotho peoples, do not observe the usual ritual separation of women and cattle, although women milk only in emergencies (Murdock, 1959).

On the other hand, there are important differences between the Hottentots and Herero, which cannot be reconciled with Murdock's theory. The Herero (Damara) formerly bred a giant-horned type of sanga, whereas the Hottentot cattle were merely long-horned. Again, the sheep of the Herero were originally of the hairy thin-tailed type; they acquired fat-tailed animals from their Hottentot neighbours less than 200 years ago; but even today the fat tails of Damara sheep are rather poorly developed, "long, straight and tapering sharply". The sheep of the early Cape Hottentots, on the other hand, were distinguished by very large broad fat tails (see also II, p. 151). The customs of the Herero and Hottentots, connected with their livestock, also differed in important particulars; the Hottentots' "practices are in many ways closer to those of the pastoral Nilo-Hamites many hundreds of miles away to the north-east, than to those of their Bantu neighbours in the east" (Forde, 1934).

Murdock (1959) dates the southward spread of milking from the interlacustrine Bantu to the middle Zambesi and thence to the south-eastern and south-western Bantu (Herero) and the Hottentots to a period after A.D. 1500. While this late date may accord with the comparative recency of Herero occupation of Damaraland, and the similarly recent westward movement of south-eastern Bantu along the south coast of Cape Province, where the Hottentot advance guard, formed by the Gonoqua tribe, for the first time encountered them at the beginning of the 18th century, it cannot be reconciled with the testimony of the Portuguese seafarers who saw Hottentots in the neighbourhood of Saldanha Bay and later on at Mossel Bay and other places already at the end of the 15th century (see p. 487). Nor is it reconcilable with the early Dutch and English accounts of the then wide distribution of the Hottentots over much of the western half of South West Africa to the Angola border and most of the Cape Province as far as the Kei river, their wealth of livestock and their deeply rooted dairying complex. As Schapera (1930) writes:—"The pastoral habits of the Hottentots, with all the customs and traditions connected with this mode of life, were already developed before their ancestors came south."

Cultural Differences between Hottentots and Bushmen

The theory that the Hottentots received their cattle and pastoral culture from a Bantu tribe is also rendered doubtful by their cultural and racial history.

Culturally the outstanding difference between Hottentots and Bushmen is that the former are pastoralists, while the latter are hunters and gatherers only. The Hottentots carve vessels from wood, weave mats and baskets from reeds and rushes, and make skin bags for holding milk and water. They differ from the Bushmen in practising the art of working copper and iron. But this does not seem to be an original trait of Hottentot culture; rather it would appear that the northern Hottentots acquired the knowledge from contact with the Bantu, while the southern tribes obtained the metal of which they had need from Dutch colonists (Clark, 1959). The Hottentots did not possess the art of painting or engraving on rock in which the southern Bushmen excelled (Seligman, 1939).

The Origin of the Hottentot Languages

The Hottentot languages are akin to those of the Bushmen, but differ from them in certain features, such as sex-gender, that present affinities to those of the Hamitic language family. They also include several words which appear to be of Hamitic origin (Clark, 1959). Therefore, Hottentot is nowadays regarded as Bushman with Hamitic admixture (Seligman, 1939).

In northern Tanzania similar Khoisan languages are spoken by the Kindiga hunters and gatherers and the Sandave pastoralists and agriculturalists of whom one section, when first observed, was still living by hunting and gathering. Kindiga and

Sandave have numerous roots, grammatical peculiarities and the three click sounds in common with the language of the Nama Hottentots, and are believed to owe their origin to a mixture of Bushman and Hamitic languages (Schapera, 1930). Dart (1966) attributes the intimate relation of the Bush-Hottentot languages to the Hamitic group of languages to the early fusion between the Hamitic and Bush races east of the Great Rift valley.

Racial Characteristics of the Hottentot People

The physical characters of the Hottentots are basically those of the Bushmen, the chief points of difference being a 10–12 cm taller stature, a longer, higher and narrower head, slightly narrower nose, rather more prognathous face, somewhat lighter skin colour, and very different blood grouping (Schapera, 1930; Seligman, 1930; Cole, 1954). Murdock (1959) attributes these differences solely to a superior diet and more secure livelihood. But all other anthropologists regard them as the result of intermixture of Bushmen with some other racial group. The prevalent theory is that, apart from Bantu admixture in certain tribes, such as the Korana, this new component was derived from an early light-skinned Hamitic stock (Huntingford and Bell, 1950). Shrubsall (1908), however, working on skeletal material alone, maintained that the non-Bushman element in the composition of the Hottentots must have been of Bantu origin, for it could be shown that the cranial and facial measurements of the Hottentots, while in some respects intermediate between those of Bushmen and Hamites, were more nearly intermediate between Bushmen and Bantu; and where they showed this least, they resembled the Sudanese negro more than the Hamite.

Against this has been argued that the Hottentots are at least as light in colour as the Cape Bushmen, whereas Bushman tribes, which have unquestionably been affected by racial intermixture with Bantu, have acquired a darker skin colour. Again, as von Luschan (1912) has pointed out, Hottentots with facial features of pronounced Hamitic type are occasionally encountered.

While Schapera (1930) held that the racially pure Hottentots now vary only slightly among themselves, recent research has demonstrated that physically the historic pastoral Hottentots show considerable variation and that the 'Hottentot type' includes all grades of hybridisation between at least three basic strains. The most characteristic of these is the small long-headed Bushman type; the second is represented by the larger, long-headed bush-boskopoid type (so called from fragments of a fossil middle stone age skull discovered near Boskop in the Transvaal and sub-fossil remains from the later stone age site of Oakhurst Shelter on the south coast of Cape Province), and the third by a long and very narrow-headed long-faced caucasoid or Erythriote (Hamitic) type which originated in north-east Africa (Clark, 1959).

On the Origin of the Hottentot People

If it be accepted that the Hottentots are a Bushman people influenced by Hamitic culture and caucasoid race, they must have originated in an area of contact between autochthonous Bushmen and Hamitic intruders with a pastoral culture. Scattered finds made in the African plateau from the eastern Sudan to South Africa indicate that the Bushman-Hottentot type once extended much farther north than it has in the historic period, while caucasoid elements extended farther south (Linton, 1956).

Sporadic advances of agricultural or pastoral Hamitic tribes from north-east Africa to the south seem to have occurred at various times and given rise to various cultures borne by bushmanoid or partly bushmanoid peoples. Thus, evidence has been collected of an early pastoral people of the giganto-pedomorphic bush-boskopoid physical type (closely resembling one of the basic Hottentot types), who inhabited the plains of northern Tanzania. They dug rain ponds and wells for their livestock.

Some bands of this people may have broken away in search of new pastures and settled at the foot of Bambandyanalo Hill and Mapungubwe in the Limpopo valley in the 10th or 11th century A.D. The skeletal remains of the early inhabitants of these sites show no negro affinities, and physically they represent a homogeneous bush-boskopoid type, closely similar to the Hottentots and akin to the post-Boskop inhabitants of Rhodesia and South Africa. However, their culture was essentially Bantu (Clark, 1959; Davidson, 1961).

For a prolonged period (Southern) Rhodesia was occupied by a pastoral stone age people, who were Hottentots proper or nearly pure Hottentots. They buried cattle in a ceremonial way that resembled the animal burials in Egypt during the neolithic age (Gardner, 1955–56). These people were driven from their homes by Bantu tribes from the north, who practised agriculture and knew the use of iron. But as late as in the 18th century A.D., the territory of the Bavenda Bantu in (Southern) Rhodesia was reoccupied by a Hottentot people that was finally dispersed by the northward-driving Matabele in 1825 (Davidson, 1961).

In 1895, Virchow described two skulls of the cattle-breeding Sandave tribe of Kondoa Irangi, northern Tanzania, which showed distinct Hottentot affinities; and Trevor (1947) also concluded that there were physical similarities between the two peoples.

A skull from the Mumbwa caves of Zambia, associated with later neolithic industries of Wilton type, combines bush-boskopoid with caucasoid features. The site of Zimbabwe in (Southern) Rhodesia was founded and first occupied in the 7th century A.D. by a people presumably Hamitic (Megalithic Cushite) (Murdock, 1959).

In South Africa the first evidence of a non-Bushman element, resulting from a very early infiltration of caucasoid peoples from the north, is represented by a skull from Tuinplaats and by fossils from the Border Cave and Fish Hoek. At Matjes river near Plettenberg Bay on the south coast of South Africa, cave deposits, associated with the later neolithic Wilton industry, have yielded remains of a long-headed people with certain caucasoid (Erythriote) features, which may reflect a migration of elements of this stock from East Africa. Skeletal remains discovered in historic graves in the Kakamas area of the Orange river show a close resemblance with the caucasoid long-headed peoples of north-east Africa, and perhaps connect also with the earlier Matjes river type (Clark, 1959).

In view of the early presence in South Africa of peoples with caucasoid features, several authors have claimed that the Hottentots originated in South West Africa from a cross between early Hamitic-speaking pastoral invaders and their Bushman forerunners. Broom (1941) regards the Korana Hottentots of the Vaal and upper

Orange rivers as a hybrid people descended from aboriginal Bushmen and long-headed caucasoid pastoralists from the north. Seligman (1939), on the other hand, writes that it is generally held that the mixed race arose in the north, perhaps in the neighbourhood of the Great Lakes, while Theal (1910) suggested that the lake district of East Africa was not the Hottentots' original home, and that it was not even likely that they had resided there very long. It is notable that the early pottery of the Hottentots is remarkably similar to that of the Lancet culture in the Rift Valley province of Kenya, which has been dated by the C^{14} method to about A.D. 1400 (Payne, 1964).

Mason (1962) suggested that there may have been a reservoir occupied by bush-boskopoid people in north-east Africa whence different bands migrated southwards at irregular intervals from about 8000 B.C. onwards; the first group leaving near the end of the stone age before the idea of food production reached them, the second, after they had become cattle and sheep owning nomads who could make pottery but not metal. The third and last wave may have reached Mupungubwe, Hatfield and elsewhere some time in the 1st millennium A.D. They were the most progressive bush-boskopoid group who gained their food-, pot- and metal-making techniques in the far north-east thousands of years after their stone age ancestors had moved south.

The view, postulated by Theal (1910), that East Africa was not the original home of the Bush-Boskopoids is shared by Coon (1968) who holds that they originated in North Africa from where they were driven out by an invasion of Caucasoids towards the end of the pleistocene and during the early post-pleistocene period.

The Hottentots' Migration to South West Africa

According to their own traditions, the Hottentots appear to have come from the lake district of central Africa whence they were driven out, at the end of the 14th or the beginning of the 15th century, by a more powerful people armed with bows and battle axes. At first they travelled with their herds and flocks between the Tanganyika and Nyasa lakes in a southern direction, then turned to the west across the high plateau of central Africa until, with their faces always towards the setting sun, they reached the 'great waters' of the Atlantic. With the ocean on their right, they slowly wandered onwards, down the west coast of Africa (Stow, 1905).

However, in common with practically all indigenous oral traditions, those of the Hottentots are not very dependable. The first Bantu invasions into (Southern) Rhodesia from across the Zambesi river appear to have taken place early in the second half of the 1st millennium A.D. It is probable that the ancestors of the Hottentots, at any rate the pastoral caucasoid component of their ancestors, crossed the upper waters of the Zambesi shortly before this date.

Origin of the Pastoral Economy of the Hottentots

The pastoral economy and culture which the Hottentots received from the Hamitic part of their ancestors must have been evolved in East Africa and carried south before the equatorial Bantu advanced into Uganda and thence entered upon their southward expansion in the first half of the 1st millennium A.D., unless we assume that the Bantu, in the course of their advance, ejected a small Hamitic tribe from its homes and followed so persistently at its heels that the Hamites had to abandon their agricultural habits and rely solely on their herds and flocks. Proceeding to the south, south-west and again to the south through Bushman territory for many centuries, a small tribe, permanently cut off from the source of its race, would be bound to lose its racial identity and speech, without however abandoning its superior economy. This would be in line with Forde's (1934) suggestion that the Hottentot "economy may indeed indicate the terminus of an early wave of pastoral advance which has been overlaid by later movements and the development of more complicated social and economic relationships in eastern Africa". It would also be in line with the fact that the cervico-thoracic-humped sanga was ubiquitous in eastern Africa, before thoracic-humped zebu cattle penetrated into the major part of this area.

On their migration from East to South West and South Africa, the Hottentots were the southern outpost of the pastoral peoples of Africa. This enabled them to withdraw from Bantu pressure to the south, where no pastoral tribes but only Bushman hunters dwelt. Here their herds were removed from outside influences until, in the most southern part of the continent, they met the vanguard of the Europeans.

3. The Long-horned Fulani Cattle of West Africa

Distribution of the Long-horned Fulani Cattle

South of the short-horned zebu belt in West Africa, comprising the Shuwa, Azaouak, Tuareg and Maure cattle, and north-west of the breeding area of the Adamawa zebu, humped cattle are bred by the Fulani, a branch of the Northern Hamites who spread their influence over the Western Sudan and Upper Senegal during the days of the ancient Ghana Empire, and had found their way into Northern Nigeria by the end of the 13th century.

Typically the Fulani cattle are long-horned, some of them giant-horned. The nomad 'Cattle Fulani', who are the purest representatives of the Hamitic element, lay more stress on horn size than do the settled Fulani who are much mixed with negro blood.

The Red Bororo Cattle

The most striking of the Fulani breeds is the Red Bororo (plural: Bororodji), also called Rahaji, Abori, Brahaza, Fogha, Gabassaé, Gadéhé, Hanagamba, M'Bororo or Bodadi. It is bred by the M'Bororo whose tribal name is derived from the fact that the tribesmen live in the Mbouroura or bush (Joshi et al., 1957). The Bororo people originally inhabited the area included in the Niger and the Sokoto Province of Northern Nigeria. To avoid islamisation in the early part of the 19th century, they fled to the east and settled in Bornu (Nigeria), Adamawa (Cameroun), Mayo Kerbi and Baguirmi (Chad), and as far east as Darfur (Republic of the Sudan) (Mornet and Koné, 1941). Bororo cattle are now found in the east of Niger, in the extreme

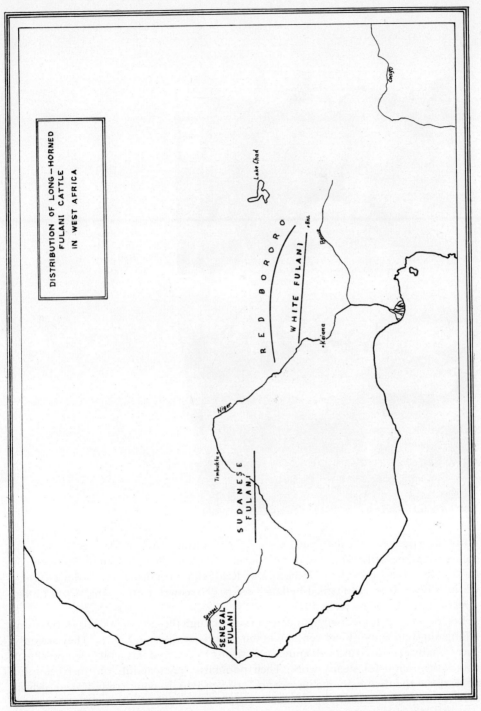

593. DISTRIBUTION OF LONG-HORNED FULANI CATTLE IN WEST AFRICA (after Mason)

594. Red Bororo (Rahaji) bull (after Mason)

595. Red Bororo (Rahaji) cattle

north of Nigeria, in Chad and the north of Cameroun; in the last-mentioned country they are called Djafoun or Djafourin (Mandon, 1948), and in the Republic of the Sudan—Red Fulbe, Red Umborora or Red Fellata (in distinction from a variety of the Shuwa Arab zebu called Fellata—see p. 399, as well as from the White Fulbe, Umborora or Fellata—see p. 501).

Red Bororo cattle are used mainly for meat although they are of poor conformation, killing out well below 50 per cent, and their meat is of inferior quality. They are only occasionally employed as pack animals, as their nervous and intractable temperament makes them unfit for steady work. Their popularity rests mainly on their beautiful appearance and their high intelligence and obedience to the commands of their owners even to the extent of charging intruders when commanded to do so or breaking and scattering in the bush at a word of command from the herdsmen or at a note from their pipes (Seligman, 1930). Owing to their uniformity in type and colour, large size, alert carriage and long graceful horns, Red Bororo cattle are most impressive animals. The live weight ranges from 350 to 450 kg, and the withers height from 140 to 150 cm.

The head is long and somewhat narrow at the forehead and muzzle. In profile it is straight or slightly concave. The almost round or oval ears are carried horizontally. The horns are 75–120 cm long, round in cross-section, with a basal diameter of about 12–13 cm. Their direction is outward and upward, with the tips again turned outwards to produce the characteristic lyre shape. The hump is thoracic or cervico-thoracic in situation, muscular in structure (Dr. D. H. Hill—personal communication), well developed and generally pyramidal in shape, but sometimes leaning over at the back. The dewlap is highly developed, often hanging loosely in pendulous folds. The prepuce and umbilical fold are also well developed, the prepuce tending to be loose and pendulous. The body of the Red Bororo is badly balanced. The animal is rather coarse of bone, narrow at the shoulders and through the body, shallow, and flat of rib. The top line rises from the hump to the high sacrum from which the rump slopes down to the tail root. The udder is poorly developed. Generally the body is insufficiently covered and lean of musculature. The limbs are long and often badly set on. The skin is loose and of medium thickness with pigmentation varying from light to dark. The hair is short and coarse (Joshi et al., 1957). The characteristic colour is a rich reddish brown varying from light to dark (chestnut to mahogany). Occasionally black animals are encountered. Colours are almost always whole save for the rather common occurrence of a light-coloured tail switch and a pale muzzle. White or pied coloration indicates a cross with the White Fulani. The horns are white or ivory in colour (Faulkner and Epstein, 1957).

The White Fulani Cattle

In the savanna zone in the Northern Provinces of Nigeria, including Kano, Katsina, Bauchi, Sokoto, Zaria, Bornu Plateau and Adamawa in the Northern Region, also in Bamenda Province of south-east Nigeria, the nomadic Fulani breed a white-coated type of humped longhorn cattle, known as White Fulani, Yakanaji, Bunaji or White Kano, the most widely scattered and probably the most numerous of all Nigerian cattle types. In Cameroun it is called Akou (Mason, 1951a), in the Republic of the Sudan—White Fulbe, White Umborora or White Fellata. In the north of the Northern Territories of Ghana the White Fulani has greatly influenced the humpless shorthorn; in Gambia its influence may also be seen in the humpless N'Dama longhorn.

The White Fulani is a medium to large type, standing approximately 130 cm at the withers; adult bulls weigh over 500 kg, adult cows about 340 kg. It is used for milk and beef and for work as pack animal and with implements. The average milk yield of the cows is about 700 kg per lactation.

The head is long but of good proportion, with a wide forehead and fairly prominent orbital arches. In profile it is straight or slightly dished. The ears are of medium size and are carried horizontally with the inner parts to the front. The horns are 80–105 cm long, round in cross-section, slender, and set high on the head, projecting forwards and upwards, often with an outward turn at the pointed tips producing the character-

596. White Fulani bull

597. White Fulani cows

istic lyre shape. The head is carried upright by the strong, short but deep neck. The hump tends to be intermediate in situation, i.e. neither clearly thoracic nor clearly cervico-thoracic. It is well developed, though more so in the male than the female. In the bull it sometimes hangs over at the back. It is usually musculo-fatty in structure (Dr. D. H. Hill—personal communication). The dewlap extends from the throat to between the front legs. It is well developed and folded but not very loose or pendulous. The prepuce and umbilical fold are also well developed without being over-pendulous or loose. The body is generally well balanced, of good depth and width. The barrel is well-sprung and of good capacity. Sometimes the body is rather shallow and lacking in width, giving the animal a leggy appearance. The top line is strong; it rises gently from the hump to the somewhat high sacrum. The back is generally of good width and fair muscular development. The rump shows good length and is fairly well muscled, but has a marked slope from hook bones to pin bones. The pin bones are often narrow. The tail is thin and long, the switch reaching almost to the ground. The udder is well developed, of good shape, and strongly attached, with well-placed and moderately large teats. The limbs are of moderate length and well placed; the upper thighs show fair breadth and fulness. The bone is clean and hard and of good quality. The skin tends to be loose, and is usually black, occasionally red in colour; the hair is soft. The coat is commonly white with black ears, muzzle and feet. Black spots on the limbs and sides of the body are very common; red markings are also frequent (Faulkner and Epstein, 1957).

The Sudanese Fulani Cattle

In western Mali, south-west of the breeding area of the short-horned Tuareg zebu, and south-east of the habitat of the Maure zebu, i.e. in the area situated in and

598. Sudanese Fulani cow

599. Sudanese Fulani bull
(after Mason)

around the flood plain of the Niger system of rivers from Ségou to Timbuktu, the long-horned Fulani cattle are known as Sudanese Fulani. They are divided into different varieties according to the tribes breeding them, such as Toronké, Samburu and Baoro. The Toronké variety came originally from Senegal and so in many ways resembles the Senegal rather than the Sudanese Fulani. During the dry season the Sudanese Fulani is found mainly in Macina, during the rainy period in Aussa in the north and Gourma in the south. In the east of its habitat the Sudanese Fulani is frequently crossed with short-horned Tuareg zebus.

In its purebred form the Sudanese Fulani is a medium-sized animal, reaching a withers height of 115–125 cm; in the bull the live weight is 300–350 kg and in the cow 250–300 kg. The head is fairly long and slender, with a flat forehead and straight profile. The horns vary in size but are generally rather long, strong at the base, and usually lyre-shaped in the bull, sickle-shaped with a forward inclination in the cow. The hump is thoracic in situation, but sometimes its position is a little more forward than it is in purebred West African zebu cattle. It is markedly developed in the bull, less prominent in the cow; frequently it hangs over at the back. The thin but well developed and folded dewlap extends from the chin to the forelegs. The umbilical fold and prepuce are less conspicuous. The chest is lacking in width; in some animals it shows sufficient depth, but cattle with a shallow chest and leggy appearance are frequent. The back is straight and moderately well muscled; the lumbar region is flat. The rump is short and markedly sloping from hook bones to pin bones. The thighs show variable development in different animals. The tail is long and fine, reaching to well below the hocks. The udder and teats are poorly developed. The limbs are long and dry of bone. The skin is soft, with pigmentation varying from light to dark (Joshi et al., 1957). The most common coat colours are light grey with black points (Ségou and Toronké cattle) and light grey with small black spots (Samburu cattle). In the northern part of the Sudanese Fulani's habitat many different colours and colour patterns are encountered (Doutressoulle, 1947).

The Senegal Fulani Cattle

West of the breeding grounds of the Sudanese Fulani, in the west and north of Senegal, extending across the Senegal river into the south of Mauritania, i.e. in the area lying between 12° and 16° W. long. and between 13.5° and 16.6° N. lat., the long-horned Fulani are known as Senegal Fulani or Gobra. The Gobra is a large-framed beast, standing 125–140 cm at the withers. Bulls weigh 300–400 kg and cows 250–350 kg; superior animals attain live weights of close to 500 kg. For meat production the Gobra is among the best of the humped cattle of West Africa. It is used also for draught and baggage work. The milk yield of the cows is small; about 450–500 kg, testing 5.5 per cent butter-fat, in a lactation period of 6 months.

The head of the Gobra is of moderate length, heavier than the head of the Sudanese Fulani, with a convex forehead and a straight or concave facial profile. The eyes are large, the ears long and pointed. The horns are lyre-shaped; long in cows and oxen, but shorter in the bull. Occasionally loose horns are encountered. The neck is short; the dewlap well developed, thin at the jaw but heavy at the chest. The hump is prominent,

600. Senegal Fulani (Gobra) cattle (after Pierre)

particularly in the bull. While essentially thoracic in situation, in many animals its position tends to be intermediate between the thoracic and cervico-thoracic. The chest is rather shallow, the back long, the loins are narrow, but the rump is fairly wide with a moderate slope to the well developed thighs which extend well down to the gaskins. The tail is long and fine, extending 5–10 cm below the hocks. The udder and teats are little developed. The skin is thick and loose, and of light pigmentation. The coat is generally white; light yellow wheat colour, red-and-white, and white with coloured spots or brindle stripes are also frequent.

Doutressoulle (1947) distinguishes between three different local varieties of the Gobra:—1) Djoloff (Jolof or Wolof), 2) Baol, 3) Dagana. The Djoloff Gobra is the largest, well muscled, with a well developed rump, and easy to fatten. The coat is commonly redand-white, with the red extending over smaller or larger areas. The Baol Gobra resembles the Djoloff, but is lighter in weight, the hindquarters are less developed, and the coat colour is whole red. The Dagana Gobra is the smallest of the three; it is poorly furnished with muscle, red-and-white in colour or white with black spots.

iv. Origin and Descent of the Humped Cattle of Africa

1. The Home of the Cervico-thoracic-humped Zebu Cattle

Records of Cervico-thoracic-humped Zebu Cattle in Egypt

The earliest records of humped cattle in Africa are found in paintings on ancient Egyptian tomb walls and other monuments. They occur, as far as can be ascertained, not until the XIIth Dynasty (c. 1990–1780 B.C.). From that period on, cervico-thoracic-humped cattle are pictured in a number of tombs at Beni Hasan, Tell el-Amarna and Thebes, particularly in the tombs of Men-kheper-Re-seneb (Thotmes III's high priest of Amen) and Amenmose (15th century B.C.) (Newberry and Griffith, 1893–1900; De

Garis Davies, 1933; Lucas, 1948). An ivory knife handle from the XVIIIth or XIXth Dynasty (Fig. 604) also has the form of a humped bull (Hornblower, 1927).

These humped cattle were probably introduced into Egypt from the south along with other objects of trade and tribute, when the rulers of the XVIIIth Dynasty conquered Nubia, Kash and the southern Sudan, and reopened the trade route to Punt (Somaliland). Apparently they had reached the Somali coast with immigrants from southern Arabia. Semitic peoples, as Davidson (1961) has pointed out, invaded Ethiopia many hundred years before the Christian era and produced, in the course of time, an Ethiopian civilisation which reflected that of their homeland. The earliest known inscription from Yeha, near Aksum, dated to the 4th century B.C., refers to the dedication of an altar to the goddesses Naura and Ashtar.

Clark (1954) does not date the introduction of humped cattle into Somaliland so early (2nd millennium B.C.) because the rock engravings in the Horn of Africa furnish no evidence to this effect. But he correctly associates this type of humped cattle with the breed represented at Mohenjo-daro in the Indus valley, and its introduction into East Africa with the Semitic migrations. He writes:—"I can find no reference to the probable date of introduction of the zebu cattle into the Horn. It appears to be in origin an Asiatic breed as it is represented in the Indus civilisation at Mohenjo-daro and it seems not unlikely that it may have been introduced during the Semitic migrations from south-west Arabia, which are said to have taken place between the first millennium B.C. and the first half of the first century A.D."

An alternative possibility, namely that some humped cattle were brought to Egypt

601. Cervico-thoracic-humped cattle and humpless cattle belonging to Nebamen, a scribe and registrar of crops at Thebes (about 1400 B.C.).

602. (left) Cervico-thoracic-humped zebu from the wall of a Theban tomb
603. (right) Cervico-thoracic-humped zebus from the period of the New Kingdom

604. Ivory figurine of a humped bull from the XVIIIth or XIXth Dynasty (after Hornblower)

direct from Asia by the pharaohs of the XVIIIth Dynasty, nearly every one of whom sent his armies into Palestine, Syria and Mesopotamia to collect loot and tribute and gain control of the caravan routes and markets, is suggested by a scene in a quarry, dated to the twenty-second year of the reign of Aahmes (Amosis) I, i.e. about 1548 B.C. This shows a large block of stone being dragged by six humped oxen which, according to the inscription accompanying the scene, had been carried off from the lands of the Fenkhu (Phoenicians?) in Asia (Breasted, 1906–07). Again, in a tomb of the XVIIIth Dynasty humped cattle are mentioned in a list of imports from Syria, from which De Garis Davies (1933) infers that "the humped bull ... is no doubt of a breed imported from Syria and farther east". However, in considering this alternative it should be borne in mind that at the time of the XVIIIth Dynasty of Egypt the cattle recorded in Syrian sculptures were of the humpless shorthorn type (Fig. 352). On Mesopotamian monuments, again, cervico-thoracic-humped cattle were no longer represented during the middle of the second pre-Christian millennium; the then prevalent zebu type was thoracic-humped (see pp. 514–515).

Humped cattle did not supersede the humpless types in Egypt; rather they seem to have been kept in small numbers as a curiosity. They are depicted with large humps cervico-thoracic in situation or, on one or two monuments, with humps of an intermediate position between the cervico-thoracic and the thoracic. Only one single beast has a clearly defined thoracic hump (Fig. 614). Their fine slender horns are of medium length and lateral direction, with one or two moderate twists; more rarely the horns are long and sickle-shaped, or short and thick and projecting upwards. The humped cattle are shown in various coat colours: red, black, white, grey, and yellow, either whole, spotted or variegated.

Records of Cervico-thoracic-humped Cattle in Mesopotamia

These humped cattle may have been brought to southern Arabia, and thence to East Africa, from the head of the Persian Gulf, where they are well represented at Tell Agrab, Ur, Susa and Arar (Larsa). Even greater is the likelihood that they reached southern Arabia direct from south Baluchistan, whence Sumer, too, appears to have received the cervico-thoracic-humped type of zebu (see pp. 511–512).

Zeuner (1963a) believes that humped cattle were introduced into northern Mesopotamia as early as 4500 B.C. But the record from the earlier Halafian period of Arpachiya, on which this assumption is based, consists of a "rather rough and somewhat doubtful figurine of a humped bull". A bull with a large cervico-thoracic hump, carved in typically Sumerian style on a green steatite vase, was found by Frankfort (1936) in a temple at Tell Agrab in Mesopotamia (Fig. 605). This fragment has been attributed by Frankfort to the predynastic period (Jamdat Nasr), about 3200 B.C., but Zeuner (1963a) dates it to the pre-Sargonic, Early Dynastic II–III period, c. 2800–2700 B.C. At Nineveh humped bulls appear on stamped seals in the Jamdat Nasr period. On a clay tablet from Larsa, dated to c. 3000 B.C., a man is depicted with raised battle axe defending an ox attacked by a lion (Duerst, 1899). The bovine has a cervico-thoracic hump and long horns similar in shape to those of the humpless longhorn cattle of ancient Mesopotamia. From the same period dates a marble amulet representing a small bull with a cervico-thoracic hump, which was found at Ur (Hornblower, 1927).

In Iran the zebu is depicted on a bitumen vessel from Susa, about 3000 B.C. or slightly earlier. A few early clay seals from Susa show cattle with transversely cleft humps which convey the impression of two humps, one situated on the neck and the other on the shoulder (Figs. 607 and 608); a similar indentation is occasionally observed in very large and elongated thoracic humps of bulls of recent zebu breeds. Another humped bull is seen on an alabaster vase from Susa, dated to the Second Period, i.e. the Early Dynastic period, c. 2600 B.C.; and a similar bull, dated to about 2500 B.C., is incised on

605. Cervico-thoracic-humped zebu bull on a vase from Tell Agrab, Mesopotamia (after Frankfort)

a clay tablet found by Frankfort. A clay figurine of a humped ox was excavated by Woolley from a pre-flood stratum (Marshall, 1931-32).

The steatite vase from Tell Agrab (Fig. 605) and several records from Susa (Fig. 606), published by De Morgan (1911), De Mecquenem and Scheil (1921) and others, show that these cattle were all of the cervico-thoracic-humped type. The hump is depicted so clearly, occasionally with such exaggerated clearness, that we may safely include this particular type of cattle among the domesticated animals of ancient Sumer.

The same cervico-thoracic-humped type may be represented by the clay figurine of a humped bull from Tell Halaf (Fig. 610); but another humped bovine bending over a small animal and facing a lion, on a seal from Susa (Fig. 609), was regarded by De Morgan (1911) as a bison, which occurred in northern Mesopotamia in a form closely allied to the recent Bison bonasus caucasius (Hilzheimer, 1926a). While this is probably correct, the occurrence of cattle with cervico-thoracic humps in the Tigris-Euphrates plain during the 3rd millennium B.C. seems to be well established on the remaining pictorial evidence.

Probable Cradleland of the Cervico-thoracic-humped Cattle

Again, this does not imply that lower Mesopotamia was the birth place of cervico-thoracic-humped cattle. For Mesopotamian records of this type are relatively few; as in Egypt, cervico-thoracic-humped cattle were never numerous enough to replace the humpless longhorn cattle common at that time. Rather they appear to have been introduced from outside.

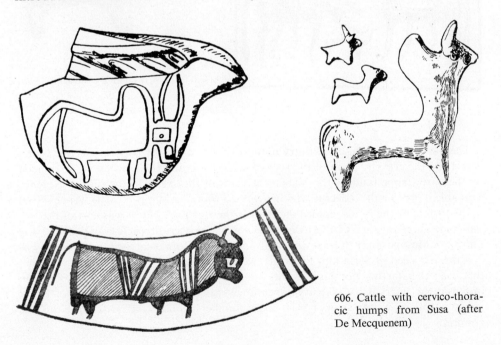

606. Cattle with cervico-thoracic humps from Susa (after De Mecquenem)

607. Bulls with indented cervico-thoracic humps on a seal cylinder from Elam (after Legrain)

608. Zebu cattle with cleft humps on clay seals from the Proto-Elamitic period of Susa (second half of 4th millennium B.C.) (after Nagel)

609. (left) Humped bovine facing a lion on a seal from Susa (after De Morgan)

610. (right) Humped bull from Tell Halaf (after Von Oppenheim)

There is steppe and desert country both west and east of Mesopotamia: the Syrian desert and the gravelly plain of El Hamed on the west, and the Lut desert surrounded by the bare steppe land of the plateau of Iran on the east. On the scanty evidence available it is difficult to decide whether cervico-thoracic-humped cattle were evolved in the former or the latter region; or whether their original home was east of the Iranian plateau, in the Great Indian desert. But as the cervico-thoracic-humped type was known in Mesopotamia at least one thousand years before its arrival in Egypt, and as it is not recorded in Syria and Palestine, we may exclude the Syrian desert. For had this been the starting point of its diffusion, the animal would have been common in Syria, the interval between its appearance in lower Mesopotamia and in Egypt would hardly have been so great, and larger numbers would have passed into the Nile country.

Records of Cervico-thoracic-humped Cattle in Northern Baluchistan

In the quest for the cradleland of cervico-thoracic-humped cattle we have therefore to turn our attention to the Great Salt and Lut deserts of Iran and to the steppe country on their eastern fringe. The proximity of this area to the Quetta–Pishin valley, situated in the semi-arid upland region between Afghanistan, southern Baluchistan, and Sind, would account also for the occurrence of cattle with large cervico-thoracic humps on Quetta pottery from the latest prehistoric (G) period (Fig. 611), contemporaneous with the Harappan civilisation (c. 2500–1500 B.C.). Bovine remains from the preceding prehistoric (H) period seem to be of the same type. In the chalcolithic site of Rhana Ghundai in northern Baluchistan humped cattle are depicted on pottery which can be correlated with Sialk III or Hissar I, and thus with the al-'Ubaid period of Mesopotamia, towards the end of the 4th millennium B.C. (Zeuner, 1963a). In the Quetta–Pishin valley the bovine remains from the prehistoric (H) period succeeded still earlier remains of cattle with larger teeth "almost identical with the teeth of Egyptian cattle" (Fairservis, 1956), while they themselves were either contemporaneous with, or followed by, zebu cattle with thoracic humps (see p. 514).

Records of Cervico-thoracic-humped Cattle in the Indus Valley

East of the Quetta–Pishin valley, cattle with well developed cervico-thoracic humps are found on seal cylinders from Mohenjo-daro and Harappa, dated to the period 2500–1500 B.C. At the present day the Siri of Bhutan is the only breed with a cervico-thoracic hump in India; but humps more cranially situated than is common in recent zebu cattle occur also in other Indian breeds (Ware, 1942; Slijper, 1951; Thorpe, 1953; Milne, 1955).

Records of Cervico-thoracic-humped Zebu Cattle in Southern Baluchistan

South of the Quetta–Pishin valley, cattle with large cervico-thoracic humps are represented on painted pottery and by baked clay figurines from the Kulli culture of the Kolwa region of South Baluchistan, which is coincident with the Amri–Nal culture of Sind and the head of the Nal valley and contemporaneous with the end of the Indus valley civilisation (Mohenjo-daro and Harappa), towards the middle of the 2nd millennium B.C. There is no trace of Kulli contacts landwards farther west than Bampur, just over the border into Persian Makran. But there is certain evidence that Baluchi

611. Neck-humped bulls on Quetta ware from the latest prehistoric period of the Quetta-Pishin valley of West Pakistan (after Fairservis)

612. Zebu bulls with cervico-thoracic humps on seal impressions from Mohenjo-daro (after Marshall)

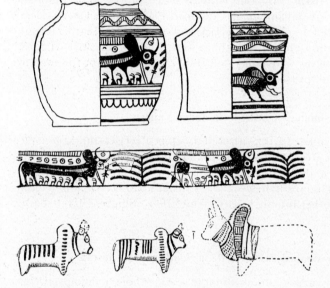

613. Painted pottery and clay figurines, showing cervico-thoracic-humped cattle, from Kulli, south Baluchistan (2nd millennium B.C.) (after Piggott)

merchants traded with Sumer by sea and that cervico-thoracic-humped cattle were introduced into southern Mesopotamia from south Baluchistan by the sea route (Piggott, 1952). Relations between Mesopotamia and India are attested very early in history, yet there is no absolute proof that they were effected by sea rather than by land (Roux, 1966). From the end of the 3rd millennium B.C. onwards, however, cuneiform texts frequently mention ships sailing from Ur to Dilmun (Bahrain), Magan and Meluhha, and there is ample evidence that the kings of Akkad, c. 2300 B.C., endeavoured to attract the countries which bordered the Persian Gulf within the sphere of their political and economic influence.

It is probable that cervico-thoracic-humped zebu cattle reached the Indus valley and Baluchistan from the semi-arid steppes on the eastern fringe of the Great Salt desert of Iran. These areas have not yet been very well explored archaeologically; and it is possible that future excavations will provide a clearer indication of the locality where the centre of evolution of the cervico-thoracic-humped cattle is to be found.

2. The Home of the Thoracic-humped Zebu Cattle

Introduction of Thoracic-humped Zebu Cattle into Africa

There appears to be only a single record of a thoracic-humped zebu in ancient Egyptian art. This consists of a bronze weight in the form of a short-horned zebu from the XVIIIth Dynasty (Fig. 614).

This ancient Egyptian figurine is followed by a gap of nearly 2000 years during which the thoracic-humped zebu is not represented in Africa. In the Horn of East Africa rock paintings and engravings of zebu cattle in semi-naturalistic and conventionalised styles appear together with those of the camel, and perhaps the horse, not earlier than the 4th century A.D., and probably appreciably later, though the weathering and patination to which some of them have been subjected assign to them a reasonable antiquity (Clark, 1954).

The recent zebu cattle of Africa are therefore comparative newcomers. Indeed, zebu cattle do not seem to have entered Africa in any considerable numbers before the Arab invasion of A.D. 669. Their distribution along the East African coast was effected primarily by Arab and Indian traders.

Similarity between East African and West Asian Thoracic-humped Zebu Cattle

The close similarity in type between the small short-horned thoracic-humped zebu breeds of Eritrea (Bahari) and Somaliland (Gasara) and the zebu cattle of southern Arabia suggests that these and similar East African breeds were introduced by Arab immigrants from Arabia in recent times.

In recent Palestine zebus cattle are absent, and in earlier times they seem to have been introduced there only occasionally from the east. The earliest Palestinian record of a thoracic-humped beast is represented by a drawing from Gerar, dated to the 8th century B.C.; Zeuner (1963a) attributes this early post-Exile record to Persian influence.

The recent zebu cattle of southern Iraq (Jenubi and Rustaqi) and of eastern Iran (Khurasani and Seistani) are of the same general type of short-horned thoracic-humped zebu that is encountered in East and West Africa, so that there can be no

614. Ancient Egyptian bronze weight in the form of a thoracic-humped zebu, XVIIIth Dynasty (1570–1305 B.C.)

15. Thoracic-humped zebu cattle in southern Arabia

doubt as to the origin of the African short-horned thoracic-humped zebu cattle from south-west Asia.

Appearance of Thoracic-humped Zebu Cattle in Mesopotamia, Baluchistan and India

In southern Babylonia the thoracic-humped zebu replaced the original humpless cattle during the middle of the 2nd millennium B.C.; a few thoracic-humped animals seem to have reached Sumer still earlier, as indicated by the occasional occurrence of this type in ancient seals and sculptures.

In the Quetta–Pishin valley, northern Baluchistan, at the head of the Bolan Pass, zebu cattle with thoracic humps and large horns (Fig. 622) occurred, along with cervico-thoracic-humped zebus (see pp. 511–512), during the latest prehistoric period, dated to approximately 2500–1500 B.C.

In the Deccan, a moderately high plateau with a temperate climate in south-central India, thoracic-humped zebu cattle appear first in a neolithic context. According to

616. Zebu ox from the synagogue at Beth Alpha (6th century A.D.)

617. Zebu bull from southern Iraq

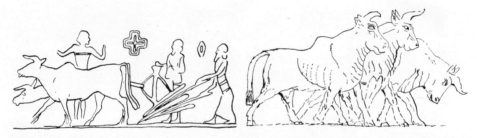

618. (left) Thoracic-humped zebu oxen on a Babylonian seal from the period of the Kassites (about 1500 B.C.)

619. (right) Sculpture of thoracic-humped zebu cattle from the Assyrian palace at Kouyunjik

620. (left) Thoracic-humped bull. Fragment of Sumerian sculpture (after King)

621. (right) Adad leading a thoracic-humped bull. Babylonian seal cylinder (after Ward)

622. Thoracic-humped zebu cattle on potsherds from the Quetta-Pishin valley, northern Baluchistan, latest prehistoric period (after Fairservis)

623. Zebu with ill-defined position of hump on a potsherd from Nal, Baluchistan (after Friederichs)

624. Rock paintings and bruisings of thoracic-humped zebu cattle in Raichur District, Mysore (after Allchin)

Allchin (1963), they were introduced by neolithic settlers. The latter apparently came from Baluchistan by way of Gujarat and Maharashtra. An early extension of the western neolithic complex is suggested also by the discovery of a neolithic tool factory, with tools out of place in the south-east Asiatic culture complex, in the state of Madras. Moreover, in north-west India, extending for some distance down the west coast there is a factor of round-headedness in the human population, which may be due to the invasion of a people from beyond the north-west frontier, a region where round heads are still common (Linton, 1956). However, in a later publication Allchin (1969) expresses the opinon that domestication of the Indian zebu "must have been from an indigenous species".

Habitation levels at Atnur have yielded a large number of bones of two types of cattle, one large and heavy, the other small and light (Allchin, 1963). The former recalls to mind the large cattle from the neolithic settlement of Fikirtepe in north-west Asia Minor (see p.249) and the cattle with large teeth reported from the lowest levels containing remains of domesticated animals in the Quetta–Pishin valley (see p.251), while the latter is doubtless of zebu stock. Figurines from the neolithic sites of Maski and Piklihal, near Mudgal, Raichur district, represent zebu cattle of a lightly built type with prominent thoracic humps and long curved horns. Bulls, bruised or painted in red ochre on rocks in the vicinity of these stations (Fig. 624), are distinguished by thoracic humps and large lyre-shaped horns some of which are even longer than those of the thoracic-humped zebu cattle on potsherds from the Quetta–Pishin valley. In view of the absence of cervico-thoracic-humped zebu cattle in the paintings, drawings and, figurines from the Deccan, in contrast with their presence in southern Baluchistan and along with thoracic-humped zebus, in the Quetta–Pishin valley, it is probable that the representations of zebu cattle in the Deccan are later than the paintings on the Kulli and Quetta wares.

A charcoal sample from a cattle enclosure in an early level of the neolithic 'ash accumulation' period in the mound of Atnur, Mahbubnagar district, has provided a radiocarbon date of 4120 ± 150 B.C. (Allchin, 1963). In a later publication, Allchin (1969) writes that at Atnur a largely pastoral people kept cattle, sheep and goats during the last quarter of the 3rd millennium B.C. With regard to the Kashmir neolithic culture, Piggott (1952) has found no valid reason for regarding this as of any proven antiquity beyond perhaps the 1st millennium B.C. On the evidence at present available, neither in central India nor in Kashmir can the local neolithic cultures be dated satisfactorily much before 500 B.C. In certain areas of southern India they are of relatively late date, surviving almost to the dawn of recorded history (Piggott, 1952). Since neither zebu nor any other cattle occur in Indian sites prior to the neolithic age, the presence of humped stock in the Neolithic of central and southern India must be attributed to relatively late introductions from the north-west.

Possible Cradleland of the Thoracic-humped Zebu Cattle

From the long interval between the arrival of thoracic-humped cattle in southern Babylonia and East Africa it may be inferred that, as in the case of the cervico-thoracic-humped type, the centre of evolution of the thoracic-humped zebu lies not west but east of Mesopotamia. This is evidenced by their early occurrence in northern Baluchistan, their introduction into the Deccan by neolithic settlers, and also by the present distributional area of the thoracic-humped zebu in Asia, which includes nearly the entire southern part of the continent, extending eastwards to the shores of the Pacific Ocean.

Marshall (1931) suggested that the breeding centre of humped cattle was India, "from which country they were introduced into Elam at a very early date". In Assyrian times they were brought into Mesopotamia from the south, or 'Sea Country' (King, 1915). In the shape of their horns and the more caudal position and lesser prominence of the hump these cattle differ from the zebus on the Indus valley seals, which portray animals with large cervico-thoracic humps and long sickle-shaped horns. "Although its original habitat has still to be found," writes Marshall (1931), "this type of bull can for the present be definitely associated with the Indus valley civilisation, for though humped cattle appear in the art of other countries they are never shown with the immensely long horns that they possess on the Indian seals." Adametz (1925) regarded the northern and north-western parts of India as the birth place of the zebu, because

India is the present centre of distribution of the type, with a large variety of different breeds; and Zeuner (1963a) says that its presence in very early prehistoric sites in India makes its Indian origin virtually a certainty.

From a geographical point of view this theory could be accepted; for the fringe of the Thar or Great Indian desert represents the kind of environment that would favour the evolution of the desert type in domesticated animals, and would encourage cattle breeders to select appropriate breeding stock. Moreover, the present distribution of the thoracic-humped zebu, as well as the chronological stages of its arrival at different localities, may well be quoted in support of the view that the evolution of this type took place in north-western India.

Introduction of Thoracic-humped Zebu Cattle into India

However, opposing views and evidence are not lacking. In northern Baluchistan as well as in Mesopotamia the appearance of humped cattle probably antedates their introduction into the Indian prehistoric sites. Col. Sewell believes that they were brought into India from the west by some immigrating offshoot of the Mediterranean race (Marshall, 1931). Friederichs (1933) shares the view that India is not the original home of the zebu. On seals from Mohenjo-daro and Harappa humpless long-horned and short-horned bulls are more frequent than humped bulls (Childe, 1935). When the Vedic Aryas invaded the Indus region, humpless cattle in the conquered territory were still superior in number to the zebu. We read in Rig–Veda (VIII. 5,37) that King Kaçu's herds included 10,000 straight-backed cattle and only 100 humped. Many recent zebu breeds, such as the Bhagnari, Gaolao, Ongole (Nellore), Hariana and related forms, are believed to have entered India through the Bolan and other northern passes between 2200 and 1500 B.C., and to have spread along the route taken by the Rig–Vedic Aryan invaders from Kalat, Baluchistan, through central India as far south as Madras (Olver, 1938; Ware, 1942). The archaeological evidence so far available likewise suggests that the ancient Indus civilisation received its first impetus from farther west (Forde, 1934). As Marshall (1931) writes:—"One thing that stands out clear and unmistakable both at Mohenjo-daro and Harappa is that the civilisation hitherto revealed at these two places is not an incipient civilisation, but one already age-old and stereotyped on Indian soil, with many millennia of human endeavour behind it."

From this it would follow that zebu cattle reached India from the west, and that the Thar or Indian desert must be discarded as the original home of the thoracic-humped zebu cattle in favour of the steppe country on the fringe of the Lut and Great Salt deserts—i.e. the same region in which the cervico-thoracic-humped type was evolved.

Probable Migration Route of the Thoracic-humped Zebu to Africa

Marshall (1931) writes:—"The evidence at present available suggests that humped cattle gradually made their way from Elam to Egypt, via Anatolia and Syria." But this is most doubtful, for records of humped cattle are not plentiful in Anatolia and Syria, where shorthorn (brachyceros) cattle have been the prevalent type for more than four thousand years; and this applies to Egypt even more strongly.

On their passage to Africa the thoracic-humped zebu took a similar course to that taken by the cervico-thoracic-humped cattle before them, namely by way of the littoral of the Arabian Sea, entering Africa at the Horn and the slave and ivory markets of the east coast.

3. The Descent of the Zebu Cattle

Peculiarities of the Zebu Skull

Klatt (1927) regarded the problem of the descent of the zebu as the most difficult in the quest for the origin of cattle; for in its extreme form the zebu skull differs far more than that of any other domestic taurine type from the skull of the wild urus.

The skull of the zebu in general displays a very high degree of variability in conformation. The typical skull may be described as long and narrow, with a convex profile, even surface, and orbits of little prominence. The convex profile and receding forehead occur in bulls more frequently than in females. In zebu breeds which have the horns directed forwards they are not encountered.

In general, the direction of the zebu horn is lateral or upright with a backward tendency, i.e. the horns are of the auchenokeratos type. In profile they are usually set well behind the face. The static forces resulting from this horn direction are the direct cause of the vaulting of the frontals and their caudal inclination. In other words, the receding forehead in the majority of zebu cattle is due to the backward direction of their horns. In heavily horned zebu cattle the effect on the conformation of the whole skull of the horns' increased moment of rotation is so great that it does not remain limited to the neurocranial profile but extends to the splanchnocranium, causing a pronounced Roman nose. However, in view of the fact that the Roman profile may occasionally occur also in polled zebus, and that the ram-like head is typical of several breeds of horses, Spöttel (1936) has suggested that, in addition to the mechanical agency of the characteristic zebu horns, certain genetic growth tendencies may be involved in the shaping of the zebu skull.

The vaulting of the frontals and the convex facial profile level all elevations and cavities of the 'normal' bovine cranium, producing the even surface by which the zebu skull is distinguished. In zebu cattle with horns pointing forwards the character of the skull approximates to either the primigenius or the brachyceros type, depending on the length and thickness of the horns. On the other hand, the likeness of orthoceros and some Apis skulls to the typical zebu skull is due to the peculiar horn direction in the former, resembling the characteristic zebu horn. The variations in the shape of the Torus frontalis in zebu skulls are likewise the direct result of the horns' weight and direction. The raised rounded poll of Africander cattle, for example, more especially of bulls, is due to the downward direction of the horns at their bases, somewhat similar to that encountered in the British Longhorn. This emphasises the vaulted appearance of the forehead (Duerst, 1926a).

Since variations in the weight, shape and direction of cattle horns are due to the effects of domestication, it stands to reason that all peculiarities in cranial conformation caused by horn weight and horn direction are secondary domestication features. For this reason the peculiar poll in many zebu skulls, their vaulted and receding frontals, convex profile, and general evenness should not be sought in the wild ancestors of the zebu cattle.

Desert Type in Cattle

Regarding the length of the skull, it should be understood that the zebu skull is not absolutely longer than the primigenius skull. It is relatively longer, owing to its narrowness, the zebus of all domesticated cattle possessing by far the greatest relative length of the splanchnocranium. The narrowness and slenderness of the skull, and the slight prominence of the orbits, agree with the general conformation of the zebu, i.e. the commonly narrow body with long legs and drooping rump, and the fine bone of the skeleton. The zebu represents the desert type of Bos taurus, analogous to the greyhound, the Nejd and similar desert breeds of sheep, and several of the savanna goats.

In all mammals, domesticated or wild, the leggy, thin-boned, slender desert type is adapted to the need of walking long distances in search of feed and water. The sloping rump and hard flinty feet and light bone of the zebu answer the requirements of speed. It is not, as Duerst (1931) suggested, that the drooping rump of the zebu cattle has been evolved through their early and long-continued use as draught animals; cattle of all types have been employed for draught. The zebu is so eminently suited for draught because selection under specific conditions has modelled its rump in horse-like fashion and fitted its frame for speed.

The Hump—a Domestication Feature

The most conspicuous zebu feature is the hump, although humps are occasionally encountered also in cattle which do not include a zebu strain. Owen (1856) recorded that the feral park cattle of Chillingham, which have no connection with the zebu group, sometimes developed humps. But this has not been confirmed by later authors; on the contrary, Whitehead (1953) writes that White cattle never develop humps. Owen's statement may therefore be due to a confusion of a prominent cervical crest in the bull with a hump. A very prominent crest approximating to a cervico-thoracic hump occurs in some of the brachyceros cattle of Egypt and western Asia (Syria, Cyprus), although there may have been no zebu cattle among their ancestors; it is also characteristic of several Italian breeds (see p. 308). Nevertheless, the hump is so typical of the zebu that we may safely generalise that the vast majority of humped cattle are either purebred zebus or carry a strain of zebu blood. Yet, while opinions may differ on other details of zebu conformation, nearly all authors on this subject are agreed that the hump—be it cervico-thoracic or thoracic in position, muscular or musculo-fatty in structure—is a typical domestication feature, similar to the fat tail and fat rump of certain sheep; that is to say, the wild ancestors of the zebu cattle had neither cervico-thoracic nor thoracic humps. Zeuner (1956), however, regards it as conceivable that a race of wild cattle existed in the drier parts of India, which was in possession of the hump and is now extinct. Were we to consider the domesticated zebu as derived from such a hypothetical southern Asiatic urus with a cervico-thoracic or a thoracic hump (no rock drawings of such an animal have ever been found), we would thereby merely shift the problem of how and why the humped form evolved from the humpless into an earlier geological period, substituting natural for directed selection.

'Hump' Formation in the Urus

Requate (1957) has pointed out that the urus displayed a tendency to hump formation. This is seen in many palaeolithic cave paintings and early historical representations of the animal (Figs. 258 and 289). However, the 'hump' of the urus, as shown in the paintings, was not cervico-thoracic, as was the hump of the earliest zebu cattle known to us, i.e. those of Sumer, India and Egypt, but thoracic in situation. Moreover, as preserved skeletons of the urus indicate, the marked elevation of the withers, suggesting a hump, was caused by the great length of the spinous processes of the thoracic vertebrae, which in turn was necessitated by the large weight of the head and horns of the animal (see p. 229).

The Bifid Spinous Processes of the Anterior Dorsal Vertebrae in Zebu Cattle

In addition to the hump, the zebu group as a whole, irrespective of the situation and structure of the hump, is distinguished by bifid spinous processes of the anterior dorsal vertebrae. The term Spina bifida has been applied to this feature in zebu cattle, but in human pathological anatomy this term is restricted to a congenital malformation in which the spinal column is cleft at its lower portion, and the membranes of the spinal cord project as an elastic swelling from the gap thus formed.

Pettit (1909) and Gans (1915) regarded the bifidity of the spinous processes as a typical racial character of zebu cattle; Klatt (1927) went further, expressing the hope that it might yet provide an important clue in the quest for the zebu's wild ancestors.

However, this view cannot be upheld since a number of humped cattle are devoid of bifid neural spines. For instance, these are absent not only in neck-humped brachyceros cattle of Egypt and western Asia (see p. 298) and in many neck-humped sanga cattle of Africa, but also in the pituitary dwarf zebu skeleton from India, in the Museum of the Royal College of Surgeons, London, which the author examined in 1936. On the other hand, Duerst (1931) mentioned the occasional occurrence of bifid spinous processes in humpless cattle (Simmentaler), where it is ascribed to the strong pull exercised by the weight of the head on the splenius and the nuchal and supraspinous ligaments. Further, no wild bovine has the neural spines of the dorsal vertebrae bifid, this character occurring only in domesticated cattle.

Bifid spinous processes are found in numerous animals in different parts of the spinal column. Mijsberg (1926) has pointed out that phylogenetically they are a specialised character in relation to the more generalised undivided processus spinosi, while ontogenetically the neural spines evolve from two separate bases. In the most primitive type (Fig. 625, 1) the neural spine develops from the cartilaginous tissue formed by the terminal union of the bilateral neural arches which are devoid of any protuberances at the tips. In the more specialised types (2a) and (2b) the tips of the

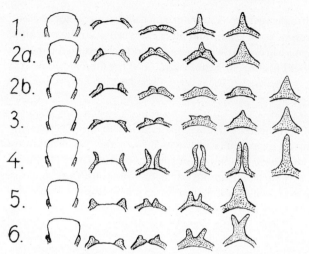

625. Sketch of different developmental types of the processus spinosi in mammals (after Mijsberg)

bilateral neural arches are thickened, and these protuberances share in the basis of the neural spine, the major part of which evolves, however, from the cartilage formed by the coalescence of the neural arches. In type (3) the protuberances at the tips of the neural arches are so strongly developed that the median tip formed by the fusion of the neural arches rises above, and absorbs, the lateral hyperapophyses only at a comparatively late stage of development. In type (4) the protuberances grow into long processes which subsequently unite into the single spine, with the median cartilage retaining only slight importance in the formation of the process. In type (5) the lateral protuberances fuse at first at the base only, the single spine evolving after a bifid stage. In type (6) the fusion of the two original halves of the processus spinosus remains restricted to the base throughout the life of the individual, with bifid neural spines as the result.

Von Eggeling (1922) has suggested that in analogy with many other skeletal crests and ridges the development of bifid spinous processes may be due to the influence of the muscles and sinews attached to the spinous processes. There may be significance in the fact that in zebu cattle—"because of the hump", as Duerst (1931) says—the nuchal ligament runs not along the top of but alongside the spines of the thoracic vertebrae. Duerst (1931) regards the bifurcation of the spinous processes as functional of the weight of the hump. But in view of the fact that certain humped zebus are devoid of the bifid character, whereas this may occur in unhumped sanga and other cattle, the validity of this theory is doubtful. In man Mijsberg (1926) has ascribed the bifurcation of the neural spines of the cervical vertebrae to a factor inhibiting the full development of the median cartilaginous tissue formed by the fusion of the bilateral neural arches, with the result that the hyperapophyses, forming the main portion of the spinous process, remain terminally separate. This would indicate a genetical basis for the bifidity of the spinal processes, such as Hammond (1940) has postulated for this character in the rabbit. While opposing the hypothesis that mechanical agency due to increased muscular development may be responsible for the evolution of bifid neural spines, Mijsberg does not deny the possibility that reduced mechanical requirements may be a contributory factor "as it is very probable that the bifurcated processes are to be regarded as less strong than those in which the halves are completely joined".

However, as Epstein (1955c) has pointed out, there is no proof that reduced mechanical pressure has any greater significance in the formation of bifid neural spines than have increased mechanical requirements. From its rather erratic occurrence in other animals, and also in man, in the most various parts of the spine it would appear that the bifidity of the neural spines of the anterior dorsal vertebrae in zebu cattle has no direct selective value. It is doubtless genetically linked to the hump; but the exceptions indicate that the linkage is not complete. The mode of inheritance of the bifurcated spinous processes in the cross-breeding of zebu and humpless cattle is still unanalysed. Along with the hump, the bifidity of the neural spines of the anterior dorsal vertebrae of zebu cattle appears to be a domestication feature, offering no clue in the quest for the wild ancestors of the zebu.

Physiological Peculiarities in Zebu Cattle

In warm regions zebu cattle have proved to be more resistant to high atmospheric temperatures and intense solar radiation than European breeds of cattle. External temperatures above 23°C greatly increase the metabolism of the latter, but only slightly that of zebus (Rhoad, 1936). At a given environmental temperature European breeds and grade cattle consume more water than zebu breeds (Lampkin et al., 1958). In the Imperial Valley of California Hereford cattle drank some 16 gallons (60 litres) of water daily, while zebu cattle drank only 10 gallons (38 litres) (Ittner et al., 1951). The large difference in water consumption, which is even more marked under high ambient temperatures, is related to differences in physiological adaptability as well as body size. The capacity of the zebu digestive system is considerably smaller than the European (Swett et al., 1961). Hence, because of limited capacity, the zebu is forced to graze more often and take less per grazing period than European cattle. Zebu cattle in the tropics lie down mostly at night (Harker et al., 1961), European (Hereford) cattle spend roughly 50 per cent of the daylight hours lying down, during most of which the animals are ruminating (Hughes and Reid, 1951).

Zebu cattle have a larger body surface—dewlap, navel flap, prepuce, large ears, hump —per unit weight, exceeding that of European breeds by approximately 12 per cent. It has been suggested that this facilitates heat dissipation by non-evaporative methods (convection, radiation, conduction)—as long as the environmental temperature does not equal or exceed the body surface temperature—as well as by evaporative (moisture vaporisation) means. But McDowell et al. (1955) and McDowell (1958) have shown that the dewlap has a low surface evaporation rate and that there are no appreciable differences in the rectal temperature or respiratory responses of Red Sindhi zebu bulls after removal of the dewlap or of the hump and about 10 cm of each ear. The sweat glands on the dewlap of zebu cattle are much less numerous per unit area, and also slightly smaller, than on the midside (Nay and Hayman, 1956).

Ferguson and Dowling (1955) have produced evidence that the apocrine sweat glands of cattle have a temperature-regulating function, and that the evaporation of sweat is the main source of heat loss in hot environments.

Brody (1948) suggested that zebu cattle had evaporative coolers, i.e. active sweat glands, while European cattle apparently had not, or sweated only slightly, so that their evaporative cooling from the skin was mostly, if not entirely, of osmotic or diffusion moisture. However, Thompson, McCroskey and Brody (1951) subsequently found that the vaporisation rate per unit surface area of zebu cows and heifers did not exceed that of Brown Swiss, Friesian and Jersey cows and heifers.

Rhoad (1940), on the other hand, has demonstrated that at a shade temperature of 30°–35°C purebred zebus vaporise nearly twice the amount of water through the skin as purebred British beef cattle. In the Sindhi zebu the sweat glands are larger and more numerous per sq. cm than in Formosa cattle and Dutch Friesians (Yamane and Ono, 1936). Dowling (1955a) found that in zebus the mean number of hair follicles per sq. cm (there is a consistent association of the apocrine gland and the hair fibre in cattle) was 1698, in zebu-Shorthorn crosses 1321, in Shorthorns on a low plane of nutrition 1064, and in Shorthorns on a high plane of nutrition 764. Similarly, Nay and Hayman (1956) concluded from the study of dairy-type zebus and dairy-type European cattle that in zebus the apocrine sweat glands were more numerous, longer and of greater diameter than in European cattle. Zebu–Shorthorn crossbred animals had larger sweat glands and a greater number per unit area than purebred Shorthorn cattle. The difference in sweat gland volume paralleled the extent to which the groups were heat-tolerant. If volume is broken down into the two components, number and size of glands, there is a suggestion that size of gland may be the important component (Nay and Dowling, 1957).

The sweat glands of zebu cattle are of the same apocrine type as those of European breeds, possessing a poor vascular supply; they do not differ in the myoepithelial cell layer nor in the chemical composition of their secretion (Yang, 1952; Goodall and Yang, 1952). In appearance the apocrine sweat glands of zebu cattle are sac-like, with few convolutions, whereas the sweat glands of European cattle are rarely sac-like, and quite convoluted. Again, the larger, more active sweat glands of zebus are situated closer to the skin surface than in European cattle (Nay and Hayman, 1956).

The zebu (like the Jersey) has a relatively thin skin; from this it may be inferred that a thick skin is not essential for adaptability to a hot environment (Dowling, 1955b). But the epidermis of the zebu is very strong and thick.

Zebu cattle generally have a short glossy coat with a high fat content; their skin is furnished with a greater number of sebaceous glands than that of European breeds; Yamane and Ono (1936) found 3181 sebaceous glands per sq. cm in an Indian (Sindhi) zebu bull as against 2253 in a Friesian.

Many zebu breeds possess protective coloration, combining a light-coloured coat, which reflects both heat rays of relatively short wave length and light rays, with a dark hide, protecting nerve and skin, and perhaps the general blood chemistry, from injurious ultra-violet radiation.

In addition, as Kibler and Brody (1951) and Kibler (1957) have shown, Brahman (zebu) heifers possess a lower basal metabolic rate than Brown Swiss, Shorthorn or even Santa Gertrudis heifers.

For these various reasons, apparently, the physiological adaptation of zebu cattle to high ambient temperatures is superior to that of European breeds, the differences in the physiological response to tropical climatic conditions being genetic in origin (Rhoad, 1940).

However, the adaptation of Texas Longhorn and Criollo cattle, for the greater part of Iberian longhorn derivation, to tropical and subtropical environments in the Americas, and of shorthorn (brachyceros) cattle to the coastal swamps and tropical rain forest of West Africa, demonstrates that such hereditary adaptation is not confined to the zebu group. The zebus have acquired genetic adjustment to high atmospheric temperatures and intense solar radiation in the course of their evolution in a warm region. There is no need to attribute the genetic factors facilitating the physiological and anatomical adaptation of zebus to ambient heat to a wild ancestor of their own, furnished with similar genetic properties, although obviously in a wild subgenus like the urus, ranging over so huge an area in Asia, North Africa and Europe, the southern representatives would be genetically better adapted to a warm climate, and the northern to a cold one.

Lack of Evidence of the Zebu's Descent from a Special Urus Race

While we distinguish at present between three geographical races of the urus, i.e. the Asiatic Bos primigenius namadicus, the North African Bos primigenius opisthonomus and the European Bos primigenius primigenius, it is probable that more than one race existed on the vast continent of Asia, and that the variation displayed by the whole urus group was not less than that encountered among the recent members of the wild genera Capra and Ovis, or Bubalus and Syncerus, for example. The small number of skulls and other skeletal fragments excavated in distant parts of Asia, which are all included in the term Bos primigenius namadicus, may in reality belong to different geographical races.

In some domestic animals it is possible to point to the wild species or subspecies from which a peculiar racial trait is derived: heteronymous horns in goats, the shoulder stripe or barred legs in domesticated asses. In others it is equally certain that peculiar racial features, such as the bulldog head in dogs, 'natism' in Niata cattle, chondrodystrophy in zebu cattle, goats and dogs, and the fat tail, fat rump or long-and-thin tail in sheep, are not directly inherited from wild races similarly distinguished. Regarding the zebu cattle, there is no evidence or likelihood that among the wild urus races of Asia one was furnished with a cervico-thoracic or a thoracic hump, bifid neural spines, horns of auchenokeratos type, cranial peculiarities due to such horns, a drooping rump, and other zebu characteristics.

Descent of the Zebu from Humpless Cattle

It would therefore appear that the various anatomical and physiological properties, held to be characteristic of zebu cattle, were evolved by selection from among ordinary cattle of Bos primigenius namadicus descent for specific purposes in a peculiar environment in south-west Asia.

Theory of the Zebu's Descent from the Banteng

However, the derivation of the zebu cattle, through the link of humpless cattle, from the Asiatic urus is not generally accepted. Until Gans (1915) published his study on the racial relationship between the zebu and the banteng (Bibos javanicus), the banteng was generally believed to be the ancestor of the zebu cattle. Even in some modern text books (Rice et al., 1957) this theory is still upheld. Still more recently Hafez (1962) notes that "... the Zebu cattle, B. indicus, are thought to descend from the Malayan banteng," adding, however:—"Despite their different origins, the behaviour patterns of European and Zebu cattle are quite similar..." And Mayr (1963), discussing the problem of 'hybridization and the origin of domestic animals', writes that "In the case of cattle there seem to have been repeated independent domestications, one starting from the (now extinct) wild Bos taurus of the western Palaearctic, another starting from the wild Bos banteng and leading to the Indian cattle". The "Indian cattle" are presumably what is commonly called zebu.

The first to suggest the banteng descent of the zebu was Rütimeyer (1865) who thought that the zebus of Asia showed a greater resemblance to the banteng than to the domesticated cattle of Europe. Keller (1902) supported this theory on observations of a number of zebu and banteng skulls in which he found several common features. Lydekker (1912a) inferred from an external resemblance that the zebu was derived from Bibos javanicus. He found a likeness in the shape of the horns between the Gujarati (Kankrej) zebu bull and the male tsine banteng, a similar development of the dewlap in the zebu and banteng, and a darker colour in the banteng bull than in the cow, a sexual dimorphism encountered also in Brahman zebus. The hump of the zebu cattle, Lydekker suggested, had developed from the high back of the banteng through the shortening of the spinous processes of the dorsal vertebrae.

The Banteng and the Balinese Cattle

The range of the banteng comprises Java, Borneo, Burma, Thailand and adjacent territories. The banteng has been domesticated on Bali island, east of Java—hence the name Balinese cattle for the domesticated type. The horns of the Balinese cattle are considerably smaller than those of the wild banteng. In the cow they are particularly small, pointing upwards and slightly backwards. In the bull they are larger than in the cow, yet much thinner, shorter and simpler in shape than in the wild banteng bull. The reduction in the weight of the horns in the domesticated Balinese cattle has considerably influenced the shape of the cranium, more especially the neurocranium. This applies to both sexes, but to the bull in particular.

Decrease in Size in Domesticated Bovini

Rejecting the theory of the zebu's descent from the banteng, Gans (1915) gave a craniological analysis of the differences between the various domesticated Bovini and their wild ancestors. In the gaur and gayal, in Bos primigenius and fossil domesticated cattle, in the wild and the domestic yak and, finally, in the wild banteng and its domes-

HUMPED CATTLE

626. Banteng bull

627. Balinese bull

628. Balinese ox

ticated descendants, the Balinese cattle, domestication has had fundamentally the same effect: the domesticated animal has become smaller than its wild progenitor. This reduction in size has affected the neurocranium more than any other part of the skeleton. For in the wild beast this part is particularly well developed, carrying powerful horns essential for attack and defence or as display structures. In the domesticated animal these weapons have become weaker, and the neurocranium has degenerated owing to the reduced static requirements.

For example, in the female gaur the length of the cranial profile, measured from the Torus frontalis to the anterior edge of the premaxillae, amounts to 115.3 per cent of the basilar length of the skull, in the male gaur to 117.0 per cent. In the domestic

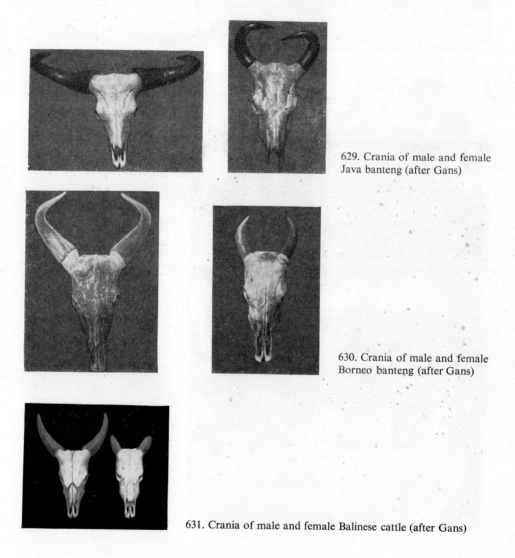

629. Crania of male and female Java banteng (after Gans)

630. Crania of male and female Borneo banteng (after Gans)

631. Crania of male and female Balinese cattle (after Gans)

gayal this percentage is 103.7 for the female, and 106.8 for the male. In the wild banteng it averages 103.5 per cent, in the domestic Balinese cattle 91.4 per cent.

In zebu cattle this ratio is 109.5 per cent. That is to say, if the zebu were actually a domesticated banteng, domestication must have had a totally different effect on its cranial conformation from that in any other bovine species. Gans considered this as sufficient evidence for rejecting the theory of the zebu's descent from the banteng.

Cranial Differences between Zebu and Banteng

In the shape of the occiput the differences between the zebu and banteng are profound. In the banteng skull the superior portion of the occiput ascends obliquely forwards, whereas in the Somali zebu skull it bulges backwards (Klatt, 1913). In Indian humped cattle the narrowest part of the occiput measures 27 per cent, and in the Somali zebu 26.4 per cent of the basilar length of the skull; in the banteng the ratio is 18.8 per cent.

In the classification of Bovini the parietal region is regarded as an important criterion. In the banteng and zebu this region, as Antonius (1922) has pointed out, differs fundamentally. The craniographic outlines of zebu and Balinese skulls illustrate these cranial differences (Figs. 634–638) (Gans, 1915).

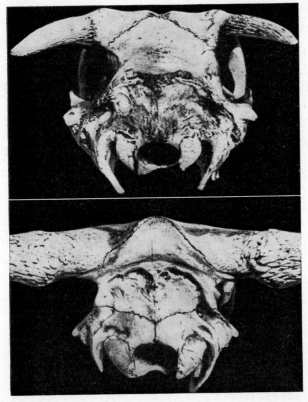

632. Skull of a banteng cow, nuchal surface (after Klatt)

633. Skull of a Somali zebu cow, nuchal surface (after Klatt)

Differences in the Anatomy of the Neural Spines between Zebu and Banteng

The anatomical differences between the zebu and banteng are not confined to the cranium. Gans (1915) has pointed out that in the zebu the bifidity of the neural spines of the anterior dorsal vertebrae is typical, whereas in the banteng and Balinese cattle the neural spines are undivided. However, as the zebu similarly differs from the urus, this cannot be considered as valid evidence against the zebu's descent from the banteng.

In zebu cattle the superior part of the spinous processes from the seventh dorsal vertebra on appears compressed fronto-caudally; in the banteng and Balinese cattle the anterior and posterior edges of the spinous processes are nearly straight and parallel, only slightly approximating at the tips. The spinous processes differ also in direction: in the zebu they form an angle of 140° with the spine, in the banteng and Bali cattle one of 125° (Fig. 639).

The Fertility of Zebu-Banteng Hybrids

The male hybrids from the crossing of zebu and banteng are usually though not always sterile (Gray, 1954); the examination of the semen of sterile males shows the absence of spermatozoa (Gans, 1915). The female hybrids of the F_1-generation, and males descended from female hybrids and banteng or zebu bulls are fertile (Kronacher, 1928).

These conditions warrant the conclusion that the zebu cannot be regarded as a descendant of the banteng.

Theory of the Zebu's Descent from the Bison

Medrano Galvis (1959) suggests that the zebu derives from the crossbred progeny of European and central Asiatic bisons, or from hybrids of the African bison and the wild cattle of Egypt, or from an intercross of these.

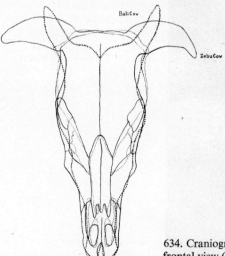

634. Craniographic outlines of female Balinese and zebu skulls. frontal view (after Gans)

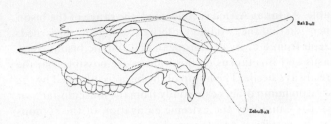

635. Craniographic outlines of male Balinese and zebu skulls, lateral view (after Gans)

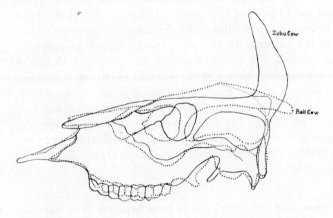

636. Craniographic outlines of female Balinese and zebu skulls, lateral view (after Gans)

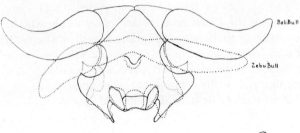

637. Craniographic outlines of male Balinese and zebu skulls, nuchal view (after Gans)

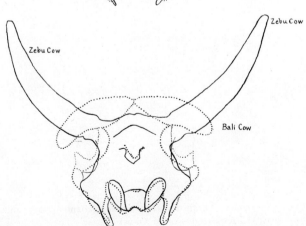

638. Craniographic outlines of female Balinese and zebu skulls, nuchal view (after Gans)

This theory is quite untenable. As far as Africa is concerned, the range of the bison never extended to that continent, so that the two latter alternatives may be discarded. As to the derivation of the zebu from a cross of European and Asiatic bisons, reciprocal crosses of Bison bonasus and Bos taurus or Bos indicus are possible, but the males of the F_1-generation are always sterile. The hump of the bison is inherited by the hybrids (Gray, 1954). But the bison hump differs essentially from the musculo-fatty or muscular hump of the zebu; it is caused by the extreme elongation of the spinous processes of the anterior thoracic vertebrae and the marked curvature of the spinal column. In addition, the bison has 14 pairs of ribs, the zebu 13 pairs. The skull of the bison is distinguished from the crania of other Bovini by the large interparietal and the shortness and great width of the frontals. In the urus, frontal length averages 57 (50–60) per cent, and frontal width 42 (38–45) per cent, of the basilar length of the skull; in the bison the respective percentages are 38 (33–42) and 55 (50–60) (Bohlken, 1958). Gans (1915) found an average frontal length, in relation to basilar length, of 50 per cent in the urus, and 49 per cent in the Indian zebu; an average frontal width of 38 per cent in the urus, and 37 per cent in the Indian zebu. In the Somali zebu the average frontal length amounts to 51 per cent, and the average frontal width to 38 per cent, of the basilar skull length (C. Keller, 1896). These skeletal and cranial conditions refute the theory of the zebu's descent from the bison.

Theory of Descent of the Ordinary and the Dwarf Zebus from Separate Wild Species

Szalay (1930) claims that the ordinary zebu is derived from a separate species of wild ox, and the dwarf zebu from yet another species. But he offers no evidence whatsoever in support of this claim. In the case of the dwarf zebu we are left in doubt whether Szalay refers to the pituitary zebu dwarf or to the achondroplastic breed, or whether he believes that the two are descended from a single or from separate wild diminutive species. Similarly, in the case of the zebu cattle of ordinary size Szalay does not specify whether he refers to the thoracic-humped or the cervico-thoracic-humped type, or if he holds that the two originate from a single or from distinct wild species.

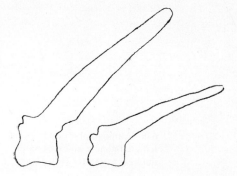

639. Ninth dorsal vertebrae of Banteng (left) and zebu (right) (after Gans)

640. Crossbred bull by cervico-thoracic-humped Africander sire out of thoracic-humped short-horned zebu dam

The Connection between the Cervico-thoracic and Thoracic Humps

This raises the question whether the cervico-thoracic-humped and the thoracic-humped zebu types were independently evolved or one gave rise to the other. Although similar variations may occur in geographically widely separated races, it is doubtful if this applies in the case of the two zebus. For in addition to the same geographical centre of origin, they appear successively in the same areas: Mesopotamia, India, Africa. Also they have too many important characteristics in common, and are too much at variance with other taurine cattle, for this likeness to be merely a matter of coincidence.

The Hump in Crossbred Cattle

The genetics of the hump, although not yet analysed, also indicates the close relationship of the two types. In crossing cervico-thoracic-humped Africander cattle with a humpless breed, the F_1-generation is furnihed with small cervico-thoracic humps. In the crossbred progeny of Africander and thoracic-humped zebu cattle the hump, as has been shown by cross-breeding experiments with Africander bulls and Iringa cows at the Government Stud Farm at Dar-es-Salaam, occupies an intermediate position. The most difficult problem is presented by the cross-breeding of thoracic-humped zebu and humpless stock: hump situation in the F_1-generation differs with different zebu and humpless parent breeds. In some instances the small hump remains thoracic, although Slijper (1951) says that "in crosses of about 50 per cent zebu blood and less, the caudal border of the hump has never been found more caudally than the longitudinal axis of the foreleg". In other instances the humps in the first crossbred generation derived from thoracic-humped zebu and humpless parent breeds is intermediate in situation; but in the majority of cases it is cervico-thoracic, contradicting Bisschop's (1937) remark that it is impossible to understand how a cross between a non-humped and a humped species could result both in a marked change of situation of the hump and the muscles involved, and in a change of the very anatomical nature of the structure. If such halfbreds are crossed back to thoracic-humped zebu cattle, the backcross generation is generally furnished with thoracic humps; if the halfbreds are bred to humpless cattle, the hump in the progeny, if present, may occupy any position. In male Santa Gertrudis cattle, bred at the ratio of three-eighths thoracic-humped zebu to five-eighths humpless Shorthorn, the hump is cervico-thoracic. Thus the change-over in hump situation is a common observation in cross-breeding.

Hump Situation in the Newborn Africander Calf

There is another remarkable phenomenon. In the newborn Africander calf the hump occupies a more caudal situation than in the adult. Soon after birth the hitherto high relative growth-rate of the thoracic part of M. rhomboideus slows down markedly, and the high allometry of its cervical part comes into play; the hump changes from the thoracic to the cervico-thoracic position. In thoracic-humped zebu breeds the allometry of the anterior part of M. rhomboideus is checked early, and the hump retains its thoracic situation. The thoracic hump in adult zebu cattle may therefore not be attributed to the retention of a visible juvenile character. It is due to the persistence into later stages of the early high relative growth-rate of the thoracic part of M. rhomboideus, and the early low relative growth-rate of the cervical part of this muscle.

Evolution of the Thoracic Hump

The cervico-thoracic-humped zebu seems to have been evolved prior to the thoracic-humped from domesticated humpless longhorn cattle in a steppe or desert environment. Here the cervico-thoracic, mainly muscular hump, developed from the prominent cervical crest of the bull, distinctly indicated in a seal impression from the protoliterate period of Babylonia (Fig. 642), was apparently valued by breeders, as it served as a store for the accumulation of reserve material; for in times of plenty a certain amount of fat is deposited in layers also in the cervico-thoracic hump. The endeavour of pastoralists to increase the quantity of fat in the hump seems to have led, by way of selection, to an over-development of the thoracic part of M. rhomboideus, capable of considerable adipose development, and to a relative under-development of the cervical part of this muscle, which forms the hump of the cervico-thoracic-humped zebu and,

641. Newborn Africander calf

642. Bulls with prominent cervical crests in a seal impression from the protoliterate period of Babylonia

apparently, does not lend itself to complete adipose 'degeneration'. Thereby the situation of the hump passed from the cervico-thoracic to the thoracic, M. trapezius simultaneously undergoing certain secondary structural and functional changes.

Whether the deposition of fat in the hump constitutes an accumulation of reserve material additional to the normal fat deposition encountered in the body of humpless cattle, or only a shift to a new location, is uncertain in view of the probability that even the fat tail of sheep may represent merely such a shift (see II, pp. 169–170). However, this is immaterial; the important factor in hump evolution is not the reality of an additional store of fat but the breeders' belief in its existence, resulting in the selection of appropriate breeding stock.

In view of the fact that dissections of Indian zebu humps have shown the majority of these to be muscular rather than musculo-fatty (see pp. 336–338), Mason (personal communication), while accepting the theory of hump development from the crest of the bull through the cervico-thoracic to the thoracic stage, does not attribute the selection for hump size involved in hump evolution to the breeders' desire for increasing the fat reserve in the hump, but to the emotional value of large humps.

Damascus and Cyprus cattle and other shorthorn (brachyceros) breeds of western Asia and Egypt, which, at maturity, develop small cervico-thoracic humps in the male, occasionally in both sexes, accompanied by undivided spinous processes of the anterior dorsal vertebrae, may represent, not crossbreds of humpless shorthorn and humped zebu cattle, as some authors believe, but the primary stage in hump evolution from the prominent crest of the bull. A primary stage in hump evolution is also found in several Italian breeds in which both sexes develop prominent crests. Whenever the selective value of the hump becomes positive in relation to environment and social or economic conditions, a cervico-thoracic-humped breed may be evolved from the primary stage by selection, and a thoracic-humped breed at a still higher stage. It is significant that both the crest of humpless bulls (Holstein) and the hump of the zebu (Red Sindhi) are rhomboideus cervical in origin (McDowell et al., 1958).

While, then, we regard the thoracic hump as evolved from the cervico-thoracic, the cervico-thoracic hump may in some instances be primary, i.e. evolved from the humpless form, in others derived, albeit not regularly, from the crossing of thoracic-humped zebus and humpless cattle.

4. The Descent of the Sanga Cattle

The great variability in general conformation, cranial character, size of horn, and situation, size and structure of hump renders it probable that the East and South African sanga cattle are derived from various ancestral stocks.

Origin of the Sanga Hump

As the two original basic types of African cattle, i.e. the longhorn (primigenius) and the shorthorn (brachyceros), are devoid of humps, the humps of sanga cattle cannot be derived from one of these two stocks. And since an independent evolution of the hump in African sanga cattle may be ruled out as improbable in view of the presence of humped zebu cattle in Africa south of the Sahara, the zebus remain the only source to which the sanga humps are traceable.

With only one known exception of an early thoracic-humped zebu in Egypt (see p. 513), zebu cattle with cervico-thoracic humps entered Africa nearly two thousand years earlier than zebu cattle with thoracic humps. Before their arrival, the cattle of North Africa, West Africa and North East Africa (Ethiopia and the Horn) were mainly of humpless longhorn stock, as demonstrated by numerous rock paintings and engravings of cattle from this early period. The interbreeding of zebu cattle with humpless longhorn stock in East Africa, more especially in Ethiopia (Fig. 643), apparently introduced the hump into the progeny. The variability in the size of the cervico-thoracic sanga hump, and the absence of the hump in individual animals of commonly humped sanga breeds, may partly be attributable to genetic segregation in a mixed type derived from humped zebu and humpless longhorn parent stocks.

However, the theory of the evolution of the cervico-thoracic sanga hump from the cross-breeding of early cervico-thoracic-humped zebu cattle with humpless cattle interprets only one of the problems involved in hump evolution, hump situation and hump structure in the sanga cattle. For the fact that a cervico-thoracic hump may be

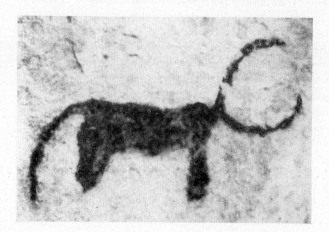

643. Rock painting in reddish-brown colour of a giant-horned sanga ox, from Saka Sharifa, Harar, Ethiopia (after Clark)

the result of cross-breeding of humpless cattle with either cervico-thoracic-humped or thoracic-humped stock renders it possible that not all cervico-thoracic-humped sanga cattle are derived from humpless longhorn and early cervico-thoracic-humped zebus. Considering no other factor but hump situation, we cannot preclude the possibility that many of the more recently formed cervico-thoracic-humped sanga breeds of Africa are derived from humpless cattle and thoracic-humped zebus, or even the possibility that *all* sanga cattle are of such origin. Indeed, the latter alternative is suggested by the synchronism of the first appearance of thoracic-humped zebu cattle in Ethiopia and the Horn of Africa and the probable period of the racial and cultural evolution of the Bantu peoples in East Africa, who, in the course of their subsequent migrations, introduced numerous sanga breeds into the southern part of the Continent (see also pp. 543–548).

The Relation between Hump Size and Hump Position

The relatively small size of the cervico-thoracic sanga hump and the generally larger size of the thoracic zebu hump have provoked the question whether there is a connection between hump size and hump position. In other words, it has been suggested that the larger the hump, the more caudal its situation. In a limited sense this is correct: high condition and consequent deposition of fat in the upper posterior portion of the hump tends to extend the upper part of the thoracic hump, if the latter is not pyramidal or dome-shaped, backwards. But the generalisation of this phenomenon is refuted by weighty evidence. Ancient representations of humped cattle from Egypt, Sumer and India show humps of very large size cervico-thoracic or intermediate in situation; they cannot all be regarded as exaggerations, at any rate not the Egyptian paintings and carvings (Figs. 691–604). Many East African zebu cattle have thoracic humps which are not larger in size or heavier in weight than the cervico-thoracic hump of recent Africander cattle. The small hump of the newborn Africander calf is not more cervical than the large hump of the adult Africander; on the contrary, its position is more caudal. The smaller hump of the East African zebu cow is not situated farther forward than the larger hump of the ox or bull. Red Bororo and White Fulani heifers, however, do have humps farther forward than in bulls. Slijper (1951) found that "there was no correlation between the position and the size (cranio-caudal diameter of the basis) of the hump". But in comparison with the zebu bull "the hump of the cow shows a distinct tendency to a shortening in the longitudinal (cranio-caudal) direction. This shortening may be brought about by a backward shifting of the cranial, a forward shifting of the caudal border or a combination of both processes."

The Impact of the Thoracic-humped Zebu on the Sanga Cattle

Since the introduction of larger numbers of thoracic-humped zebu cattle into East Africa from approximately the 7th century A.D. on, these have superseded the sanga, owing to their superiority in functional properties, over wide areas in East Africa north of the Zambesi river. In other areas of East Africa the impact of thoracic-humped zebu on cervico-thoracic-humped sanga cattle has given rise to new breeds, in many ways similar to the original sanga but differing in the situation of the hump which owing to the increased share of thoracic-humped zebu blood has become mainly thoracic. How far the structure of the hump has been affected by the additional thoracic-humped zebu component has not yet been investigated; but hump structure appears to be basically a function of hump situation, that is of the variable participation of the anterior and posterior parts of M. trapezius and M. rhomboideus in hump development.

In sanga cattle with thoracic humps, which are encountered among the Nilotic and Bashi breeds, for example, but occasionally occur also in other sanga breeds north of the Zambesi river, zebu features generally predominate over traits derived from the humpless longhorn ancestry, as may be expected in a cross consisting of two different zebu strains to only one of longhorn.

The Spinous Processes of the Anterior Dorsal Vertebrae in Sanga Cattle

Zebu cattle are distinguished by the bifid tips of the spinous processes of the anterior dorsal vertebrae. In humpless longhorn and humpless shorthorn cattle the neural spines are generally single. In sanga cattle with cervico-thoracic humps the spinous processes of the anterior dorsal vertebrae display considerable variability. In some the fissure is moderately deep, although commonly not quite so deep as in zebu cattle; in others it is hardly perceptible; and even within one and the same breed the differences are considerable. The variability of this character in cervico-thoracic-humped sanga cattle supports our contention that these are derived from a mixture of zebu and humpless longhorn stocks. In thoracic-humped sanga cattle the neural spines of the anterior dorsal vertebrae have not yet been studied; therefore it is unknown whether they are generally bifid or of variable conformation.

Cranial Variability in Sanga Cattle

The cranial character of sanga cattle varies greatly. In some the cranium is long and relatively narrow, coffin-shaped, with the convex frontals and receding forehead characteristic of the typical zebu. This skull conformation is more general in sanga breeds in which thoracic humps are frequent; but it is by no means restricted to these, for it occurs also in sanga cattle with exclusively cervico-thoracic humps, as among the Nguni cattle. In other sanga breeds, such as the Tswana, the cranium more often

644. Ninth dorsal vertebrae of an Ovambo sanga cow and Zulu sanga ox

shows pronounced primigenius (humpless longhorn) character. With regard to the Ankole sanga of East Africa, Antonius (1943-44) writes:—"Among the giant-horned cattle of the lake district ... we can recognise along with pure primigenius types—in head formation and topline—also those showing distinct zebu influence." Again, the primigenius skull conformation is not limited to sanga cattle with cervico-thoracic humps, although in these it is commoner. It is encountered also in thoracic-humped sanga cattle in East Africa. This variability in cranial type has misled several authors to present opposite extreme generalisations with regard to the racial character of the sanga group. While Stegmann von Pritzwald (1924) regarded the sanga cattle as of pure zebu stock, Antonius (1922) classed them with the primigenius (humpless longhorn) type.

Horn Gigantism in Sanga Cattle

The horns of sanga cattle vary greatly in length, shape and basal circumference. This applies to the cervico-thoracic-humped as well as to the thoracic-humped type. In giant-horned sanga cattle the cores are often hollow and of such a porous material that they may not exceed those of long-and-slender-horned animals in weight.

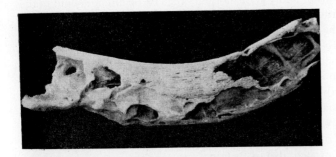

645. Horn core fragment of a giant-horned sanga

646. Texas Longhorn

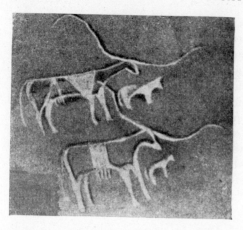

647. Rock drawing of giant-horned cows from Abrak, Nubian desert

Various interpretations of the causes of megaloceraty in sanga cattle have been offered. Adametz (1920) believed that it was derived from the African urus, a skull of which, with gigantic cores, was found in the Fayum (Fig. 259). But the validity of this theory is doubtful, for the porous structure of the cores and the tremendous basal girth characteristic of the horns of many giant-horned sanga cattle did not occur in the urus. Again, horns of gigantic size also occurred in domesticated cattle beyond the confines of Africa. Prior to the introduction of British beef breeds, giant-horned cattle, which were descended from Iberian stock, first landed by Columbus in Santo Domingo on his second voyage to America (1493–96), and brought over from Santo Domingo to the mainland by Gregorio de Villalobos in 1521, were bred in Brazil (Franqueiro) and Texas (Texas Longhorn) (Dobie, 1943). In Crete giant-horned cattle are known from the Third Middle Minoan of Knossos (c. 1730–1550 B.C.); in the basement of a small house of that period, Evans (1921–35) found the heads of "two large oxen of the *urus* breed, the horn-cores of one of which were over a foot in girth at the base". During the neolithic age such cattle existed also in Britain.

Moreover, megaloceraty was not restricted to the African race of the urus; some European Bos primigenius primigenius and Asiatic Bos primigenius namadicus skulls are furnished with longer and thicker cores than those of the African Bos primigenius opisthonomus (Epstein, 1958). At Çatal Hüyük in southern Anatolia the clay head of a bull, into which real horns had been inserted, formed the principal ceremonial object in one of the temples, dated to the 6th millennium B.C. The horn sheaths are no longer preserved, but one of the horn cores measures more than 100 cm in length, indicating that the animal to which it belonged must have carried horns of truly gigantic dimensions (Mellaart, 1964a).

In Egypt giant-horned cattle seem to have been unknown; had they existed, the artists of the Nile country would doubtless have preserved their memory in sculptures and paintings. But in the deserts to the west and east of the Nile, in Ethiopia (Fig. 241), and on Mount Elgon on the Kenya-Uganda border (Fig. 240), ancient rock drawings and paintings indicate that at an early time Hamitic pastoralists had evolved a giant-horned type from their humpless longhorn stock. The gigantic horns characteristic of many sanga breeds offer additional evidence of the share of the humpless cattle of the Hamites in the evolution of the sanga.

The reasons for the directed selection and propagation of inheritable horn gigantism in cattle are social and ritual rather than economic, and are connected with the cattle complex of the pastoral peoples of Africa (Epstein, 1955b). This may still be observed among the Masai and several other pastoral tribes of East Africa. In Tanzania short-horned zebus are encountered in addition to giant-horned Ankole sanga cattle. Economically, the zebu is definitely the more useful animal; hardier, maturing earlier, milking better, working better, and furnishing a superior carcase. It maintains its condition when feed is scarce and innutritious, and water available only every other day. But the owner of an Ankole beast judges altogether differently. He will not exchange a giant-horned Ankole cow of poor conformation and production for a better but shorter-horned zebu. In some parts of Africa, e.g. in the Congo, the interest of the pastoralists in gigantic horns in their cattle is so great that occasionally horn growth is stimulated by irritation of the horn buds. Comparative study has suggested that the cattle complex of the pastoral peoples of East Africa developed farther north among the Hamitic Galla of Ethiopia, and has been carried south by the advances of Nilo-Hamitic cattle breeders (Forde, 1934).

With this we have not exhausted the subject of megaloceraty in the sanga cattle. Duerst (1926a; 1931) has presented an entirely different theory as to its causes, considering these to be environmental. Originally he held the climate, more especially the humidity of the air, in the lake district of Africa responsible; but subsequently he ascribed megaloceraty to the lack of phosphate and a high degree of acidity in the soil. These, he suggested, caused acidosis in the organism, resulting in an increased production of horn substance through hyperaemia and hypertrophy of the horn epidermis, associated with a blood supply of insufficient alkalinity; this tendency was further enhanced by the effects of endocrine secretion.

A theory which completely excludes the hereditary disposition to megaloceraty is unacceptable. For it offers no explanation for the occurrence of giant-horned, long-horned, short-horned and polled cattle in the same area, the same herd, even the same family. It offers no explanation for the facts that herds of giant-horned sanga cattle are maintained under widely differing environmental conditions, and that in districts where the giant-horned sanga is replaced by short-horned zebu cattle the latter do not

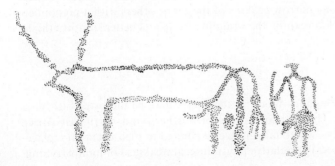

648. Rock engraving of a giant-horned bull from Aswan, Roman period (after Schweinfurth)

commence to grow gigantic horns. Nor does it offer an explanation for the fact that megaloceraty in sanga herds can be maintained only by continuous careful selection of giant-horned breeding stock.

For these reasons megaloceraty in sanga cattle should be regarded as a hereditary feature. This does not exclude the possibility of its endocrine basis. It appears, however, that Duerst (1926a; 1931), while conceding a connection between megaloceraty and endocrine constitution, considers the latter, as far as horn gigantism is concerned, to be dominated by feeding and environment rather than by hereditary factors.

Environment undoubtedly does influence horn development in cattle. It influences horn development along with the skin, hair and bone, along with the whole endocrine constitution of the animal, with the entire physiological and conformational complex. For example: Jersey cattle bred by the author in the subtropical environment of Israel grew markedly longer and coarser horns than those of their parents imported from the island of Jersey.

We have to distinguish between two different types of megaloceraty: one induced by environment, leading to acidosis of the organism, and the other independent of environment. In the giant-horned sanga cattle the principal factor involved is not natural environment but artificial selection of a heritable disposition over a very long period of time. Probably megaloceraty in African longhorn cattle originally occurred only sporadically and not in its present extreme forms. The selection of animals with the largest horns for breeding purposes propagated this feature and created the present giant-horned breeds.

That the giant-horned sanga is merely a variation of the long-horned sanga is readily seen from a comparison of their skulls. Cranially the giant-horned Tswana cattle are distinguishable from the long-horned Ambo cattle by little more than the gigantic horn necks and their effect on the conformation of the neurocranium.

Variability in Size and Conformation in Sanga Cattle

In body size, we have seen, sanga cattle vary from dwarfed to large-bodied animals. They are, as a rule, well covered, as could be expected from a cross of zebu and Hamitic longhorn cattle (Bisschop, 1937). The majority of the sanga cattle have well developed dewlaps, similar to those of zebu cattle; but some breeds have small dewlaps, more like those in humpless longhorn cattle. The hindquarters vary considerably. Breeds with a larger share of zebu blood show the typical sloping rump, low-set pin bones, rounded upper thighs and long narrow gaskins of the zebu; others with a preponderance of humpless longhorn blood exhibit the straighter rump and better let-down thighs of their longhorn ancestors.

Adaptability of Sanga Cattle

The presence of humpless longhorn blood in the sanga may be indicated also by a physiological phenomenon. The longhorn cattle of northern and north-east Africa seem to have lacked hardiness, which may have contributed to their eventual replacement by more adaptable types. This relative physiological weakness of humpless longhorn cattle has apparently been inherited by some of the sanga cattle. In Uganda, in 1935, Carmichael (1939) diagnosed bovine tuberculosis in 33.3 per cent of the carcasses of Ankole sanga cattle as against 0.6 per cent in those of zebu cattle. Bahima (Ankole) cattle grow weak and perish when taken from their fertile and healthy highland pastures into an inferior environment. Here the sanga cattle cannot compete with the zebu, and their owners' reluctance to part with their giant-horned stock is of no avail. Thus the Bahima between Lake Albert and the northern shore of Lake Victoria now no longer possess giant-horned sanga cattle but breed short-horned zebus. Bahima pastoralists, who had moved with their giant-horned herds into the vicinity of Tabora, were found a few years later in possession of zebu cattle only (Adametz, 1894). As zebus in general are distinguished by a high degree of adaptability to tropical and subtropical environments, the lack of resistance characteristic of many sanga breeds may reasonably be suspected to be a heritage of their humpless longhorn parent stock.

Referring to this physiological factor, Bisschop (1937) remarks that "the question of blood ratio in the Sanga is a very difficult one to answer. In size and in general resistance the "Sanga" sub-types vary considerably, but it is suggested that this is due, not so much to blood ratio, as to natural selection and adaptation to the very different environment complexes, in which they finally became separate sub-types or breeds. Thus, e.g. the Makalanga cattle have managed to survive under some of the worst conditions in Southern Africa. They are exceptionally hardy cattle, but due to their environment have become dwarfed in comparison to, for instance, the Sanga cattle of Bechuanaland."

Environment undoubtedly is the principal factor in imparting hardiness to livestock. The relative hardiness of the entire zebu group, and the comparatively lower degree of resistance displayed by the cervico-thoracic-humped sanga cattle of Africa, are probably due to the different environmental conditions under which these cattle types were originally evolved. But in the course of time these physiological characteristics have become racial characteristics, the zebus as a group having acquired greater hardiness under tropical and subtropical conditions than cattle wholly or partly of humpless longhorn type.

On the Origin and Dispersal of the Bantu

The dating of the evolution of the sanga cattle is a problem closely connected with the origin and dispersal of the Bantu, who were the most important agent in the diffusion of the sanga type in East and South Africa.

Bantu-speaking peoples occupy today approximately one-third of the African continent. Until the first centuries of the Christian era practically all this vast territory was occupied by hunters and gatherers of pygmoid people in the west and bushmanoid people in the east and south. Howells (1967) has suggested that the original territory of the negroes comprised the southern Sahara and the open country and grasslands on its southern edge, from where they moved into the forests before the beginning of the Christian era when the acquisition of the knowledge of iron made efficient clearing of the forest for planting possible. However, it would appear that at least some of the Negro peoples inhabited the forest zone already at an earlier date. For

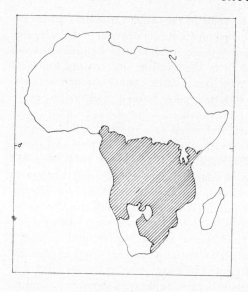

649. Bantu territory in Africa

Greenberg (1955) has demonstrated on linguistic grounds that the Bantu originate from a very small area near the Cameron-Nigerian border. Here their differentiation from a single ancestral speech community began approximately at the beginning of the 1st millennium B.C.; and after a thousand years or so of linguistic differentiation in their homeland they launched forth upon their explosive expansion (Olmsted, 1957). During this period of linguistic differentiation the Bantu were primarily agriculturalists, cultivating the indigenous Sudanic crops in addition, possibly, to some of those brought under cultivation in south-west Asia. They probably also possessed dogs, pigs, hairy thin-tailed sheep, goats and cattle; but the latter were not milked, for the limit of early diffusion of the milking complex in Cameroun did not extend south beyond Adamawa.

Murdock (1959) has claimed that the expansion of the Bantu to the south and east was rendered economically possible by their acquisition of the Malaysian food plants, banana, taro and yam, which are adapted to an equatorial environment. According to the author of the 'Periplus of the Erythraean Sea', an anonymous Greek merchant mariner who made a voyage around the Horn and down the African coast, these plants were cultivated on the Azanian coast (the coast of Kenya and of adjacent Somalia and Tanzania, from Kisimayu in the north to Kilwa in the south and including the islands of Zanzibar and Pemba, was known to Mediterranean antiquity as Azania) by A.D. 60; they might have been introduced several centuries earlier, but not before the earliest southward expansion of a Hamitic people with a true neolithic culture from south-western Ethiopia into western Kenya, central Uganda, northern Tanzania and down to the Azanian coast. The neolithic cultures established by these immigrants in East Africa are dated to 850 B.C. (Cole, 1954); according to Murdock (1959), waves of caucasoid (Hamitic) peoples, bringing with them a full neolithic culture with agriculture and domestic animals, began to fan out from southern Ethiopia at some time before 1000 B.C. None of these settlers on the Azanian coast and its hinterland, where they occupied elevated locations with substantial precipitation, could have received the Malaysian food plants and transmitted them to the forest peoples in the west. Hence their diffusion from the Azanian coast across the continent to the Bantu homeland near the Cameroun-Nigerian border appears to have taken place in the course of the second part of the 1st millennium B.C., which would agree with the dating of the commencement of the Bantu migrations to the end of the pre-Christian era.

When the vanguard of the equatorial Bantu emerged from the tropical forest into Uganda in the early centuries of the Christian era, they encountered and eventually absorbed the Hamitic inhabitants of this region, from whom they received the principal elements of East African agriculture—the Ethiopian cultigens eleusine and sorghum (durra and Kafir corn), and the sanga type of cattle with the Hamitic cattle-and-dairying complex. Subsequently this cultural pattern appears to have been passed on from tribe to tribe around the eastern and southern periphery of the tropical forest zone, enabling the various Bantu groups to leave the forest and continue their expansion into the adjacent savanna zone (Murdock, 1959).

However, it is rather doubtful if the primary factor in the rapid dispersal and expansion of the equatorial Bantu was, as Murdock suggests, their acquisition of the Malaysian food plants and the elements of East African agriculture—important as these were. The most powerful impetus to the expansion of the early Bantu from their original homeland on the Cameroun-Nigerian border to the east and south was probably provided by their acquisition of iron and the knowledge of smelting and working metal. It is significant that the great spread of the negroids over Africa does not seem to have begun before the iron age (Cole, 1966). The Bantu are invariably associated with an iron age pottery-using culture of mixed farming. They thus had a greatly superior economic, social and military structure to that of the palaeolithic hunters and gatherers whose territory they invaded. Iron and the working of iron appear from good evidence to have been transmitted to the savanna negroes of West Africa, and through these to the equatorial Bantu, by the southern Berbers towards the end of the pre-Christian era or early in the 1st millennium A.D. (Clark, 1959). Chronologically this fits in well with the emergence of the Bantu vanguard from the tropical forest into Uganda in the early centuries of the Christian era.

The rapid dispersal and expansion of the Bantu-speaking peoples from West Africa into territory occupied by other racial types is believed to have been facilitated by the relatively high tolerance for malignant subtertian malaria characteristic of West Africans. Their superior fitness in areas with an excessively high rate of morbidity due to subtertian malaria is attributed, in addition to factors of secondary importance, to the prevalence of the gene for sickle-cell anaemia (haemoglobin S) and the protection this confers against the invasion of the red blood cells by Plasmodium falciparum by which this most dangerous form of malaria is caused (Allison, 1955; Livingston, 1958). The largest number of people with sickle-cell (haemolytic) anaemia are found in Central and West Africa, where tertian malaria was or is endemic. It is especially in the early years of life that malaria is a very killing condition in these parts of Africa, and it is at this time that heterozygosity for the sickle-cell gene protects and confers a selective advantage. Raper (1956) has published evidence in support of the idea of balanced polymorphism with malaria as the balancing condition: among 818 children admitted to hospital during a certain time period, the incidence of sickle-cell trait was largely uniform for various admitting diseases (17 per cent on an average), but was zero in a group of 47 children suffering from cerebral malaria (Ingram, 1963). Linton (1956) has pointed out that most West Africans are malaria carriers, and whenever they have penetrated into a region infested with Anopheles, they have introduced this disease, with catastrophic results for the local populations.

Early Pastoral Iron Age Invaders of South Africa

The use of iron appeared in south-central Africa at about the same time as in the forest belt of West Africa, or not much later. The formative period of the East and South African iron age can be narrowed down to the middle centuries of the 1st Christian millennium (Davidson, 1961). The first iron age people migrated to Rhodesia early in the 1st millennium A.D. By A.D. 90, the 'Channelled Ware' (type of pottery decoration) peoples, breeders of livestock, were already settled north of the Zambesi. They were followed by the 'Stamped Ware' peoples, still in the earlier half of the 1st millennium A.D. It is possible that iron working had reached Rhodesia, via the Horn of Africa, already prior to the arrival of the Bantu, sometime before A.D. 500, from the kingdom of Meroe on the middle Nile, where an intensive iron industry flourished for a full millennium, from 650 B.C. to A.D. 350 (Clark, 1959). Recent radiocarbon research has dated the occupation of Zimbabwe in (Southern) Rhodesia by the 'Stamped Ware' peoples to approximately A.D. 600. Skeletal remains of negroid type at Zimbabwe confirm that the early Bantu invasions across the Zambesi river took place at about this time.

The early Bantu invaders of southern Africa, of Sotho, Shona and Venda stocks, were in possession of herds and flocks and a pastoral and agricultural economy. They were not nomads, but settled cattle folk and hoe-cultivators—nomad only when compelled to seek new pasture grounds for their animals. It has been pointed out that their racial evolution from a mixture of negro and Hamitic elements, and the evolution of their cultural pattern from a combination of agricultural and pastoral elements occurred in central East Africa (Uganda). This must have happened some time before their arrival in (Southern) Rhodesia, that is not later than the middle of the 1st millennium A.D. On the other hand, it could not have occurred very much earlier since not a single trace of negroid peoples appears archaeologically anywhere in the vast region of East Africa from western Ethiopia and the Horn south to the Cape of Good Hope, until well after the time of Christ (Murdock, 1959).

As the cattle of the recent Bantu peoples of South Africa are all of sanga type, and as stone bowls from Zimbabwe are decorated with pictures of cattle with long lyre-shaped horns such as were common in north-east Africa, and a clay figurine from the Zimbabwe excavations, dated to about A.D. 900, represents an ox with a cervico-thoracic hump (Robinson, 1961), and, further, as there is no evidence of the presence of any other type of ox in South Africa in former times, it may be assumed that the cattle which the Bantu obtained from the early Hamitic inhabitants of East Africa were of sanga stock. This is confirmed by the historical evidence which shows that sanga cattle were the common type in East Africa before the short-horned zebu filled the vacuum caused in many areas by repeated pandemics of rinderpest or superseded the sanga owing to its superior functional properties (see also p. 409).

Period of Evolution of the Sanga Cattle

While the early Hamites of East Africa must have brought a developed cattle and dairying complex from southern Ethiopia, it is uncertain if their cattle, by about 1000 B.C., were of long- or giant-horned humpless type or already of humped sanga stock. In view of the importation of cervico-thoracic-humped cattle into Egypt by the middle of the 2nd millennium B.C., they may already have been humped; but this possibility cannot be reconciled with the absence of zebu cattle in rock paintings in the Horn of Africa until after A.D. 400, unless this absence is attributed to a failing of the archaeological record. The alternative possibility is that the early neolithic Hamites of East Africa, on their descent from southern Ethiopia, were in possession of humpless longhorn cattle, and that their brethren, settled on the Azanian coast, received, along with the Malaysian food plants, zebu cattle from the coastlands of the Arabian Sea via the Sabaean Lane, the great monsoon trade route from India via southern Arabia to Azania, in the course of the 1st millennium B.C. This alternative is supported by the presence of cave paintings of humpless giant-horned cattle, at least 2000 years old, as far south as Mount Elgon on the Kenya-Uganda border (Fig. 240), and of cervico-thoracic-humped zebu cattle in the Kulli culture of south Baluchistan, bordering the Sabaean Lane. From the ports of importation the zebu cattle were passed on, together with the Malaysian plants, to the Hamitic agriculturalists and pastoralists of northern Tanzania, Kenya and Uganda where the intermixture of the autochthonous humpless long-horned or giant-horned cattle and the recently introduced zebu produced the sanga type. From this alternative it would follow that the sanga was first evolved in central East Africa.

The evolution of the sanga in the Horn of Africa and Ethiopia may have occurred independently in the course of the first pre-Christian millennium and the early centuries of the Christian era, when the Himyaritic Arabs of southern Arabia exercised political and mercantile control over the Eritrean and Somali coasts, and an extensive trade existed between the two centres of complex agricultural civilisations in southern Arabia and highland Ethiopia across the narrow strait which connects the Red Sea with the Gulf of Aden.

During the second half of the 1st millennium A.D., sanga cattle were introduced into southern Somalia from the west when the north-east coastal Bantu pressed into Kenya and thence into Somalia, where the valley of the Shibeli river represents the limit of their penetration. Whether the pastoral Galla, who descended from the southern edge of the Ethiopian plateau into the savanna and steppe country of Somalia at about A.D. 1000, were in possession of giant-horned humpless or humped cattle is uncertain. But when, 400 or 500 years later, they were displaced in this area by Somali tribes, and were forced, in search of suitable grazing grounds for their livestock, to invade highland Ethiopia, where they occupied extensive tracts of territory in the west and north-west during the 16th century A.D., their cattle were of humped giant-horned sanga type.

Giant-horned sanga cattle were introduced into Uganda, Rwanda-Burundi and the shores of Lake Kivu, and north-western Tanzania by the pastoral Nilotic Hima (Bahima) and Watusi. The Bahima infiltrated the Bantu (Bahera) of Uganda in or around the 14th century A.D. (Davidson, 1961). Posnansky (quoted by Payne, 1964) states that there is evidence that the Bahima settled in western Uganda as early as A.D. 1200. In the Rwanda region the local agricultural Bantu peoples were subjected by the Watusi invaders, who still form the ruling aristocracy in this area. The Bahima and Watusi belonged to one of the comparatively recent waves of pastoral folk with a Hamitic element, forced to leave their homes in the north in search of new pastures—far more recent than those early waves of Hamites who, mixing with the negro Bantu, gave rise to the agricultural and pastoral peoples of Bantu speech in East and South Africa (Seligman, 1930).

We have to regard the evolution of the sanga breeds, not as a solitary phenomenon during an isolated period of history, but as a prolonged process that occurred first in Uganda or Ethiopia and the Horn of Africa, and subsequently in all those parts of the continent where zebu cattle, first, probably, with cervico-thoracic humps, later with thoracic humps, came into contact with humpless long-horned or giant-horned stock or with sanga cattle previously evolved from a mixture of humpless long-horned or giant-horned and humped stocks. We have to consider the evolution of the sanga breeds as a process that may have begun with the first introduction of cervico-thoracic-humped or thoracic-humped zebu cattle on the Azanian coast 3000, or into Ethiopia more than 3500 years ago, that received its main impetus with the increasing importations of thoracic-humped zebu cattle since the early centuries of the Christian era, and which continues with the extension of the range of the thoracic-humped zebu in Africa to this day (Epstein, 1957a).

5. *The Status of the Africander and Hottentot Cattle*

Previous Classifications of the Africander Cattle

In previous publications Epstein (1933; 1937) classed the Africander and Hottentot cattle with the zebu group. This classification has been accepted by several authors who have qualified the term 'zebu' with reference to the Africander cattle by the addition of different attributes: Longhorned Zebu (Curson and Epstein, 1934; Curson, 1935; Curson and Bisschop, 1935; Bisschop, 1937); Lateral-horned Zebu (Curson, 1936; Curson and Thornton, 1936); Flesh-humped Zebu (Epstein, 1955b); Cervico-thoracic-humped (neck-humped) Zebu (Epstein, 1955a; 1956; Faulkner and Epstein, 1957).

Reasons for the Classification of the Africander with the Zebu

The classification of the Africander with the zebu, in distinction from the sanga group was based on several facts and assumptions. In general conformation the Africander conveys the impression of a true and unadulterated zebu. This impression derives chiefly from the prominent hump, high sacrum (in the former draught type) and drooping zebu-like rump, the rounded buttocks, slender legs and fine bone; further, the conspicuous folded dewlap, well developed navel fold and prepuce, and the loose skin of the body. It is enhanced by the long head with the convex profile, the rounded poll and backward sweep of the horns. Anatomically the deeply cleft spinous processes of the anterior dorsal vertebrae are wholly zebu-like, ranging the Africander with the zebu rather than the sanga group. With regard to the anatomy of the skin and hair and certain physiological properties facilitating adaptation in warm regions, the Africander also approaches the zebu cattle, albeit not completely, its skin characteristics being intermediate between those of the temperate and the true tropical breeds (Walker, 1957).

A number of representations of humped cattle from the period of the New Kingdom of Egypt (Figs. 601 and 603), which show a likeness to the Africander in the position of the hump, the shape of the laterally twisted horns and the general zebu conformation, suggested a close racial connection between these two types over a distance of the whole length of Africa and a time interval of nearly three-and-a-half millennia.

The preservation of the original zebu type in the Africander was attributed to the derivation of the Hottentot people from a southern offshoot of Hamitic pastoralists and autochthonous Bushman hunters and gatherers in the latter's territory where the Hottentots remained isolated from agricultural and further pastoral influences. They were removed from such influence also later on their migration to South West and South Africa in the van of the pastoral and agricultural Bantu.

Objections to the Classification of the Africander with the Zebu

The main argument against this theory, brought forward by Slijper (1951) and strongly supported by Mason (personal communication), is based on the cervico-thoracic situation of the Africander hump, which separates the Africander from practically all present zebu cattle of Africa and Asia, placing it in a group with the recent African and Asian cervico-thoracic-humped cattle, the majority or possibly all of which are derived from thoracic-humped zebu and humpless parent stocks. In addition, Jones (1955) has adduced pictorial evidence to show that the cattle of the Hottentots were of sanga rather than of zebu type.

The Status of the Cervico-thoracic-humped Cattle of Ancient Mesopotamia, India and Egypt

The author acknowledges the validity of this criticism. But before going into this question, the status of the ancient cervico-thoracic-humped cattle of Mesopotamia, Baluchistan, India and Egypt requires clarification.

If it be accepted that the thoracic hump did not appear of a sudden but was evolved by breeders from the cervico-thoracic hump, and that the cervico-thoracic hump was evolved by selection of bulls with the most prominent neck crests, a term for the cervico-thoracic-humped stage, intervening between the crested but still humpless primary condition and the specialised thoracic-humped zebu stage, has to be found. The term 'sanga' would not seem suitable, as its use is restricted to African cervico-thoracic-humped cattle derived from a definite and peculiar racial mixture. Representations of the ancient cervico-thoracic-humped cattle, which are historically and anatomically the forerunners of the thoracic-humped zebus, already show the characteristic conformation of the zebu type. In the literature on the cultures of ancient Sumer, Baluchistan, the Indus valley and Egypt they are generally referred to as zebu. There seems to be no valid reason to change this terminology and restrict the term 'zebu' to the thoracic-humped type.

Reasons for the Africander's Zebu Conformation and Similarity to Ancient Egyptian Cervico-thoracic-humped Zebus

The general zebu conformation of the Africander cattle, and their likeness to some of the ancient Egyptian paintings of cervico-thoracic-humped zebu cattle may indeed entitle them to be classed with the cervico-thoracic-humped zebu. But the characteristic zebu feature in the recent Africander cannot be attributed to immediate derivation from a pure zebu type, nor the likeness to the ancient Egyptian prototype to direct descent from the latter. The prominence of many characteristic zebu features in the Africander is due to the emphasis placed by the standards of excellence of 1912 and 1926, and probably also by many of the earlier breeders, on certain features characteristic of the zebu type, while the likeness to the ancient Egyptian neck-humped zebu must be attributed to coincidental convergence.

The Sanga Type in Early Africander and Hottentot Cattle

While some of the photographs of Africander cattle from the first decade of the present century already show the pronounced zebu conformation characteristic of purebred Africander cattle following the introduction of the first standard of excellence in 1912, others (Figs. 650 and 651), though they may not be representative of the then prevalent general type, do not; they rather represent animals that may be classed with the sanga.

Again, in some paintings of Hottentot cattle the horns anticipate the characteristic horn shape of the recent Africander, in others this applies to the hump or to coat colour. But practically none of the paintings, drawings and photographs of Hottentot

650. Africander cow (after Burton, 1903)

651. Africander bull (after Molhuysen, 1911)

cattle shows the distinct zebu type and peculiar conformation typical of the recent Africander. Rather they show cattle of sanga type, in which the mixture of humpless longhorn and humped zebu stocks is discernible.

Reasons for the Classification of the Africander with the Sanga Group

From these conditions we may draw the conclusion that the breeders of Africander cattle, more especially the more recent breeders (but prior to the present emphasis on beef type), bred away from the humpless longhorn strain present in the Hottentot and early Africander cattle, and concentrated on the zebu component. However, they did not breed for the thoracic hump.

Curson and Bisschop (1935) have pointed out that the cervico-thoracic hump of Africander (= longhorned zebu) cattle resembles the sanga hump in situation and structure:—"The similarity of the humps of the Africander and sanga types, being cervico-thoracic and muscular, is noteworthy in view of Epstein's theory that the latter is derived from the intermixture of the long-horned zebu and Hamitic longhorn. The dissimilarity between the above humps and that of the shorthorned zebu is striking."

The implied interpretation that the cervico-thoracic sanga hump resembles the Africander's in situation and structure because one of the components in the ancestry of the sanga cattle was derived from the Africander's sole ancestral stock, i.e. the long-horned or cervico-thoracic-humped zebu, cannot be upheld. Rather the similarity in hump position and structure between the Africander and sanga cattle indicates that the Hottentot cattle, from which the Africander is descended, were a sanga breed.

In the classification of the Africander breed of cattle, particularly in a work dealing with racial origins, the sanga type of their ancestors and the retention of the cervico-thoracic sanga hump are decisive factors. Accordingly, the Africander breed has been included in the sanga, and not in the zebu group.

The classification of the Africander with the sanga group is supported by their approximation in the distribution of the two electrophoretically different adult bovine haemoglobins. In the frequencies of the genes controlling adult bovine haemoglobin the Africander is even nearer to the humpless longhorn (N'Dama), and farther removed from the zebu, than the Ankole sanga:—

Frequencies of Genes controlling Adult Bovine Haemoglobin

Breed	Haemoglobin Types		Reference
	A	B	
N'Dama	1.00	0.00	(Bangham and Blumberg, 1958)
Africander	0.91	0.09	(Singer and Lehmann, 1963)
Ankole Sanga	0.80	0.20	(Lehmann and Rollinson, 1958)
White Fulani	0.79	0.21	(Bangham and Blumberg, 1958)
Nganda	0.77	0.23	(Lehmann and Rollinson, 1958)
Zebu—Uganda	0.68	0.32	(Lehmann and Rollinson, 1958)
Zebu—Sudan	0.53	0.47	(Bangham and Blumberg, 1958)
Zebu—Indian Gir	0.50	0.50	(Lehmann, 1959)

6. The Descent of the Long-horned Fulani Cattle

Distinctive Features of the Long-horned Fulani Cattle

The long-horned Fulani cattle form a group of their own. From the zebu cattle of West and East Africa they are distinguished by the length of their horns, and from the cervico-thoracic-humped sanga cattle of South and East Africa (but not from the thoracic-humped sanga cattle of East Africa) by the thoracic, occasionally intermediate, situation of their humps. The principal question with regard to their racial origin is whether they should or should not be classed with the zebu cattle of West Africa, considering that zebu cattle with long lyre-shaped horns (e.g. Kenwariya) occur also in certain parts of India, especially of northern India.

Descent of the Long-horned Fulani Cattle from Thoracic-humped Zebu and Humpless Longhorn Stocks

Obviously the long-horned Fulani cattle received their thoracic humps and general zebu conformation from zebu cattle introduced into West Africa from the east. Zebu cattle are believed to have first arrived in the present breeding area of the Sudanese Fulani in the course of the Arab invasions in the 7th century A.D. (Joshi et al., 1957). Into the lower part of Senegal in the Fouta Toro basin they penetrated during the latter part of the 8th century; from there they spread to the plateau area of Ferlo and farther westwards (breeding area of the Senegal Fulani) in the 9th century (Doutressoulle, 1947). Larger numbers followed in their steps when, after the fall of the Christian kingdom of Dongola in the 14th century A.D. and of Alwa to the south in 1504, the Arabs surged unchecked up the Nile and thence westwards into the central Sudan, where their advance guard penetrated as far as Lake Chad.

The typical zebu cattle of East Africa, like those of West Africa, are short-horned. Also short-horned are the recent zebu cattle of Arabia, Iraq, Iran, Baluchistan and the western coastlands of India. We may, therefore, assume that the zebu cattle introduced into East Africa from Arabia and the littoral of the Arabian Sea were for the greater part short-horned. It seems most unlikely that the nomad Fulani, of all West African cattle breeders, received a special type of long-and-lyre-horned zebu from India, although in view of the presence of such zebu cattle in the Neolithic of the Deccan and during the latest prehistoric period (c. 2500–1500 B.C.) of the Quetta-Pishin valley, this is not entirely impossible. However, it seems far more likely that the long horns (White Fulani, Sudanese Fulani, Senegal Fulani) or the gigantic horns (Red Bororo) by which their humped cattle are distinguished from the typical zebu cattle of West and East Africa are a heritage of humpless long-horned or giant-horned cattle bred by the Northern Hamites before humped cattle were introduced into their breeding grounds at a relatively recent date.

The Northern Hamites, like the Eastern Hamites, paid special attention to the large size of the horns in their cattle. This is shown not only by the recent Kuri cattle of Lake Chad, but also by numerous rock drawings and paintings of giant-horned cattle in pasture oases in the Sahara and Libyan desert (Fig. 652).

652. Rock drawing of a giant-horned cow from the pasture oasis of Uweinat, Libyan desert (after Winkler)

The humpless Hamitic Longhorn (primigenius) strain in the long-horned Fulani breeds can be recognised not only in the length of the horns but also in the position of the hump, which in many animals is more forward than in zebu cattle. It can be recognised also in cranial conformation. Antonius (1943–44) has added the following legend beneath the photograph of a White Fulani (he calls it Bororo) herd en face:—"The head types show all transitional stages from pure primigenius to zebu." Of the Sudanese Fulani he says:—"The Niger region is the home of the large-horned Macina breed which combines pure primigenius skull conformation with marked hump development."

In the frequencies of genes controlling adult haemoglobin in cattle, the White Fulani is intermediate between the Ankole sanga and the Nganda zebu-sanga type (see table p. 551).

Whether the different hump structure in the Red Bororo and White Fulani cattle (muscular in the former, and usually musculo-fatty in the latter) may be quoted as evidence for their origin from different zebu types or from different proportions of zebu and humpless longhorn blood is uncertain in view of the variability of hump structure in typical short-horned zebu breeds (e.g. muscular in the Azaouak and Shuwa, and musculo-fatty in the Sokoto and Adamawa zebus).

Historical Evidence for the Descent of the Fulani Cattle from Humpless Longhorn and Thoracic-humped Zebu

The history of the Fulani people confirms the diphyletic descent of their cattle. The Fulani are generally divided into two groups differing in racial, social and economic characteristics. One, strongly negroid, consists for the major part of sedentary village dwellers; the other, non-negroid, of pastoral nomads. The latter depend for their livelihood on their cattle, which are, however, never killed for food except at festivals, the staple article of diet being milk and butter (Seligman, 1930).

The origin of the nomad Fulani has been subject to various theories: Delafosse (1912) believed that they were descended from Syrians of Aramaic speech, who allegedly penetrated negro Africa from Cyrenaica about A.D. 200, while Meinhof (1912) derived them from Hamites of the Horn of Africa. According to Seligman (1939), they are a branch of the Northern Hamites. But Murdock (1959), on the basis of Greenberg's (1955) demonstration that they speak a Nigritic language of the Atlantic subfamily of that stock, traces the origin of the ancestral Fulani to the middle region of the Senegal valley and the savanna region of Fouta Toro immediately south thereof. The infiltration of Northern Hamitic elements into this area began in the 11th century when the Berbers of Morocco expanded southwards under Arab pressure from the rear. They occupied the savanna country of Fouta Toro which offered excellent pasturage for their herds of humpless longhorn cattle. Thus the pastoral population became predominantly Hamitic in race, while the sedentary tillers and townspeople along the Senegal remained predominantly negroid. During the following centuries the pastoral Fulani spread across the entire breadth of the western Sudan. In the 12th and 13th centuries they infiltrated into Ferlo and eastwards into Kaarta; during the 14th century they moved eastwards through Soninke and Bambara country to Macina, and in the 16th century from Senegal southwards into Fouta Djallon. The main movement continued eastwards, forming an enclave in Liptako and penetrating the Hausa country, where they are first mentioned historically in the 15th century. Their vanguard reached Adamawa by at least the early 18th century, and smaller groups extended as far as Wadai and penetrated south-east into parts of Cameroun during the 19th century (Murdock, 1959).

Murdock (1959) suggests that, upon the Berber intrusion into their country on the middle Senegal, the incipient Fulani people (Tukulor) replaced their older humpless cattle with the new humped, or zebu, breed from the north. But this conjecture is entirely unfounded, for the Berbers could not have brought zebu cattle from their homes in Morocco where this type has never been bred. Rather it would appear that the Fulani, on obtaining thoracic-humped zebu cattle in the course of their eastward expansion in payment for service as herdsmen, by trade, or in raids or ordinary warfare, incorporated these with their original humpless long-horned or giant-horned herds, but continued to select their breeding stock with a view to horn length.

Similar Racial Origin of the Long-horned Fulani and the East and South African Sanga Cattle

The origin of the long-horned Fulani cattle of West Africa must therefore be regarded as similar to that of the sanga cattle of East and South Africa. Indeed, regarding the giant-horned Bororo breed, Gates (1952) suggested that it might have had its origin in sanga cattle introduced from Upper Egypt. But whereas the zebu cattle initially (that is before the introduction of thoracic-humped zebu cattle) involved in the development of the sanga type in Uganda or the Horn of Africa and Ethiopia may have been cervico-thoracic-humped, the zebus sharing in the evolution of the long-horned Fulani breeds were thoracic-humped. Moreover, the zebu element in the long-horned Fulani breeds is probably much stronger than it is in the sanga—stronger, at any rate, than in the sanga cattle south of the Zambesi river.

V. Summary—Classification, Distribution and Origin of African Cattle

African cattle are divided into two main groups: humpless and humped. The humpless cattle are sub-divided into longhorn and shorthorn, while the humped cattle are classed into zebu proper and zebu crossbred types.

Humpless longhorn cattle occur in the south-west of West Africa (N'Dama) and in the Chad region (Kuri). Formerly they were ubiquitous in North Africa, Ethiopia, Kenya and Uganda; and in the mountains and oases of the Sahara, Hamitic pastoralists had evolved a type with gigantic horns. Longhorn cattle were originally introduced into Egypt from south-west Asia in the second half of the 5th millennium B.C.; from Egypt they spread to the west and south. During the period of the early Hamitic cattle breeders, the wild North African urus (Bos primigenius opisthonomus) may occasionally have been incorporated into domestic longhorn herds. The original domestication of longhorn cattle took place in south-west Asia, probably in the vicinity of Mesopotamia. The wild parent stock was furnished by the Asiatic urus (Bos primigenius namadicus).

Humpless shorthorn (brachyceros) cattle occur in the northern half of Africa in a narrow belt skirting the Sahara desert. They are bred in Egypt, with a vanishing offshoot in the Nuba mountains of the Sudan, and extend from Libya through the Atlas countries into Morocco. In West Africa they are found mainly in the coastal regions from Gambia to Cameroun, with a recent offshoot in the western extension of the Congo. Shorthorn cattle were introduced into Egypt from south-west Asia in the 2nd millennium B.C. The original type appears to have been evolved through selection from domesticated longhorn cattle in Elam during the 4th millennium B.C.

Both zebu proper and zebu crossbred types are classed according to hump situation into cervico-thoracic-humped and thoracic-humped stocks.

Cervico-thoracic-humped zebu cattle are known only from records from ancient Egypt, Mesopotamia, Persia, Baluchistan and India. They were introduced into ancient Egypt from the Somali coast, less likely from Syria, from the XIIth Dynasty (c. 1990 to 1780 B.C.) on, and are represented in wall paintings in tombs of the 15th century B.C. This zebu type has not survived in a pure form into recent times.

Thoracic-humped zebu cattle are distributed over a wide area in East and West Africa, namely from the Sudan south to the Zambesi river, and from Chad west to Mauritania. Only a single record of this zebu type is known from ancient Egypt

(XVIIIth Dynasty—1570–1305 B.C.). Subsequently thoracic-humped zebu cattle were introduced from south-west Asia into East Africa where they appear in rock drawings in Ethiopia and the Horn of Africa in the 4th century A.D. and later.

All recent cervico-thoracic-humped cattle of Africa are of mixed racial stock. To these the term 'sanga' is applied. Sanga breeds are either long-horned or giant-horned. They are found, along with thoracic-humped zebu cattle, in East Africa, from northern Ethiopia and the southern Sudan to Zambia (Northern Rhodesia). Before the introduction of European breeds into South Africa, the sanga was the only type of cattle south of the Zambesi river. The cervico-thoracic-humped Africander cattle of South Africa, derived from the cattle of the Hottentots, are also classed with the sanga group. The sanga type has been evolved from a mixture of cervico-thoracic-humped or thoracic-humped zebus and humpless cattle distinguished by either long or gigantic horns. Its evolution probably began in Ethiopia or Uganda after the introduction of cervico-thoracic-humped zebu cattle from south Baluchistan and southern Arabia during the 2nd millennium B.C. and continued with the later importations of thoracic-humped zebu cattle and their rapid extension in East Africa until recent times. It cannot be ascertained whether the zebu stocks involved in the evolution of the various sanga breeds were cervico-thoracic- or thoracic-humped, because in the crossbred progeny of humpless and thoracic-humped parent stocks the hump may be either cervico-thoracic or thoracic in situation.

The thoracic-humped zebu crossbred type is represented by the Fulani cattle of West Africa, which extend from Northern Nigeria to Senegal. With the exception of one giant-horned breed (Red Bororo), the Fulani cattle are long-horned. They are derived from a mixture of thoracic-humped zebu and humpless long-horned or giant-horned cattle.

It is suggested that cervico-thoracic-humped zebu cattle were evolved in the semi-arid steppe on the eastern fringe of the Great Salt desert of Iran in the 4th millennium B.C. From here they were introduced into northern Baluchistan and the Indus valley in the east, southern Baluchistan in the south, and south-western Iran, southern Mesopotamia, the Somali coast and Egypt in the west. Thoracic-humped zebu cattle were probably developed in the same region during the 3rd millennium B.C. All anatomical and physiological peculiarities of the zebu type are attributed to artificial selection from domesticated humpless longhorn cattle in a hot and semi-arid environment. In particular, the thoracic hump is believed to have been evolved from the cervico-thoracic hump, and the latter from the high crest of the bull of a humpless longhorn breed.

Chapter III

BUFFALO

I. The Classification of Wild Buffaloes

Pliocene and Pleistocene Buffaloes

The earliest wild buffalo of which we have any knowledge is Bubalus platyceros Lyd. (= Bubalus sivalensis Rüt.) from the Pliocene of the Siwalik Hills of India. The Siwalik buffalo survived in a closely allied form, which has been named Bubalus (Bos) palaeindicus, in the Pleistocene of the Narbada valley of India (Falconer, 1859). While larger in size, the skull of Bubalus palaeindicus Falc. resembles in conformation the skull of the recent Indian buffalo (arnee).

A number of geographic races nearly allied to the Narbada buffalo occur in the Pleistocene of Africa: in North Africa—Syncerus (= Bubalus) antiquus Duvernois; in East Africa—Syncerus (= Bubalus) nilssoni Lönnb.; in the Transvaal—Syncerus (= Bubalus) baini Seeley; a similar pleistocene race is known from the swamps of Ambolisatra, Madagascar.

The North African Pleistocene Buffalo (Syncerus antiquus)

The North African pleistocene buffalo (Syncerus antiquus = Bubalus antiquus) is of particular interest. It was contemporaneous with pleistocene man by whom it was hunted and featured in numerous rock engravings all over the Atlas countries (Fig. 656). It was a gigantic species, more generalised than either the recent African or Asiatic forms. In its cranial conformation it shows also a certain similarity to the recent Asiatic buffaloes, which, if factual, should not be surprising since the pleistocene fauna of North Africa in many instances represents an extension of the contemporaneous Indian. However, Bate regarded this resemblance as purely superficial and attributed it to convergence in horn conformation (Zeuner, 1963a).

The giant-horned pleistocene buffaloes were typical pluvial forms which became extinct with the desiccation of their ranges (Lönnberg, 1933). None of them was domesticated.

The Classification of the Recent Wild Buffaloes of Africa and Asia

The recent wild buffaloes of Africa and Asia are assigned to two different genera: Syncerus Hodgson and Bubalus H. Smith. In the African buffaloes (Syncerus) the hair

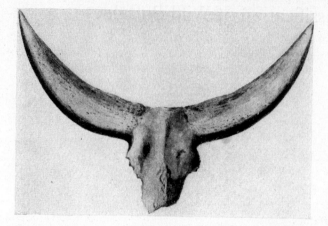

653. Restored skull and horn Cores of the pliocene Siwalik buffalo

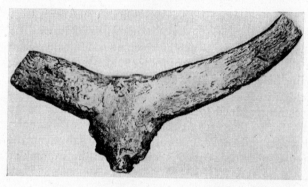

654. Skull and horn cores of the p'eistocene Narbada buffalo

on the middle line of the back is normally directed backwards from nape to rump; in the Asiatic buffaloes (Bubalus) it grows forwards from the haunches to the nape. The ears of the African buffaloes are large and heavily fringed, those of the Asiatic buffaloes are relatively small. The Syncerus skull is short and wide, the vomer (ploughshare bone) is free from the palatines and does not separate the posterior nares (choanae). The Bubalus skull is long and narrow, the large vomer is fused to the palatines and completely divides the posterior nares. Differences between Syncerus and Bubalus skulls exist also in the shape and position of the interparietal, and the situation of the nasal process of the premaxillae. Moreover, the horns of Bubalus are slenderer than those of the African genus (Bohlken, 1958).

The African buffaloes (genus Syncerus) range over nearly the whole of Africa south of the Sahara. Only one species is recognised, i.e. Syncerus caffer, which is classed into three subspecies: Syncerus caffer caffer Sparrman, S.c. aequinoctialis Blyth, and S.c. nanus Boddaert.

The Asiatic buffaloes (genus Bubalus) are classed into two species: Bubalus arnee and Bubalus depressicornis. Bubalus arnee occurs in four subspecies, namely, Bubalus arnee arnee Kerr, B.a. fulvus Blanford, B.a. hosei Lydekker, and B.a. mindorensis Heude. These are distinguished from one another mainly by size and coat colour,

B. a. mindorensis (tamarao) being the smallest form (100–120 cm withers height) owing to the restriction of its range to a small island.

Bubalus depressicornis (anoa or dwarf buffalo) is found on Celebes. It is the smallest of all recent wild Bovini, reaching a withers height of only 60–100 cm. The recogni-

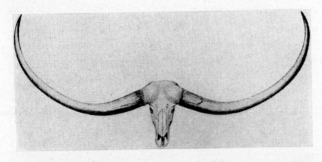

655. Skull and horns of the North African pleistocene buffalo

656. Syncerus antiqueus. Rock drawing from Tel Issaghen I, Fezzan (after Frobenius)

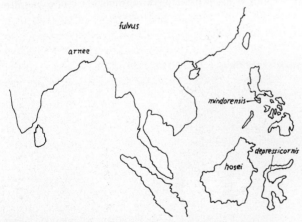

657. Distribution of the Asiatic buffaloes (after Bohlken)

tion of the anoa as a separate species, comprising three subspecies or geographic races, is based mainly on horn direction and the similarity of the anoa's cranial proportions to those of Bubalus arnee— a similarity which from an allometric point of view is significantly at variance with the great difference in size.

The Indian Buffalo

The range of the wild Indian buffalo (B.a. arnee and B.a. fulvus) is now restricted to the grass jungles of the Nepal Terai and the plains of the Ganges and Brahmaputra in Assam; a few herds survive in parts of Orissa, adjoining the Raipur district of

658. Anoa (Bubalus depressicornis) (above), tamarao (Bubalus arnee mindorensis) (below) (after Lydekker)

659. Indian wild buffalo (after Lydekker)

Madhya Pradesh (Central Provinces), and in the south-eastern districts of Madhya Pradesh (Prater, 1947). In former times the range of the wild buffalo comprised a large part of India and extended west into Mesopotamia. A buffalo hunt by Chosroes (Khosru) II of Persia (A.D. 591–628), depicted on a silver plate in the National Library of Paris, indicates that at that time the range of the animal still included Iran.

Typically the Indian buffalo is a large animal, reaching a shoulder height of over 180 cm. The hair on the middle line of the back and neck is directed forwards from the haunches, and the relatively small tubular ears are only sparsely fringed. The hair on the body is coarse and scanty; old animals are almost bare. The tail reaches about to the hocks where it ends in a small tuft. The colour of the typical subspecies is dark ashy grey, approaching black, while in B. a. fulvus it is brownish dun. The horns are commonly black in colour, very large, transversely rugose, and widely separated at the bases, either curving in a crescentic form or directed mainly outwards with an upward and slight forward curvature near the tips, departing throughout their length but little from one plane, although typically there is a distinct recession behind the plane of the centre of the forehead. Like the hair, the horn sheath is commonly rectangular at the base, turning triangular in cross-section at some distance from the base and elliptical towards the tip (Duerst, 1931). In cows the horns are more slender and often longer than in bulls, and not so strikingly angular, pointing also more straightly outwards. The horns of 8–10-year-old bulls from Assam measure 100–107 cm in length and 40–46 cm in basal girth, those of cows of a similar age 119–132 cm in length and 31–34 cm in basal circumference (Duerst, 1926a). The skull of the Indian wild buffalo is relatively light, long and narrow, with a convex forehead and nearly straight or slightly concave facial profile, slender elongated nasals which are narrow near the middle and project on either side considerably beyond the middle suture, with their longest diameter laterally. The vomer is fused with the palatines, the tympanic bullae are small, and the crowns of the upper molars nearly square (Lydekker, 1913).

II. Distribution and Characteristics of the Domestic Buffalo of Egypt

The Domestic Buffalo in the Nile Valley

Domestic buffaloes are met with throughout the Nile valley, the number of animals found in Egypt amounting to approximately one million. Commonly they are kept in the neighbourhood of pools and lagoons where they can wallow in the mud, as the proximity of water is essential to their existence. Grass constitutes their chief nutriment. They are sluggish animals, feeding in the early morning and evening and resting during the heat of the day. In Egypt they are kept for ploughing, frequently yoked together with camels; on small holdings most of the farm work is done by them. The buffalo is the main dairy animal of Egypt. It gives far more milk than the native cow, with fat ranging from 6 to 9 per cent. The meat is rather stringy and coarser than that of ordinary cattle, but Zaki (1951) noted that young male buffaloes have finer and firmer meat than that of Egyptian cattle. Early in the 19th century A.D. the buffaloes survived a rinderpest epidemic which killed nearly all Egyptian cattle (Zeuner, 1963a).

Domestic Buffaloes in North, East and Central Africa

Outside the Nile valley, domestic buffaloes are only occasionally met with in Africa. In North Africa the Arabs, on their westward expansion, did not take buffaloes beyond the confines of Egypt. Johnston mentioned a herd, about 50 in number, that was kept in a semi-wild state in the marshes of Ischkeul near Matur in northern Tunisia. This herd was apparently descended from a few domestic buffaloes presented by a king of Naples to the Bey of Tunis about a century ago (Lydekker, 1898). More recently the animal was re-introduced into Tunisia for agricultural work (Boettger, 1958). These herds have now been exterminated.

In East Africa, Emin Pasha (1840–92) endeavoured to import buffaloes into Lado, his residence in the former Equatorial Province of Egypt in the Sudan, but was unsuccessful (Reinhardt, 1912). An import of buffaloes into East Africa at the time of the German administration has left no trace (Kaleff, 1942). In Zanzibar dairymen used to keep Indian buffaloes for milk. A small herd of domestic buffaloes, derived from breeding stock imported from India in 1926 and 1928, was kept at Mpwapwa,

660. Egyptian buffalo bull (after Brehm)   661. Egyptian buffalo cow

Tanzania, for experimental purposes, but they did not thrive and the remnants were slaughtered in 1932. After the first World War a few buffaloes were also introduced into Mozambique. In 1956 four males and twenty-six females were imported from India into Madagascar (I. L. Mason, 1970 – personal communication).

In 1910 twelve Italian buffaloes were taken to Zambi Station in the Lower Congo. It is assumed that these died of trypanosomiasis to which buffalloes are very sensitive. In 1953 two male and ten pregnant female Kundi buffaloes were imported from Pakistan into the Congo; one bull and four cows were taken to Nioka in the north-east, and one bull and six cows to Yangambi on the Congo river near Kisangani (Stanleyville) (Gillain, 1953; 1955; Mammerickx, 1961).

Characteristics of the Egyptian Buffalo

Four different types of buffalo are recognised in Egypt; three of these in Lower Egypt and one in Upper Egypt (Mason, 1969). The Lower Egyptian varieties are the Minufi, called after its breeding area in Minufiya province, which comprises the southern part of the Nile Delta; Beheri (= towards the sea), in Beheira province in the north-western part of the Nile Delta; and Baladi in the remaining parts of Lower Egypt. Khishin (1951) wrote that these three types can be distinguished by their general appearance, while body measurements do not show any actual differences. But Zaki (1951) claimed that the Minufi buffalo has a shorter body than the Beheri, although nearly the same height at withers. The Saidi, found mainly in Es Said (Upper Egypt), is smaller than the Lower Egyptian varieties, but it is hardy and well adapted to the climatic and feeding conditions of its breeding area. The colour of its skin is black, while that of the Beheri has been described as slate grey and that of the Minufi as ashy blackish-grey.

Generally, the domestic buffaloes of Egypt range from 120–150 cm in withers height. The following average measurements have been taken on 200 adult Minufi buffalo cows from different parts of the breeding area (Zaki, 1951): height at withers 134 cm, length of body from shoulder point to pin bone 150 cm, heart girth 235 cm. The live weight of adult cows is slightly above 600 kg. The head, more especially the facial part, is relatively long—50–59 cm in the Minufi—and the muzzle moderately broad. The facial profile is nearly straight, and the forehead only slightly convex. The horns are black in colour, of medium length, widely separated and rectangular to triangular in cross-section at the base, tapering regularly from base to tip, with irregular transverse ridges and grooves for the greater part of their length. They distinctly recede from the plane of the forehead, curving outwards and backwards, with the tips turning inwards. In cows the horns are longer and more slender than in bulls. The ears are comparatively small and tubular, with slight fringes of hair on their margins. The neck is fairly long, the hock and pin bones are prominent, the rump is steeply drooping, the hooves are large and broad, and the tail is short, ending in a small tuft. The sparse coarse hair almost completely disappears in the adult. On the middle line of the neck and, to a variable extent, on the back the hair is reversed and directed forwards to the occiput, forming a whorl on the line of parting. The colour is usually ashy grey throughout, rarely dun. Occasionally the legs below the knees and hocks are dirty white, and patches of the same colour may be found on the forehead, face and lower lip, frequently accompanied by blue eyes. White animals with blue eyes and albinotic buffaloes are occasionally observed.

III. The Origin of the Domestic Buffalo of Egypt

Although the domestic buffalo is relatively a new-comer in Africa, a similar degree of obscurity shrouds its origin to that which conceals the beginnings of most of the other domestic animals in Africa.

Tamed African Buffaloes in Ancient Ethiopian Rock Paintings

At Genda-Biftou, Sourré, on the Harar plateau in Ethiopia, Breuil (1934) identified buffaloes among herds of humpless longhorn cattle painted on rock faces (Fig. 239). Clark (1954) confirms that the illustrations leave little doubt that the animals represented are indeed buffalo. He writes:—"The earlier, naturalistic painters specialized in pastoral scenes with long-horned, humpless cattle, and buffalo." In rock paintings at Saka Sharifa, not far from Kondoddo mountain, Harar, Von Rosen (1949) observed heavily built and long-horned humpless bovids—either buffalo or domestic cattle, most probably the latter.

With this meagre evidence at present at our disposal it is difficult, if not impossible, to be certain whether in pre-Christian millennia breeders of longhorn cattle in Ethiopia occasionally incorporated African buffaloes into their herds. There is no suggestion that the animal is represented also in the rock art from a later period. We may infer that the early cattle breeders of Ethiopia either did not attempt or did not succeed in their attempt to domesticate the African buffalo.

The Introduction of the Domestic Buffalo into Egypt

The dynastic Egyptians were not in possession of domestic buffaloes, nor was the animal known to the ancient Israelites or to the Greeks and Romans. Alexander the Great encountered buffaloes in Persia and India. Aristotle mentioned the occurrence of domestic buffaloes, with horns curved back on to their necks, in Arachosia, the region of Khokand, Uzbekistan, north of the Afghanistan–India border. The animal may have been introduced into the Nile country from Syria towards the end of the pre-Christian era; for in Palestine it is represented in the ruins of Sebastiye which date from Herodian times, i.e. the last decades B.C. (Bodenheimer, 1935). According to Stuhlmann (1909), the buffalo entered Egypt in A.D. 596; but Boettger (1958) writes that the Mohammedan Arabs first brought the animal into Palestine and Egypt. In A.D. 723, Saint Willibald observed buffaloes in the Jordan valley; he called them 'armenta mirabilia'. Zaki (1951) gives a still later date for their introduction into Egypt, namely the 9th century A.D.

Erroneous Assumption of the Introduction of Buffaloes into Italy by the Lombards

According to the testimony of the monk and historian Paul Warnefried, recorded by the Langobardian (Lombard) writer Paulus Diaconus in his 'Historia Langobardorum', domestic buffaloes reached Italy in the year A.D. 595 during the reign of the Langobardian king Agilulf (591–611). In connection with this record, Hehn and Stallybrass (1885) remarked:—"It seems probable, as they appear in company with wild horses, that they were a present to the Longobardian kings from the Khan of the Avars, for this Turkish race of nomads, who at that time dwelt near the Danube and scourged the Roman Empire with fearful devastations, were on friendly terms with the Longobardian court. If King Agilulf sent shipbuilders to the Avarian Khan to supply the vessels necessary to taking an island in Thrace, that Khan may well have sent presents from the heart of Asia in return."

This seems to have been an erroneous conclusion. Hahn (1896) has pointed out that the animal to which Warnefried referred with the name bubalus—"tunc primum caballi silvatici et bubali in Italiam delati Italiae populis miracula fuerunt"—was not the buffalo; for Gajus Secundus Plinius (A.D. 23–79) employed the term bubalus with reference to an antelope, and at a later time the same term was used with regard to Bison bonasus; hence Hahn believed that the Lombards brought bisons over the Alps. Boettger (1958), on the other hand, says that Pliny referred to the urus, yet it is unlikely that wild uri were brought by the Lombards from Pannonia (formerly a Roman province bounded north and east by the Danube) to Italy.

Introduction of the Domestic Buffalo into Europe

Be this as it may, the buffalo was brought by the Arabs to Sicily after their conquest of the island in 827, and from Sicily it spread into Italy (Campania).

Into the Balkan peninsula the animal has been introduced since the 12th century A.D. There its range extends approximately to the limits reached by the Turks at the height of their expansion, including Greece, Yugoslavia (Macedonia), Albania, Bulgaria, Romania, Transylvania. It is bred also in Transcaucasia (U.S.S.R.).

The Domestic Buffalo in Western Asia

In Asia the domestic buffalo was common in the marshes of Lake Huleh in northern Palestine until 1948. Formerly it was bred also in several river swamps near the Mediterranean coast; but with the drainage and amelioration of the low-lying, malaria-infested lands it disappeared there. Owing to the unfavourable environmental conditions, combined with the relatively close inbreeding practised within the restricted area

662. (left top) Sumerian seal portraying a buffalo and human figures (about 2800 B.C.)
663. (right) Ancient Assyrian clay tablet with a buffalo en face (about 2000 B.C.)
664. (left bottom) Mesopotamian seal cylinder showing the watering of buffaloes (c. 2500–2100 B.C.)

of distribution, the Huleh buffalo was smaller than either the Egyptian or Iraqi breed, although not as small as the dwarf breed of Albania.

In Syria the breeding of buffaloes is as unimportant as it was in Palestine, and restricted to a few marshy districts (Hirsch, 1932). But in Turkey the buffalo plays an important part in agriculture in many regions.

The centre whence the domestic buffalo was diffused to Turkey, Syria, Palestine and Egypt, is Iraq where the animal is bred in large herds in the river swamps.

The domestic buffalo was well known in ancient Mesopotamia, as shown by numerous seal cylinders and clay tablets from the Sumerian as well as the Semitic periods—on cylinder seals of the kings of Shirpula and Ur, and of Sargon, king of Akkad. A seal impression was found below the level of the royal cemetery at Ur, which implies that the buffalo was known prior to 2500 B.C. On a cylinder seal of the Akkadian dynasty of Mesopotamia (c. 2500–2100) two buffaloes are depicted on a river bank, being watered by men or gods (Fig. 664). A seal from the time of Shar-kali-sharri, the successor of Naram–sin (c. 2150 B.C.), shows two buffaloes being fed by two men (Zeuner, 1963a). Watering and feeding suggest that the buffaloes were domesticated. The presence of the domestic buffalo in Mesopotamia in the later part of the 3rd and the beginning of the 2nd millennium B.C. is also indicated by an ancient Assyrian clay tablet (Fig. 663), dated to about 2000 B.C., on which the head of a buffalo is depicted with horns curved in a manner that is occasionally encountered in domestic buffaloes but never in the wild beast.

The wild buffalo is frequently portrayed in Mesopotamian mythological scenes, showing its subjugation to Gilgamesh. Great numbers were killed by the Assyrian king Ashur-nasir-pal in the hunting grounds near the Euphrates.

Theory of the Domestication of the Buffalo in Mesopotamia

Klatt (1948) suggested that the buffalo was the first bovine domesticated in Mesopotamia, and that the reason for its domestication was the nearly perfect crescent shape of its horns, which made it a suitable epiphany of the lunar deity. As the buffalo became extinct in Mesopotamia already in early historical times, the urus was probably domesticated as a substitute.

However, it is unlikely that the buffalo was first domesticated in Mesopotamia. Had this been the centre of the buffalo's domestication and dispersal, the animal would have spread westwards at a much earlier date than it actually did. Also, the principal breeding areas of the buffalo are situated not in the west of Asia but in the central and eastern parts.

665. Buffalo on a seal from Mohenjo-daro, Indus va ley (2500–1500 B.C.)

666. Indian domestic buffalo, Nagpuri type (after Brehm)

667. White buffalo, south-east China

Domestication of the Buffalo in India

In ancient India the buffalo was among the earliest domesticated animals, being well represented at Mohenjo-daro and Harappa (3rd millennium B.C.). The remains of domesticated buffaloes from Mohenjo-daro do not appear to differ appreciably from those of the modern buffalo, although a comparison of the foldings of the enamel of the cheek-teeth of the ancient Mohenjo-daro breed with those of recent domestic animals from Bengal suggests the possibility that there has been some degree of simplification in the modern breeds (Marshall, 1931–32). The occurrence of the domestic buffalo, together with the tamed elephant and the domesticated hen, in the early Indus valley civilisation indicates that the domestication centre of the buffalo has to be sought in one of the jungle or river regions of India.

At the present time numerous local breeds are found in India, some of which are used mainly for dairy purposes, and others for work. The Toda of the Nilgiri Hills, Madras, live an almost exclusively pastoral life dependent on their buffalo herds (Forde, 1934). The various Indian breeds are distinguished in size, colour and conformation, in addition to horn size and shape. Generally the skin is greyish black, more rarely dun, and only sparsely haired, sometimes with white on the face, lower part of the legs, and tail switch. In calves the coat is often reddish brown. Occasionally white buffaloes with blue eyes or albinotic animals with red eyes, a pink skin and whitish hair are encountered. In some Indian breeds, e.g. the Murrah, Kundi, Mehsana, Nili and Ravi, the horns are short and coiled; in the Nagpuri, they are long and scimitar-shaped. Long-horned and short-horned varieties are found also in south and south-east China, Indo-China and Indonesia. Some buffalo breeds are distinguished by very large horns closely approximating to those of the wild arnee. The conformation of the cranium, particularly the neurocranium, varies considerably in accordance with the size, weight and direction of the horns. In buffaloes with horns of small or medium size directed mainly backwards, the forehead is markedly vaulted. In large-horned animals in which the horns frequently project from the head in the plane of the forehead, the latter is commonly flat.

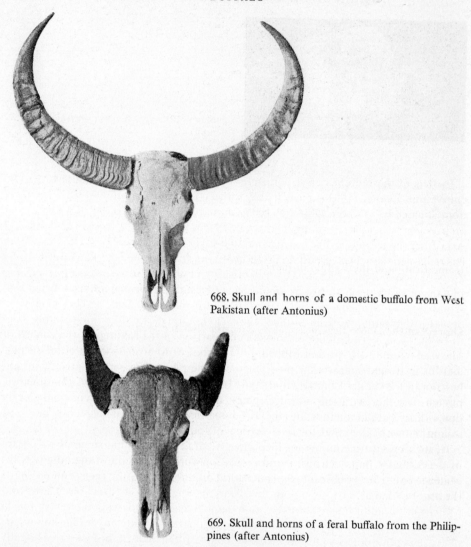

668. Skull and horns of a domestic buffalo from West Pakistan (after Antonius)

669. Skull and horns of a feral buffalo from the Philippines (after Antonius)

Theory of the Buffalo's Domestication in South-East Asia

In south-east Asia the buffalo, being far better suited than cattle to hot humid regions, very largely replaces the latter for draught, and in this area the use of cattle for dairy purposes fades out. The fully irrigated rice complex involves the use of the water buffalo. Tropical swamps are the animal's natural habitat. It can be used as a source of traction, meat and milk in regions where cattle cannot survive. Its domestication made it possible to extend the use of the plough and wheel into the south-east Asiatic region (Linton, 1956).

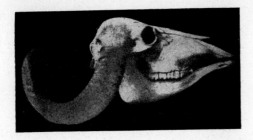

670. Skull and horns of a female domestic buffalo, Surti type, Bombay, lateral view (after Duerst)

In view of the buffalo's close connection with south-east Asiatic wet rice (paddy) cultivation, Zeuner (1963a) regards it as conceivable that the animal was originally domesticated in southern China or Indo-China rather than in India. But there is no archaeological evidence to support this view. None of the buffalo breeds of south-east Asia differs essentially from the wild Indian buffalo (Sanders, 1925). To this day India has remained the centre of buffalo breeding. In the arnee's range in India tamed arnees are still occasionally introduced into domestic buffalo herds, while domesticated buffalo cows are permitted to be served by wild arnee bulls in the jungle.

Descent of the Domesticated Buffaloes from the Indian Arnee

The wild buffaloes of Africa are excluded from the ancestry of the domesticated buffaloes for anatomical reasons. For in all domestic buffaloes, in common with the wild buffaloes of Asia but distinct from the African buffaloes (Syncerus), the vomer or ploughshare bone is fused to the palatines. Again, the wild buffaloes of Asia and all domesticated buffaloes have the hair on the middle line of the back directed forwards, in contrast with the wild buffaloes of Africa in which the dorsal hair grows backwards.

It has been shown that the wild buffaloes of Africa are excluded from the ancestry of the domestic buffaloes also for historical reasons. The anatomical and historical evidence points to the descent of all domestic buffaloes, including those of Egypt, from the arnee of India.

POLAND, DOUGLAS
COOPERATIVE EQUILIBRIA IN PHY
000304424

HCL   QP521.P76

LIVERPOOL UNIVERSITY LIBRARY
WITHDRAWN FROM STOCK

STOCK
FROM
WITHDRAWN
LIVERPOOL UNIVERSITY LIBRARY